The origins and development of African livestock

The origins and development of African livestock

Archaeology, genetics, linguistics and ethnography

Edited by Roger M. Blench and
Kevin C. MacDonald

Routledge
Taylor & Francis Group

LONDON AND NEW YORK

First published 2000
By Routledge
2 Park Square, Milton Park, Abingdon, Oxon, OX14 4RN
711 Third Avenue, New York, NY 10017

The name of Routledge is a registered trade mark used by
Routledge with the consent of the owner.

Routledge is an imprint of the Taylor & Francis Group

First issued in paperback 2011

British Library Cataloguing in Publication Data
A catalogue record for this book is available from the British Library

Library of Congress Cataloging in Publication Data
A catalogue record for this book has been requested

ISBN13: 978-1-841-42018-9 (hbk)
ISBN13: 978-0-415-51577-1 (pbk)

Contents

Preface

The present book developed from a meeting held at the Institute of Archaeology in London, 22–24 September 1995. The convenors, Roger Blench and Kevin MacDonald, felt that the quantity of new information which had become available concerning the origin, diffusion and current situation of African domestic animals was such that a new synthesis was needed. The last overview was effectively that of Epstein (1971), whose massive two-volume work combining studies of present-day races with archaeological and historical material set the agenda until the mid 1990s. However, Epstein excluded poultry and made little use of what small amount of genetic data was available during the 1960s. Recent DNA studies of Old World *Bos* species have caused a major rethink of conventional views on the origins of African cattle. Also, archaeological data on the origins of African livestock, which was sparse in 1971, has accumulated and now allows the temporal mapping of the spread of domesticates across the continent, and opens a window onto the development of "breeds" in prehistory. In addition, although Joseph Greenberg's pioneering studies of the classification of African languages appeared in the 1960s, few Africanist linguists began to make use of them to propose hypotheses concerning prehistory until the 1980s.

For all of these reasons a new overview of the topic seemed well overdue. It should be emphasized that the present volume is *not* a proceedings. Speakers were invited to the conference to present as rounded a picture of recent scholarship as possible, but some papers have been dropped, some papers have been re-focused and several additional papers have been commissioned or added by the editors to ensure completeness.

The nature of research institutions are such that they do not lend themselves well to interdisciplinary scholarship. Archaeologists and linguists do not often meet in seminars and even more rarely does DNA research move from the laboratory bench into the more uncertain world of socio-historical hypotheses. Development agencies with substantial resources and voluminous databases of recent information rarely communicate with academic anthropologists and livestock specialists. Although lip-service is often paid to the concept of working together, in practice it is all too rare.

This book aims to change this situation, at least in one small corner of academia. The authors in this volume have all been chosen for their interest in straying beyond the boundaries of their official discipline. This has been particularly fruitful in relation to cattle. Since the late 1980s, data from archaeozoology had suggested an indigenous origin for African cattle. Similarly, linguists wondered why cattle seemed to be so well embedded in African language phyla if they were introduced at the comparatively late dates then suggested. So, DNA research in the mid 1990s which indicated that cattle were domesticated not once, nor twice but three times (in Europe, Africa and the Indian region) came as an exciting confirmation for what had previously been only a controversial hypothesis (see, for example, Smith 1995 for the previous conventional view).

Beyond these major results, detailed research continues on aspects of the intro-duction and diffusion of various species. This volume presents for the first time discrete syntheses of what archaeozoology can tell us with some certainty about the prehistory of domestic animals in Africa. While such findings remain patchy for certain species, they go a long way towards providing a benchmark for live-stock distributions over time, and include significant information on the develop-ment of specific cattle size classes or "breeds" (*sensu lato*) in prehistory. Following on from Gautier's (1987) archaeozoological synthesis of early African cattle re-mains, this work comes up to the present millennium and includes all major types of livestock.

In some ways, it must be admitted that linguists and geneticists are ahead of archaeology: words and blood samples can be collected far more rapidly and at much less cost than bones. For instance, a linguistic paper in the present volume discusses the spread of the chicken in Africa – both within the continent and on the routes whereby it reached Africa from southeast Asia. These suggest that chickens may have reached sub-Saharan Africa far earlier than can be substantiated from archaeozoological findings, and cause a certain amount of interdisciplinary dis-agreement. Such conflicts are inevitable when different disciplines draw upon varied corpora of data; data which indicate and data which confirm, data that may be freely gathered and data that only slowly become available. Only through negotiat-ing the divides between disciplines can a multidimensional picture of the past be constructed.

Conference Acknowledgements
The editors would like to acknowledge with gratitude the advice and assistance that they received from Professor David Harris and Barbara Brown of the Institute of Archaeology, University College London, in bringing together the 1995 confer-ence. The meeting also received kind financial support from the British Academy. Finally, Rachel Hutton MacDonald and Helen Cook are due thanks for their organ-izational assistance in the early stages of this project.

Radiocarbon Dating Conventions used in this Book
Throughout the present volume, uncalibrated dates and time ranges are referred to as bp (before present) and bc/ad. Calibrated dates and time ranges are denoted by upper case letters (BP, BC, and AD).

References

Epstein, H. 1971. *The origin of the domestic animals of Africa* [2 volumes]. New York: Africana Publishing.

Gautier, A. 1987. Prehistoric men and cattle in North Africa: a dearth of data and a surfeit of models. In *Prehistory of arid North Africa*, A. Close (ed.), 163–87. Dallas: SMU Press.

Smith, B.D. 1995. *The emergence of agriculture*. New York: Scientific American Library.

R.M. Blench
K.C. MacDonald
London, April 1999

Acknowledgements

The editors would like to gratefully acknowledge permission from the following persons and organizations for the reproduction of original drawings and photographs: the editors of Antiquity for Figures 10.2 and 10.3, Professor Desmond Clark for Figure 10.4, the South African Archaeological Society for Figure 10.5, and the Cape Archives for Figure 11.2. We would also like to thank Mr Richard Freeman and The Institute of African Studies (Ibadan) for permission to reprint the poem preceding Chapter 23.

List of Illustrations

List of Tables

List of Contributors

Nigatu Alemayehu Adami Tulu Research Station, Institute of Agricultural Research, Ethiopia

Workneh Ayalew FARM Africa, Ethiopia

Marianne Bechhaus-Gerst Universität zu Köln, Germany

Roger M. Blench Overseas Development Institute, UK

D. Bourzat CIRAD/EMVT Laboratoire de recherches zootechniques et veterinaires, Tchad

Dan Bradley Genetics Department, Trinity College, Ireland

Juliet Clutton-Brock Department of Zoology, The Natural History Museum, UK

Patrick Cunningham Genetics Department, Trinity College, Ireland

Caroline Grigson Odontological Museum, Royal College of Surgeons, & Institute of Archaeology, University College London, UK

Sian Hall Natal Museum, South Africa

Stephen J.G. Hall Department of Animal Science, de Montfort University, UK

Fekri A. Hassan Institute of Archaeology, University College London, UK

Rachel Hutton MacDonald Department of Anthropology, University College London, UK

Samuel M. Kahinju National Museums of Kenya, Kenya

G. Kana CIRAD/EMUT, Laboratoire de Recherches Vétérinaire et Zootechnique de Farcha, Tchad

Andrew D. Kidd Institute for Social Sciences of the Agricultural Sector, University of Hohenheim, Germany

Paul N. Kunoni Kenya

J.J. Lauvergne INRA, Jouy-en-Josas, Laboratoire de Génétique Factorielle, France

Ronan Loftus Genetics Department, Trinity College, Ireland

Kevin C. MacDonald Institute of Archaeology, University College London, UK

David MacHugh Genetics Department, Trinity College, Ireland

Fiona Marshall Department of Anthropology, Washington University, USA

Ciaran Meghen Genetics Department, Trinity College, Ireland

F. Minvielle (INRA, Jouy-en-Josas), Laboratoire de Génétique Factorielle, France

Karega Munene National Museums of Kenya, Kenya

Alfred Muzzolini France

François Paris Institut Français D'Archéologie Orientale, Egypt

Christie Peacock FARM Africa, UK

Alemayehu Reda Oromiya Bureau of Agriculture, Ethiopia

Bernard Rey CIRAD/EMVT (Departement d'Elevage et de medicine veterinaire) Centre de Cooperation Internationale en Recherche Agronomique pour le Developpement, France

Kathleen Ryan University of Pennsylvania Museum, USA

B. Sauveroche France

Berthold Schrimpf Germany

Andrew B. Smith Department of Archaeology, University of Cape Town, South Africa

Paul Starkey Centre for Agricultural Strategy, University of Reading, UK

Wim Van Neer Royal Museum of Central Africa, Belgium

Kay Williamson Department of Linguistics and Nigerian Languages, University of Port Harcourt, Nigeria

PART ONE
Introduction

CHAPTER ONE

The origins of African livestock: indigenous or imported?

Kevin C. MacDonald

Knowledge of cultivation of cereal crops was, on existing archaeological evidence, transmitted across the Sahara from South-West Asia via the Nile and perhaps the Maghrib, and with agricultural knowledge must have come domestic stock – long- and short-horned cattle, sheep and goats.   (Clark 1962:227)

Since there were wild cattle in North Africa and the Nile valley north of the tropic of Cancer during the Pleistocene and up to Roman times, the possibility of an indigenous domestication of cattle in North Africa must not be ruled out.   (Shaw 1977:109)

Early neolithic sites in the southern part of the eastern Sahara, even those with radiocarbon dates earlier than 9000 BP, usually yield bones believed to be those of domestic cattle. The age and location of the sites mean that cattle must have been independently domesticated in Africa, probably in the tenth millennium BP.   (Close & Wendorf 1992:69)

1. Introduction

The origins and development of African livestock – issues long ignored or marginalized – have come dramatically to the fore over the past decade. This is primarily the result of fresh evidence for an indigenous origin for African cattle. New data has mostly taken the form of osteological material from northeast Africa (Close & Wendorf 1992, Wendorf & Schild 1994), although tantalizing results from DNA studies on modern cattle now also seem to support an independent domestication of African cattle (Bradley et al. 1996, Bradley & Loftus, Ch. 13 in this volume). This is not to say that indigenous domestication has been unanimously accepted (see, for example, Muzzolini, Ch. 6 in this volume and Grigson, Ch. 4 in this volume). Indeed, it must be admitted that neither arguments for indigenous origin nor for importation are absolutely conclusive. But whichever hypothesis is eventually accepted, it is becoming increasingly apparent that contrary to expectation, pastoralism preceded cereal agriculture in most of Africa. This hypothesis, discussed in more detail below, rests upon Sahelo–Saharan studies that have repeatedly produced evidence for livestock-keeping, without contemporary evidence for domestic cereals or field clearance (see also Hassan, Ch. 5 in this volume).

The notion that pastoralism preceded the domestication of cereals in some parts of the world is hardly a new one. In the evolutionary schema of the Enlightenment it was commonly assumed that shepherding of animals occurred before the advent of agriculture. Certainly Adam Smith, the Marquis de Condorcet, and Turgot took this view (cf. Meek 1976), and the tendency continued into the Victorian era with the work of early archaeologists such as Hodder Westropp (1872). These models, like those that later gave priority to cultivation, were highly speculative. Indeed, it can be argued that the Enlightenment prioritization of pastoralism stems from Roman historical philosophy and Lucretius' *De Rerum Natura*, where pastoralism is depicted as preceding the sowing of crops (Meek 1976:8–10). That such an argument can be made again today, after nearly a century of scholarship has held a contrary view (e.g. Childe 1928, 1956, Cohen 1977) or at least posited that animal-keeping and cereal agriculture began relatively contemporaneously (e.g. Reed 1977), is a tribute to the ability of new data to surprise us and confound the most tidy of models.

Cattle are, of course, not the only (probable) indigenous domesticates of Africa. There are some "minor species" such as the guinea-fowl, the cat, and the donkey that are also in need of consideration (see below, and Blench, Ch. 20 in this volume). Also, there are some more controversial cases for indigenous domestication which have been made by some researchers for the sheep, the pig, the camel and the dog. Of the probable African domesticates, however, it is undeniable that cattle are the most important, as they are associated with an entirely new form of economic life: pastoralism. Coupled with this new lifestyle is the concept of capital accumulation (the very word capital itself derives from the French word for livestock *cheptel*). Add to this changes in mobility patterns made to protect cattle from disease and to keeping them watered and fed, ranging from the stall-raising of animals by sedentists, to semi-nomadic seasonal displacements, to full nomadism; and one begins to appreciate the behavioural richness engendered by this economic change. Thus Africa, long a source of case studies on contemporary livestock management and historical ecological change and economic response, must now also be considered as a subject suitable for the study of the origins of pastoralism itself (Fig. 1.1).

In considering the prehistoric origins of cattle and other types of livestock, it should be remembered that the archaeologist/archaeozoologist is not the only arbiter of academic opinion. Increasingly, microbiological studies, initially of the blood-serum typing variety, but now of DNA, are becoming central to the formation of prehistoric models (see below, and Cunningham, Ch. 12 in this volume). Additionally, historical or comparative linguistics supplies a continuing input in terms of the temporal ordering of animal acquisitions by different language speakers, and the directionality of livestock transfers (Blench 1993, and Ch. 20 in this volume). Ethnography has also long played a role in the formation of analogical models of prehistoric herd management strategies (e.g. Dahl & Hjort 1976, Ryan et al., Ch. 25 in this volume). As locators of "hard evidence" for the presence or absence of

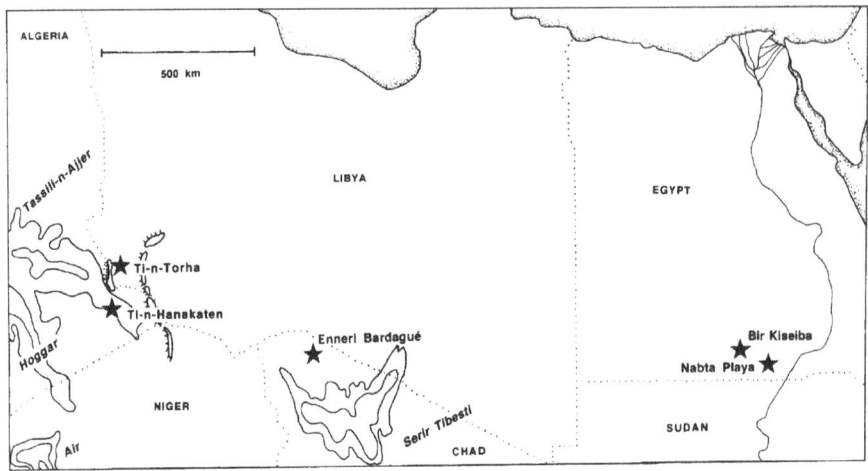

Figure 1.1   Map showing the location of African sites with osteological evidence for domest-icated cattle dating to around or before 7000 bp. The major central Saharan escarpments and mountain ranges are also featured.

specific animals and other types of phenomena, archaeologists may (justifiably?) feel a certain confidence in the reliability of their arguments, but it would behove us not to automatically discount hypotheses from other disciplines that run counter to our findings. Or, for that matter, we must also be cautious not to accept too uncritically the work of scholars from other disciplines without digging sufficiently deep into their own literature to verify the feasibility of their assertions. In essence, the diverse contributions and opinions expressed in this volume teach us the need for wide reading and an ongoing dialogue with other disciplines so as to reach a better understanding and increasing consensus in a "multi-vocal" academic world.

As the chapters of this volume feature the current state of knowledge for the interpretively interwoven fields archaeology, linguistics, ethnography and biology, rather than the historical narrative of their inquiries, it is worthwhile here to con-sider in some detail the development of the domestic cattle origins debate. Addi-tionally, the origins of the pastoral way of life in Africa will be considered in relation to evidence for the beginnings of plant domestication on the continent. Finally, other putative African domesticates of varying likelihood will be consid-ered, inasmuch as their origins have been placed by some researchers in Africa.

2. The Large Bovids of Bir Kiseiba and Nabta Playa: wild or domestic?

In 1980, Gautier (1980) and Wendorf & Schild (1980) announced the probable presence of domestic cattle at the site of Nabta Playa dating to *c.* 8840±90 bp. As Muzzolini (1983i:189) has noted, this was *"suprenant, car le site présente actuellement les dates les plus anciennes connues pour de possibles bouefs domestique, non seulement en Egypte – òu il éclipse définitivement le vénérable Fayoum A – mais meme dans le monde*". Subsequently, Gautier (1984) made known still earlier remains, tentatively attributed to *Bos taurus*, from the site of Bir Kiseiba dating to *c.* 9500 bp. Needless to say, this new case for a precocious, and virtually world-leading African pastoralism, was rapidly challenged (Smith 1984, 1986, Muzzolini 1983, 1989).

The argument for the domestic status of the Bir Kiseiba and Nabta Playa rests primarily on palaeoenvironmental factors and only secondarily on osteological criteria. Morphologically, the large bovid remains from these sites have now been definitively identified as *Bos* – and not the African buffaloes *Syncerus* or *Pelorovis* (Wendorf & Schild 1994). Metrically, Gautier (1980, 1984) notes that the few measurable remains are somewhat smaller than one would expect for wild cattle (*Bos primigenius*), falling well within the upper range of the well-studied European neolithic cattle of Manching (Bavaria, cf. Boessneck et al. 1971). But as the critics have asserted, none of this physical evidence is conclusive: there are few specimens, those dating to between 9500 bp and 7900 bp numbering less than 30 (cf. Wendorf et al. 1987:Table 1; Wendorf & Schild 1994:119–20), and it is thus possible that the metric sample is skewed (Smith 1986). Indeed, examination of the metric data using the "standard animal" method, appears to demonstrate that the Nabta Playa and Bir Kiseiba cattle were – on the basis of their large size – wild rather than domestic (see Grigson, Ch. 4 in this volume). But, it is worth remembering that the size of regional wild cattle populations might have varied considerably in the early Holocene. This would seem particularly true for Egypt, where recent research has shown that late Pleistocene Nile Valley wild cattle were unusually large (see Wendorf & Schild 1994:127).

As a result of this rather equivocal osteological evidence, ecological arguments have remained central to the case: if cattle could not have existed in the Nabta Playa/Bir Kiseiba area without human intervention, then even one bone in context becomes evidence for "cattle-keeping". To this same end, it is primarily the harsh and changeable ecology of the eastern Sahara that has been invoked as a potential trigger for an independent African cattle domestication: with wild cattle initially protected and conserved in drought years, by mobile populations, as a critical source of protein and fat (Close & Wendorf 1992; see also Hassan, Ch. 5 in this volume). If this were true, it would again defy conventional archaeological wisdom, which normally predicates increasing sedentism as a precondition for domestication (e.g. Flannery 1969, Reed 1977). But, to properly evaluate these claims, we must first review the evidence for the early and middle "neolithics" of the eastern Sahara.

To begin, the terrestrial fauna of the "early neolithic" of El Adam type (9500– 8900 bp) includes two species of gazelle (*Gazella dorcas and dama*), as well as hares (*Lepus capensis*), jackals (*Canis aureus*), and tortoises (*Testudo sp.*). None of these taxa require a year-round supply of standing water. Thus, it has been argued that the 20+ cattle bones in the El Adam type assemblages, must be from managed animals – led by humans from waterhole to waterhole in a low biomass landscape (Wendorf et al. 1984, Wendorf & Schild 1994; see also Hassan, Ch. 5 in this volume). Smith (1986) has noted, however, that these and later assemblages are curiously deficient in large mammals (especially other larger bovids), even for such an arid environment. To this end, both Smith (1986, 1992) and Muzzolini (1983i:193, 1989) assert that even the sparse steppes that would have surrounded the ephemeral playa lakes should have supported populations of widely ranging grazers including giraffes, rhinos, elephants, hartebeests, addax, oryx, and (conceivably) wild cattle. Since "arid landscape" archaeological assemblages from Dakleh Oasis in Egypt and from South Africa are known to possess such diverse taxa, they are led to doubt the ecological, and thus interpretative, validity of the Nabta Playa/Bir Kiseiba faunal samples. Close & Wendorf's (1992) riposte to this criticism has been to stand by the representativeness of their bone collections, and to invoke palaeobotanical studies indicating very low rainfall during this period (e.g. Neumann 1989).

Strangely, the two subsequent phases in the "early neolithic" of the eastern Sahara seem to feature a smaller ratio of cattle remains (only three out of eight sites with preserved fauna dating between 8600 bp and 7900 bp having cattle remains, and then only a handful). An interesting addition is that the "early neolithic" of the El Nabta type (8100–7900 bp) begins to feature 1.5–2.5 m deep wells associated with house structures, a good indicator of either more arid conditions or dry season occupations (Wendorf & Schild 1994). It is difficult to fix a definite date to the original digging of these wells, and they may well be slightly more recent than the El Nabta phase. As an explanation for the small quantity of cattle remains in this period, Close & Wendorf (1992:69) have argued for a "milk and blood" exploitation practice with no intentional culling.

The "middle neolithic" period (7700–6500 bp) sees more definitive evidence for cattle-keeping with larger samples and more extensive settlements. There is even a recently investigated megalithic complex dating to the later part of this period, situated in the Nabta Playa Region (site E-75-8), and associated with a "relative abundance of cattle remains" (Wendorf et al. 1996:131). But, of course, by this time we cross the threshold into a period when contact with western Asiatic herders would have become much more likely. From the "middle neolithic" onwards, and certainly after the definite appearance of ovicaprines in northeast Africa around 7000 bp (cf. Vermeersch et al. 1996), it becomes possible to speak of "stimulus diffusion" as a trigger for the domestication of wild African cattle. Indeed, ovicaprines are present in the "middle neolithic" of the eastern Sahara (Wendorf & Schild 1994). It would seem that if there was a "pristine" cattle domestication in Africa we must look to the "early neolithic" of the eastern Sahara or to similarly early sites farther to the west.

When were domesticated sheep and goats, and particularly cattle, available in the southern Levant (i.e. Jordan, Israel/Palestine) for import into Africa? The goat (*Capra hircus*) was undoubtedly the first bovid domesticate in the southern Levant, probably domesticated locally by *c.* 9000 bp (Legge 1996). Sheep would appear to have been introduced into the area later, by *c.* 8600 bp, from the north and east (ibid.). The first appearance of cattle, however, is a much more thorny issue. In her 1989 synthesis on the subject, Grigson indicates that the earliest good metric evidence for cattle domestication begins around 8000 bp, particularly in Anatolia. There also appears to be good evidence from *c*. 8000–7000 bp from northern Syria, although any claim for domestic cattle in the southern Levant during this period would seem far more shaky; Grigson's (1989) only data point being the site of Ashkalon (with only eight elements attributable to *Bos* and, seemingly, no radiometric dating, cf. Ducos 1968). However, recent analysis of finds from the Jordanian site of 'Ain Ghazel, may yet indicate the presence of cattle during the final prepottery neolithic (PPNC) period (*c*. 8000–7500 bp) (Köhler-Rollefson 1989, Rollefson & Köhler-Rollefson 1993). Unfortunately, the basis for the identification of domestic cattle in the 'Ain Ghazal PPNC is not clear. Köhler-Rollefson (1989:203) states that cattle from this period "are now apparently domesticated", yet no metric data is referred to, and *Bos* actually decrease in quantity from the (presumably) wild cattle of the preceding phase (12.2 per cent to 5.9 per cent of total faunal specimens). This assertion may be linked to evidence for the domestication of caprines during this same period, but it is clear that we are at this time without definite evidence for domestic cattle in the southern Levant before 7000 bp. After 7000 bp, there are more plentiful and reliable finds (e.g. Tel Dan, cf. Grigson 1989).

Still closer to Africa, there are only a few preserved or documented faunal assemblages from the Negev/Sinai region for the period of interest to us (*c*. 8500–6500 bp) (Goring-Morris 1993). Such assemblages as there are for the PPNB (*c*. 8500–8000 bp) seem to consist principally of gazelle and ibex, and for later periods there would appear to be no faunal evidence at all, simply material culture traces, which might argue for the seasonal presence of hunters or herders (ibid.). For the similarly arid region of eastern Jordan, at least, there is good evidence for the presence of domestic sheep and goats from 8000 bp at the site of Asraq (Garrard et al. 1988, 1995). As a consequence, Byrd (1992) has argued that although faunal evidence is scarce for other arid zones adjoining the Levant (i.e. the Negev and Sinai) it is possible that sheep and goat herds could have also reached hunter–gatherers in these areas by 8000 bp. Thus, for northeast Africa "cut-off dates" for independent domestication would have to fall before *c*. 8000 bp for there to be no knowledge of domestic sheep and goats, and before *c*. 7000 bp for the purely indigenous development of wild cattle stock.

Outside of the eastern Sahara there is no good data for the presence of domestic cattle in Africa before or around 8000 bp, and little good data for domestic cattle before 7000 bp (see Fig. 1.1). The core of the existing evidence are the enigmatic *steinplätze* or stone places of eastern Egypt and Sudan, Libya, Chad, and Algeria (Gabriel 1973, 1987). The *steinplätze* consist of stone concentrations, hearth rings, and (frequently) waisted tethering stones (Gabriel 1987, Pachur 1991). Elements of material culture and faunal remains are rare. They thus appear to be ephemeral campsites, arguably of pastoral economic orientation. The over 50 radiocarbon dates available for these sites range between 10,500 bp and 1500 bp, but if we accept only those periods that have at least four dates per 800 years, we have three principal phases of their accumulation: 8500 bp to 5800 bp (five dates per 800 year period), 5800 bp to 5000 bp (20 dates per 800 year period), and 5000 bp to 3800 bp (four dates per 800 year period) (Gabriel 1987:97). For the earliest of these phases the only site with identified cattle remains is that of Enneri Bardagué in the Serir

Tibesti (northern Chad). There cattle, and possibly sheep or goat, occur with a fauna containing gazelle, African buffalo, elephant, and giraffe (Gautier 1987:174). There is a direct date on animal bone of 7455±180 bp, and another on human bone of 6930±370 bp (ibid.). Meanwhile, in southwestern Libya and southeastern Algeria, there is evidence for the appearance of cattle sometime between 7000 bp and 6500 bp. There are dates for deposits with cattle remains (and ovicaprines) at Wadi Ti-n-Torha North (Libya) ranging from 7070±60 bp to 5260±130 bp, with a median date in the region of 6000 bp (Barich 1987). Similarly, there is a range of dates for the cattle containing S4 layer at Ti-n-Hanakaten (Algeria) ranging from 7220±140 bp to 5800±120 bp, with a median date of 6650±90 bp (Aumassip & Delibrias 1982–3). On the basis of this limited evidence, cattle-keeping progressed – probably with introduced sheep and goat stock – from western Egypt to the Tibesti region (Libya/Chad) by at least 7000 bp, and into the Tassili and Tadrat Acacus (Libya/Algeria) by 6500 bp. Unfortunately, such dates are neither too early to make the conventional diffusionist hypothesis untenable, nor too late to rule out a slightly delayed movement from the putative "early neolithic" of western Egypt.

The early campsites and tethering stones scattered across the plains of the central Sahara may or may not be decisive, depending on the predisposition of the individual researcher. For example, Pachur (1991) believes that tethering stones, which are associated with radiocarbon dates ranging from 7500 bp to 4000 bp in the central and eastern Sahara, relate to the management of domestic cattle or even the taming of wild cattle. In rock art from southwestern Libya, cattle (seemingly always male animals) are clearly represented tethered to these waisted stones (Castiglioni & Negro 1986). However, Morel (1982) has argued that the primary function of the "tethering stones" was to act as part of trap lines for hunting (even possibly for the hunting of wild cattle). Pachur's arguments are seductive but not proven.

In terms of archaeological evidence alone, the issue of African cattle domestication was deadlocked by 1994. Then, a genetic survey of modern cattle by the Department of Genetics at Trinity College, Dublin, added a new dimension to the debate. Initial research using mtDNA sequences showed no maternal connection between modern African humped cattle and east Asian *Bos indicus*/zebu – indicating an origin either through imported zebu stud bulls or (notionally at least) from indigenous humped African stock (Loftus et al. 1994). This was interesting, but held only slight portent for issues of indigenous domestication (i.e. only if one were to follow Grigson's arguments for the African wild cattle being humped, cf. Grigson 1991, and Ch. 4 in this volume). However, further research produced conclusions that were far more dramatic: mtDNA sequencing indicated the genetic separation at *c*. 275–117 kyr for Indian (*B. indicus*) and Euro–African cattle, with a last common ancestor for African and European cattle at 26–22 kyr (Bradley et al. 1996, see also Bradley & Loftus, Ch. 13 in this volume). Thus, Bradley et al. (1996:5131) could justifiably state that "as cattle domestication is thought to have occurred at approximately 10,000 BP, these estimates suggest the domestication of genetically distinct *aurochsen* strains as the origins of each continental population". The implications of this research for the more tenuous data from the eastern Sahara are plain to see – suddenly an African origin for African cattle is an expected event supported in terms of time and place by the archaeological data, not indicated by the archaeological data alone. But, confident prehistoric extrapolations from this initial genetic data alone might prove dangerous. As Bradley et al. (1996) have noted, other possibilities remain, including the option that small Euro–Asiatic founder herds were augmented and in time replaced by the incorporation of local wild cattle. Although they go on to state: "the evidence is most suggestive of two domestic origins that were either temporally or spatially separate and that involved divergent strains of taurine progenitors" (ibid.:5135).

3. The chronological relationship of pastoralism and cultivation in Africa

Whatever conclusion can be drawn from this, cattle were present in most of North Africa (including the Sahara) by at least c. 6500 bp. A great deal of tortuous reasoning has been advanced to explain the absence of evidence of domestic cereals in Africa until well after that date. The now long-withdrawn claims for 18,000-year-old domestic barley at Wadi Kubbaniya in the eastern Sahara (Wendorf & Schild 1984), have given way to widely disputed claims of cultivated (but not domesticated) sorghum in the middle neolithic of Nabta Playa on the basis of lipid analysis (Wasylikowa et al. 1993), to broader claims of cultivation without domestication in northeastern Africa by c. 6000 bp (Haaland 1995). The notion of cultivation without domestication (because of harvesting methods and re-crossing with local wild progenitors) is a reasonable hypothesis, although the timing of this development remains largely uncertain (ibid.; Blench 1997). But, the persistent absence of domestic cultigens augurs for a real lag between the adoption of pastoral economy and the origins of cereal agriculture in Africa.

At present, accepted, physical evidence for domesticated crops in Africa is patchy and recent. Indeed, curiously, the earliest good evidence comes from western rather that eastern Africa. East of Lake Chad, sorghum has long been suspected as an early domesticate. There is some possible indication of domestic sorghum at Middle Nile sites dating to c. 5500 bp on the basis of potsherd impressions (Stemler 1990), but even their identifier doubts their domestic status. Ironically, the best evidence for (prior) sorghum domestication or cultivation in Africa comes from Yemen with dates for sorghum by 3700 bp (Constantini 1990). Grains of a similar age are also known from India (Kajale 1991). Within Africa the earliest "sure" dates are those from Jebel el Tomat (Sudan) at c. 1700 bp (Clark & Stemler 1975), Meroë at c. 2000 bp (Rowley-Conwy 1991), and Jenné-Jeno (Mali) at c. 2050 bp (McIntosh 1995).

In the Sahara, wild grain collection, rather than domestication, would seem to be universal in early pastoral sites (Barich 1992). Indeed, the domestication of cereals in the West African Sahel is well supported from c. 3500 bp onwards – but not before (MacDonald 1994, Breunig & Neumann 1996). Identifications of domestic millet from well-dated contexts include: carbonized remains from Windé Koroji (Mali) c. 3600–3100 bp (MacDonald 1996b, Capezza, in press); potsherd impressions from numerous sites of the Tichitt-Oualata complex (Mauritania) dating to *c.* 3100–2700 bp (Munson 1971, 1976, 1989, Amblard & Pernès 1989, MacDonald 1996a); and carbonized remains from Kursakata (Nigeria) dating to *c.* 2500 bp (Neumann et al. 1996). Interestingly, pollen work at the Mare de Oursi (Burkina Faso) indicates that field clearance, and thus cultivation, only began in that part of the Sahel from *c.* 3500 bp (Ballouche & Neumann 1995).

Thus, on the basis of present evidence, pastoralism preceded cereal agriculture in northeastern and arid West Africa by at least two to three thousand years. The increasing weight of current regional research supports the argument that this is not a sampling error, but a reflection of prehistoric economic reality. The implications of this for modelling pastoral origins in Africa, and early African pastoral economies has not yet been fully considered either within or outside of Africanist scholarship. It is an important challenge to deterministic conceptualizations favouring sedentism as the crucial factor in the shift to food production.

4. Other African animal domestication events

Aside from cattle, there are a number of other possible indigenous domestication events on the continent. These may be graded from certain indigenous domestications, to probable ones, to merely speculative or possible ones. Few of these have much supporting documentation, and all require further research.

4.1. A certain African domestication: the guinea-fowl (Numida meleagris)

The only certain African domestication is that of the guinea-fowl. Unfortunately, there is as yet scant osteological evidence to provide evidence of the place or timing of this development (see MacDonald & MacDonald, Ch. 8 this volume, and Blench, Ch. 20 in this volume). Indeed, it is not certain whether it was first domesticated in North or West Africa. We know from artistic evidence that domestic guinea-fowl were present in Greece by the fifth century BC, and throughout the Roman Empire in slightly later times (Monghin & Plouzeau 1984). Ancient Egypt lacks evidence for guinea-fowl exploitation, so one must assume a domestication event somewhere in the Maghreb, or in the Sahelian belt, within the range of the wild bird, sometime during or before the early first millenium BC.

4.2. Probable African domestications: the donkey (Equus asinus) and the cat (Felis sylvestris)

The donkey was almost certainly first domesticated in northeast Africa from the Nubian ass, although there remains some insistence that an earlier western Asiatic domestication cannot be ruled out (Groves 1986, Clutton-Brock 1987). Osteological and artistic evidence from Egypt strongly indicates that domestic donkeys were in use there by the pre-Dynastic period (*c.* 3500–4500 BC) (Epstein 1984). Outside of Egypt and Nubia early evidence for the domestic donkey in Africa is almost non-existent. This may be attributable to a genuinely slow diffusion of the animal, or it may be a factor of depositional tendencies (with carcasses of non-food animals left to rot outside of settlement areas). Several lines of evidence for the origins and development of the domestic ass are considered elsewhere in this volume (Blench, Ch. 21 in this volume).

The cat is long thought to have been domesticated in dynastic Egypt, although the possibility exists that is was first domesticated in western Asia and then imported by the Egyptians. Morrison-Scott (1952) who examined the crania of the mummified cats of dynastic Egypt, found that they conformed principally with *Felis sylvestris libyca* (the wild sub-species of most of Africa, the Levant and the Arabian peninsula), rather than other wild sub-species. However, some skulls conforming to *Felis sylvestris chaus* (the wild sub-species of southwestern and southeastern Asia) were also identified. Archaeologically, Baldwin (1975) has argued for a multi-phase domestication of the cat by the Egyptians ranging from commensality in the pre-Dynastic and early Dynastic periods (with cats feeding on agricultural pests around settlements) to domestication and religious incorporation from the time of the New Kingdom (approximately 1600 BC onwards). However, both Zeuner (1963) and Clutton-Brock (1987) suggest that the cat may well first have been domesticated in the Levant, with possible commensal animals recovered from Jericho dating to *c.* 7000 BC. The key notion here is that cats were most probably domesticated as a result of their commensal association with cultivating peoples – keeping rodents away from grain storage bins and fields. The rather late appearance of cultivation in Africa might thus push forward the timing of domestication on the continent.

4.3. Possible African animal domestications: the dog (Canis familaris), the sheep (Ovis aries), the pig (Sus scrofa) and the camel (Camelus dromedarius)

The dog (*Canis familiaris*) is generally accepted to have been first domesticated from the wolf (*Canis lupus*) somewhere in southwestern Asia during the Terminal Pleistocene (e.g. 14,000–10,000 bp). In Africa the earliest definite evidence for domestic dog remains that from Merimde in Egypt at *c.* 6000 bp/5060 BC (Hawass et al. 1988, von den Driesch & Boessneck 1985). Thus, there is certainly no indication that any African domestication event *preceded* developments in Asia. However, there remains the possibility that African jackals (particularly *Canis aureus*) played a roll in the ancestry of modern African dog populations. Golden jackals, wolves, and domestic dogs are all interfertile – thus in the low management sphere of the African pariah dog, the possibility of genetic input from jackal stock is high (cf. Gray 1972). Jackals and wolves even share some behavioural particularities, including the adoption of an erect tail posture in tamed animals when in a state of anticipation (Brewer et al. 1994:116). So there remains the remote possibility of an indigenous "jackal domestication" somewhere in Saharan Africa during the early to mid Holocene, even if most modern African "breeds" derive from Asian imports. However, the small brain size of jackals (even when compared to similar sized dogs) and morphological differences in the first upper molar have caused most scholars to exclude the jackal from the African pariah dog's ancestry; attributing it instead to a haphazard mixture of incoming greyhound and other domestic types (Epstein 1971i:138–47). Osteological differentiation of dog, wolf, and jackal relies primarily on measures of relative robusticity (see MacDonald & MacDonald, Ch. 8 in this volume), and although relatively straightforward with dental and cranial specimens, is still imperfect. As there are few specimens, particularly from early sites, it will be some time before archaeozoology can paint a convincing picture of the origins of African dogs. More widespead genetic characterizations of the diverse pariah breeds of Africa would shed some important new light on this matter.

One of the most contentious postulated African animal domestications is that of the sheep (*Ovis aries*). This theory's chief proponent is Muzzolini (1987, 1993, and Ch. 6 in this volume). In outline, this argument first derived from the observations of rock art depictions of the "ornamented rams" of the Atlas. Muzzolini (1987:136–7) observed the curious fact that these animals seem to resemble wild rather than domestic sheep and from there began to question traditional assumptions of their introduction from western Asia. The principal difficulty stems from the definition of the wild range of *Ovis orientalis* at the Pleistocene/Holocene transition. Muzzolini argues that if wild sheep could have recolonized the (northern) Levant by this time, then there is no reason that they could not also have spread into North Africa. Unfortunately evidence for this hypothesis (osteological evidence pre-dating the traditional diffusionist arrival time of 8000 bp to 7000 bp) is tenuous. In most instances old identifications of *Ammotragus lervia* (the Barbary Sheep), such as those at Haua Fteah, are questioned on the basis that the criteria used for their separation from domestic sheep is essentially contextual (e.g. too early to be imported domestic sheep) rather than strictly osteological. However, none of the instances cited by Muzzolini (1993) are sufficiently old enough, well-dated enough, or even properly identified enough to support his claims on their own. To validate or discard this interesting, though doubtful, hypothesis a good deal more osteological evidence from the early Holocene of Africa, the Levant and Sinai, must be excavated and scrutinized.

The ancestor of the domestic pig, the wild boar (*Sus scrofa*), is native to Africa north of the Sahara, and wild populations continue to exist in Morocco and Algeria (Kingdon 1997). There have been only a few assertions that these populations might have independently given rise to domestic populations (see Blench Ch. 22 in this volume, for a detailed consideration of this issue). The origins of pig-keeping are usually placed in Anatolia by *c.* 7000 BC and in the early neolithic of China, with subsequent spread across the rest of the Old World (Epstein & Bichard 1984). However, evidence for domestic pigs in Egypt and Morocco is also early (*c.* 5000–4000 BC), and it is likely that local *Sus scrofa* populations at least contributed genetically to imported stock (Gilman 1975, Brewer et al. 1994). A programme of DNA study, similar to that already executed on cattle, could prove very illuminating on pig origins. Unfortunately, much of the native domestic pigs of North Africa have been exterminated or left to go feral since the coming of Islam, although some small populations kept by Coptic Christians remain in Egypt and Sudan (Epstein & Bichard 1984).

In 1979, Brent Shaw put forward a detailed, if contestable, case for the African origin of the domestic camel (*Camelus dromedarius*). The conventional "introductionist" hypothesis places the domestication of the camel in Arabia, with its introduction via Egypt into the Sahara. In essence his claim is that the camel did not go extinct in North Africa after the lower and middle Palaeolithic periods – from which it is well attested in the faunal sequence – but remained in place and was eventually domesticated locally during the "neolithic" (Shaw 1979). Shaw's hypothesis rests upon three factors: (a) several controversial camel bone finds of Capsian (*c.* 9000–6000 bp) or "neolithic" date from the Maghreb; (b) abundant rock art depictions of the camel in North Africa; and (c) a matter of fact textual mention of camel use in the African interior at 46 BC during the campaigns of Caesar – a date thought by Shaw too early for the "introductionist" hypothesis. The third argument is now probably irrelevant, given new data from Qasr Ibrim (southern Egypt) placing animals (presumably) introduced from Asia on the continent by the early first millenium BC (Rowley-Conwy 1991). The second argument supplies with it no dating evidence – all such depictions could be (as currently believed by most scholars) of the late first millenium BC/first millenium AD "Cameline" period. The first argument might still hold some weight, but post middle Palaeolithic finds are few, and none are uncontested (i.e. they may be intrusive into the earlier layers of these sites and have not been directly dated).

In short, although these three controversial species are not in any way proven African domesticates, they deserve further serious examination before they are discounted.

5. Final comments

Although this chapter has examined the possible indigenous origins of some modern domestic African animals, it should not be thought that "first past the post" scenarios dominate African archaeozoology and livestock studies. Processual inquiries into economic change relative to livestock and their breeding are ultimately of greater interest and contemporary relevance. However, it is important to convey the message that one should not be overly doctrinaire about the presence/absence or status of domestic animals in the early Holocene sites of Africa. There is much that remains to surprise us, if we combine an open mind with rigorous data recovery and analysis.

References

Amblard, S. & J. Pernès 1989. The identification of cultivated pearl millet (*Pennisetum*) amongst plant impressions on pottery from Oued Chebbi (Dhar Oualata, Mauritania). *African Archaeological Review* 7, 117–26.

Aumassip, G. & G. Delibrias 1982–3. Ages des dépôts néolithiques du gisement de Ti-n-Hanakaten (Tassili-n-Ajjer, Algérie). *Libyca* 30–31, 207–11.

Baldwin, J.A. 1975. Notes and speculations on the domestication of the cat in Egypt. *Anthropos* **70**, 428–48.

Ballouche, A. & K. Neumann 1995. Pollen from Oursi/Burkina Faso and charcoal from NE Nigeria: a contribution to the Holocene vegetation history of the West African Sahel. *Vegetation History and Archaeobotany* **4**, 31–9.

Barich, B.E. (ed.) 1987. *Archaeology and environment in the Libyan Sahara. The excavations in the Tadrat Acacus, 1978–1983*. BAR International Series 368. Oxford: BAR.

Barich, B. 1992. Holocene communities of western and central Sahara: a reappraisal. In *New light on the northeast African past*, F. Klees & R. Kuper (eds), 185–204. Africa Praehistorica 5. Köln: Heinrich-Barth Institut.

Blench, R.M. 1993. Ethnographic and linguistic evidence for the prehistory of African ruminant livestock. In *The archaeology of Africa: foods, metals and towns*, T. Shaw, P. Sinclair, B. Andah, A. Okpoko (eds), 71–103. London: Routledge.

Blench, R.M. 1997. *Neglected species, livelihoods and biodiversity in difficult areas: how should the public sector respond?* Natural Resource Perspective Paper 23. London: Overseas Development Institute.

Boessneck, J., A. von den Driesch, U. Meyer-Lemppenau, E. Wechsler von Ohlen 1971. *Die Tierknochenfunde aus dem Oppidum von Manching*. Wiesbaden: F. Steiner.

Bradley, D.G., D.E. MacHugh, P. Cunningham, R.T. Loftus 1996. Mitochondrial diversity and the origins of African and European cattle. *Proceedings of the National Academy of Sciences, USA* **93**, 5131–5.

Breunig, P. & K. Neumann 1996. Archaeological and archaeobotanical research of the Frankfurt University in a West African context. *Berichte des Sonderforschungsbereichs 268 (Frankfurt am Main)* **8**, 181–91.

Brewer, D.J., D.B. Redford, S. Redford 1994. *Domestic plants and animals: the Egyptian origins*. Warminster: Aris & Phillips.

Byrd, B.F. 1992. The dispersal of food production across the Levant. In *Transitions to agriculture in prehistory*, A.B. Gebauer & T.D. Price (eds), 49–62. Madison: Prehistory Press.

Capezza, C. in press. The macrobotanical remains from Windé Koroji Ouest. In *The southwestern Gourma project (vol. 1): the Windé Koroji Complex*, K.C. MacDonald (ed.). London: Institute of Archaeology.

Castiglioni, A. & G. Negro 1986. *Fiuma di Pietra. Archivo della preistoria sahariana*. Varese: Lativa.

Childe, V.G. 1928. *The most ancient East: the Oriental prelude to European prehistory*. London: Kegan Paul.

Childe, V.G. 1956. *Man makes himself*, 3rd edn. London: Pitman.

Clark, J.D. 1962. The spread of food production in sub-Saharan Africa. *Journal of African History* **3**, 211–28.

Clark, J.D. & A. Stemler 1975. Early domesticated sorgum from central Sudan. *Nature* **254**, 588–91.

Close, A.E. & F. Wendorf 1992. The beginnings of food production in the eastern Sahara. In *Transitions to agriculture in prehistory*, A.B. Gebauer & T.D. Price (eds), 63–72. Madison: Prehistory Press.

Clutton-Brock, J. 1987. *A natural history of domesticated mammals*. London: British Museum of Natural History.

Cohen, M. 1977. *The food crisis in prehistory*. New Haven: Yale University Press.

Constantini, L. 1990. Ecology and farming of the protohistoric communities in the central Yemeni Highland. In *The Bronze Age cultures of Hawlan, At-Tyial and Al-Hada*, A. De Maigre (ed.), 187–204. Rome: Ismeo.

Dahl, G. & A. Hjort 1976. *Having herds: pastoral herd growth and household economy*. Stockholm Studies in Social Anthropology, No. 2. Stockholm: University of Stockholm.

Ducos, P. 1968. *L'origine des animaux domestiques de Palestine*. Publications de l'Institut de Préhistoire de l'Université de Bourdeaux No. 6.

Epstein, H. 1971. *The origin of the domestic animals of Africa* [2 volumes]. New York: Africana Publishing.

Epstein, H. 1984. Ass, mule, onager. In Mason (1984), 174–84.

Epstein, H. & M. Bichard 1984. Pig. In Mason (1984), 145–62.

Flannery, K. 1969. Origins and ecological effects of early domestication in Iran and the Near East. In *The domestication and exploitation of plants and animals*, P.J. Ucko & G.W. Dimbleby (eds), 73–100. London: Duckworth.

Gabriel, B. 1973. Steinplätze: Feuerstellen neolithischer nomaden in der Sahara. *Libyca* 21, 151–8.

Gabriel, B. 1987. Palaeoecological evidence from neolithic fireplaces in the Sahara. *African Archaeological Review* 5, 93–104.

Garrard, A., S. Colledge, C. Hunt, R. Montague 1988. Environment and subsistence during the late Pleistocene and early Holocene in the Azraq Basin. *Paléorient* 14, 40–49.

Garrard, A., S. Colledge, L. Martin 1995. The emergence of crop cultivation and caprine herding in the "marginal zone" of the southern Levant. In *The origins and spread of agriculture and pastoralism in Eurasia*, D.R. Harris (ed.), 204–226. London: UCL Press.

Gautier, A. 1980. Contributions to the archaeozoology of Egypt. In *Prehistory of the eastern Sahara*, F. Wendorf & R. Schild (eds), 317–44. New York: Academic Press.

Gautier, A. 1984. Archaeozoology of the Bir Kiseiba region, eastern Sahara. In *Cattle-keepers of the eastern Sahara: the neolithic of Bir Kiseiba*, A.E. Close (ed.), 49–72. Dallas: SMU Press.

Gautier, A. 1987. Prehistoric men and cattle in North Africa: a dearth of data and a surfeit of models. In *Prehistory of arid North Africa*, A. Close (ed.), 163–87. Dallas: SMU Press.

Gilman, A. 1975. *The later prehistory of Tangier, Morocco*. Cambridge, Mass.: Peabody Museum of Archaeology and Ethnology.

Goring-Morris, N. 1993. From foraging to herding in the Negev and Sinai: the early to late neolithic transition. *Paléorient* 19, 65–89.

Gray, A.P. 1972. *Mammalian Hybrids*. London: Commonwealth Agricultural Bureaux.

Grigson, C. 1989. Size and sex: evidence for the domestication of cattle in the Near East. In *The beginnings of agriculture*. A. Milles, D. Williams, N. Gardner (eds), 77–109. BAR International Series 496. Oxford: BAR.

Grigson, C. 1991. An African origin for African cattle? – some archaeological evidence. *African Archaeology Review* 9, 119–44.

Groves, C.P. 1986. The taxonomy, distribution, and adaptations of recent equids. In *Equids in the ancient world*, R.H. Meadow & H.-P. Uerpmann (eds), 11–51. Wiesbaden: Ludwig Reichert.

Haaland, R. 1995. Sedentism, cultivation, and plant domestication in the Holocene Middle Nile Region. *Journal of Field Archaeology* 22, 157–74.

Hawass, Z., F. Hassan, A. Gautier 1988. Chronology, sediments and subsistence at Merimde Beni Salama. *Journal of Egyptian Archaeology* 74, 31–8.

Kajale, M.J. 1991. Current status of Indian Paleoethnobotany: introduced and indigenous food plants with a discussion of the historical and evolutionary development of Indian agriculture and agricultural systems in general. In *New light on early farming*, J. Renfrew (ed.), 155–91. Edinburgh: Edinburgh University Press.

Kingdon, J. 1997. *The Kingdon field guide to African mammals*. London: Academic Press.

Köhler-Rollefson, I. 1989. Resolving the revolution: late neolithic refinements of economic strategies in the eastern Levant. *Archaeozoologia* 3, 201–8.

Legge, T. 1996. The beginning of caprine domestication in southwest Asia, In *The origins and spread of agriculture and pastoralism in Eurasia*, D.R. Harris (ed.), 238–62. London: UCL Press.

Loftus, R.T., D.E. MacHugh, D.G. Bradley, P.M. Sharp, P. Cunningham 1994. Evidence for two independent domestications of cattle. *Proceedings of the National Academy of Sciences, USA* 91, 2757–61.

MacDonald, K.C. 1994. *Socio-economic diversity and the origins of cultural complexity along the Middle Niger (2000 BC to AD 300)*. Unpublished PhD dissertation, University of Cambridge.

MacDonald, K.C. 1996a. Tichitt-Walata and the Middle Niger: evidence for cultural contact in the second millenium BC. In Pwiti & Soper (1996), 429–40.

MacDonald, K.C. 1996b. The Windé Koroji Complex: evidence for the peopling of the eastern Inland Niger Delta (2100–500 BC). *Préhistoire Anthropologie Méditerranéennes* 5, 147–65.

Mason, I.L. (ed.) 1984. *Evolution of domesticated animals*. London: Longman.

McIntosh, S.K. (ed.) 1995. *Excavations at Jenné-Jeno, Hambarketolo, and Kaniana (Inland Niger Delta, Mali). The 1981 Season*. Berkeley: University of California Press.

Meek, R.L. 1976. *Social Science and the ignoble savage*. Cambridge: Cambridge University Press.

Mongin, P. & M. Plouzeau 1984. Guinea-fowl. In Mason (1984), 322–5.

Morel, J. 1982. Les Pierres à gorge du Sahara. Inventaire provisoire et essai d'interprétation. *Journal des Africanistes* **52**, 68–94.

Morrison-Scott, T.C.S. 1952. The mummified cats of Ancient Egypt. *Proceedings of the Zoological Society of London* **121**, 861–7.

Munson, P.J. 1971. *The Tichitt tradition: a late prehistoric occupation of the southwestern Sahara*. PhD thesis, Department of Anthropology, University of Illinois at Urbana-Champaign.

Munson, P.J. 1976. Archaeological data on the origins of cultivation in the southwestern Sahara and their implications for West Africa. In *Origins of African plant domestication*, J. De Wet & A. Stemler (eds), 187–209. The Hague: Muton.

Munson, P.J. 1989. About *Economie et Société Néolithique du Dhar Tichitt* (Augustin Holl). *Sahara* **2**, 107–9.

Muzzolini, A. 1983. *L'Art rupestre du Sahara central: classification et chronologie. Le bouef dans la préfihistoire Africaine* [2 volumes]. Aix-en-Provence: Université de Provence.

Muzzolini, A. 1987. Les premiers ovins et caprins du Sahara d'après l'Art rupestre. *Archaeozoologia* **1**, 129–48.

Muzzolini, A. 1989. Les debuts de la domestication des animaux en Afrique: faits et problemes. *Ethnozootechnie* **42**, 7–22.

Muzzolini, A. 1993. L'origine des chèvres et moutons domestiques en Afrique. Reconsidération de la thèse diffusionniste traditionnelle. *Empuries* **48–50**, 160–71.

Neumann, K. 1989. Zur vegetationsgeschicte der Östsahara im Holozän: Holzkohlen aus prähistorischen fundstellen. In *Forschungen zur Umweltgeschichte der Östsahara*, R. Kuper (ed.), 13–181. Africa Praehistorica 2. Köln: Heinrich-Barth Institut.

Neumann, K., A. Ballouche, M. Klee. 1996. The emergence of plant food production in the West African Sahel: new evidence from northeast Nigeria and northern Burkina Faso. See Pwiti & Soper (1996), 441–8.

Pachur, H.-J. 1991. Tethering stones as palaeoenvironmental indicators. *Sahara* **4**, 13–32.

Pwiti, G. & R. Soper (eds) 1996. *Aspects of African archaeology. Papers from the 10th Congress of the Pan-African Association for Prehistory and Related Studies*. Harare: University of Zimbabwe Publications.

Reed, C.A. (ed.) 1977. *Origins of agriculture*. The Hague: Mouton.

Rollefson, G.O. & I. Köhler-Rollefson 1993. PPNC adaptations in the first half of the 6th millennium BC. *Paléorient* **19**, 33–42.

Rowley-Conwy, P. 1988. The camel in the Nile valley: new radiocarbon accelerator (AMS) dates from Qasr Ibrim. *Journal of Egyptian Archaeology* **74**, 245–8.

Rowley-Conwy, P. 1991. Sorgum from Qasr Ibrim, Egyptian Nubia, *c.* 800 BC–AD 1811: a preliminary study. In *New light on early farming*, J. Renfrew (ed.), 191–212. Edinburgh: Edinburgh University Press.

Shaw, B. 1979. The camel in Roman North Africa and the Sahara: history, biology, and human economy. *Bulletin de l'IFAN (séries B)* **41**, 663–721.

Shaw, T. 1977. Hunters, gatherers and first farmers in West Africa. In *Hunters, gatherers and first farmers beyond Europe*, J.V.S. Megaw (ed.), 69–126. Leicester: Leicester University Press.

Smith, A.B. 1984. The origins of food production in northeast Africa. *Palaeoecology of Africa* **16**, 317–24.

Smith, A.B. 1986. Cattle domestication in North Africa. *African Archaeological Review* **4**, 197–203.

Smith, A.B. 1992. *Pastoralism in Africa: origins and development ecology*. London: Hurst.

Stemler, A. 1990. A scanning electron microscopic analysis of plant impressions in pottery from the sites of Kadero, El Zakiab, Um Direiwa and El Kadada. *Archéologie du Nil Moyen* **4**, 87–106.

Vermeersch, P.M., P. Van Peer, J. Moeyersons, W. Van Neer 1996. Neolithic occupation of the Sodmein area, Red Sea Mountains, Egypt. See Pwiti & Soper (1996), 411–19.

von den Driesch, A. & J. Boessneck 1985. *Die Tierknochenfunde aus der Neolithischen Siedlung von Merimde-Benislama am Westlichen Nildelta*. Cairo: Deutsches Archäologisches Institut.

Wasylikowa, K., J.R. Harlan, J. Evans, F. Wendorf, R. Schild, A.E. Close, H. Korolik, R.A. Housley 1993. Examination of botanical remains from early neolithic houses at Nabta Playa, western Desert,

Egypt, with special reference to Sorgum Grains. In *The archaeology of Africa. Foods, metals and towns*, T. Shaw, P. Sinclair, B. Andah, A. Okpoko (eds), 154–64. London: Routledge.

Wendorf, F., A.E. Close & R. Schild 1987. Early domestic cattle in the eastern Sahara. *Palaeocology of Africa* **18**, 441–8.

Wendorf, F. & R. Schild 1984. The emergence of food production in the Egyptian Sahara. In *From hunters to farmers: the causes and consequences of food production in Africa*, J.D. Clark & S.A. Brandt (eds), 93–101. Berkeley: University of California Press.

Wendorf, F. & R. Schild 1994. Are the early Holocene cattle in the eastern Sahara domestic or wild? *Evolutionary Anthropology* **3**, 118–28.

Wendorf, F. & R. Schild (eds) 1980. *Prehistory of the eastern Sahara*. New York: Academic Press.

Wendorf, F. & R. Schild (assemblers) & A.E. Close (ed.) 1984. *Cattle keepers of the eastern Sahara: the neolithic of Bir Kiseiba*. Dallas: SMU.

Wendorf, F., R. Schild, N. Zodeno 1996. A late neolithic megalith complex in the eastern Sahara: a preliminary report. In *Interregional contacts in the later prehistory of northeastern Africa*, L. Krzyzaniak, K. Kroeper, M. Kobusiewicz (eds), 125–32. Poznan: Poznan Archaeological Museum.

Westropp, H.M. 1872. Pre-historic phases, or introductory essays on pre-historic archaeology. London: Bell & Daldy.

Zeuner, F.E. 1963. *A history of domesticated animals*. London: Hutchinson.

CHAPTER TWO

A survey of ethnographic and linguistic evidence for the history of livestock in Africa

Roger M. Blench

1. Introduction

In reconstructing the history and evolution of domestic animals, interdisciplinary studies relating archaeozoological materials with iconography, historical texts, genetics, animal production data, contemporary ethnography, and linguistics are essential to create a rich and convincing description of the past and its links with present production systems. Results from these different disciplines are, however, not always easily synthesized. This is partly because sampling strategies are generally not co-ordinated, leading to patchy availability of data. However, the different styles of data presentation and even types of argumentation are often difficult for disciplinary scholars to integrate. Nonetheless, recent developments especially in molecular genetics have made the process of synthesis essential if coherent models are to be developed.

Table 2.1 shows the different disciplines used for the reconstruction of prehistory of domestic animals in Africa and tabulates various features associated with both their collection and availability. It gives impressionistic estimates both of the type and amount of data available in specific disciplines and also the extent to which such data has been exploited. This introduction considers the nature of the evidence for their potential and actual contribution to livestock prehistory drawn from different disciplines, their relative progress and likely future developments.

2. Archaeology

Archaeology deals in point data, namely archaeological sites. For every site excavated, many more have been identified; their excavation depends on the availability of resources, both human and financial, and a stable political and administrative framework within which to operate. The likely finds are also important; it is no

Table 2.1 Types of data and its uses in reconstructing livestock prehistory in Africa.

	Samples	Precision	Dating	Degree of Exploitation in			
				North Africa	East and South Africa	Central Africa	West Africa
Archaeology	Small number of point samples	High	High	High	Medium	Low	Low
Iconography	Highly variable	High	Medium/ Low	High	Medium	Low	Medium
Textual	Very small, chronologically limited	Medium	High	High	Medium	Low	Medium
Genetics	Large Number	High	Medium	Low	Low	Low	Low
Animal production	Large Number	Low	None	Low	Low	Low	Low
Ethnography	Very small number	Very low	None	Medium	Medium	Medium	Medium
Linguistics	Very small number	Low	Low	Medium	Medium	Medium	Medium

accident that Egypt has seen a greater concentration of resources than the rest of the continent aggregated. Egypt has produced and continues to yield remarkable art objects, texts and iconography that allow almost unparalleled access to patterns of subsistence as much as 5,000 years ago. It has a hold on popular imagination quite unlike any other region, ensuring a continuing flow of resources. Striking monuments such as Axum and Zimbabwe create a public profile of a region that enables funding of archaeology at nearby sites even where they have little or nothing to do with the monuments in question. An extensive and well-resourced university and museum system also encourages archaeology as the concentrations of sites in South Africa demonstrates.

Preservation is also a significant feature. The Sahara is more likely than other regions to produce well-preserved complete remains and preferentially attracts researchers. The acid soils of the humid forests, by contrast, make sites harder to find and the results of excavations less spectacular. The result is that excavations are extremely unevenly distributed; northen Africa has been extensively sampled, whereas west–central Africa has an extremely limited number of sites.

Fashion and the salience of particular questions in a given region are an important determinant of how much archaeozoology is carried out. Marshall (Ch. 10 in this volume) points out that pre-Iron Age sites in eastern Africa have a substantial amount of faunal analysis related to their presumed pastoral subsistence strategies. However, nearer to the present, where the questions relate to the expansion of the

Bantu and the identification of sites that are presumed to correlate with this, faunal assemblages are treated casually and published either in summary form or not at all.

3. Iconography

The major sources of iconography are the rock-paintings and engravings found throughout the continent, and best preserved in the arid and semi-arid regions. In the case of Egypt and North Africa, wall-paintings and artefacts are an important source of data, especially as they can usually be dated precisely. Model animals in clay occasionally turn up in excavations and these can be used to determine the presence of a species at a given period or establish the existence of gross anatomical features such as the humps on cattle. Early drawings and engravings can also be of value, especially where they record animals or a production system now vanished (for example, the Khoi of southern Africa documented by Andrew Smith, Ch. 11 in this volume).

Rock art has two problematic aspects, dating and the selectivity of the artists. Rock art cannot usually be dated directly, although techniques are becoming available to do this. It is therefore dated on stylistic grounds or occasionally by associated artefacts. Superposition and patina enable the establishment of rather general chronologies, but the considerable debate within the scholarly world on coherence of style must imply that these can be used at only the most general level.

Rock art creates positive evidence; the representation of animals and practices suggest their presence and importance in the mind of the artists. However, it does not necessarily mean that the item represented in the locale of the rock art actually existed there. The schematic representation of wheeled vehicles in the southern Sahara suggests that the painters may have heard about such vehicles rather than seen them directly. Similarly, absence on rock art does not imply absence in reality. In the case of animal representation, cultural or economic salience is all. The majority of images are of large and medium-sized mammals, principally those hunted. As domestic stock became more important, cattle, camels, sheep and goats are represented. Chickens, pigeons, dogs, cats and other species are rare or absent in the Sahara although there is strong evidence that they crossed the desert.

The same reservations apply to wall-paintings and historical iconography, although the presence of co-occurring texts often act as a check on the visual representations. Rock art may be presumed to be a collective art, somehow synoptic of the desire and imagination of the people who made it. Wall-paintings, however, are very strictly political, expressive of the status and authority of those who commissioned them. As such they must be read with additional care when generalizing from image to narrative.

Some of the same reservations apply when using images from early travellers' narratives and ethnographies. A primary bias of such literature is its coastal emphasis; until the middle of the nineteenth century, few travellers were able to sustain long journeys into the interior. This means that there are fewer records of early pastoral systems since most of the coastal regions are not arid or semi-arid. Travellers tended to be fascinated by the exotic and recorded unusual practices more than the everyday.

4. Textual

Historical textual material on African livestock can be divided into three main categories: a) ancient North African texts (Egyptian, Greek, Latin); b) Arabic texts; c) early texts in European languages. Although there is some small inscriptional material in other North African languages (Phoenician, Old Libyan) it is mostly too fragmentary to provide more than names of species. There are also references to livestock in the classical texts of Ethiopia in the Ge'ez language. Many texts have been published, but there is no research systematically combing them for information about livestock, although some information is given in Pankhurst (1968). Most of the classical and Egyptian sources have been published and analyzed at some length and are usefully summarized in Epstein (1971), Boessneck (1988) and Brewer et al. (1994, but see also MacDonald & Nesbitt 1995) as well as individual articles in the *Lexicon der Aegyptologie*.

The principal Arabic sources for information on West African from the eighth century onwards are the writings of geographers and travellers. Almost all of these exist in some edition, although not necessarily a modern one, and have been translated into a major European language. In the case of West Africa, the corpus of sources has been conveniently assembled in a single volume of translations (Levtzion & Hopkins 1981). Lewicki (1974), meanwhile, is a useful synthesis of everything related to food in these sources, which inevitably collect most material on livestock. For East Africa, the Arabic corpus is more scattered and texts must be searched individually.

Descriptive texts in European languages are again summarized in Epstein (1971) and more briefly in individual contributions to Mason (1984). As with iconography, there is a bias towards the coastal and the exotic, compounded often by the sometimes limited descriptive vocabulary of authors not specialized in livestock terminology.

5. Ethnography

Scattered throughout travellers' accounts and professional anthropological monographs describing Africa are extensive materials on systems of livestock production. Most references do not usually include the detail that a professional zoologist would prefer, but they still represent a major source of information concerning both the species, races and management systems of domestic animals.

Much of this information is biased towards large animals. Evans-Pritchard (1940) has much to say about Nuer cattle-keeping, but makes only passing reference to small ruminants and completely ignores other types of backyard stock. Anthropologists, in particular, are prone to be influenced by the values and symbolic system of those among whom they work. Again, taking the example of the Nuer, the high symbolic value of cattle among Nilotic peoples determines that much of the investigation should focus on cattle. Agriculture, which the Nuer also practise, was given a lower status by the anthropologist and so much less space was allotted to describing it. Traditional anthropologists would usually defend this approach as "actor-focused", i.e. if the people being studied attach symbolic value to the cattle then the task of the anthropologist is to describe preferentially that system. A more modern economic anthropology, however, would attempt to rectify the balance by taking as its starting point different contributions to subsistence. In other words, cattle-focused accounts of Nilotic society would not provide a convincing account of Nuer subsistence strategies. They are, moreover, strongly biased by male ideology, while it is women who make major contributions to agriculture.

Although unfashionable among social scientists, the type of descriptive ethnography written by colonial officers and missionaries who attempted to cover all aspects of a society without any very explicit theoretical framework are often of much greater value to the historian of livestock. Although the colonial ideology is easily deconstructed in fashionable seminar rooms, this does not affect their concise descriptions of pigeon-keeping or the place of pigs in a system of sacrifice.

Such accounts were commonly collated by the cultural geographers of the North European traditions. In Germany and Sweden, in particular, "ethnology" was held to consist of the collection of accounts of particular practices or cultural items and their mapping. In the case of Frobenius, and in particular of the folios of the *Atlas Africanus*, this was to illustrate rather wayward theories of cultural strata in Africa. However, scholars such as Lagercrantz (e.g. 1950), who did most towards mapping livestock-related practices, appear to have had no very explicit agenda. In some ways this had slightly unfortunate results. Lagercrantz painstakingly plotted the distribution of references to, for example, geese, combining and conflating all occurrences of geese whether they were recent European introductions or records of Ancient Egypt.

Nonetheless, with a more explicit historical agenda, it is possible to re-analyze this data by including only occurrences of a species, breed or practice falling within a specific category. What is usually revealed are highly skewed distributions in need of interpretation. The domestic pigeon is a good example of this. Although its ancestor, the rock-pigeon, *Columba livia*, is part of the indigenous fauna of Africa, keeping pigeons in cotes only occurs in specific geographical regions (see Blench Ch. 20 in this volume). With this in mind, it is possible to ask archaeologists, zoologists and linguists for possible correlates of such a distribution.

6. Genetics

Genetics can be divided into two categories, corresponding to the categories of phenotypic and molecular. Determining the races, species and wild antecedents of African domestic animals through comparative anatomy has a venerable history, going back at least to Darwin's identification of the rock-pigeon, *Columba livia*, as the ancestor of the domestic pigeon. In a more elaborate form it is represented by a series of monographs combining comparative anatomy with ethnography, beginning with Hahn (1896) via Doutressoulle (1947), Boettger (1958), Mason & Maule (1960) and reaching a climax in Epstein's (1971) 2-volume masterpiece *The origin of the domestic animals of Africa*.

Phenotypic work of this type continues, but since the early 1980s the growing availability of molecular techniques, in particular the ability to compare DNA between species and breeds, has led to a radically different approach to evolutionary genetics. This has been most apparent in the case of human genetics and has led to some major revisions of models of human phylogeny. Approaches to the phylogeny of livestock have initially concentrated on breeds of importance to Europe and America, but a more global approach encouraged by FAO has flourished, especially in Dublin (see papers by Cunningham, Bradley & Loftus, Meghen et al., and references therein, Chs 12, 13, 15, respectively in the present volume). The announcement that domestic cattle had two distinct evolutionary origins was dramatic enough to make the non-scientific press. The hypothesis that zebu and humpless cattle were of different stock had in fact a venerable history among German anatomists. However, it had remained a speculation largely discounted by mid-century scholars. DNA, however, produces quantified and reproducible answers and so generates results of a different order of certainty than comparative anatomy.

The potential of molecular biology to resolve many of the troubling questions of African livestock history is undoubted. The calibration of molecular clocks which might allow the dates of splitting into particular races to be established has only just begun. Sampling procedures for clarifying particular breeds and species need to be more coherent and the questions asked by archaeologists, cultural geographers and linguists need to be unified more effectively with the procedures of geneticists. Nonetheless, the techniques are in principal available; the major constraint will henceforth be resources.

7. Animal production

A literature that is often ignored or only mentioned in passing by archaeozoologists is the animal production literature, much of it related to development agencies. The FAO, for example, has made substantial contributions to the description and categorization of livestock breeds, and this work has accelerated with the growing realization of the loss of genetic diversity. For example, the single most complete account of African humpless cattle, is ILCA (1979) (with FAO 1987), which collated everything known about the status and conformation of these cattle in west–central Africa. As with the ethnological literature, this enables the compilation of a type of breed geography, and in turn suggests a potential for historical interpretation.

Beyond simple occurrence, however, is the distribution of production systems and the actual productivity of animals within those systems. One of the hard science models of animal productions is represented by experimental station data. In Africa, this was classically represented by the dedicated research centre or the university farm. Livestock were bought from local producers, fed on a variety of experimental diets, crossed with imported breeds, and their productivity in milk, meat and fecundity carefully recorded. This produced a great variety of research papers, but was almost completely useless for any other purpose, since no livestock were likely to be kept under those conditions elsewhere in the continent.

In recent years there has been a widespread recognition that productivity under conditions resembling those of traditional management is a key element in understanding African livestock production whether past or present. A major text in terms of anthropological understanding of pastoralism is Dahl & Hjort (1976), which argued that pastoralism cannot be understood without a more sophistic-ated model of the constraints that the productivity and behaviours of particular spices impose upon producers. Such insights lie behind detailed studies of spe-cific systems such as the pastoral systems of northern Niger described in Swift (1984).

Despite the best efforts of developers, most animals and animal products that reach the market continue to be produced by traditional sector, especially in west–central Africa. As a result, there has been a major switch from attempting to acclimatize exotic races to documenting and measuring the productivity of animals within traditional production systems. Some examples of the methods used are given in Hall (Ch. 16 in this volume) and in RIM (1992) and typical results are cited in Blench (Chs 20, 21 in this volume).

In general, the productivity of animals away from farms is lower than those under controlled conditions. However, various demonstrations suggest that overall productivity (milk + meat + calves) may be higher than on commercial farms emphasizing one product. As with any type of ethnoarchaeology, it is crucial to understand the extent to which results from the present can be read back into the past. Bone assemblages in Africa are usually too small to make more than educated guesses at herd structure (see MacDonald & MacDonald Ch. 8, and Marshall, Ch. 10, this volume) and under certain conditions the bones are too fragmentary to identify a significant percentage. Nonetheless, present-day data on the viability of a herd (i.e. the number of animals required for stock to be replaced while supporting a sustainable offtake) can help understand past sites and thus production systems. Similarly, the different herd structures implied by herds kept for milk, milk and blood, meat or mixtures of these can be modelled in the present and used to interpret the past.

8. Linguistics

The use of linguistics in reconstructing African livestock history is represented by a long scholarly tradition beginning with the speculations of Barth (1862) on the diffusion of the pigeon and the names for "dog" in Africa (see discussion in Blench Ch. 20 in this volume). In the 1880s, Harry Johnston (1886) set out clearly the method of reconstructing Bantu culture history through linguistics and used chicken as one of his examples.

Linguistics can be used in two distinct ways to shed light on the otherwise undocumented history of domestic animals: (a) animals of ancient establishment but uncertain antiquity can be reconstructed historically with particular language phyla; (b) recent introductions can be traced through the movement of loanwords from one language to another.

The tradition of linguistic reconstruction is well traced out in other parts of the world for other language phyla. The reconstruction of "horse" in Indo–European has long been held to be crucial to the understanding of the identity of the proto-Indo–Europeans. The reconstruction of "pig", "dog" and "fowl" in Oceanic (Austronesian) enables us to establish the subsistence strategies of the colonizers of the Pacific. Attempts to use similar methods in Africa exist, notably a series of papers by Ehret (1967, 1968) and more recently Bender (1982) and Blench (1993a, 1995). Within the present volume, the papers by Blench (Chs 20–22), Bechhaus-Gerst (Ch. 24) and Williamson (Ch. 23) use linguistic data.

A key element of this type of historical linguistics is its reconstructions, usu-ally denoted by an asterisk * and often referred to as "starred forms". These are abstract forms, derived from attested languages that a researcher claims are part of a hypothetical proto-language. Thus an author citing * plus a formula for a word is implying that it formed part of the proto-language spoken by the particular recon-structed group. For example, if it is claimed that "dog" can be reconstructed to proto-Afroasiatic, it means that wherever and wherever proto-Afroasiatic was spoken, that society was familiar with the dog.

For the archaeologists, trying to reconcile the results of linguists with the more concrete evidence of radiocarbon dates, claims by linguists can be perplexing. For example, not all linguists agree on the classification of African languages (see Blench 1993b, 1999, for a description of both mainstream and speculative views), and the homelands of its major phyla constitute a major arena of disagreement. Ehret (1993) made major claims for the internal reconstruction of Nilo–Saharan and the anti-quity of both cultivation and livestock production among its speakers. Bender (1996) has strongly questioned the reconstruction and its implications.

The lesson is almost certainly to be wary of grandiose claims. Small, local-level reconstructions or tracing the progress of loanwords across a region are far more likely to be trustworthy than continent-wide reconstructions. Linguists will un-doubtedly continue to launch speculations and these can create useful tools for thinking. Unlike cereals and other domestic plants, livestock are older and are apparently more linguistically stable; it is certainly tempting to reconstruct them in advance of local-level reconstructions. As dates and sites become more numerous it should be that linguistic and archaeological results gradually being to show more harmony.

9. Conclusion

Calls for greater interdisciplinary scholarship have become something of a cliché in this type of literature. Yet if any area of research demands a subtle and thoughtful integration of past and present, of the ethnographic and the archaeological, and of hard science and sensitive cultural description, it is surely the history of domestic animals. So much that is still practised today is reflected in the archaeological record and so many ideas and conceptions about animals persist in the genetics and phenotypes of livestock today. Western conceptions of livestock development have come close to destroying traditional breeds and systems of production in many areas. Only the rational resistance of African farmers to inappropriate and unwanted exotics has allowed the conservation of a rich genetic heritage.

The last decade has seen a gradual turnaround in this area, from pejoration to respect. It has become clear that pastoral production systems are highly effective in using marginal land and that the survival qualities of traditional breeds under conditions of extreme moisture stress must be conserved. This in turn has led to a greater respect and understanding of the historical point of view from animal production experts and geneticists. There is every reason to hope that academic traffic will also flow in the opposite direction, that archaeozoologists will also make use of the descriptions of geneticists and ethnographers to make sense of their own finds. Similarly, the large database of information that linguists now command in relation to African livestock terminology can be more effectively analyzed and interpreted as the domestic animals of Africa past and present become better known and the white spaces on the map cease to be filled merely by speculation. Under all circumstances there can be no return to the previous situation where each discipline worked in almost conscious ignorance of the others.

References

Barth, H. 1862. *Collection of vocabularies of Central African languages*. Gotha: Justus Perthes.

Bender, M.L. 1982. Livestock and linguistics in North and East African ethnohistory. *Current Anthropology* **23**, 316–18.

Bender, M.L. 1996. *The Nilo-Saharan languages: a comparative essay*. Munich: Lincom Europa.

Blench, R.M. 1993a. Ethnographic and linguistic evidence for the prehistory of African ruminant livestock, horses and ponies. See Shaw et al. (1993), 71–103.

Blench, R.M. 1993b. Recent developments in African language classification and their implications for prehistory. See Shaw et al. (1993), 126–38.

Blench, R.M. 1995. A History of domestic animals in northeastern Nigeria. *Cahiers de Science Humaine ORSTOM* **31**(1), 181–238.

Blench, R.M. 1999. The languages of Africa: macrophyla proposals and implications for archaeological interpretation. In *Archaeology and Language IV*. R.M. Blench & M. Spriggs (eds). London: Routledge.

Boessneck, J. 1988. *Die Tierwelt des Alten Ägypten*. München: C.H. Beck.

Boettger, C. 1958. *Die Haustiere Afrikas*. Jena.

Bradley, D.G., D.E. MacHugh, E.P. Cunningham, R.T. Loftus 1996. Mitochondrial diversity and the origins of African and European cattle. *Proceedings of the National Academy of Sciences, USA* **93**, 5131–5.

Bradley, D.G., D.E. MacHugh, R.T. Loftus, R.S. Sow, C.H. Hoste, E.P. Cunningham 1994. Zebu-taurine variation in Y chromosomal DNA: a sensitive assay for genetic introgression in West African trypanotolerant cattle populations. *Animal Genetics* **25**, 7–12.

Brewer, D.J., D.B. Redford, S. Redford 1994. *Domestic plants and animals: the Egyptian origins*. Warminster: Aris & Phillips.

Dahl, G. & A. Hjort 1976. *Having herds: pastoral herd growth and household economy.* Stockholm Studies in Social Anthropology, No. 2. Stockholm: University of Stockholm.

Doutressoulle, G. 1947. *L'élevage en Afrique occidentale française.* Paris: Editions Larose.

Ehret, C. 1967. Cattlekeeping and milking in eastern and southern African history: the linguistic evidence. *Journal of African History* **8**(1), 1–17.

Ehret, C. 1968. Sheep and central Sudanic peoples in southern Africa. *Journal of African History* **9**(2), 213–21.

Ehret, C. 1993. Nilo–Saharans and the Saharo–Sudanese neolithic. See Shaw et al. (1993), 104–25.

Epstein, H. 1971. *The origin of the domestic animals of Africa* [2 volumes]. New York: African Publishing.

Evans-Pritchard, E.E. 1940. *The Nuer: a description of the modes of livelihood and political institutions of a Nilotic people.* Oxford: Clarendon Press.

FAO 1987. *Trypanotolerant cattle and livestock development in West and Central Africa* [2 volumes]. Rome: FAO.

Hahn, E. 1896. *Die Haustiere and ihre Beziehungen zur Wirtschaft des Menschen.* Leipzig.

ILCA 1979. *Trypanotolerant livestock in West and Central Africa* [2 volumes]. ILCA Monograph 2. Addis Ababa: ILCA.

Johnston, H.H. 1886. *The Kili-manjaro expedition; a record of scientific exploration in eastern equatorial Africa.* London.

Lagercrantz, S. 1950. *Contributions to the ethnology of Africa,* Studia Ethnographica Upsaliensia, I. Lund: Håkan Ohlssons.

Levtzion, N. & J.F.P. Hopkins 1981. *Corpus of early Arabic sources for West African history.* Cambridge: Cambridge University Press.

Lewicki, T. 1974. *West African food in the Middle Ages.* Cambridge: Cambridge University Press.

MacDonald, K.C. & M. Nesbitt 1995. Review of: Brewer, D.J., Redford, D.B. and S. Redford. 1994. *Domestic plants and animals: the Egyptian origins.* Warminster: Aris & Phillips. *Antiquity* **69**(264), 637–39.

Mason, I.L. (ed.) 1984. *Evolution of domestic animals.* London: Longman.

Mason, I.L. & J.P. Maule 1960. *The indigenous livestock of eastern and southern Africa.* Technical Communication No. 14 of the Commonwealth Bureau of Animal Breeding and Genetics, Edinburgh. Farnham Royal: Commonwealth Agricultural Bureaux.

Muzzolini, A. 1995. *Les images rupestres du Sahara.* Toulouse: Préhistoire du Sahara, 1.

Pankhurst, R. 1968. *Economic history of Ethiopia, 1800–1935.* Addis Ababa: Artistic Printing Press.

RIM 1992. *Nigerian National Livestock Resource Survey* [6 volumes]. Abuja, Nigeria: Report by Resource Inventory and Management Limited (RIM) to FDL&PCS.

Shaw, T., P. Sinclair, B. Andah, A. Okpoko (eds) 1993. *The archaeology of Africa. Food, metals and towns.* London: Routledge.

Skinner, N.A. 1977. Domestic animals in Chadic. In *Papers in Chadic linguistics,* P. Newman & R.M. Newman (eds), 175–98. Leiden: Afrika-Studiecentrum.

Swift, J. (ed.) 1984. *Pastoral development in central Niger: report of the Niger Range and Livestock Project.* Niamey: USAID and Republic of Niger Ministère de Développement Rural.

Zeuner, F.E. 1963. *A history of domesticated animals.* London: Hutchinson.

PART TWO
Archaeology

CHAPTER THREE

Cattle, sheep, and goats south of the Sahara: an archaeozoological perspective

Juliet Clutton-Brock

1. Introduction

The spread of domestic animals southwards from the Sahara to the Cape Province probably had many similarities with the neolithic spread of the same species west and north across Europe. People who had lived for millennia by hunting and gathering slowly altered their ways of life to encompass the ownership of tamed animals for whom they were responsible, leading to fundamental changes in human and animal communities that today provide central points of discussion for archaeologists and anthropologists. In both continents, it is envisaged that the spread occurred by many different agencies, including barter and trade, bride-price, thieving, warfare, and the migration of people themselves with their stock. In Europe, the spread of dogs and domestic livestock was complete by 5,000 years ago while in Africa south of the Sahara it was only just beginning.

The ways in which humans enfolded tamed animals into their societies may have been the same wherever it occurred, but here the similarities end, because in so many other respects the problems faced by the peoples of Africa were very different from those of the first farmers of northern Europe (home of many of the archaeologists who study African prehistory). The first agriculturists and livestock herders in Africa had to keep their plants and animals alive in environments that were usually diametrically opposite to those of Europe: being hot, drought-ridden, and teeming with dangerous predators. But perhaps the greatest enemy of the early farmers would have been the parasites and endemic diseases that are fatal to allochthonous immigrants both human and animal. These remain today and include horse sickness, trypanosomiasis, malaria, schistosomiasis, tick-borne fevers, and so on.

The domestic horse has long been established along the North African littoral and in the Sahel, but along the equator and southwards, African horse sickness was the prime factor in limiting their spread. In the early days of British colonialism it was not unusual for 80–90 per cent of imported horses to succumb to this disease, from which they died a few hours after the first signs of visible symptoms (see, for example, the descriptions by Lord Randolph Churchill 1895:112, 176–7). African horse sickness, which is endemic south of the Sahara, is a peculiarly virulent disease caused by orbiviruses. Very few horses survive infection but if they do they usually gain lifetime immunity. Donkeys and zebras can catch the disease but have a degree of immunity as they usually show few clinical symptoms, and mules inherit a partial immunity (Hine 1988). Livestock in the tsetse fly belts were similarly affected with the fatal trypanosomiasis, as in recent times were the oxen of the *voortrekkers* travelling north from the Cape.

The large mammals of Africa escaped most of the wide-scale extinctions that occurred in Europe, Asia, and North America at the end of the Pleistocene, so that even today (at least in some wildlife reserves) we can see much of the range of species of large mammals that populated the earth before the advent of human hunters. Almost a hundred large ungulates, including elephant, rhinoceroses, giraffe, hippopotamus, suids, around 78 species of bovids, and many species of carnivores give the African fauna a unique character, described by Bigalke (1978) as "pre-Pleistocene" in composition (by which I assume he meant pre-glacial). It was in this environment that the first agriculturists had to establish themselves, south of the Sahara.

I find it strange that in all the writings of archaeologists about the spread of farming in Africa there is hardly a mention of the huge problems that elephants and all the other crop-robbers must have caused, and the vast losses that the human populations must have sustained not only to their crops and animals but also to their own numbers. Before the arrival of the Europeans the grasslands of Africa were teeming with ungulates and their predators, and to me it is truly amazing that human settlements managed to develop and flourish against such odds. It would take a brave group of humans to live among great clans of elephants, as described for example by the "big game" hunter, Cornwallis Harris (1839:203) in the Cashan mountains (latitude south 26 degrees – not far from Johannesburg):

The whole face of the landscape was actually covered with wild elephants. There could not have been fewer than three hundred within the scope of our vision. Every height and green knoll was dotted over with groups of them, whilst the bottom of the glen exhibited a dense sable living mass.

In a recent review of the conflict between elephants and people in the Central African forests, Barnes (1996) gives details of how farmers claimed that their forefathers lived on the brink of starvation as their food supply was regularly devastated by elephants. The conflict between people and elephants has ancient origins as described by the Greek historian Diodorus Siculus (80–20 BC) in his writings about ancient Egypt:

In that part of the country which lies along the Nile in Libya there is a section which is remarkable for its beauty ... consequently this region is a bone of contention between the Libyans and the Ethiopians who wage unceasing warfare for its possession. It is also the gathering place for a multitude of elephants from the country lying above it because, as some say, the pasturing is abundant and sweet. ... Consequently they remain there and destroy the means of subsistence of the human beings; and because of this the inhabitants are compelled to flee from these regions, and to live as nomads and dwellers in tents (book III, 3–4).

Even today, elephants can cause persistent damage to crops and drive villagers away from their homes. I consider that archaeologists, in seeking to interpret pre-historic sites, should take much greater account of the ways in which buildings and walls would have been constructed to keep away carnivores and wild ungulates including elephants. For example the wide, low stone walls that surround the ruined circular buildings at the site of Ziwa (formerly called the Van Niekerk ruins) in Nyanga, Zimbabwe, could well have been built by early farmers as a protection against elephants whose prying and marauding trunks would not be able to reach across the walls. Similarly, at the same site, the purpose of the maze of low stone walls encircling small patches of ground and leading up to the circular buildings could have been to protect growing vegetables from grazing animals, both wild and domestic. However these thoughts must remain as speculation since, with the demise of these communities in Zimbabwe several hundred years ago, there are no memories or ethnographic parallels to divulge the purpose of the massive stone walls.

2. The spread of livestock herding southwards

It is certain that the early agriculturists in Africa had much to contend with that was absent in the neolithic of Europe where, broadly, there was a temperate climate and where the danger from large carnivores was limited to bears, wolves, and the occasional lion and hyena in the east. In Africa, humans struggled against much greater adversities and that the spread of agriculture and herding took a long time is to be expected. It began with the southward movement of domestic cattle from the Mediterranean fringe and the eastern Sahara, perhaps as long as 8,000 years ago, and in some places it is even now incomplete: for there are still peoples who live largely or entirely by hunting, fishing, and gathering (see MacDonald Ch. 1 and Hassan Ch. 5 in this volume).

2.1. Cattle

It is generally assumed that, around 5,000 years ago, as a result of increasing desiccation of the Sahara, people and cattle began to move southwards, together with the tsetse fly barrier. This southward spread of cattle has been dated archae-ologically and by radiocarbon dates, as cited in Chapters 5, 8, 9, 10 and 11 and in Smith (1992). The approximate dates are summarized as follows:

- Eastern Sahara to Sudan (from indigenous *Bos primigenous africanus*) perhaps as early as 9000 bp
- Pre-Dynastic Egypt – *c*. 6000 bp
- Hoggar Mountains – 6500–4500 bp
- Kadero, north of Khartoum – *c*. 6000 bp
- Adrar Bous in Niger – *c*. 5760 bp
- West African river basins – *c*. 4500 bp
- Gaji 4, Kenya – *c*. 4000 bp
- Zambia and south to the Cape province – *c*. 1600 bp

It appears from the recent findings of molecular biologists, which support the osteological evidence presented by Grigson (Ch. 4 in this volume), that African cattle may have had a separate and autochthonous evolution, both from the taurine cattle of Europe and from the zebu cattle of south Asia (see Loftus et al. 1994, and Chs 13 and 14 in this volume).

It is evident from the dates above that the domestication of cattle occurred in North Africa around the same period as in Europe and Asia, and it can be postulated that there was a consequent rather rapid and dramatic change in many human social systems from the hunting, gathering and foraging culture to that of herding domestic animals. This new economy gathered momentum as it spread and evolved in different ways in adaptation to local conditions of climate and environment. Wherever the soil conditions and the necessary humidity allowed, settlements developed and eventually people began to grow crops as well as herd livestock. But it is assumed that in the arid zones and in the grasslands of Africa south of the Sahara the soil was too poor and the grazing insufficient to allow anything except shifting agriculture coupled with pastoralism.

2.2. Sheep and goats

In recent years there have been dramatic finds of naturally mummified sheep remains from the site of Kerma in lower Nubia, ranging in date from *c*. 4500–3700 bp (Chaix & Grant 1987). As cited by Marshall (Ch. 10 in this volume), these sheep were hairy and most were single-coloured black or white. They were thin-tailed and without the Roman nose of modern desert sheep. In general, the archaeological record shows that sheep together with some goats followed cattle in their spread southwards.

At what period and in what region the dwarf goat first appeared is not known, although the remains of small breeds of both sheep and goat have been identified by Van Neer from the forested region of Central Africa (Ch. 9, in this volume), and MacDonald & MacDonald (Ch. 8 in this volume) have identified remains of dwarf goats from the Middle Niger region of West Africa, with some remains dating to as early as 3,000 years ago. MacDonald & MacDonald (ibid.) believe the mutation for dwarfism was associated with trypanosome-tolerance. Similarly, the origins of fat-tailed sheep are unknown. Muzzolini (Ch. 6 in this volume) believes that they were introduced into Egypt during the Middle Kingdom (*c*. 4050–3786 bp) and into Maghreb during the historic period. He states that there are no fat-tailed sheep in the Saharan rock art.

3. The nature of pastoralism in sub-Saharan Africa

Khazanov (1983) in his classic work on nomadism has reviewed the many different theories on the origins of pastoralism. One of his main tenets was that a specialized pastoral economy could only evolve from a more sedentary form, although he points out that nomadism in Africa was never as homogeneous and specialized as in north Eurasia. His conclusions, however, are in my view only partially relevant to explaining the spread of domestic livestock in Africa south of the Sahara: first, because no primary domestication of wild species of ungulates was initiated in sub-Saharan Africa as it was in western Asia. Secondly, because pastoralists in Africa could not employ the horse so they had no rapid means of migration, and thirdly, because, unlike Eurasia, there remained in sub-Saharan Africa enormous numbers of wild large mammals that could provide meat and other raw materials. Why then did domestic cattle, sheep, and goats become so valuable? Smith (1992:57) suggests that it was because the carcasses of these animals can provide larger amounts of fat than the wild herbivores and this must certainly have been an important factor, as evinced by the prevalence of the fat-tailed sheep.

4. Migration of Bantu-speaking farmers

By 4,000 years ago livestock herding was well-established across Africa, north of the equator, but it did not reach further south than southern Kenya and northern Tanzania (Phillipson 1993:157). As discussed by Vansina (1996:17) a consensus of opinion among linguists and archaeologists had, until recent rethinking, claimed that the first proto-Bantu farming began in the forested region of the Cameroons at around this date of 4,000 years ago. It was presumed that settlement, the adoption of agriculture, and livestock-herding, together with the assets of the new technologies of pottery and iron-smelting, led to a build-up of population that made migration of the Bantu-speaking peoples inevitable. This assumption probably followed from the popular theory that nomadic pastoralism can develop in a society as a result of an increase in the size of herds beyond the grazing capacity of the land around settlements. To outsiders, pastoralists appear to increase the numbers of their livestock beyond their needs for subsistence. This is assumed to be so that the long-term survival of the herds can be assured as well as for economic and social reasons. If unchecked the herds will lead to degradation of the environment, and even desertification, and the need to continually find new grazing. African pastoralists, however, have probably always been well aware of such dangers, as evidenced in this book by the work of Ryan et al. (Ch. 25 in this volume).

Until recently the general belief was that the Bantu-speaking people moved rapidly southwards, in eastern and western streams (see for example Phillipson (1980:344). The herders migrated, taking their cattle, sheep, and goats with them, planting their crops with their new iron tools, and building nation states across the continent. According to the archaeological record these appear to be the facts for the first five centuries of the Christian era, and they support the speculation of Klein (1986) that while cattle moved south with the eastern stream, sheep moved with the Bantu-speakers down the western side of the continent to arrive at the Cape around AD 500. However, in a study of animal remains from sites in Central Africa, Van Neer (Ch. 9 in this volume) has shown that there is very little evidence for domestic livestock from the early Iron Age sites in Cameroon and he does not believe that early Bantu-speaking populations from there were responsible for the propagation of food production, which he believed occurred in the interlacustrine region. This fits with the revised theory of Phillipson (1993:188), who sees the origins of iron-using Bantu-speakers not in Cameroon but in the Chifumbaze complex which first appears in sites around Lake Victoria dated to around 2,500 years ago. Furthermore, the very concept of streams of Bantu-speaking peoples migrating southwards was contested by many contributors to the conference on the growth of farming communities in Africa, held in Cambridge, UK, in July 1994 (Sutton 1996).

Archaeologists, like other scholars, have to begin with simple premises, so it is not surprising that the theory was postulated of Bantu-speaking peoples emerging from their heartland in the Cameroons, taking up pastoralism, and moving southwards fast. This idea seemed to fit the few facts available from sparse excavations and linguistic studies in much the same way as, in an earlier period, did the concept of Indo–European migrations. However, as in all things to do with human cultural systems and ecology, the truth must have been vastly more complicated, as cogently discussed by Smith (1992, and Ch. 11 in this volume). Kinahan (1996) claims that, "The spread of pastoralism into south-western Africa through the migration of ethnically-distinct pastoral groups is no longer tenable as a general explanation for the archaeological evidence". He and other authors in Sutton (1996) also put forward the linguistic and archaeological evidence for the possession of sheep by the Khoisan people of southern Africa before the immigration of the Bantu-speakers.

But what of the pastoralists themselves – were they already present as sparse inhabitants of the continent south of the equator? Did they migrate, or did they simply move from place to place tending their flocks and moving on when their settlements were marauded by other groups of people, by crop-raiding animals, or by an unbearable load of parasitic fleas, lice, and ticks? For what purposes were the herds of cattle, sheep, and goats kept and tended with such care? They would have had to be given regular water and enclosed in a *boma* every night to protect them from carnivores. It is hard to believe, at least in the early phase of livestock herding, that the animals were important as a source of meat, considering the abundance of wild ungulates that could be killed at will. To me it seems more likely that the herds were a basis of wealth and status.

5. Discussion

In propounding views about pastoralism in Africa it is easy for Eurocentric archaeologists to forget the multitude of wild ungulates that inhabited all the grasslands of Africa before the incursions of white people. In the nineteenth century the massacre of wildlife in South Africa was even more devastating than the massacre of bison in North America. In the words of Churchill (1895:75), writing about wildlife in the Transvaal:

> Persons whose word can be implicitly relied upon have informed me that within the last fifteen years they can remember these plains being covered as far as the eye could reach with thousands of wildebeest, blesbok, springbok, and other varieties of the deer [sic] and antelope tribes. So desolate and lifeless is the appearance of these plains now that it is difficult to credit the assertion.

This slaughter was not only carried out by European "sportsmen" who visited the country in droves, but also by the white inhabitants to fulfil a well-paid demand for hides. According to Churchill (1895:75) large shooting parties:

> shot the beasts down by scores, by hundreds, and by thousands, leaving the carcasses to be devoured by the vultures, and going a few days afterwards to gather up the skins which the vultures had neglected, and which the sun had dried and tanned.

In all aspects of study of the complex prehistoric societies in Africa, it can be seen that archaeologists need to break away from the discipline of North European attitudes and scholarship. They should see the southern African continent, before European contact, as a pre-glacial environment in which there was no division into wilderness and human settlement but only a continuum in which humans and animals struggled to survive and develop their separate cultures.

Over the past 2,000 years, within the manifold ecosystems of southern Africa, human societies developed into nations, while their herds of livestock, by a combination of natural and artificial selection, evolved into separate breeds that were perfectly adapted to overcome the varying environmental hazards and many of the diseases that assaulted them. It has been, sadly, another result of colonialism that great efforts have been made to "improve" these "indigenous"[1] livestock by crossbreeding in the light of knowledge about European husbandry systems. Thus, for example, the heat-resistant Afrikander cattle of South Africa, which had most probably been bred by both Khoisan people and Bantu-speakers for nearly two millennia, have been interbred with European breeds, especially the shorthorn and the Aberdeen Angus. This procedure seems to lack all reason as agricultural policy on the African continent. Fortunately, these efforts at "improvement" have been more or less curtailed and a society has been set up in Pretoria to conserve the remaining breeds of "indigenous" livestock in South Africa; this includes approximately six breeds of cattle, four of goats, and six of sheep.

It now seems reasonably certain from their molecular biology that the domestic cattle of North Africa had a separate origin from those in Eurasia. I consider that the next stage of research in this field should be on the fat-tailed, hairy sheep, of which there are many breeds throughout the African continent. Perhaps such studies will be able to divulge their ancestry and show whether these fat-tailed breeds of sheep are descended from those in the Middle East or whether they had a separate origin in the arid regions of Africa (see Muzzolini Ch. 6 in this volume).

I believe it is time for archaeology, and perhaps even more so, archaeozoology, to enter a new phase of synthesis in Africa and to escape from the strictures of thought that are implicit even in post-colonial attitudes. As an ending, I can do no better than to quote de Maret (1996:321):

We have tended to oversimplify highly complex issues, to consider events rather than process, to construct tree-like models rather than networks, to imagine homogenous areas rather than patchwork, and to rely too much on natural rather than cultural factors to explain social evolution.

Note

1. The term "indigenous" is used here to denote presence of the domestic species in Africa before European contact.

References

Barnes, R.F.W. 1996. The conflict between humans and elephants in the Central African forests. *Mammal Review* **26**, 67–80.

Bigalke, R.C. 1978. Present day mammals of Africa. In *Evolution of African mammals*, V.J. Maglio & H.B.S. Cooke (eds), 1–16. Cambridge, Mass.: Harvard University Press.

Chaix. L. & A. Grant 1987. A study of a prehistoric population of sheep (*Ovis aries* L.) from Kerma (Sudan). *Archaeozoologia* **1**, 93–107.

Churchill, Lord Randolph, S. 1895. *Men, mines and animals in South Africa*. London: Sampson Low, Marston.

Cornwallis Harris, W. 1839. *The wild sports of southern Africa*. London: John Murray.

de Maret, P. 1996. Pits, pots and the far-west streams. See Sutton (1996), 318–23.

Hine, R.S. (ed.) 1988. *Concise veterinary dictionary*. Oxford: Oxford University Press.

Khazanov, A.M. 1983. *Nomads and the outside world*, J. Crookenden (transl.). Cambridge: Cambridge University Press.

Kinahan, J.A. 1996. A new archaeological perspective on nomadic pastoralist expansion in southwestern Africa. See Sutton (1996), 211–26.

Klein, R.G. 1986. The prehistory of Stone Age herders in the Cape Province of South Africa. In *Prehistoric pastoralism in southern Africa*, M. Hall & A.B. Smith (eds), 5–12. Vlaeburg: South African Archaeological Society.

Loftus, R.T., D.E. MacHugh, D.G. Bradley, P.M. Sharp, P. Cunningham 1994. Evidence for two independent domestications of cattle. *Proceedings of the National Academy of Sciences, USA* **91**, 2757–61.

Phillipson, D.W. 1980. Iron Age Africa and the expansion of the Bantu. In *The Cambridge encyclopedia of archaeology*, A. Sherratt (ed.), 342–7. Cambridge: Cambridge University Press.

Phillipson, D.W. 1993. *African archaeology*, 2nd edn. Cambridge: Cambridge University Press.

Smith, A.B. 1992. *Pastoralism in Africa origins and development ecology*. London: Hurst.

Sutton, J.E.G. (ed.) 1996. *The growth of farming communities in Africa from the equator southwards*. British Institute in eastern Africa, *Azania* [special volume] **29/30**.

Vansina, J.A. 1996. A slow revolution: farming in subequatorial Africa. See Sutton (1996), 15–26.

CHAPTER FOUR

Bos africanus (Brehm)? Notes on the archaeozoology of the native cattle of Africa

Caroline Grigson

1. Introduction

In a paper originally given at a colloquium at the British Museum in 1988 and in another published in 1991 – "An African origin for African cattle?" – I advanced the hypothesis that modern Sanga cattle in Africa might have originated in the domestic cattle of early Dynastic, or even pre-Dynastic, Egypt, and that modern Sangas are taxonomically distinct from both taurine cattle (*Bos taurus*) and humped – zebu – cattle (*B. indicus*) (Grigson 1991, 1996). The rather scanty evidence presented for this view was largely pictorial and based on the similarity between some modern longhorn Sanga cattle and those depicted in ancient Egypt. This view is in contradistinction to the widely-held belief that Sanga cattle *originated* as crosses between zebu and taurine cattle, and contends that their zebu-like appearance is attributable to relatively recent crossing with imported zebu. Interbreeding with both taurines and zebu has been so intense since the decimation of cattle populations in Africa caused by rinderpest and other diseases in the late nineteenth and early twentieth centuries, that in many places the original local cattle have been either eliminated or outbred into new forms that bear little, if any, resemblance, either genetically or anatomically, to those of the past. The paper also suggested that Sanga cattle were the original cattle over the entire continent and that they were often quite small. It implied that in West Africa the humpless, short-horned cattle of West Africa are not derived from imported taurines, but like the larger, longer horned N'Dama cattle, are of the same ultimate origin as Sangas. The conclusion was that African cattle are, or were until recently, basically African, arguably comprising a separate taxon, *B. africanus*.

The purpose of the present paper[1] is to extend these arguments further back in time, in order to investigate the hypothesis that African cattle are descended not only from a common domestic progenitor, but from a wild ancestor that was different from both the supposed ancestors of taurines, the wild ox or aurochs (*B. primigenius primigenius*) and of zebus, the wild ox of the Indian subcontinent

(*B. primigenius namadicus*), and that this ancestor was the African form of the aurochs (*B. primigenius mauritanicus*), also sometimes referred to as *B. opisthonomus*.

A brief discussion of what has been established about the osteology of both early and modern cattle in Africa and of the genetics of modern cattle is followed by an analysis of the archaeozoological information that is beginning to emerge from archaeological excavations in various parts of Africa.

Five scenarios suggest themselves:

(a) that the wild populations of Levantine and African cattle were isolated from one another as the result of an environmental episode, such as a period of hyperaridity affecting North Africa or Sinai or both;

(b) that domestic cattle originated in Africa from the local wild aurochs *B. opisthonomus* and spread first to the Levant, and then both eastwards towards Iran and to the west and north around the Mediterranean to Greece and beyond;

(c) that domestic cattle originated in the Levant from *B. primigenius* and spread to Africa;

(d) that domestic cattle originated simultaneously all around the western Mediterranean, including North Africa;

(e) that although descended from a common ancestor the native cattle of Africa have had different histories in different areas.

2. Osteology

2.1. The osteology of modern cattle in Africa

Bos indicus can be distinguished from *B. taurus* on the basis of many cranial characters, including the long, narrow skull, the shape of the ridge between the horns, the convexity of the forehead, the concavity of the occipital region, the less prominent orbits, the persistence of flat orbital rims into old age and the diagonally upward direction of the horns at the base (Grigson 1980). Among the postcranial differences are the elongated dorsal spines on the anterior thoracic vertebrae, bifid dorsal spines on *some* of the *posterior* thoracic vertebrae (Stallibrass 1983, Grigson 1984) and relatively slender limb bones. The differences between the taxa are more marked in large than small individuals.

Sanga cattle have long faces, straight intercornual ridges, flat foreheads, prominent orbits and long legs. The horns are usually directed diagonally outwards and upwards at the base, and tend to be crescentic in males and lyre-shaped in cows and castrates. The horns of some Sanga breeds are disproportionately large in both girth and length. Whether these characters hold good for small cattle that are thought to be of Sanga origin is uncertain.

2.2. The osteology of Egyptian cattle in Dynastic and pre-Dynastic times

Little is known about the osteology of African cattle in the past, despite the fact that many complete skulls and skeletons have been excavated in Egypt. When

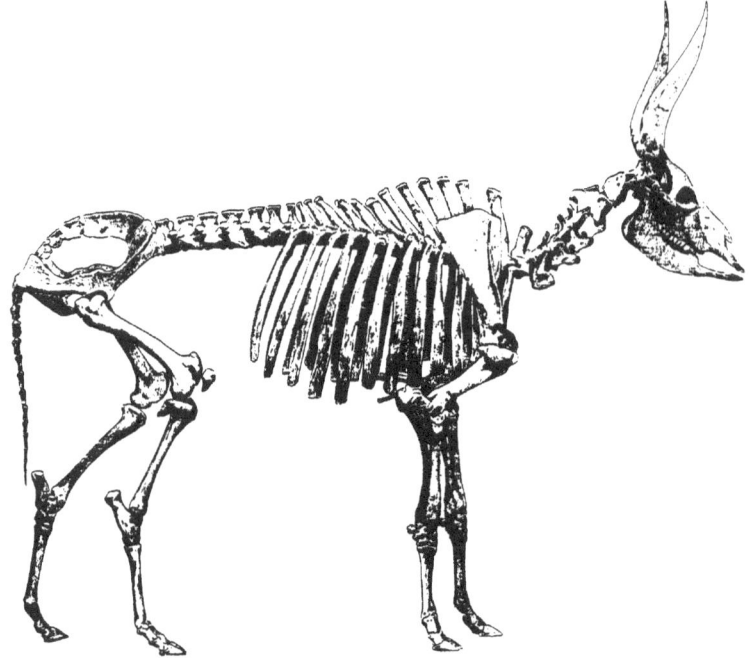

Figure 4.1 Egyptian longhorn. A skeleton of a bull from Saqqara, described by Lortet & Galliard (1903, 1905) as the taxon *Bos africanus*, distinct from both zebu *B. indicus* and taurine cattle *B. taurus* (after Lortet & Galliard 1903, Fig. 23).

describing the (Egyptian) "Munich Ox", Boessneck & von den Driesch (1987) did not invoke the features that Grigson (1974, 1975, 1976, 1978, 1980) found could be used to distinguish *indicus* and *taurus*, nor are all their measurements comparable with hers. However, their account of the skeletal morphology is detailed enough to show that the Egyptian longhorn was quite unlike *indicus* and, although closer to *taurus*, was quite distinct from that taxon as well. The same is true of the skulls from Saqqara, Giza, Illahun, Tel el-Amarna, the Baqaria and the Bucheum (Dürst 1899, Jackson 1934, Burleigh & Clutton-Brock 1980), the cow skeletons from the Baqaria (Jackson 1934) and the skeletons from Saqqara and Abousir (Lortet & Gaillard 1903, 1905).

The numerous complete cattle skeletons from Saqqara (Figs 4.1–4.3) and Abousir, were described in detail by Lortet & Gaillard (1903, 1905) as a different taxon, *Bos africanus*. Although this is probably not a valid taxonomic category, the fact that they invoked it does emphasize the marked difference between Egyptian cattle, taurines and zebus. They state that Egyptian longhorns are the same as the modern cattle living on the plains of the Upper Nile, that is the Hamitic longhorns, which they also refer to as *B. africanus*.

Boessneck & von den Driesch (1987) do not state whether any of the vertebral spines of the "Munich Ox" were bifid, but presumably if they had been they would

Figure 4.2 Egyptian longhorn. Front view of a bull's skull from Saqqara (after Lortet & Galliard 1903, Fig. 25).

Figure 4.3 Egyptian longhorn. Side view of a bull's skull from Saqqara (after Lortet & Galliard 1903, Fig. 24).

have said so. None of those illustrated by Lortet & Gaillard (1903) or by Jackson (1934) are bifid, but one of the bulls from Saqqara which was not illustrated did have bifid vertebrae, Lortet & Gaillard (1903:43) wrote; "The last dorsal processes of the dorsal vertebrae are weakly bifid at the summit, thereby tending to link this bovid with the zebu form. I believe however that one should not attach even the least importance to this character which seems to be little constant" (my translation). They also found that the shape of the sacrum in the Egyptian ox was similar to that in *B. taurus* and differed from that in *B. indicus* (ibid.:48; see Fig. 4.4).

Lortet & Gaillard (1903:46) said that the Egyptian longhorn stood high on its legs and had long, slender limb bones like those of the zebu. Comparison of the available measurements shows that the metacarpals of African cattle from archaeological sites in Africa and *B. indicus* from archaeological sites in the Indian subcontinent were indeed very long and slender and quite distinct from those of *B. taurus* from archaeological sites in the Near East (Fig. 4.5A). Those of African cattle may have been even more slender than those from India, but the sample is too small for certainty. The same is true of the metatarsals (Fig. 4.5B).

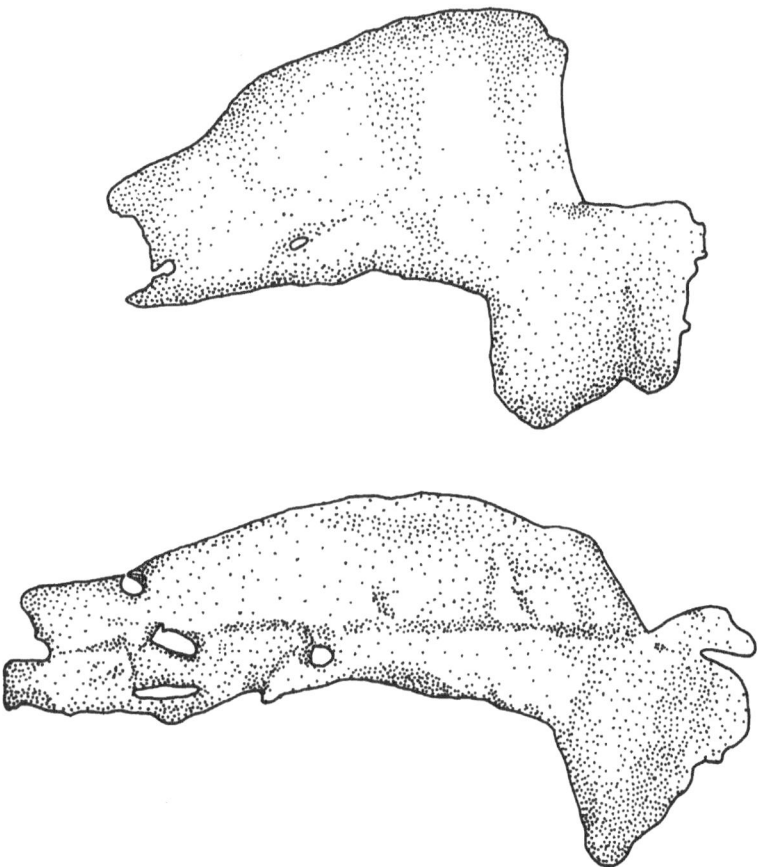

Figure 4.4 Sacrum of an Egyptian longhorn (below) compared with that of a zebu *Bos indicus* (above). Note that the zebu sacrum slopes down in a characteristic manner clearly visible in living animals (after Lortet & Galliard 1903, Figs 28, 29).

Although it has been claimed that there were three distinct types of cattle in ancient Egypt, that is short-, medium- and long-horned cattle (e.g. Boessneck 1988), no osteological evidence has yet been published for the presence of more than one type, except for some polled cattle. Indeed Lortet & Gaillard (1903, 1905) maintained that despite the numerous depictions of short- and medium-horned cattle, only longhorns were represented among the material examined by them. It may well be that only long-horned cattle were buried in monuments, smaller cattle may have been utilized in everyday life and therefore their remains would be present in the domestic archaeozoological assemblages that have only recently begun to be retrieved in Egyptian excavations. Boessneck (1988) maintained that long-, medium- and short-horned cattle were represented in the faunal assemblage

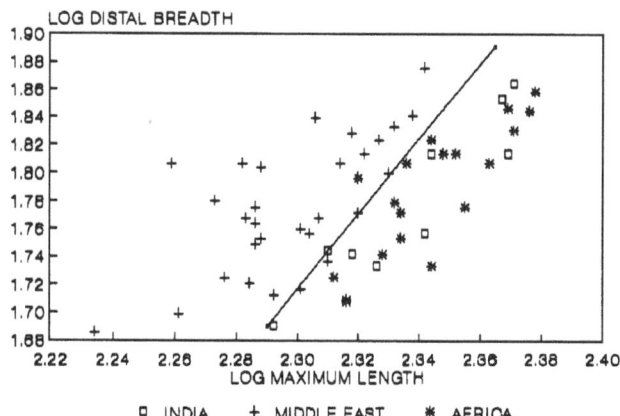

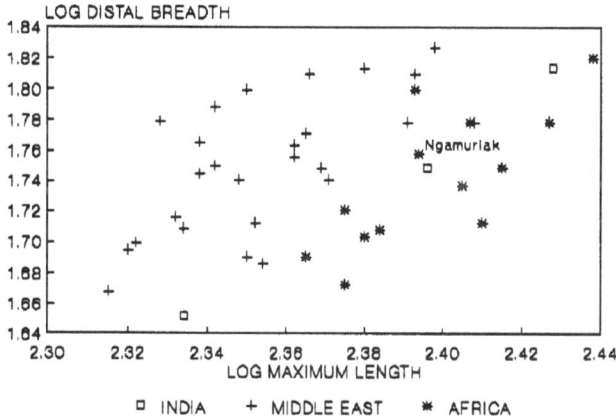

Figure 4.5 Metapodial shape in cattle from archaeological sites. Log plots of measurements of maximum length and distal breadth. A: Metacarpals. Comparison of the available measurements show that those of *Bos indicus* from India tend to be longer and more slender than those of *B. taurus* from the Middle East. Those from Africa appear to be as long as those of *B. indicus*.

B: Metatarsals. The metatarsals of the African cattle seem to be as long, if not longer, than those of *Bos indicus*, although for this bone the *B. indicus* sample is very small.

African data from the Munich Ox (Boessneck & von den Driesch 1987), the Baqaria and the Bucheum (Jackson 1934), Elephantine (Boessneck & von Den Driesch 1982), the Roman level at Qasr Ibrim (Grigson, unpubl. data), and Ngamuriak (Marshall 1990).

at Ma'adi III in the early fourth millennium bc, but the recent detailed report by Boessneck et al. (1989), which is discussed further below, does not confirm this conclusion.

Cattle are variable in size (Grigson 1989) and there is a wide, but overlapping, sexual dimorphism, complicated by the likelihood that some of the domestic males were castrated. It is difficult to imagine that the presence of three size groups could easily be demonstrated, though presumably the total range of measurements would be wider than in a single population. Also, it must be remembered that as all cattle are born with no horns, long-horned cattle go through short- and medium-horned stages in the course of maturation.

There is so far no osteological evidence for the presence of zebu in ancient Egypt, but the possibility that true zebu cattle were occasionally imported into Egypt cannot be entirely ruled out (Nicolotti & Guérin 1992). As Epstein (1971:513) pointed out, they can have had little or nothing to do with the much later appearance of zebu cattle in most of sub-Saharan Africa (Grigson 1996), which is discussed in Section 3 (below).

2.3. The osteology of wild cattle in Africa – Bos opisthonomus and Bos ibericus

It is well known that there was a North African form of the aurochs, *B. opisthonomus*, also sometimes referred to as *B. primigenius mauritanicus*; its range extended into the Sahara and south down the Nile to Nubia (Pomel 1894, Perkins 1965, Gautier 1988b). It was present during the Upper Pleistocene and early Holocene and it is not known when it became extinct, but recently published measurements of bones of *Bos* from archaeological sites suggest that it may have survived until at least the second millennium bc in Egypt (Boessneck & von den Driesch 1992). The extent to which it differed from the wild ox of Europe, Asia north of the Himalayas and the Near East (*B. primigenius primigenius*) or the wild form of Iran and the Indian subcontinent (*B. primigenius namadicus*) is unknown. Very few finds have been made and the skulls described are incomplete. It is said that the horncores of *B. opisthonomus* were directed outwards, inwards and downwards, rather than outwards inwards and upwards as in the other taxa, but to judge from photographs this appears to be true of only one of the skulls. The skulls appear to have straight frontal profiles. This is a subject that badly needs researching; the main difficulty lies in locating the material, most of which was described in the nineteenth or early part of the twentieth century (Gautier 1988b).

Gautier (1988b) has shown that the smaller form, sometimes identified in North Africa as *Bos ibericus*, is misnamed, misunderstood and misidentified; it is not a valid taxon, being no more than the female form of the North African aurochs or, in some cases, having been confused with domestic cattle. A similar taxon of a small and supposedly wild *Bos – B. longifrons*, or *B. brachyceros* – was effectively demolished, on much the same grounds, by the German, von Leithner, in 1927; but its ghost lingers in the literature, even in works by such eminent writers as Epstein (1971) and Bouchud (1981, 1987).

3. Genetics

3.1. Protein polymorphism

Manwell & Baker (1980) and Baker & Manwell (1980) have amalgamated the results of many studies on protein polymorphism to show that zebu and taurine cattle are quite distinct, even though they are completely interfertile. In their opinion *Bos indicus* and *Bos taurus* are valid species, which accords with the proposition based on their craniology, which is that they are descended from different wild ancestors, namely *B. primigenius* and *B. namadicus* (Meadow 1984, 1986, Grigson 1985). However they found African and Indian zebu to be closely related and the differences that exist between them explicable in terms of outbreeding of zebu in Africa with Sanga cattle. Although the Sanga group is distinct from their taurine group and closer to *B. indicus*, it is not clear to what extent this similarity is the result of cross-breeding in at least some of the breeds classified as Sangas, particularly the Afrikander. Thus the results of the study of protein polymorphism could be interpreted as meaning that Sanga cattle were originally intermediate between taurines and zebus, and that they arose either as a result of crossing between taurines imported from the Middle East and the local wild cattle, or by the domestication in North Africa of the local form of wild cattle (*B. opisthonomus and B. primigenius*). In the latter case it could be argued that African domestic cattle were originally also a separate taxon (*B. africanus*).

Although it is doubtful whether the division of domestic cattle into three (or even two) separate species is taxonomically valid, it has been suggested that African domestic cattle are phylogenetically distinct from both *B. indicus* and *B. taurus* (Grigson 1991). This view is also supported by the fact that the humpless cattle of West Africa are tolerant of several local diseases, including trypanosomiasis. It would be of interest to know whether the tolerance existed in the past in any of the purer Sanga breeds, particularly the small Sanga-type cattle, which were present from the Sudan to the Cape, most of which are either extinct or have been absorbed into other breeds.

3.2. Mitochondrial DNA

New work on sequencing of mitochondrial DNA by Loftus, Bradley and their colleagues has confirmed that *B. taurus* and *B. indicus* have a very separate ancestry, having diverged from their common ancestor at least 200,000 or even 1,000,000 years ago (Loftus et al. 1994). They found that African and European taurine cattle are quite close to one another and quite distinct from *B. indicus*. (Bradley et al. 1996, Bradley & Loftus, Ch. 13 in this volume). This is surprising because, although highly variable, many individuals of three of the African breeds chosen (White Fulani, Kenana and Butana) have obvious zebu characteristics. They suggest that the lack of *B. indicus* mitochondrial DNA in the African cattle is attributable to the fact that the zebu element in their make-up derives largely from crossing from imported males. Bradley and his colleagues found that the African cattle breeds chosen were more closely related to one another than to taurines, suggesting divergence from them at about 26,000–22,000 BP. This confirms the view that African and taurine cattle have been separate for a long time and, unless cattle were domesticated far earlier than presently thought, implies a genetic separation between these two wild cattle (*B. primigenius opisthonomus and B. primigenius primigenius*) during the late Upper Pleistocene, rather than a more recent divergence of the domestic forms.

4. Archaeozoological comparisons

4.1. Wild cattle in the Levant and North Africa

It is first necessary to attempt to establish the size range of the wild cattle in North Africa in the early Holocene. This can be done by comparing measurements of its limb and foot bones with those of a standard animal as previously done for cattle in the Near East and northern Europe by Grigson (1989) and in Pakistan by Meadow (1984). A large set of data, produced by making logarithmic comparisons of the dimensions of postcranial bones from archaeological sites with those of a standard animal – a complete skeleton of a *Bos primigenius* cow found at Ullerslev in Denmark (Buitenhuis 1985) – was plotted in the form of simple histograms.

The first histogram, of Danish archaeological material, included dimensions of bones of animals of known sex and it was clear that the bimodality could be attributed to sexual dimorphism, the bones of aurochs bulls being larger than those of cows (Fig. 4.6A). Comparisons of histograms of measurements of cattle bones from archaeological sites suggested that the size of wild cattle in the Near East remained constant from about the tenth to the seventh millennia bc. Amalgamation of the data from the various millennia into a single histogram (Fig. 4.6B) produces a bimodal distribution similar to that obtained on Danish material and is a useful baseline for comparisons with material from later sites. It also shows that the animals were slightly smaller than those of northern Europe (Fig. 4.6A, B).

There are very few data available on the size of *B. primigenius* in the Holocene in North Africa, but some can be culled from Higgs's work on the relevant levels at Haua Fteah (Higgs 1967). The results suggest that wild cattle of North African cattle *may* have been slightly smaller than those of the Near East, but the sample sizes are much too small for certainty (Fig. 4.6C).

The main point of these comparisons is to test whether the analysis of size indicates a discontinuity between the wild cattle of Africa and the Levant in the early Holocene; the results suggest that this could have been the case. However, it is also possible that the size reduction simply exemplifies Bergmann's Rule, although this is less likely as Haua Fteah is on much the same latitude as the Levant.

4.2. Cattle keepers at Bir Kiseiba and Nabta Playa (eighth–fourth millennia bc)?

A few cattle bones retrieved from shallow deposits and the surface of several sites at Bir Kiseiba and Nabta Playa in the eastern Sahara have been claimed as domestic and their measurements have been published (Gautier 1980, 1984). This claim is spelt out in detail in a recent paper by Wendorf & Schild (1994). As a

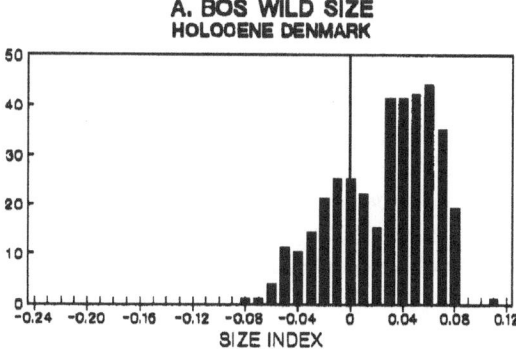

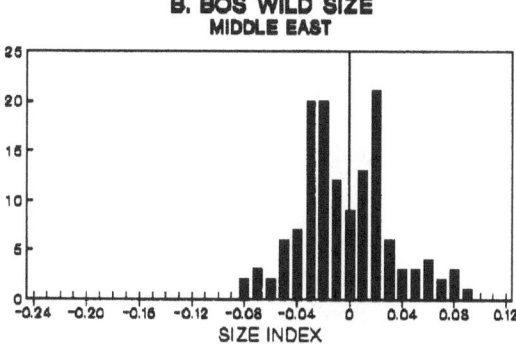

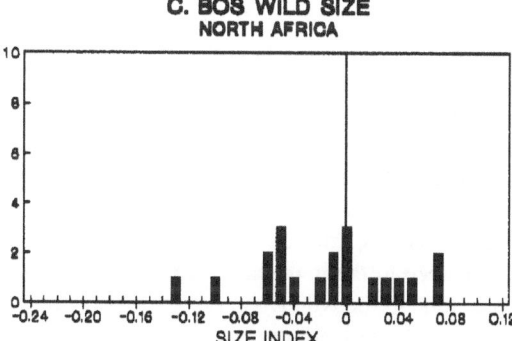

Figure 4.6 Wild Size. Comparisons of the size of wild cattle from Holocene sites in northern Europe, the Levant and North Africa, with those of a standard animal (see Buitenhuis 1985).

A: Denmark. The data include dimensions of bones from complete skeletons of known sex, so the bimodality is attributed to sexual dimorphism. Data from Grigson (1989) and Degerbøl & Fredskild (1970).

B: The Levant. 10th–7th millennia bc. The graph shows the same bimodal distribution as the Danish material, and indicates that the animals were slightly smaller than those of northern Europe. Data from Grigson (1989).

C: North Africa. Early Holocene sites. The few data available suggest that wild cattle in North Africa *may* have been slightly smaller than those of the Levant. Data from Higgs (1967).

result, the period of occupation, formerly referred to as "Terminal Palaeolithic", has been renamed "early neolithic". The main reason given is that the area is considered to have been too arid for wild cattle at the time of occupation (eighth to seventh and sixth to fourth millennia bc). This line of evidence is supposed to be confirmed by the fact that hartebeest (*Alcelaphus buselaphus*), and other fauna which apparently normally accompanies wild *Bos* in North Africa, is absent from the site. This, however, cannot be sustained – in such small samples absence of certain animals cannot be taken as significant – absence of proof is not the same as proof of absence. It is difficult to believe that domestic animals could survive in an environment that was too inimical for their wild progenitors. It is also doubtful whether people with a mobile lifestyle had a level of social organization complex enough to allow them to achieve such a major co-operative step, bearing in mind that domestication of herbivores is usually thought to have been associated with sedentism.

Comparisons of the size of the Bir Kiseiba and Nabta Playa cattle bones (Fig. 4.7B) with those of wild and domestic cattle of Egypt (Fig. 4.7A, C) do not support the notion that these cattle were domestic. The matter will not be finally settled until much larger well-dated faunal collections from the relevant period have been studied in detail.

4.3. Early domestic cattle in the Levant and North Africa

Grigson (1989) used the standard animal method to show that there was a marked decrease in body size from the seventh to the sixth millennia bc in the Levant, Turkey and Greece and suggested that this was evidence for the first domestication of cattle in the Near East (Fig. 4.8A, B). Similar plots for the fifth millennium bc suggest a further decrease in size (Fig. 4.8C).

Was some similar process occurring in North Africa? Unfortunately, there is a dearth of material from the eighth to the sixth millennia bc in the area. Some sites exist, some material exists, but it has not been published to anything approaching adequate standards. For example, faunal remains were retrieved in quantity from the sixth millennium bc site of Capeletti and are said to indicate that domestic cattle were present but there is next to no quantification in the report and no adequate presentation of measurements (Carter & Higgs 1979).

A site recently published to the high standard that characterizes the work of the Munich School is Merimde in the Nile Delta (von den Driesch & Boessneck 1985). It dates from the late fifth millennium bc and, depending on the terminology employed, belongs to either late neolithic or early pre-Dynastic. As well as domestic cattle, the site has domestic pigs, sheep, goats, and dogs. As the size of the cattle from Merimde (Fig. 4.7C) is much less than that of the early Holocene of Haua Fteah (Fig. 4.7A) and almost identical to that of the cattle of the fifth millennium bc in the Levant (Fig. 4.8C), there is little doubt that they stem from domestic animals. The similarity to contemporary cattle in the Levant means that the analysis of size has not produced evidence to indicate different stocks or different management practices in the two areas; but again it must be remembered that absence of evidence is not the same as evidence of absence.

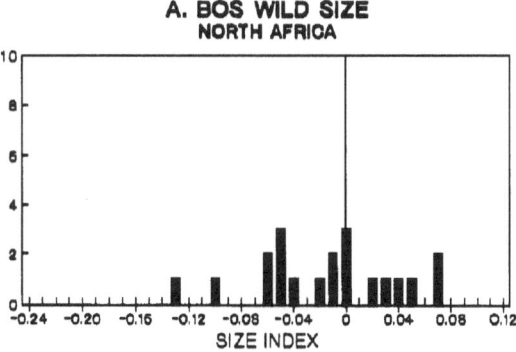

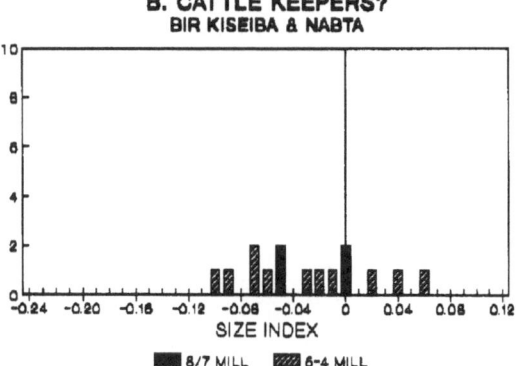

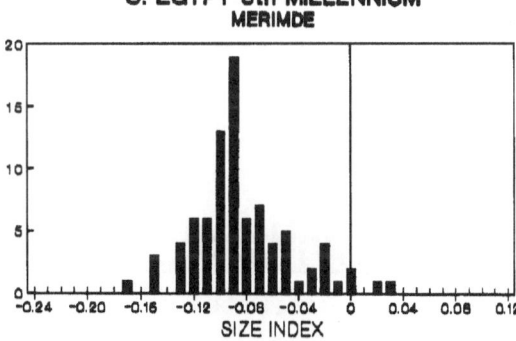

Figure 4.7   Size change from wild to domestic cattle in North Africa.
A: Wild cattle in North Africa. Early Holocene (data as Fig. 4.6C).
B: Bir Kiseiba and Nabta Playa. 8th–7th and 6th–4th millennia bc. Comparison of the distribution and range of cattle bone dimensions from these sites with those of wild and domestic cattle from Egypt does not support the notion of Wendorf & Schild (1994) that these cattle were domestic. Data from Gautier (1980, 1984).
C: Domestic cattle from the 5th millennium site of Merimde. Data from von den Driesch & Boessneck (1985).

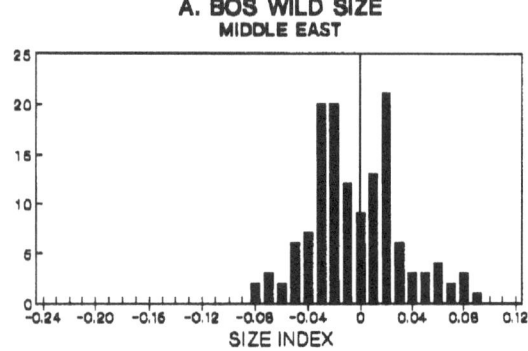

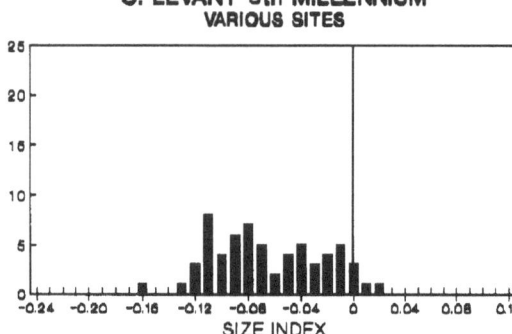

Figure 4.8   Size change from wild to domestic cattle in the Levant, 10th–5th millennia bc. A: 10th–7th millennia bc, wild cattle. (Data as Fig. 4.6B).

B: 6th millennium bc, domestic cattle. Comparison with Figure 4.8A suggests that there was a marked decrease in cattle body size from the 7th to 6th millennia bc in the Levant – the most likely explanation for which is that the animals had been domesticated. Data from Grigson (1989).

C: 5th millennium bc, domestic cattle. The plot suggests a further decrease in size. Data from Grigson (1989).

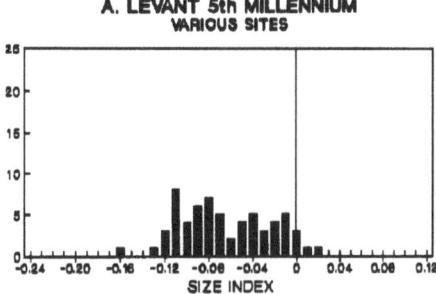

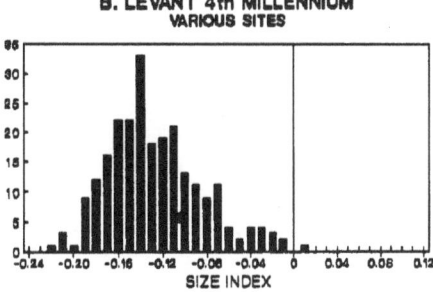

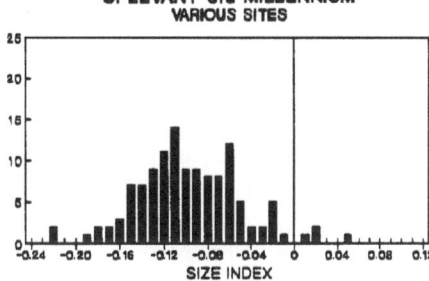

Figure 4.9   Size change in domestic cattle in
the Levant, 5th–3rd millennia bc.
A: 5th millennium bc. (Data as Fig. 4.8C)
B: 4th millennium bc. Showing a further
diminution. Data from Grigson (1989).
C: 3rd millennium bc. Showing a slight
recovery in size. Data from Grigson (1989).

However, the histograms do suggest that something very different was happen-
ing in the two areas in the fourth millennium bc. In the Levant, cattle have undergone yet another marked decrease in size (Fig. 4.9A, B), whereas in the fourth
millennium (pre-Dynastic) site of Ma'adi (Boessneck et al. 1989), and also in the
Nile Delta (Fig. 4.10A, B), the size has increased slightly, and indeed size continues

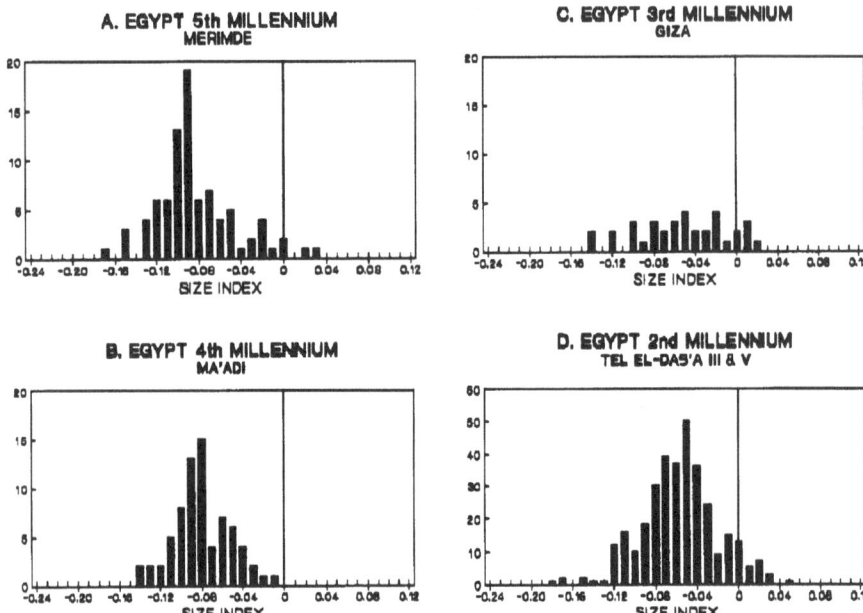

Figure 4.10   Size change in the domestic cattle of Egypt, 5th–2nd millennia bc.

A: 5th millennium bc, Merimde. (Data as Fig. 4.7C)

B: 4th millennium bc, Ma'adi. The size has increased slightly. Data from Boessneck et al. (1989).

C: 3rd millennium bc, Giza. Showing a further increase in size. Data from Kokabi (1980).

D: 2nd millennium bc, Tel el-Dab'a III & VII. Showing yet a further increase in size. Data from Boessneck (1976) and Boessneck & von den Driesch (1992).

The different pattern of size change in Egypt and the Levant (Fig. 4.9) suggests that from the 4th millennium onwards the cattle of Egypt were genetically isolated from those of the Levant, and distinctive in terms of size and probably also of management.

to increase into the third and second millennia at other sites in Egypt (Fig. 4.10C, D). In the Levant size recovers only slightly in the third millennium (Fig. 4.9C). This suggests that from at least the fourth millennium cattle of Egypt were genetically isolated from those of the Levant, and distinct in terms of size and probably also of management.

4.4. Travelling south – the neolithic of the Sudan (Shaheinab and El-Kadada)

Re-examination of the faunal remains from the fourth millennium bc site of Esh Shaheinab by Peters (1986) showed that most of the cattle bones originally identified by Bate (1953) as being of African buffalo were in fact of domestic cattle. Domestic sheep or goats or both were also present. This is in contrast with the sixth millennium bc site of Khartoum Hospital which had a rich fauna with no trace of any domestic animals. Only one measurement of a *Bos* bone from Esh Shaheinab is suitable for the present comparisons, it has an index of −0.117, rather small, but well within the range of the domestic cattle from the third millennium bc. According to Gautier (1988a) domestic cattle were also present at the early neolithic sites of Geili and El-Kenger East and El-Kenger Middle, but no details of the cattle bones have been published.

The earliest site in the Sudan from which measurements of domestic cattle bones have been published is the fourth millennium bc site of El-Kadada (Gautier 1986), northeast of the sixth cataract of the Nile on the east side of the river. I had assumed that in spreading southwards up the Nile Valley cattle would have declined in size, as they appear to have done in crossing the Sahara. However comparison of the size data from El-Kadada (Fig. 4.11B) with those of Ma'adi (Fig. 4.11A) show that the cattle from the two sites were equally large and it is reasonable to suppose, since horn size is usually directly proportional to body size that these were long-horned cattle, similar in appearance to Egyptian cattle, from which they were almost certainly descended.

4.5. Further south – the neolithic of Kenya (Ngamuriak)

Archaeozoological information from sites in East Africa is scarce. By far the most comprehensive, detailed and thoughtful report is that by Marshall (1990) on the neolithic site of Ngamuriak, dated to about the year 0 (BC/AD). Again one might have expected a reduction in size, but as at El-Kadada (Fig. 4.11C) the cattle seem to have been at least as large as those from Ma'adi in the fourth millennium bc. Again we can assume that these were long-horned cattle, fitting for people in whose lives cattle played an important role, probably in terms of prestige as well as subsistence.

I disagree with Marshall that the Ngamuriak cattle were zebu (*Bos indicus*). She reached this conclusion using two criteria. The first was the large size of the one complete cattle metatarsal. But, as Figure 4.5B shows, this is of similar size and shape to those of other African cattle, as well as those of zebu. She also drew on my work on the distinction between *B. taurus* and *B. indicus*, in which I showed that the normal effect of age on the shape of the lower orbital rim was that the shape of the rim changes from being flat in young animals to being pointed anteriorly in old ones, and I suggested that this particular aspect of the ageing process happens later in life in zebu than in taurines (Grigson 1980). This is a matter of the inherent variability of cattle – few characters are absolute – and it is important not to use age-dependent data to draw hard and fast conclusions, since such data are by definition liable to gradual change. This means that the presence of flattened orbital rims on the cattle skull fragments from Ngamuriak cannot be taken as indicating zebu unless it can be shown that they come from elderly animals. An additional

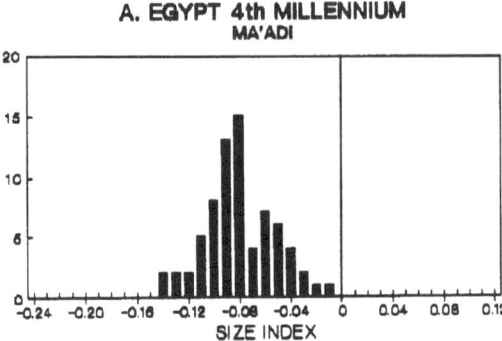

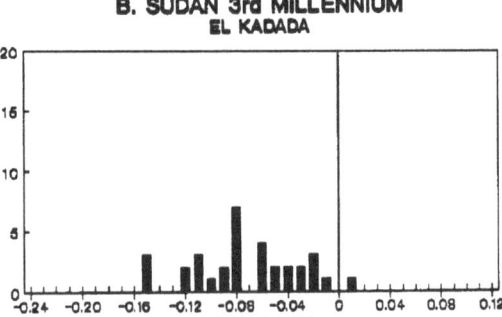

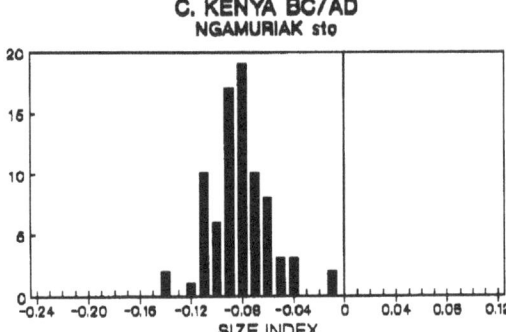

Figure 4.11   Cattle size and latitude in Africa.

A: Egypt in the 4th millennium bc. (Data as Fig. 4.10B).

B: The Sudan in the 4th millennium bc, El-Kadada. Data from Gautier (1986). The cattle are as large as those from Egypt (Fig. 4.11A).

C: Kenya at about BC/AD, Ngamuriak. Data from Marshall (1990). The cattle are as large as those from Egypt (Fig. 4.11A) and the Sudan (Fig. 4.11B).

As horn size is usually proportional to body size it is reasonable to assume that the cattle from El-Kadada and Ngamuriak were longhorns, similar in appearance to Egyptian cattle, from which they were probably descended.

problem is that the proper comparison for African cattle is not with *B. taurus*, but with the native cattle of Africa, about whose craniology we know little and about whose orbital rims we know nothing at all.

4.6. Diachronic change in cattle size outside Egypt

4.6.1. Across the Sahara to West Africa   It is widely accepted that cattle spread south across the Sahara from the fifth millennium bc onwards, reaching West Africa during the second millennium bc and getting smaller as they went (Shaw 1980, Smith 1980, Clutton-Brock 1989, 1993; and particularly Gautier 1987, Grigson 1991, Blench 1993). The data supporting this view are scanty, and few individual measurements have been published up until now (but see MacDonald & MacDonald, Ch. 8 in this volume). Thus, histogram plots using the present logarithmic method of comparison have not been made. However in some cases ranges, or means, or both, are given and simple graphical plots of these confirm the general view (Fig. 4.12A). Nevertheless, far more data are needed to establish this properly and to investigate whether the remains can be related to the present cattle of the area.

4.6.2. Nubia and the Sudan   The plots of size data comparing the sizes of cattle at El-Kadada (Gautier 1986) from the fourth millennium bc, Meroë (Carter & Foley 1980) from about AD 0, and from the Roman levels at Qasr Ibrim (Grigson unpubl.) shown in Figure 4.12B, suggest that the cattle there have suffered a decline in size over the years, similar to that established for the Sahara, though less extreme.

4.6.3. Kenya, Zambia, Mozambique and the Transvaal   The plots of size data of cattle from sites in eastern and southern Africa (Fig. 4.12C), which range in time from Ngamuriak in Kenya (*c.* AD 0; Marshall 1990), to the Behrens site (eighteenth and nineteenth centuries) (Fagan et al. 1969) and as far south as Mapungubwe in the Transvaal (Voigt 1983), give a similar picture of diminution in size, perhaps more extreme than that in the Sudan, but extending over a longer time period. One would assume that these small cattle had short horns and, as Fagan (1967) pointed out, those from the eighteenth and nineteenth centuries in Zambia may well be the same small, short-horned cattle which David Livingstone encountered in Tonga.

5. Conclusions

Although the present study has not succeeded in pinpointing the place or date of origin of the domestic cattle of Africa, some patterning has emerged that suggests that domestic cattle in the Levant and North Africa have been distinct from one another since at least the fourth millennium bc and probably well before that. Although it is likely that the native cattle of Africa are descended from a common

SITE

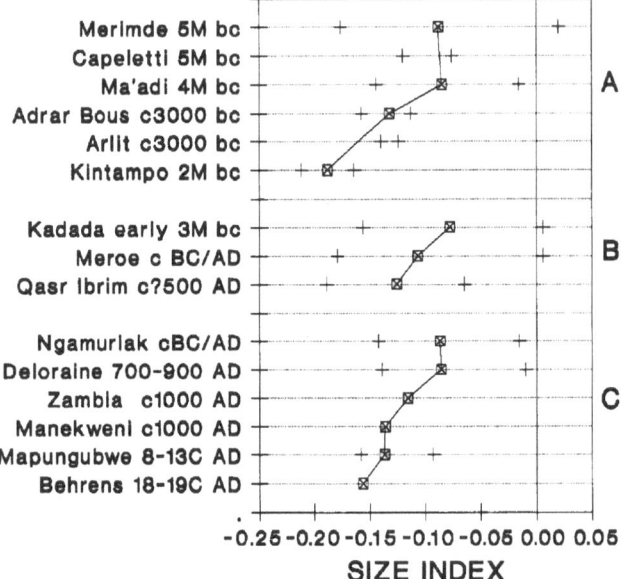

Figure 4.12   Size change over time in different parts of Africa.
A: Egypt, the Sahara and West Africa. The data, though scanty,
confirm the view that as cattle spread southwestwards across the
Sahara from the 5th millennium BC onwards, they suffered a diminution in size. Data from von den Driesch & Boessneck (1985),
Carter & Higgs (1979), Boessneck et al. (1989), Carter & Clark
(1976), Smith (1980), and Carter & Flight (1972).
B: Nubia and the Sudan. Cattle seem to have suffered a decline in
size over the years, similar to that established for the Sahara, though
less extreme. Data from Gautier (1986), Carter & Foley (1980),
and Grigson (unpubl. data).
C: East Africa. Cattle here seem to have continued to decline in
size from about AD 0 to the nineteenth century. Data from Marshall
(1990), Fagan (1967), Fagan et al. (1969), and Barker (1978).

ancestor the analysis of their size shows that they have had very different histories
in different areas. It also sheds light on the later history of cattle, indicating that
in crossing the Sahara they underwent a reduction in size, whereas the earliest
populations of cattle on the eastern side of Africa were initially as large as their
supposed Egyptian forebears. Yet in the course of time, it appears that cattle throughout Africa did get smaller.

The results are by no means conclusive; far more work will need to be done
before it can be established that these suppositions are correct, and no doubt the
situation will be found to be far more complex. Nevertheless, the patterning found
so far is suggestive and indicates that continuation of this line of enquiry may

well prove fruitful in the future, with the usual proviso that useful results will only be forthcoming if large samples of bones are collected systematically from a large number of sites and then analyzed and reported upon in detail, with the publication of individual measurements of individual bones. If such archaeozoological research were to be co-ordinated with work on the DNA of modern populations, on DNA extracted from cattle material from archaeological sites, and on the comparative osteology of modern cattle and Egyptian cattle, much information could be gained not only about the relationships of African people and their livestock in the past, but also about patterns of resistance to trypanosomiasis, which could influence cattle breeding programmes of the future.

Note

1. I thank Kevin MacDonald for inviting me to participate in the Conference "African Livestock: Archaeology, Linguistics and DNA" at the Institute of Archaeology, University College London, 22–24 September 1995; and Achilles Gautier, Angela von den Driesch and Andrew Smith for help with the literature.

References

Baker, C.M.A. & C. Manwell 1980. Chemical classification of cattle. 1. Breed groups. *Animal Blood Groups and Biochemical Genetics* 11, 127–50.

Barker, G. 1978. Economic models for the Manekweni, Mozambique. *Azania* 13, 71–100.

Bate, D.M.A. 1953. The fauna. In *Shaheinab*, A.J. Arkell (ed.), 10–19. Oxford: Oxford University Press.

Blench, R. 1993. Ethnographic and linguistic evidence for the prehistory of African ruminant livestock, horses and ponies. In *The archaeology of Africa. Foods, metals and towns*, T. Shaw, P. Sinclair, B. Andah, A. Okpoko (eds), 71–103. London: Routledge.

Boessneck, J. 1976. *Tell el-Dab'a III. Die Tierknochenfunde 1966–1969*. Österreichischen Akademie der Wissenschaften. Denkschriften der Gesamtakademie, 5. Vienna: Österreichischen Akademie der Wissenschaften.

Boessneck, J. 1988. *Die Tierwelt des Alten Ägypten*. Munich: C.H. Beck.

Boessneck, J. & A. von den Driesch 1987. Zoologisch-häustierkundliche Befunde an der Rindermumie. In *Die Münchner Ochsenmumie*, J. Boessneck et al. (eds), 55–71. Hildesheimer Ägyptologische Beiträge, 25. Hildesheim: Pelizaeous Museum and Gerstenberg Verlag.

Boessneck, J. & A. von den Driesch 1992. *Tell el-Dab'a VII. Tiere und historische Umwelt im nordost-Delta im 2. Jartausend v. Chr. anhand der Knochenfunde der Ausgrabungen 1975–1986*. Österreichischen Akademie der Wissenschaften. Denkschriften der Gesamtakademie, 11. Vienna: Österreichischen Akademie der Wissenschaften.

Boessneck, J., A. von den Driesch, R. Ziegler 1989. Die Tierreste von Maadi und Wadi Digla. In *Maadi III. The non-lithic small finds and the structural remains of the predynastic settlement*, I. Rizkana & J. Seeher (eds), 87–128 + 5 plates. Mainz am Rhein: Philipp von Zabern.

Bouchud, J. 1981. Les mammifères de la Mauritanie occidentale. In *Mélanges offerts au Doyen Lionel Balout*, C. Roubert, H.-J. Hugot, G. Sonville (eds), 239–50, Paris: Editions ADPF.

Bouchud, J. 1987. *La Faune du Gisement Natoufien de Mallaha Eynan Israel, Jerusalem*. Mémoires et Travaux du Centre de Recherche Francais, 4. Jerusalem: CNRS.

Bradley, D.G., D.E. MacHugh, E.P. Cunningham, R.T. Loftus 1996. Mitochondrial diversity and the origins of African and European cattle. *The Proceedings of the National Academy of Sciences, USA* 93, 5131–5.

Buitenhuis, H. 1985. Preliminary report on the faunal remains of Hayaz Hüyük from the 1979–1983 seasons. *Anatolica* **12**, 61–74.

Burleigh, R. & J. Clutton-Brock 1980. A sacrificial bull's head from Illahun. *Journal of Egyptian Archaeology* **66**, 151–3.

Carter, P.L. & J.D. Clark 1976. Adrar Bous and African cattle. In *Proceedings of the PanAfrican Congress of Prehistory, Addis Ababa 1971*, A. Berhanou, J. Chevaillon, J.E.G. Sutton (eds), 487–93. Addis Ababa: Provisional Military Government of Socialist Ethiopia, Ministry of Culture.

Carter, P.L. & C. Flight 1972. A report on the fauna of Ntereso and Kintampo Rock Shelter Six in Ghana: with evidence for the practice of animal husbandry during the second millennium BC. *Man* 7, 277–82.

Carter, P.L. & R. Foley 1980. A report on the fauna from the excavations at Meroë, 1967–1972. In *The Capital of Kush 1*, P.L. Shinnie & R.J. Bradley (eds), 298–312. Meroitica, 4. Berlin: Akademie Verlag.

Carter, P.L. & E.S. Higgs 1979. A study of the faunal remains from "La Grotte Capeletti Khanguet si Mohamed Tahar" Aures, Algeria. In *Economie pastorale préagricole en Algerie Orientale: le Néolithique de Tradition Capsienne*, C. Roubet (ed.), 411–14. Paris: CNRS.

Clutton-Brock, J. 1989. Cattle in ancient North Africa. In *The walking larder: patterns of domestication, pastoralism and predation*, J. Clutton-Brock (ed.), 200–6. London: Unwin Hyman.

Clutton-Brock, J. 1993. The spread of domestic animals in Africa. In *The archaeology of Africa. Foods, metals and towns*, T. Shaw, P. Sinclair, B. Andah, A. Okpoko (eds), 61–70. London: Routledge.

Degerbøl, M. & B. Fredskild 1970. The Urus *Bos primigenius* Bojanus and neolithic domesticated cattle *Bos taurus domesticus* Linné in Denmark. *Det Kongelige Danske Videnskabernes Selskab Biologiske Skrifter* **17**, 1–177.

Dürst, J.U. 1899. *Die Rinder der Babylonien, Assyrian und Ägypten und ihr Zusammenhang mit den Rindern der alten Welt*. Berlin: Georg Reimer.

Epstein, H. 1971. *The origin of the domestic animals of Africa* [2 volumes]. New York: Africana Publishing.

Fagan, B. 1967. *Iron Age cultures in Zambia Kaloma and Kangila, Vol. 1*. London: Chatto & Windus.

Fagan, B., D.W. Phillipson, S.G. Daniels 1969. *Iron Age cultures in Zambia Dambwa, Ingombe, Ilede and the Tonga, Vol. 2*. London: Chatto & Windus.

Gautier, A. 1980. Contributions to the archaeozoology of Egypt. In *Prehistory of the eastern Sahara*, F. Wendorf & R. Schild (eds), 317–44. Dallas: Southern Methodist University Press.

Gautier, A. 1984. Archaeozoology of the Bir Kiseiba region, eastern Sahara. In *Cattle-keepers of the eastern Sahara: the neolithic of Bir Kiseiba*, F. Wendorf & R. Schild (assemblers), A. Close (eds), 49–76. Dallas: Southern Methodist University Press.

Gautier, A. 1986. La faune de l'occupation néolithique d'el Kadada secteurs 12-22-32 au Soudan Central. *Archéologie du Nil Moyen* **1**, 59–111.

Gautier, A. 1987. Prehistoric men and cattle in North Africa: a dearth of data and a surfeit of models. In *Arid North Africa. Essays in Honor of Fred Wendorf*, A.E. Close (ed.), 163–87. Dallas: Southern Methodist University Press.

Gautier, A. 1988a. Notes on the animal bone assemblage from the early neolithic at Geili Sudan. In *El Geili, the history of a Middle Nile environment 7000 BC–AD 1500*, I. Caneva (ed.), 57–63. BAR International Series, 424. Oxford: BAR.

Gautier, A. 1988b. The final demise of *Bos ibericus*? *Sahara* **1**, 37–48.

Grigson, C. 1974. The craniology and relationships of four species of *Bos*. I. Basic craniology: *Bos taurus* L. and its absolute size. *Journal of Archaeological Science* **1**, 353–79.

Grigson, C. 1975. The craniology and relationships of four species of *Bos*. II. Basic craniology: *Bos taurus* L. Proportions and angles. *Journal of Archaeological Science* **2**, 109–28.

Grigson, C. 1976. The craniology and relationships of four species of *Bos*. III. Basic craniology: *Bos taurus* L. Sagittal profiles and other non-measurable characters. *Journal of Archaeological Science* **3**, 115–36.

Grigson, C. 1978. The craniology and relationships of four species of *Bos*. IV. The relationship between *Bos primigenius* Boj. and *Bos taurus* L. and its implications for the phylogeny of the domestic breeds. *Journal of Archaeological Science* **5**, 123–52.

Grigson, C. 1980. The craniology and relationships of four species of *Bos*. V. *Bos indicus* L. *Journal of Archaeological Science* **7**, 3–32.

Grigson, C. 1984. Bifid dorsal vertebral spines and *Bos indicus* – a reply to Stallibrass. *Journal of Archaeological Science* **11**, 177.

Grigson, C. 1985. *Bos indicus* and *Bos namadicus* and the problem of autochthonous domestication in India. In *Recent advances in Indo-Pacific prehistory*, V.N. Misra & P. Bellwood (eds), 425–8. New Delhi: Oxford & IBH Publishing House.

Grigson, C. 1989. Size and sex – morphometric evidence for the domestication of cattle in the Near East. In *The beginnings of agriculture*, A. Milles, D. Williams, N. Gardner (eds), 77–109. BAR International Series, 496. Oxford: BAR.

Grigson, C. 1991. An African origin for African cattle? – some archaeological evidence. *African Archaeological Review* **9**, 119–44.

Grigson, C. 1996. Early cattle around in the Indian Ocean. In *The Indian Ocean in antiquity*, J.E. Reade (ed.), 41–74. London: Kegan Paul.

Higgs, E.S. 1967. Environment and chronology – the evidence from the mammalian fauna. In *The Haua Fteah, Cyrenaica*, C.B.M. McBurney (ed.), 16–44. Cambridge: Cambridge University Press.

Jackson, J.W. 1934. The osteology. In *The Bucheum 1*, R. Mond & O.H. Myers (eds), 137–42, The Egypt Exploration Society, 41. London: Egypt Exploration Society.

Kokabi, M. 1980. Tierknochenfunde aus Giseh/Ägypten. *Annalen der Naturhistorische Museum Wien* **83**, 519–37.

Loftus, R.T., D.E. MacHugh, D.G. Bradley, P.M. Sharp, P. Cunningham 1994. Evidence for two independent domestications of cattle. *The Proceedings of the National Academy of Sciences, USA* **91**, 2757–61.

Lortet, A. & C. Gaillard 1903. La faune mommifiée de l'ancienne Égypte. *Archives du Muséum d'Histoire naturelle, Lyon, 8, Memoire* **2**, 41–71.

Lortet, A. & C. Gaillard 1905. La faune mommifiée de l'ancienne Égypte; deuxieme seriés. *Archives du Muséum d'Histoire naturelle, Lyon, 9, Memoire* **2**, 54–68.

Manwell, C. & C.M.A. Baker 1980. Chemical classification of cattle. 2. Phylogenetic tree and specific status of the zebu. *Animal Blood Groups and Biochemical Genetics* **11**, 151–62.

Marshall, F. 1990. Cattle herds and caprine flocks. In *Early pastoralists of south-western Kenya*, P. Robertshaw (ed.), 205–60. British Institute in eastern Africa, Memoir 11. Nairobi: BIEA.

Meadow, R.H. 1984. Animal domestication in the Middle East: a view from the eastern margin. In *Animals and archaeology: 3. Early herders and their flocks*, J. Clutton-Brock & C. Grigson (eds), 309–37. BAR International Series, 202. Oxford: BAR.

Meadow, R.H. 1986. Faunal exploitation in the Greater Indus Valley: a review of recent work to 1980. In *Studies in the archaeology of India and Pakistan*, J. Jacobson (ed.), 43–64. New Delhi: Oxford & IBH Publishing House.

Nicolotti, M. & C. Guérin 1992. Le zébu (*Bos indicus*) dans l'Egypte ancienne. *Archaeozoologia* **5**, 87–108.

Perkins, D. 1965. Three faunal assemblages from Sudanese Nubia. *Kush* **13**, 56–61.

Peters, J. 1986. A revision of the faunal remains from two central Sudanese sites: Khartoum Hospital and Esh Shaheinab. *Archaeozoologia*, **Mélanges** [special volume], 11–33.

Pomel, A. 1984. *Les Boeufs-Taureaux*. Carte Géologique de l'Algérie. Paléontologie–Monographies. Algiers: Fontana.

Shaw, T. 1980. Agricultural origins in Africa. In *The Cambridge encyclopaedia of archaeology*, A. Sherratt (ed.), 179–84. Cambridge: Cambridge University Press.

Smith, A.B. 1980. Domesticated cattle in the Sahara and their introduction into West Africa. In *The Sahara and the Nile*, M.A.J. Williams & H. Faure (eds), 489–501. Rotterdam: Balkema.

Stallibrass, S. 1983. A bifid thoracic vertebral spine from a bovine in the Roman fenland. *Journal of Archaeological Science* **10**, 265–6.

Voigt, E.A. 1983. *Mapungubwe: an archaeozoological interpretation of an Iron Age community*. Transvaal Museum Monograph, 1. Pretoria: Transvaal Museum.

von den Driesch, A. & J. Boessneck 1985. *Die Tierknochenfunde aus den neolithischen Siedlung von Merimde-Benisalame*. Munich: Institut für Paläoanatomie, Domestikationsforschung und Geschichte der Tiermedizin der Universität München.

Von Leithner, O.F. 1927. Der Ur. *Bericht der internationalen Gesellschaft zur Erhaltung des Wisents. Berlin* **2**, 1–140.

Wendorf, F., A. Close, R. Schild 1987. Early domestic cattle in the Sahara. *Paleoecology of Africa* **18**, 441–8.

Wendorf, F. & R. Schild 1994. Are the early Holocene cattle in the eastern Sahara domestic or wild? *Evolutionary Anthropology* **3**, 118–28.

CHAPTER FIVE

Climate and cattle in North Africa: a first approximation

Fekri A. Hassan

1. Introduction

The end of the last glacial period, heralded by global warming, led to dramatic oscillations in climate superimposed on the global shift from cold to warmer conditions. Some of these oscillations were rather abrupt, severe and short lived. They triggered changes in vegetation and the distribution and biomass of animals as well as the availability and distribution of surface and subsurface water inevitably creating new opportunities for subsistence while foreclosing certain options.

It is important to bear in mind the nature of the exploitation of cattle in different communities during the Holocene. I will suggest that the incorporation of cattle was initially a means of supplementing the subsistence hunting–foraging base around the lakeshores of large desert playas in the northeastern corner of Egypt. The spread of cattle-keeping from this centre into other parts of North Africa was a response to severe climatic events and interannual variability and occurred by 8500 bp or/and again by 7500–7000 bp.[1]

Between 7500 bp and 7000 bp, flocks of small stock (sheep and goats) were also introduced from the Near East, along the Mediterranean coast, and down the Red Sea Hills. Ovicaprines are recorded in the Sodmein Cave near Qusseir, Red Sea Hills, Egypt, dated to just after 7000 bp (Vermeersch et al. 1996).

In the central Sahara, where cattle and livestock spread by 6500 bp, specialized cattle pastoralism developed in certain habitats associated with mountain massifs and depressions. Cattle pastoralists then spread westward from the central Sahara and became integrated with hunting, foraging, and fishing economies.

Cattle and small livestock also reached the Nile Delta and central Nile Valley by 6000 bp. From the central Sudan, where specialized livestock herding was soon developed, cattle, sheep and goats spread into Ethiopia and the rest of East Africa *c.* 4500 bp. There, initially, domestic cattle were kept only as an economic supplement by foraging and hunting peoples. After 3700 bp, the southward movement of the Intertropical Zone (ITCZ) and the tsetse barrier permitted the movement of cattle to northern Nigeria south of Lake Chad, as well as southward into Ghana.

Coincident with the final onset of Holocene dry conditions at 4500–3000 bp, combinations of specialized herding (pastoralism) and cultivation began to emerge in East and West Africa. This was particularly the case in the savanna zone, limited in the north by the Sahara and in the south by the tsetse belt, and stimulated by severe droughts at 3700 bp and 2500–2000 bp. The presence of appropriate cultigens and the suitability of the climate for cultivation were key factors in promoting the shift to cultivation. This emergence of specialized herders and farmers, and the spread of iron metallurgy after 2700 bp, created new conditions linked with their further spread southward, as well as powerful dynamics of ethnic, linguistic, and political change that still shape the cultural landscape of Africa.

It would be a grave error to reduce the myriad processes and events that shaped the later prehistory of Africa to a simple model of climatic change and the spread of cattle – although these were two of the fundamental forcing mechanisms. Additionally, we cannot understand the cultural history of Africa without clarifying the temporal stages of change and the regional context of interaction between existing societies and newcomers. There is no evidence of mass invasion, master races, or population replacement. Even an analogy with the precolonial "present" is deceiving because it represents only a later stage in the cultural dynamics of Africa, responding both to internal economic and political developments and external factors.

The spread of cattle, before the development of specialized pastoralism in the Sahara and East Africa, was most likely the result of numerous dispersals by very small groups who were then integrated into pre-existing host communities. The inventiveness, dynamics, and particularity of each situation was different creating a colourful kaleidoscope of developments. The richness of Africa's past lies in this diversity and can hardly be attributed to deterministic causes such as climate or cattle.

It should be noted that in a review published while this work was in final preparation, Bower (1996) covers many of the topics discussed here. Although the main conclusions are similar, the reader is advised to refer to Bower's synthesis for differences in approach and argumentation.

2. Climatic variability

Although there is no space for the details of Holocene climatic events (see Pachur & Braun 1980, Petit-Maire 1990, Grove 1993, and Hassan 1996), the following aspects of Holocene climatic change should be emphasized:

(1) The transition from global glacial to post-glacial conditions commencing 14,000 years ago was associated with abrupt climatic fluctuations in the tropics. Lake levels in East Africa and the Sahel show minimal stands at 11,200–10,100 bp, 8200–7100 bp, 4200–3000 bp and 1000–1 bp (Street-Perrot & Perrot 1990:609; see also Roberts et al. 1994). Gasse & Van Campo (1994) recognize several abrupt dry events simultaneously in Asia and Africa within the monsoonal belt. In the Egyptian Sahara, there are abrupt short-term episodes of aridity at c. 8700–8600 bp, 7200–7000 bp,

6100–5900 bp, 4500–3900 bp with the general onset of arid conditions after 4500 bp (cf. Hassan 1986a, see also Hassan 1986b on Nile flood levels over the past 10,000 years). A survey of Holocene climatic events in Africa also reveals possible abrupt arid events at 9500 bp, 8600 bp, 7900–7700 bp, 7000 bp, 6000 bp, 3700 bp, 2500 bp and 2000 bp. These events are neither equally certain nor sufficiently precise (Hassan 1996).

(2) One of the key elements of African climate is the Intertropical Convergence Zone, ITCZ (Rognon & Williams 1977, Flohn & Nicholson 1980). Displacement of this zone may span distances of 700–1,000 km causing major environmental changes in a broad belt and triggering a domino effect if people are displaced or a novel subsistence or social coping mechanism comes into play.

(3) One of the major aspects of climate relevant to culture change in Africa is the magnitude of interannual variability, especially in sub-arid regions, as clearly shown by the survey of recent rainfall fluctuations in Africa (Nicholson 1994, Lamb et al. 1995).

(4) The effects of the ITCZ create similar and almost simultaneous events in various parts of North Africa from Mauritania to the Egyptian Sahara. Nicholson (1994:124) recognizes the strong spatial coherence of African rainfall anomalies and an overall similarity on a continental scale. Climatic changes in the Sahara are correlated with variations in Nile floods (fed by rainfall in Ethiopia and Equatorial East Africa) and lake levels in East Africa (Hassan 1981, Said 1993).

(5) Local feedback mechanisms in the Sahel prolong and intensify major rainfall anomalies triggered by large-scale meteorological factors (Nicholson 1994).

(6) In addition to abrupt climatic events, major changes on a scale of thousands of years eventually create limiting conditions or special opportunities. For example, the onset of hyper-arid conditions in the Sahara by 4500–3000 bp created a different landscape that has profoundly influenced cultural events since that time.

(7) The transition from the moist/wet intervals of the early–middle Holocene to the present conditions of aridity was apparently a result of a migration of the wetting front southward and perhaps westward (cf. Haynes 1987). The effects of this shift are thus likely to be different along the clinal gradients.

(8) The coastal Mediterranean zone has a different meteorological regime from equatorial Africa (Bryson 1992).

(9) In contrast to certain general similarities in the climatic conditions of North Africa, there are localized climatic conditions and differences in the topographic–hydrographic settings that can create different habitats under the same climatic regimes. Such areas include the Red Sea coast, the major depressions in the Sahara, the Atlantic coast and the mountainous ranges in the central Sahara.

(10) The impact of climatic changes must be assessed in terms of the link between climate, water resources, vegetation and animals. In addition, we have to allow for an understanding of environmental perception and attitudes

towards movements, innovation and inertia. The mobility and demographic flux of hunter–gatherers, for example, provides a different cultural milieu for change in comparison to large, sedentary, agrarian communities.

The relations between climate and Holocene cultural events in North Africa have been explored by many other scholars (e.g. Clark 1967, 1980, 1984, Shaw 1976, Smith 1979, 1992, Banks 1984, Clutton-Brock 1989, McIntosh 1993). These authors, mostly focusing on specific regional developments, recognize many of the causal mechanisms and climatic events incorporated in this contribution. My own interests in climate and cattle are in conjunction with the origins of Egyptian civilization. My original hypothesis was that the beginning of food production in Egypt was attributable to the eighth millennium droughts that promoted a population movement from the eastern Sahara to the Lower Nile Valley (Hassan n.d., 1981). Here, I expand the canvas to show the effect of that critical climatic event, and others, on North Africa as a whole.

3. Cattle-keeping and pastoralism: concepts and terminology

To understand the emergence of cattle pastoralism in Africa it is important to realize the plethora of terms often used in the literature. These concepts include "husbandry", "domestication", "breeding", "exploitation", and "management", in addition to the ubiquitous "pastoralism".

It is important to differentiate between taming, keeping and domesticating cattle. Taming involves a change from a wild or savage state to a docile state, which may be an initial step towards management or domestication, but implies neither. However, cattle may be kept more or less wild and still be managed. Management involves improvement by selective feeding and protection. To manage implies to handle, direct, control, or train. Domestication, on the other hand, is an advanced state of management involving selective breeding. Archaeologically, domestication can lead to discernible morphometric changes while taming or limited management may not be visible.

The variables involved in recognizing varieties of the "exploitation" of cattle may include:
- subsistence mix (hunting, foraging, fishing, horticulture, farming)
- size of group
- daily mobility
- seasonal mobility
- periodic mobility
- magnitude of dependence on cattle
- magnitude of interference with cattle feeding and watering
- magnitude of interference with cattle breeding and selection
- size of herd
- animal varieties and requirements
- land use
- organizational strategy.

Harris (1996) has addressed this confusing issue and proposed a terminology that recognizes a stage of "taming" or "protective herding" (free range or management) followed by a stage of "livestock raising by settled farmers, nomadic transhumance, or pastoralism". There is no implication in his scheme of unilinear evolutionary trajectory. The distinctions made by Harris (1996) are essential, since it is very unlikely that either domestication or pastoralism were short-term events. Hunters did not become nomadic pastoralists overnight. The process was most likely protracted, involving beginnings, false beginnings, setbacks, and repeated trials.

The fluidity and reversibility of subsistence modes is suggested by the shift in the east–central Sudan in the easternmost part of the Sahel (rainfall 200–400 mm) where subsistence strategy changed from a sedentary mixed economy to specialized nomadic pastoralism from 1500 BC to AD 400 (Sadr 1993). Marshall (1994) suggests that pastoralists in East Africa between 5,000 and 4,000 years ago were characterized by a broad subsistence strategy including wild and domestic animals, fishing, and presumably domestic plants; only much later becoming specialized pastoralists. At present there is also great variability in African pastoralism including specialized, camel-breeding full nomads, as in northern Somalia, and sedentary and/or partial nomads holding cattle as their major economic and social focus.

4. Controversial beginnings

The claim for an independent origin of domesticated cattle in Africa has been heatedly debated (Smith 1984b, 1986, Clutton-Brock 1989, Muzzolini 1989). It was based on the discovery by the Combined Prehistoric Expedition of bones at Nabta Playa and Bir Kiseiba in the Egyptian Sahara, identified by Gautier as *Bos* sp. (Gautier 1984a, Wendorf et al. 1987). In this region of the southeastern corner of the western Desert of Egypt (also known as the eastern Sahara), sites associated with ephemeral lakes (playas) yielded extensive faunal collections (Gautier 1987). Sites, before and after 8000 bp, from which there is a moderately large faunal collection, yielded a few bones of "large bovids" (Close & Wendorf 1992). However, the most common animals are Dorcas gazelle (*Gazella dorcas*) and hare (*Lepus capensis*). Other species include Addax (*Addax nasosulcatus*), oryx (?), and wild carnivores. The list also includes seasonal aquatic elements such as molluscs, frogs, and turtles. In addition, bones of birds, desert hedgehog, ground squirrel, field rat, and porcupine were identified (Gautier 1987:177).

One of the problems is the difficulty in identifying "domestic cattle" from small, fragmentary samples (Grigson 1989). Smith (1986) also contended that there is a lack of good information on the ecology and morphology of African forms of *Bos primigenius*. However, Gautier (1987), Wendorf et al. (1987), Close & Wendorf (1992), and especially Wendorf & Schild (1994) have maintained their position and substantiated their claim.

Close & Wendorf (1992:64) indicate that large bovids in Saharan Holocene contexts in northeastern Africa may be African buffalo (*Syncerus caffer*), giant buffalo (*Pelorovis antiquus*), wild cattle (*Bos primigenius*) or domestic cattle

(*B. primigenius f. taurus*). The bones from the Egyptian Sahara partially fall within the range of African buffalo but are morphologically distinct from them. Additionally, they are morphometrically and osteometrically different from giant buffalo. It would thus appear from Gautier's work that these large bovid remains are mostly *Bos*. Nevertheless, "In terms of size, they could be either wild or domestic ... size is a treacherous criterion and our argument that they were domestic is based primarily on ecological considerations" (Close & Wendorf 1992:64).

The ecological argument for domestic cattle rests on the assumption that the climate was too dry to support wild cattle. Rainfall was meagre and estimates vary from 25 mm to 200 mm per annum, not enough to support cattle, especially if the lakes were ephemeral and intermittent (Neumann 1989, 1993, Wendorf et al. 1984). If gazelle and hare are regarded as exclusively dry desert forms, the presence of cattle would be anomalous. Under wetter climate, large bovids were associated with other large animals such as antelope, giraffe, camel, and rhinoceros (Close & Wendorf 1992:67).

Recent work on mitochondrial DNA suggests that African and European cattle diverged from one another at about 22,000 bp (Loftus et al. 1994, Bradley & Loftus, Ch. 13 in this volume). Although Grigson (Ch. 4 in this volume) admits that cattle in Africa were clearly domesticated before the fifth millennium, she suggests that there is not enough data to say when or where domestication occurred. She also casts doubt on the domestic status claimed for the "early neolithic" cattle of Bir Kiseiba and Nabta Playa (see also Grigson 1989).

Although there is no definitive osteometric evidence for domestic cattle before the eighth millennium bp,[2] I suggest that the domestication of African cattle emerged as a result of a sequence of events beginning with the management of cattle within communities associated with desert lakes during the early Holocene. As I hope to show here, cattle were *kept* in the Nabta Playa and Bir Kiseiba region by 8500 bp, if not already by 9000 bp. Note that the dating for the earliest kept cattle is based on a date of 8290±80 bp at site E-75-6 from Nabta Playa. However, it is noted that cattle "may go back to 9840 BP ±380 years at Site E-79-8, but since the remains were mostly collected from the surface, they may be related to the later phases of occupation around 9000 BP" (Gautier 1984a: 69).

5. Before cattle

The Nabta Playa and Bir Kiseiba region is not too far from the Nile Valley. Thus, it may be useful to give a brief overview of the fauna represented in archaeological sites in Egypt and the Sudan before and contemporaneous with the emergence of cattle-keeping in the Egyptian Sahara.

In Egypt, the fauna from final Palaeolithic sites until 7000 bp are characterized by many species. However, wild cattle (*Bos primigenius*) and hartebeest (*Alcelaphus buselaphus*) are the most frequently encountered species at archaeological sites. Red-fronted gazelle (*Gazella rufifrons*) and the Nubian wild ass (*Equus asinus africanus*) are less frequent, but still common at many sites. Dorcas gazelle (*Gazella dorcas*) was recognized at Kom Ombo, but not elsewhere. Other species include

Cape hare (*Lepus capensis*), kob (*Adeonta kob*) and possibly Cape buffalo (*Syncerus caffer*), as well as carnivores such as the striped hyaena (*Hyaena hyaena*), spotted hyaena (*Crocuta crocuta*), African wild cat (*Felis libyca*), the common jackal (*Canis aureus*) and the black-backed jackal (*Canis mesomales*). Rodents are represented by the Egyptian bandicoot or pest rat (*Nesokia indica*). Aquatic resources include African hippopotamus, Nile catfish (*Clarias anguillaris*) and Nile perch (*Lates niloticus*), Nile soft-shelled turtle, Nile clams, and molluscs. Bird remains include a variety of shore, wading and diving waterfowl. Birds include both resident and migratory species (Hassan 1974, Gautier 1984b).

Sudanese sites (El Damer, Abu Darbein and Aneibis) dating from 8500 bp to 7000 bp yielded a diverse fauna of fish and remains of many species of mammals; Cape buffalo (*Syncerus caffer*), topi, kob, kudu, dikdik, African elephant, giraffe, black rhinoceros, hippopotamus, porcupine, mongoose, golden jackal, wild cat, and caracal/serval. The mammals have a high proportion of grazing animals often associated with alluvial grasslands. The environment is interpreted as thorn savanna and scrub with annual precipitation at the junction of the Nile and the Atbara of 300–400 mm (Peters 1993).

6. Climate, ecology, and cattle-keeping: a model

Following the late glacial hyperaridity, the advent of greater rainfall at about 14,000 bp with global warming began to replenish the water table. A rising water table and higher rainfall especially around 12,500 bp coincided with high flooding of the Nile flood plain and a rise in the levels of lakes in East Africa.

The post-glacial warming was associated with the advance of a wetting front eastwards and northwards bringing rain to the Sahara. The depressed water table began to be replenished. Vegetation began to recover. Desert conditions gave way to an environment that supported large mammals such as elephants and giraffes. Populations confined to refugia in the central Saharan massifs, the northern Sahel and the riverine regions began to disperse into the Sahara (Hassan 1997). The number of early desert pioneers were small and given that the desert was still recovering from the hyperaridity of the glacial maximum, generalized foragers and hunters were mobile. Their traces were negligible and the droughts of the "Younger Dryas" probably deflated their scanty remains.

The greening of the desert and the availability of surface water after the initial wet periods from *c.* 12,700 bp to 11,700 bp was interrupted by a cold, dry phase equivalent to the Younger Dryas in Europe or 11,350 bp to 10,250 bp. Grass and shrubs withered; game animals died or migrated. The dry, cold conditions contemporaneous with the Younger Dryas, an event of great global significance, led to the emergence of responses that would have included the establishment of communities around accessible water resources in the vicinity of desert lakes. The droughts would have severely reduced the biomass, distribution and size of large animals, including cattle. It is likely that this event led to the confinement of desert communities to the better watered habitats, especially around lakes in large depressions with large catchment areas. The demise of large mammals during the droughts might have initiated the first attempts towards cattle-keeping. There were also adjustments in foraging and hunting strategies. Hare and gazelle figured prominently in the diet of the communities of the eastern Sahara.

By c. 10,500 bp the return of the rains revived vegetation and supported a wide variety of game animals, providing opportunities for hunting and foraging. The replenishment of the water table and the rain led to the formation of desert lakes in depressions with large catchment areas, and wadi activity and flowing streams and springs in mountainous regions and the Saharan massifs. The repetition of spells of drought c. 9500 bp and high interannual variability, especially in the southeastern margin of the wetting zone in front of the ITCZ is likely to have led to attempts to keep cattle. Management consisted primarily of protecting the animals from predators and other hunters, as well as ensuring that the animals were provided with food and fodder. Calves were probably captured by lassoing and then tethered. They were led to pastures and drinking pools. Wells were dug when necessary to get to the water near the surface. Even today water is still available close to the surface.

Situated in the southeastern corner of the Sahara, the Nabta–Kiseiba region was a potential cradle for cattle-keeping. Receiving about 50–100 mm of rain annually, by contrast to areas to the south that received more rain, this area was far more susceptible to interannual variability than other areas. It was also the first to experience the severity of drought spells and to suffer longer from such spells. The playas near Nabta and Bir Kiseiba with very large catchment areas and precipitation probably held water for most of the year, if not for the whole year, during years of plentiful rain. It would thus seem that the Nabta–Kiseiba region, where rainfall was neither too low (as farther north as Kharga) or too high (as in the Sahel or western Sahara), that permanent or quasi-permanent lakes existed. Short-term interannual fluctuations in rainfall as well as the episodic recurrence of droughts in this region were critical factors that would have prompted people to experiment with a variety of subsistence strategies. Fishing, collecting molluscs, frogs, and turtles were perhaps already common. Fowling and capture of small animals such as lizards, porcupines, and rats was not unknown. Similarly, wild carnivores were also probably trapped. These communities would have also attempted to keep some cattle during those years when rainfall was not sufficient to maintain large mammals.

From 9500 bp to 8500 bp, relatively wet conditions led to the emergence of semi-sedentary communities that resided on the shores of desert lakes. These groups intensively utilized a wide variety of available wild plants, fish and other aquatic resources, as well as the variety of animals that are attracted to the water and vegetation of the lakes and ponds. Among the large animals available, cattle were probably the most important source of meat. Captured calves were probably tethered and sheltered within the encampments as a supplement to hunting–fishing–fowling–foraging. They were probably no more than an insurance policy – an occasional resource kept for slaughtering at times of famine.

The absence of wild ancestors of sheep and goats may explain why cattle were chosen. Cattle were preferred over relatively unmanageable larger mammals such as elephants and giraffes, and were also apparently chosen in preference to oryx, ibex, gazelle, and hartebeest. Smith (1986:200) suggests that cattle were preferentially domesticated because of their exceptional ability to store fat during dry years and droughts. Smith argues that fat is important for health and grave metabolic problems can occur when it is not available – a person can starve to death even when the amount of lean meat is unlimited.

In the Nabta–Bir Kiseiba region, the management of cattle could not have become a viable specialized subsistence activity on its own considering their ecological requirements. The biomass of vegetation and the available amounts of water curtailed the number of kept animals. The animals probably became smaller because of ecological and nutritional stress. The small size of the herd prevented it from becoming a biologically viable population that allows for selective breeding. Wild cattle continued to be lassoed and captured. This would explain both the paucity of cattle bones at the sites as well as their ambiguous wild/domestic characteristics. It would also explain why the proportions of cattle bones are very small in comparison to those of gazelle and hare which were the major sources of animal protein, in addition to fish, birds, rats, turtles and other delicacies.

7. The earliest spread of cattle

An episode of droughts from *c.* 8500 bp was most likely responsible for a new stage in the history of people and cattle in North Africa. The responses to these droughts consisted of territorial localization and intensive utilization of local resources, retreat to refugee areas, and dispersal by small, highly mobile groups to take advantage of the ephemeral and unpredictable resources associated with erratic rainfall. It must be noted that a reduction in rainfall would have included greater interannual variability that might approach 80 per cent, as well as greater spatial unpredictability of humidity.

It seems from the spatial distribution of the early sites with domestic or putative domestic cattle that their spread could have indeed begun from the Nabta–Kiseiba region *c.* 8500 bp coincident with the prevalence of droughts at that time. This first phase of dispersal is documented by the early dates on *Steinplätze* compiled by Gabriel (1984) dating back to 8550 bp. The radiocarbon age determinations of domestic or putative domestic cattle in North Africa reveal that the early dates of 9500 bp at Bir Kiseiba (E-79-8) and 8840 bp at Nabta (E-75-3) are not matched by sites with early "domestic" cattle anywhere else. The majority of sites with "domestic" cattle cluster between 7000 bp and 3500 bp. The paucity of sites from 8500 bp to 7000 bp is probably owing to the small size of encampments, high mobility, and a high dependence on hunting and foraging.

The earliest good evidence for domestic cattle outside the Kiseiba–Nabta area is reported from the Acacus (southwestern Libya) *c.* 7440±220 bp (Gautier 1984a, Barich 1987). There is also a date of 5950±120 bp associated with a domestic *Bos* skull. The earliest sites also include Grotte Capeletti in northern Algeria on the Mediterranean littoral where the earliest date is 6530±250 bp (Roubet 1979, Roubet & Carter 1984).

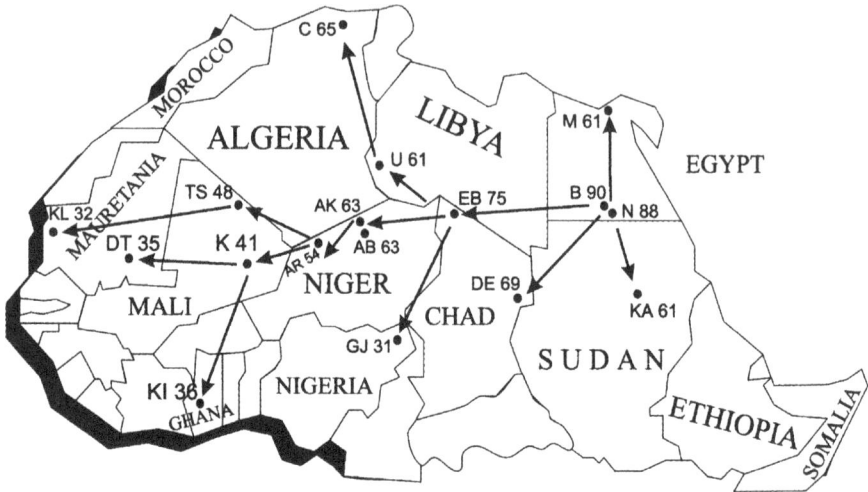

Figure 5.1 Map showing sites with earliest radiocarbon age measurements of "domestic" cattle in North Africa. The earliest dates shown are only suggestive of the probable date of the presence of domestic cattle in the area of the site.

NB. The earliest age measurements are often far less probable than an average date of several concordant age measurements. A re-examination of the radiocarbon chronology to provide a better basis for interpreting the spread of cattle in North Africa is now underway by the author. Preliminary results suggest the majority of sites with domestic cattle in the central Sahara most probably date to 6300–6000 bp. The emerging pattern suggests an initial spread after 7000 bp from the eastern Sahara to Enndi and Tibesti. To maintain continuity with Gautier (1987), the earliest age measurements are shown after the name of the site in hundred years.

Key: AB = Adrar Bous; AK = Adrar n'Kiffi; AR = Arlit; B = Bir Kiseiba; C = Grotte Capeletti; DE = Délébo; DT = Dhar Tichitt; EB = Enneri Bardagué; GJ = Gajiganna; K = Karakarchinkat; KA = Kadero, Geili, Saggai, Rabak; KI = Kintampo; KL = Khatt Lemeitag; M = Merimde Beni Salama; N = Nabta; U = Uan Muhuggiag; TS = Tessalit.

It appears that an initial movement could have taken cattle, or the idea of domesticating cattle, from the Nabta–Kiseiba region soon after 8500 bp to the Ennedi, Tibesti, and the Acacus (Fig. 5.1). This was rapidly followed by the spread of cattle to the Aurès where Grotte Capeletti is situated. This route, as well as others later, take advantage of orographic rain and the tectonic-topographic relief permitting surface or near surface water availability.

8. The Near Eastern connection: the coming of ovicaprines

In the Levant, sheep and goats were domesticated by 9000 bp, but they were mostly restricted to a small area before they spread at 8500 bp (Legge 1996).

However, Garrard et al. (1996) suggest that there is no evidence for domesticated ovicaprines before 8500 bp in the Levant (see also Bar-Yosef & Meadow 1995).

The area for domestication of cattle in the Near East is difficult to pinpoint. Uerpmann (1996:236) suggests three possible centres: the southern Levant, Anatolia, and the eastern margins of southwest Asia. In Greece, domestic cattle appear from c. 8000 bp following a level with pigs and goats dated to 9000 bp (Halstead 1996:296).

However, neither domestic ovicaprines nor domestic cattle are recorded in the Sinai or the Negev before the seventh millennium bp (Smith 1989). The vast majority of Holocene sites in the Sinai are Chalcolithic or later, but a site with pottery and domestic cattle and ovicaprines at Qatif, near Gaza, indicates that domestic sheep, goats and cattle arrived close to Africa at c. 7000 bp, and perhaps earlier (Oren 1979). However, according to Rosen (1984:49), the earliest certain evidence for animal domesticates in the Levant dates to the early Bronze Age (c. 3000 BC) although strong circumstantial evidence suggests that the transition to a pastoral nomadic way of life took place as early as the fifth millennium BC.

Sheep and goats, and possibly some cattle, might have thus made it along the Mediterranean coast from the Sinai to coastal Cyrenaica and then to the Aurès. However, the status of domestic sheep and goats at Haua Fteah in coastal Cyrenaica is problematic. A re-investigation of the Haua Fteah fauna first identified by Higgs (1967) could not verify the date of early domesticated cattle because of stratigraphic problems (Klein & Scott 1986).

The fauna at Grotte Capeletti is dominated by sheep and goats. Cattle remains range from 7.3 per cent at base to 24.7 per cent in the upper levels. The faunal assemblage at Capeletti suggests a specialized, transhumant, pastoralist economy with an emphasis on small livestock (Roubet 1979:384–404). If we exclude the remote possibility that cattle were independently domesticated at Capeletti (see suggestion by Smith 1984a:90), and assume that cattle did reach Capeletti from Nabta–Kiseiba via the central Sahara, it is possible that they might have come along the coast with ovicaprines from the Near East (see Smith 1984a:90–91). Given that cattle and ovicaprines are so far dated at c. 7000 bp or slightly earlier in the Sinai, the movement along the coast would have been fairly rapid.

However, even if domestic cattle arrived at Haua Fteah during the seventh millennium bp, there would be no problem fitting this with a source in the Nabta–Kiseiba region. Cattle from the Egyptian Sahara could have arrived at about the same time as sheep and goats to Cyrenaica and the Aurès. Alternatively, the spread of cattle from the Egyptian Sahara did not reach either Haua Fteah or Capeletti. Since mitochondrial DNA data shows that modern North African cattle cluster with the European group (Loftus et al. 1994), this possibility is not unlikely. However, this may be a function of later introductions of cattle from the Near East.

It seems also that herders moving from the Levant into North Africa (once they crossed the Sinai) could have travelled along the Red Sea coast southwards to the well-watered wadis of the Red Sea Hills. It is in one of those wadis that ovicaprines were recently discovered in a cave near Qusseir dating to about 7000 bp (Vermeersch et al. 1996).

Farther south in the eastern Desert, Sadr et al. (1994) discovered, along one of the tributaries of Wadi Alaqi, a low circular tumulus. Although the grave was looted they recovered a gold wire bracelet, pottery, a quantity of charcoal, several beads, a stone pendant, and two animal horns: one of cattle and the other of sheep or goat. A radiocarbon age determination of the charcoal dates the bones to the mid seventh millennium bp.

Sheep and goats also spread rapidly into the interior of the Sahara because of their greater adaptation to desert conditions and because of the spread of arid conditions between 7500 bp and 7000 bp. Sheep or goat first appear in Nabta Playa at 6700 bp or somewhat earlier (Close 1992:172).

9. Colonization of the Nile Valley

The droughts of the period from 7500 bp to 7000 bp both in the Levant and the Egyptian Sahara seem to have been also responsible for an initial colonization of both the Lower and Upper Nile Valley. Sheep and goats as well as wheat and barley were introduced into the Nile Valley in Egypt where they are first recorded at Merimde Beni Salama *c.* 5900 bp, and perhaps by 6100 bp (Hassan 1988). Domestic cattle and ovicaprines appear in the central Sudan within the Shaheinab–Khartoum tradition at *c.* 6000 bp (Fig. 5.1). Those who settled in Egypt were integrated with the local fishermen and foragers and began a course of agrarian life as a mode of subsistence. The fauna from Merimde Beni Salama and the Fayum (dated to 5900–5800 bp) include bones of cattle, ovicaprines and pigs (Gautier 1987). By 5000 bp at Nagada, bones of wild game are almost negligible suggesting a devotion to farming supplemented with some herding, although some fishing and fowling continued (Hassan 1988).

In the Sudan, both cattle and ovicaprines were herded. Gautier (1989) reviews the prehistoric fauna of the central Sudanese Nile Valley, confirming that the specimens previously identified as African buffalo from Shaheinab are in fact domestic cattle. Unlike the Lower Nile, wheat and barley were not favoured. The Sudanese communities, taking advantage of the higher local rainfall and perhaps prevented from productive cultivation by the ecological conditions of the Gezira, opted for herding as a more productive and secure pursuit. Most samples show that cattle were a major subsistence item. By contrast, sites dating before the advent of livestock (early Khartoum tradition) yielded bones of kob (*Kobus kob*), molluscs, and fish. It is thus likely that the adoption of livestock was a result of the arrival of groups who kept cattle and small livestock (Gautier 1987).

Examination of pottery, decorated with various dotted wavy line motifs (Caneva & Marks 1990), confirms the introduction of Saharan ceramics into the central Sudan around 6000 bp. According to Caneva (n.d.), three main types of dotted wavy line pottery dating from *c.* 9370 bp to 6950 bp in the Sahara appear together in the Sudanese Nile Valley at the same time. This may suggest that the introduction of livestock from the central Sudan was via Ennedi. At Délébo in the Ennedi, the third type of the dotted wavy line appears at *c.* 7000 bp (ibid.). Caneva also suggests that the early Khartoum complex in the northernmost part of Sudanese Nubia was introduced from the south, perhaps ultimately from the Atbara region, where it dates from the ninth to the seventh millennium bp. In the intermediate Khartoum region, the majority of the early Khartoum sites date from the second half of the eighth millennium to the end of the seventh millennium bp. This presumed northwards movement contrasts with the movements eastwards from the Sahara, and appears to be either a result of the transmission of the early Khartoum ceramic types as a welcome innovation, or the displacement of groups from the south northwards following the wetting front during the interval from the tenth to the sixth millennium bp.

Since there is so far no evidence for herding or farming communities in the Nile Valley (Lower or Upper) before 6000 bp or at the earliest 6500 bp, and since both ovicaprines and cattle were introduced by about 7000 bp in nearby regions, it seems that herders were not keen on inhabiting the Nile Valley, preferring instead habitats similar to those of the Sinai desert. Those who kept cattle in the Egyptian Sahara do not seem also to have been attracted to the Nile Valley, moving instead initially towards similar habitats farther west in the Sahara. The penetration of the Nile Valley might thus have been a result of the severe arid spells between 7000–6000 bp.

10. The western frontier: cattle in the central Sahara

The earliest date for domestic cattle at Délébo, Chad, is 7180 bp (Gautier 1987). This suggests a movement before that time from the Nabta–Kiseiba area. If we accept that this date represents the earliest occupation of this region and fits dispersal after *c.* 8500 bp, much younger dates at Daima and Gajiganna (see below) suggest that the spread to the south was not favoured until much later. This may be because the tsetse zone was much farther north than it is now (see below).

Sites in the central Sahara are concentrated around the well-watered massifs. The earliest dates with domestic cattle range in age from 6130 bp to 5400 bp. The dates fall within a time gradient from the Nabta–Kiseiba region with a rate of spread 0.6 km per year from there. This rate corresponds to about 12 km per generation. The spread was westwards across the desert via the Tibesti, Acacus, and Aïr massifs. Cattle also spread farther north in a northwestern direction and later westwards to Dhar Tichitt. Synchronous with the movement westwards to Dhar Tichitt, cattle spread from the central Sahara southwards towards the Lake Chad Basin (e.g. Gajiganna).

11. Emigrants, pioneers or refugees?

The spread of cattle and small livestock into the central Sahara, the Nile Valley, and the Mediterranean littoral between 7000 bp and 5900 bp is extraordinary. It calls for a trans-regional explanation, and is most likely not independent of the severe aridity that swept much of North Africa and the Levant at that time.

The almost instantaneous and simultaneous appearance of small livestock in a broad region suggests a very rapid rate of movement. This was hardly a demic (population) expansion of an advancing "wave" or a "frontier", but a "leap-frog" movement by small groups. In a demic expansion, the dispersal is a result of a build-up of population in the parent region, which is followed by fission and the establishment of daughter colonies. Even with a fast rate of population growth (2 per cent per year) the rate of advance would have been 0.6 km/year (Hassan 1981). At the more likely rate of 0.01–0.1 per cent, the rate would have been much slower. This suggests that the movement was not a result of population increase.

Muzzolini (1993) has suggested that the transition to food production in North Africa was attributable to demographic pressure. He postulates a demographic pressure against an inflexible ceiling. Groups were limited by water and could not migrate to the "territories" of other groups. This led to the "first crisis", which created fixed territories. According to Muzzolini (1993) this was followed by intensification and specialized hunting, which in turn led to accelerated population increase leading to the "second crisis". This crisis led in some cases to warfare and in others to either cultivation of cereals or animal husbandry. This fanciful scenario has no substantive support. Theoretically, the appeal to unbridled population increase in view of the universal practice of cultural checks is unfounded (Hassan 1981:143–75). Moreover, the Muzzolini scenario would have led to the independent development of cultivation and animal husbandry everywhere in North Africa, which is clearly not the case.

Alexander (1984) provides a "frontier" model with the spread of food production in northeastern Africa recognizing several husbandry complexes and two phases of spread with various implications for the existing hunting-gathering populations. The model presented here favours a dispersal by small groups who are too few to overwhelm local communities demographically. The newcomers are then integrated with the existing societies both culturally and physically. Only in certain ecological settings does specialized husbandry emerge. These include:

(1) the appearance of specialized pastoralism in the central Sahara and the Mediterranean littoral between 6000 bp and 5000 bp,

(2) the development of wheat-barley farming and raising herds of cattle and ovicaprines in the Lower Nile Valley beginning *c*. 6000 bp,

(3) the adoption of herding small livestock and cattle in the Middle Nile Valley (Nubia and the Sudan) beginning at 6000 bp,

(4) the development of specialized cattle/ovicaprine pastoralism in the Sahel by the fourth millennium bp, and East African savanna during the third millennium bp,

(5) and the emergence of plant food production in association with livestock herding in the savanna belt *after* 4500 bp, and perhaps even after 3700 bp.

If we assume that the original inhabitants in the Nabta–Bir Kiseiba region were 200–500 persons, those who "emigrated" were probably no more than 100–200. The emigrants were probably single families or groups of families (say up to 50 persons, most likely with a high ratio of young adult males). Some of those "bands" would settle after a few days journey, while others would bypass them to settle elsewhere with short sojourns before they moved away to prevent deterioration of their habitats from over-exploitation by sheep and goats. A distance of as much as 5,000 km can be covered in 500 years if they moved only 10 km per year. The rate estimated for the spread of farmers in Europe is 15 km per generation (Ammerman & Cavalli-Sforza 1973).

We assume that the groups moved along certain routes where water and pastures would have been available to sustain several days of travel. The fastest rate of movement would have been of about 5–10 km per week if they resumed travel after say a week of residence in one spot then relocating to another area about one day farther away before resettling. The group would cover a distance of 200–400 km per year. This is far greater than the estimated rate, suggesting that the movements probably did not occur continuously, but might have been a result of episodic movements interrupted by long-residence in hospitable habitats. It is thus likely that the spread of cattle did not follow a regular linear pattern, but consisted of a series of moves at a faster rate in the wake of successive episodes of spells of droughts or over-exploitation of local resources.

We hardly envision a take-over by pastoralists as cattle spread. Instead, we can picture a dynamic interaction between the struggling newcomers and existing foragers, fishers, and hunters with cultural exchanges leading in different contexts even within the same region to a diversity of syncretic, assimilated and transformed new cultural modalities (cf. Zvelebil 1996 and Thomas 1996 on the spread of food production in the circum-Baltic region and continental and northwest Europe).

A generalized, flexible, opportunistic economy characterized by a spectrum of alternate or complementary subsistence activities including cattle-keeping/herding, foraging, fishing, and hunting, especially in arid and semi-arid regions would appear to be a long-term, sustainable system. It explains the composition of numerous faunal assemblages and fits well with the archaeological data on land-use. This model can also explain the dispersal of cattle-keeping as a combination of "drift" and "leap-frog" spatial movements under conditions of high interannual variability in rainfall and episodic droughts. The model presented here fits with the "trickle-and-splash" pattern suggested by Bower (1991). Instead of a "bow wave" frontier, Bower suggests an interstitial frontier where small groups of diverse origins coalesce. During a drought a trickle of people from different areas may converge on a favourable region where, through demographic and cultural fusion and amalgamation, new practices would emerge. In this case, it is difficult to distinguish a single "parent" industry or tradition (Hassan 1988). This model is antithetical to the migration of ethnically-distinct pastoral groups with all its implications for the spread of language and genetic mapping. The "ethnicity" model, in a re-casting of racialism, still permeates and inhibits archaeological thinking and redirects attention from more deserving issues (see for example Kinahan's (1993) thoughtful examination of the rise and fall of nomadic pastoralism in the central Namib desert).

12. Meanwhile, back at the ranch

In the eastern Sahara, those who remained behind after the ninth millennium droughts (*c.* 8600 bp) are likely to have congregated around certain well-watered resources. The emergence of clusters of dwellings by 8100 bp at Nabta suggests that sedentary or semi-sedentary communities were established by that time. Mediation of conflicts at times of stress probably prompted the emergence of religious ideologies and modes of political organization that resemble ranked societies or tribal chiefdoms. This appears to have been what happened for those who stayed in the Nabta–Kiseiba region. Close (1992:169) summarizes the situation by 8000 bp as follows:

> ... in spite of the fickleness of the local climate, there was a significant human occupation of the eastern Sahara. People were collecting wild plants, herding their cattle across the grasslands after the rains, digging wells in the playa bottoms during the dry season, and cherishing their rare pottery vessels (wherever they might have been made). Sites are known from the Egyptian eastern Sahara (although curiously, they are exceedingly rare in the Sudan) and occupation of the desert may have been year-round. Most sites are not very large, but the structures found at some may imply semi-permanent occupations. Sites known near Nabta (but not studied) are considerably larger than any other known (several hundred meters in diameter), and might reflect more social complexity than has hitherto been suggested for the early neolithic.

Recently, Wendorf & Schild (1995) and Wendorf et al. (1992/3) presented evidence dating to *c.* 6500 bp for a number of large megalithic alignments (calendar) circles, and stone tumuli. Excavation of two of the tumuli disclosed that one covered a pit in which there was deliberate burial of a young bull; the other tumulus contained most of a disarticulated bull together with bones of sheep or goat. This is interpreted as evidence for a cattle cult and the emergence of ranked society in the eastern Sahara before the emergence of social complexity in the Nile Valley. The elaborate burials suggest that a cattle complex involving ritual performances was already in existence by 6500 bp in the eastern Sahara. It also hints at the possibility of "tribal chiefdoms" that may have emerged in conjunction with an integration of several Saharan communities in a political organization that brought them together for ceremonial events.

The developments in the eastern Sahara after 8500 bp, also involved intensive utilization of grasses. The recent discovery of a spectrum of wild plants from contexts dating to 7500–7000 bp at Nabta (Close & Wendorf 1992:66) analyzed by Wasylikowa (Wasylikowa et al. 1993) and at Farafra Oasis (Barich et al. in press) analyzed by Ahmed G. Fahmi, include an exceptionally high frequency of *Panicum* sp. grass and wild sorghum. There are also herbs and fruits. Conspicuous among the herbs and shrublets are the cucurbits and *Arnebia* sp. There are also abundant elements of leguminosae and tubers. These plants would have been mostly associated with the wetter margins of the desert lakes, sedge-marshes, and ponds, which would have also supported tamarix. A high frequency of *Zizyphus* is also noted.

13. The spread of cattle after 6000 bp

The numerous *steinplätze* (scatters of stone, presumably related to transient camps) surveyed by Gabriel (1984), and most frequently documented for the period from 5800 bp to 5300 bp, are probably related to generalized foragers and hunters who probably kept cattle. Their movement was a result of the unstable climatic conditions between 7800 bp and 5900 bp, with severe droughts at 7000 bp and 6000 bp, with pronounced interannual variability. Specialized pastoralism was also developed within the Mediterranean coastal biome as at Capeletti after 6500 bp, and in the central Sudan also probably at about the same time (see below). The development of a different, highly specialized strategy of pastoralism also developed within the savanna and other ecologically favourable areas in East Africa after 3000 bp. Similarly, specialized herding (nomadic pastoralism) was a seemingly late development in West Africa.

Those who stayed at Nabta and those who moved away into northwestern Africa were faced by the recurrence of droughts and a progressively increasing aridity. The droughts of *c.* 7000 bp and again probably at *c.* 6000 bp were followed by another climatic crisis at about *c.* 4500 bp, which heralded the onset of recent hyper-arid conditions. By *c.* 3000 bp the transition to the modern desert landscape was complete. In addition to the abrupt, short-term spells of aridity, the change from the wetter conditions of the early Holocene was gradual from 4500 bp to 3000 bp, with a retreat of the rains southwards and westwards. Haynes (1987) estimates a rate of 0.25–0.83 km/year. At an average rate of 0.5 km/year, the onset of hyper-arid conditions in the northeastern region would have become evident farther south and east (a distance of 1,000 km) in about 2,000 years. If we assume a minimal rate of 0.25 km/yr it would have required 4,000 years. The fast rate would require 1,200 years. The estimate of a retreat between 4500 bp and 3000 bp is thus probably reasonable. Because conditions with the life span of generations would have been wetter to the northwest and south, we would thus expect that this period brought about significant slow dispersal of human groups. In addition, the recurrence of severe aridity at *c.* 4500 bp and 3700 bp and again 2500–2000 bp are likely to have led to rapid population movements.

The changing climatic conditions during the late Holocene may thus explain the appearance of domestic cattle within hunting–foraging–fishing subsistence regimes farther west and south of the zone where cattle were introduced 7500–6500 bp. Domestic cattle in northern Niger at Agorass-in-Tast in the Valley of Adrar Bous have been dated to 5780 bp and 5740 bp (each with a fairly high sigma of 500 years) on a sample from a complete skeleton. *Bos* from the same site yielded a date of 5050±140 bp (Clark et al. 1973, Smith 1974). At the Adrar Bous sites, cattle are dated to 5130±300 bp and 4910±135 bp (Smith 1976), and 5140±300 bp (Hugot 1962). However, a recent date from the latter site yielded a date of 6325±300 bp (Roset 1987). At Arlit, domestic cattle are dated to 5400–4800 bp, with one date as recent as 2700 bp (Gautier 1987). The high frequency of domestic cattle at Arlit (cattle 84.9 per cent, ovicaprines 6.3 per cent) suggests that fully specialized cattle pastoralism was established. Domestic cattle have also been dated from Adrar n'Kiffi in the Adrar Bous region to 6250±250 bp and 6325±300 bp (Paris, Ch. 7 in this volume). Farther south at Daima, a mound-site in the Chad Basin, south of Lake Chad in Nigeria (Connah 1976), evidence for fishermen and herders with small and large livestock was dated to *c.* 2500 bp, but new investigations in the region revealed domestic cattle at Gajiganna dating to *c.* 3100 bp (Breunig et al. 1996). There are no potential wild ancestors for cattle in Nigeria, and cattle were introduced from the north. The cattle were small and may well have belonged to the same stock that gave rise to the West African dwarf shorthorn (Shaw 1984).

Farther west at Dhar Tichitt (Mauritania), "pastoralists" with cattle and small livestock appear around 3500 bp (Gautier 1987). It thus seems that expansion westwards continued well into the third millennium bp. In Central Africa, small livestock (sheep and goats) were recorded at 2600–2400 bp in the secondary forest region of Cameroon (Van Neer, Ch. 9 in this volume). However, pottery was being used and oil palm and *Canarium* were being exploited by 4000 bp (de Maret 1996:320).

Smith (1984a:86) has suggested that deterioration of climate after 4500 bp pressured herders to move out of the Sahara, as surface water became scarce. The southwards movement of the ITCZ brought rain to the interior of West Africa and the tsetse limits of the 500–700 mm rainfall isohyet also moved southwards. Accordingly, the more southerly parts of the Sahel became free from tsetse infestation, opening up new areas for colonization by cattle-keepers. In the meantime, cattle-keepers followed the river systems draining southwards to the Niger River. By 3300 bp, according to Smith (1984a:86) the sites of Karkarichinkat, with evidence of domestic cattle, had been abandoned. At that time, cattle were brought south of this zone into the savanna via the Gourma region at Windé Koroji (MacDonald 1996), through tsetse-free corridors open during dry seasons or droughts. This is indicated by the early (*c.* 4000–3500 bp) presence of "dwarf" domestic cattle and dwarf goats at Kintampo Tradition sites in Ghana (Carter & Flight 1972). According to Posnansky (1984) northern immigrants were integrated with existing Ghanaian societies in different degrees, as reflected in the regional variations of the Kintampo Tradition.

The spread of livestock into East Africa probably from the Red Sea Hills or the Sudan (where it is recorded from Kadero at 6500 bp) is first documented at 4200 bp at Dongodien, East Turkana (Marshall et al. 1984) apparently as a consequence of increasing aridity in the north (Bartheleme 1984, Marshall 1994, and Ch. 10 in this volume). Small domestic stock in hunter-gatherer levels at *c.* 4500 bp at Enkapune Ya Muto and cattle at *c.* 3100 bp (Maréan 1992) suggests, as in other parts of North Africa, that livestock were an element in a generalized economy.

Bartheleme (1984:205) notes that the archaeological evidence from northeastern Lake Turkana provides a geographical and a chronological link between the early food producers along the Sudanese Nile and sites later in time located in the central Rift of Kenya and farther to the south. Bartheleme (1984) takes account of the droughts of the fifth millennium bp as a factor in the introduction of new economic and cultural traits as well as population movements. These movements led to the adoption of domestic cattle in the southern area of Kenya and northern Tanzania.

In an early interpretation, Clark (1962, 1967) suggested that the spread of cattle-herding was introduced to the highlands of Ethiopia by "C-Group" immigrants from Nubia some 4,000 years ago. The movement of the herders was instigated by mid-Holocene aridity, which resulted in famines and drought, forcing people to search for new pastures, such as the temperate Ethiopian plateau. The herders moved into northern Ethiopia introducing cattle-keeping to the indigenous populations. Cattle were then dispersed farther south down the Red Sea into the Afar Rift and finally up onto the Ethiopian plateau (Clark 1980). The inhabitants of the low-lying plains of the Sudan and western Ethiopia were also subjected to long periods of drought by 4500 bp, resulting in movement towards the Ethiopian Plateau.

Evidence for a subsistence economy devoted to herding cattle and small live-stock as the primary subsistence activity is recorded at Ngamuriak, Kenya, dated to 2000 bp. According to Marshall (1994:24), although wild game were common in the area, only 22 bones of wild animals were recovered from a collection of some 60,000 bones. Other sites from 3000–2000 bp also revealed the specialized use of animal resources. However, Ambrose (1984) and Marshall (1994) distinguish between two ecological patterns – Elmenteitan and savanna Pastoral Neolithic (originally Gumban A). According to Ambrose (1984), the Elmenteitan groups co-existed with the Eburran (originally Kenya Capsian) and were primarily hunter–gatherers.

The ecological issues involved in the management and exploitation of livestock in Kenya is admirably illustrated by Gifford-Gonzalez's (1984) study of the site of Prolonged Drift, southwest of Lake Nakuru, dating to 2530–2300 bp. Gifford-Gonzalez recognized the presence of many species of wild animals along with cows and caprines. The animals present included wildebeest, kongoni, zebra, Thompson's gazelle, Grant's gazelle, impala, eland, buffalo, warthog, giraffe, white rhino, mole rat, mouse, and leopard. The overall representation of wild species resembles that of wild game in the Nairobi Park and a "low risk/high yield per kill take". According to Bower (1991, 1996:137) from about 4000 bp to 3000 bp, the remains of domestic animals at sites in Kenya and northern Tanzania are minor by comparison to wild species. This suggests that domestic livestock were perhaps kept either as insurance against food shortages or as prestige items. By 3000 bp, an increase in the frequency of domestic livestock suggests the emergence of specialized pas-toralism (Bower 1996:137). Marshall (1994:34) attributes the appearance of spe-cialization in East Africa to the establishment of a bimodal rainfall in Africa about 3500 bp.

Robertshaw (1993:370), who examined the beginnings of food production in southwestern Kenya, notes that "in South Nyaza a broad-spectrum, gathering–hunting–fishing adaptation is replaced by, or perhaps assimilated into, the Elmenteitan cultural tradition of immigrant pastoralists some two thousand years ago". By con-trast, in the Mara region, "pioneer herders settled in the Lemek valley, from which they would appear to have been displaced around 400 bc or somewhat later by the makers of Elmenteitan pottery and stone artefacts". The Mara region sustained viable pastoral production and thus, hunting, and possibly cultivation, were shunned. The shift to "pure" pastoralism in the first centuries AD probably occurred when agricultural products could be obtained through exchange networks.

Bender (cited in Blench 1993:73) suggests that the linguistic evidence for Nilo–Saharan and other branches of Afroasiatic implies that cattle were brought into East Africa twice; initially from the Ethiopian highlands and then by the ancestors of the present-day Nilotic speakers moving in from the northwest. Blench argues that the primary introduction from Ethiopia was the all but vanished humpless shorthorns and the secondary introduction was of the zeboid type. It is thus likely that the primary introduction was that of the humpless shorthorns before 2000 bp.

It is still not clear what role climatic change might have played in the dispersal of cattle into southern Africa. However, the attenuation and fragmentation of the zone of tsetse infestation after 3000 bp when modern aridity was established is likely to have facilitated a southwards movement. The motivation for this might also have been triggered by the effect on domestic and wild game of arid spells, as well as the emergence of tribal agro-pastoralists (cf. Bower 1996:137). A revised chronology for pastoralism in southernmost Africa (Henshilwood 1996:945) indicates that sheep were present at Blombos Cave, southernmost Cape, at 2000 bp. However, the route by which livestock were introduced is still controversial (Sutton 1996).

14. Summary

The emergence and spread of domestic cattle in Africa during the Holocene is a controversial subject because of a sparsity of archaeozoological and archaeological data. Many areas remain to be explored. The present information poses certain puzzles and hints at some tantalizing possibilities. Regardless of the place and timing of domestication, and even of what we call domestic, the changes in climatic conditions during the Holocene were influential. On a large scale, the early and middle Holocene periods were wetter than the present. Aridity began to set in by 4500 bp in the north and was complete by 3000 bp. During the moist and wet period interannual variability was great. In addition, episodes of droughts and hyperaridity that spanned decades or a few centuries had devastating effects on water, vegetation, and the distribution, quality, and biomass of game animals.

By 6000 bp, bones of domestic cattle and sheep and goat are found in many sites in North Africa. Sheep and goats were introduced from the Levant and spread rapidly into the Sahara. They entered North Africa between 7000 bp and 6500 bp perhaps in response to the droughts and unstable climatic conditions between 7800 bp and 5900 bp. It is not likely that cattle were also introduced from the Levant at that time. Near Eastern sheep spread southwards along the Red Sea coast and westwards along the Mediterranean coast. By 6500 bp they were already common in the Egyptian Sahara and by 6000 bp or earlier in the central Sudan.

African cattle were probably tamed and managed in the Nabta–Kiseiba region between 9500 bp and 8000 bp following in association with an economy exploiting the resources of desert lakes. The droughts of 7500–7000 bp are likely to have led to population movements, as well as the spread of cattle management to the well-watered areas of central Sahara and the Sahel. Specialized pastoralism in the central

Sahara and the Mediterranean was established by *c.* 6000 bp and much later in West Africa and East Africa.

Cattle and livestock were introduced to the western Sahara and the Sahel after 3700 bp. Early domestic cattle remains in these areas are often found in association with remains of generalized hunting–fowling–fishing subsistence regimes. Small groups apparently moved rapidly during the drought years and mingled with the local hunting–gathering communities. By that time the Sahara was changing rapidly into the current desert conditions, which were established between 4500–3000 bp.

At that time interval (especially between 3700 bp and 3000 bp), communities in the savanna zones and East African highlands began to develop specialized pastoral activities. In both East and West Africa the first attempts to cultivate plants were also underway during that interval. The shift to specialized pastoralism and cultivation might have been triggered by droughts at *c.* 3700 bp. In the regions farther north, desert conditions made herding or cultivation difficult. The central Sudanese Nile Valley was all but depopulated during the period between 3500 bp and 2500 bp (Marks et al. 1985). Tsetse prevented the spread of cattle and livestock farther south until the late Holocene when drier conditions prevailed. The tsetse barrier was then broken by corridors which permitted the spread of cattle into central and southern Africa.

The panorama of the spread of cattle across the African continent is a testimony to the remarkable ingenuity of human responses to climatic adversities, and to the capacity to seize new opportunities and create new worlds of labour, social relations, and thought. The role of climate emphasized here is only one ingredient in the exceedingly complex ecological and cultural situations that were faced in so many different regions.

15. Concluding remarks

The model and various hypotheses posited here rest on a set of data of variable quality. Anyone who attempts to put together a comprehensive model of events associated with the appearance of cattle in North Africa from the Atlantic Coast to the Nile Delta and the central Sudan over 5,000 years faces numerous difficulties. This contribution is an initial formulation of such a comprehensive model with all its shortcomings and inadequacies. I continue to examine the chronological evidence proposed by various authors for the earliest dates of domesticated cattle wherever such evidence has been advocated. The continual reliance on the oldest date from any given site is questionable, and future work should focus on estimating the most probable age determinations based on an average of several calibrated dates. Unfortunately many sites are dated by one or two radiocarbon dates, some with large standard deviation. The status of some cattle bones in several cases also remains inconclusive.

I hope that this work will stimulate an interest in trans-regional examinations of the ramifications of the spread of cattle from a core area encompassing the eastern Sahara, Ennedi and Tibesti to the rest of the central Sahara and the Nile Valley, and the subsequent spread after 4500 bp to the western Sahara and thereafter to the area occupied by the Sahel today. The spread of cattle as an all-encompassing explanation for cultural developments in North Africa is inadequate, and we should aim instead to elucidate the creative dynamics by which local communities dealt with a changing world.

Notes

1. All years reported here as bp are in uncalibrated radiocarbon years before present.
2. In India (Meadows 1996:403 in Harris) domesticated goats date to the seventh millennium bp (5300–4700 cal BC) and cattle bones, comprising more than 50 per cent of the faunal remains, date to the second half of the sixth millennium bp (5500–5000 bp).

References

Alexander, J.A. 1984. The end of the moving frontier in the neolithic of northeastern Africa. See Krzyzaniak & Kobusiewicz (1984), 57–63.

Ambrose, S.H. 1984. The introduction of pastoral adaptation to the highlands of East Africa. See Clark & Brandt (1984), 212–39.

Ammerman, A.J. & L.L. Cavalli-Sforza 1973. A population model for the diffusion of early farming in Europe. In *The explanation of culture change*, C. Renfrew (ed.), 343–58. London: Duckworth.

Banks, K.M. 1984. *Climate, cultures and cattle: the Holocene archaeology of the eastern Sahara*. Dallas: Southern Methodist University Press.

Barich, B.E. 1987. Adaptation in archaeology: an example from the Libyan Sahara. See Close (1987), 189–210.

Barich, B.E., F.A. Hassan, A. Fahmi, in press. Early to mid-Holocene occupation at Farafra, western Desert, Egypt. *Proceedings of the XIII Congress of the International Union of Prehistoric and Protohistoric Sciences*, Forli, Italy.

Bartheleme, J.W. 1984. Early evidence for animal domestication in eastern Africa. See Clark & Brandt (1984), 200–205.

Bar-Yosef, O. & R.H. Meadow 1995. The origins of agriculture in the Near East. In *Last hunters, first farmers: new perspectives on the prehistoric transition to agriculture*, T. Douglas Price & B. Gebauer (eds), 39–94. Santa Fe: School of American Research Press.

Blench, R.M. 1993. Ethnographic and linguistic evidence for the prehistory of African ruminant livestock, horses and ponies. See Shaw et al. (1993), 71–103.

Bower, J. 1991. The pastoral neolithic of East Africa. *Journal of World Prehistory* 5, 49–82.

Bower, J. 1996. Early food production in Africa. *Evolutionary Anthropology* 5, 130–39.

Breunig, P., K. Neumann, W. Van Neer 1996. New research on the Holocene settlement and environment of the Chad Basin in Nigeria. *African Archaeological Review* 13, 111–45.

Bryson, R.A. 1992. A macrophysical model of the Holocene intertropical convergence and jetstream positions and rainfall for the Saharan region. *Meteorology and Atmospheric Physics* 47, 247–58.

Caneva, I. n.d. *The influence of Saharan prehistoric cultures on the Nile Valley*. Paper presented at the XIII International Congress of Prehistoric and Protohistoric Sciences, Forli, Italy, 8/14 September 1996.

Caneva, I. & A.E. Marks 1990. More on the Shaqadud pottery: evidence for Saharo–Nilotic connections during the 6th–4th millennium BC. *Archéologie du Nil Moyen* 4, 11–35.

Carter, P.L. & C. Flight 1972. A report on the fauna from the sites of Ntereso and Kintampo Rock Shelter Six in Ghana with evidence for the practice of animal husbandry during the second millennium BC. *Man* 7, 277–82.

Clark, J.D. 1962. The spread of food production in sub-Saharan Africa. *Journal of African History* 3, 211–28.

Clark, J.D. 1967. The problem of neolithic culture in sub-Saharan Africa. In *Background to evolution in Africa*, W.W. Bishop & J.D. Clark (eds), 601–27. Chicago: University of Chicago Press.

Clark, J.D. 1980. Human populations and cultural adaptations in the Sahara and Nile during prehistoric times. In *Late prehistory of the Nile Basin and the Sahara*, M.A.J. Williams & H. Faure (eds), 387–410. Rotterdam: Balkema.

Clark, J.D. 1984. The domestication process in northeast Africa: ecological change and adaptive strategies. See Krzyzaniak & Kobusiewicz (1984), 23–41.

Clark, J.D. & S.A. Brandt (eds) 1984. *From hunters to farmers: causes and consequences of food production in Africa*. Berkeley: University of California Press.

Clark, J.D., M.A.J. Williams & A.B. Smith 1973. The geomorphology and archaeology of Adrar Bous, central Sahara: a preliminary report. *Quaternaria* 17, 245–96.

Close, A.E. 1992. Holocene occupation of the eastern Sahara. In *New light on the northeast African past*, F. Klees & R. Kuper (eds), 155–83. Köln: Heinrich-Barth Institut.

Close, A.E. (ed.) 1987. *Prehistory of North Africa*. Dallas: Southern Methodist University Press.

Close, A.E. & F. Wendorf 1992. The beginnings of food production in the eastern Sahara. In *transitions to agriculture in prehistory*, A.B. Gebauer & T.D. Price (eds), 63–72. Monographs in World Archaeology, No. 4. Madison, Wisconsin: Prehistory Press.

Clutton-Brock, J. 1989. Cattle in Ancient North Africa. In *The walking larder: patterns of domestication, pastoralism and predation*, J. Clutton-Brock (ed.), 200–206. London: Unwin.

Connah, G. 1976. The Daima sequence and the prehistoric chronology of the Lake Chad region of Nigeria. *Journal of African History* 17, 321–52.

de Maret, P. 1996. Pits, pots and the far-west streams. *Azania*, 29–30, 318–23.

Flohn, H. & S. Nicholson 1980. Climatic fluctuations in the arid belt of the "Old World" since the last glacial maximum: possible causes and future implications. *Palaeoecology of Africa and the Surrounding Islands* 12, 3–21.

Gabriel, Baldur 1984. Great plains and mountain areas as habitats for the neolithic man in the Sahara. See Krzyzaniak & Kobusiewicz (1984), 393–8.

Garrad, A., S. Colledge, L. Martin 1996. The emergence of crop cultivation and caprine herding in the "Marginal Zone" of the southern Levant. See Harris (1996), 204–26.

Gasse, F. & E. Van Campo 1994. Abrupt post-glacial climate events in West Asia and African Monsoon domains. *Earth and Planetary Science Letters* 1256, 435–56.

Gautier, A. 1984a. Archaeozoology of Bir Kiseiba region, eastern Sahara. In *Cattle keepers of the eastern Sahara: the neolithic of Bir Kiseiba*, Wendorf & R. Schild (assemblers), A.E. Close (ed.), 49–72. Dallas: Southern Methodist University Press.

Gautier, A. 1984b. Quaternary mammals and archaeozoology of Egypt and the Sudan: a survey. See Krzyzaniak & Kobusiewicz (1984), 47–56.

Gautier, A. 1987. Prehistoric men and cattle in North Africa: a dearth of data and a surfeit of models. See Close (1987), 163–87.

Gautier, A. 1989. A general review of the known prehistoric faunas of the central Sudanese Nile Valley. See Krzyzaniak & Kobusiewicz (1989), 353–7.

Gifford-Gonzalez, D. 1984. Implications of a faunal assemblage from a pastoral neolithic site in Kenya: findings and a perspective on research. See Clark & Brandt (1984), 240–51.

Grigson, C. 1989. Size and sex: evidence for the domestication of cattle in the Near East. In *The beginnings of agriculture*, A. Milles, D. Williams, N. Gardner (eds), 77–109. Oxford: BAR.

Grove, A.T. 1993. Africa's climate in the Holocene. See Shaw et al. (1993), 32–42.

Halstead, P. 1996. The development of agriculture and pastoralism in Greece: when, how and what? See Harris (1996), 296–310.

Harris, D. 1996. Themes and concepts in the study of early agriculture. See Harris (1996), 1–9.

Harris, D. (ed.) 1996. *The origins and spread of agriculture and pastoralism in Eurasia*. London: University College London Press.

Hassan, F.A. n.d. *Holocene Palaeoclimate in North Africa*. Paper presented at the 4th Meeting of the Society of Africanist Archaeologists in America, New Orleans.

Hassan, F.A. 1974. *The archaeology of the Dishna Plain, Egypt: a study of a late Palaeolithic settlement*. The Geological Survey of Egypt, Paper No. 59.

Hassan, F.A. 1981. *Demographic archaeology*. New York: Academic Press.

Hassan, F.A. 1986a. Desert environment and origins of agriculture in Egypt. *Norwegian Archaeological Review* **19**, 63–76.

Hassan, F.A. 1986b. Holocene lakes and prehistoric settlements of the western Faiyum. *Journal of Archaeological Science* **13**, 483–501.

Hassan, F.A. 1988. The predynastic of Egypt. *Journal of World Prehistory* **2**, 135–85.

Hassan, F.A. 1996. Abrupt Holocene climatic events in Africa. In *Aspects of African archaeology*, G. Pwiti & R. Soper (eds), 83–9. Harare: University of Zimbabwe Publications.

Hassan, F.A. 1997. Climate, famine, and chaos: Nile Floods and political disorder in early Egypt. In *Third millennium BC climate change and Old World collapse*, H.-N. Dalfes, G. Kukla, H. Weiss (eds), 1–23. Berlin: Springer.

Haynes Jr, C.V. 1987. Holocene migration rates of the Sudano-Sahelian wetting front, Arb'in Desert, eastern Sahara. See Close (1987), 69–84.

Henshilwood, C. 1996. A revised chronology for pastoralism in southernmost Africa: new evidence of sheep at *c.* 2000 bp from Blombos Cave, South Africa. *Antiquity* **70**, 945–9.

Higgs, E.S. 1967. Domestic animals. In *The Haua Fteah (Cyrenaica) and the Stone Age of the southeast Mediterranean*, C. McBurney (ed.), 313–19. Cambridge: Cambridge University Press.

Hugot, H.J. 1962. Premier aperçu sur la préhistoire du Ténéré du Tefassasset. In *Missions Berliet: Ténéré-Tchad*, H.J. Hugot (ed.). Paris: Berliet.

Kinahan, J. 1993. The rise and fall of nomadic pastoralism in the central Namib desert. See Shaw et al. (1993), 372–85.

Klein, R. & K. Scott. 1986. Re-analysis of faunal assemblages from the Haua Fteah and other Late Quaternary archaeological sites in Cyrenaican Libya. *Journal of Archaeological Science* **13**, 515–42.

Krzyzaniak, L. & M. Kobusiewicz (eds) 1984. *Origin and early development of food-producing cultures in northeastern Africa*. Poznan: Polish Academy of Sciences.

Krzyzaniak, L. & M. Kobusiewicz (eds) 1989. *Late prehistory of the Nile Basin and the Sahara*. Poznan: Polish Academy of Sciences.

Krzyzaniak, L., M. Kobusiewicz, J. Alexander (eds) 1993. *Environmental change and human culture in the Nile Basin and northern Africa until the second millennium BC*. Poznan: Poznan Archaeological Museum.

Lamb, H.F., F. Gasse, A. Benkaddour, N. El Hamouti, S. van der Kaars, W.T. Perkins, N.J. Pearce, C.N. Roberts 1995. Relations between century-scale Holocene arid intervals in tropical and temperate zones. *Nature* **373**, 134–7.

Legge, Tony 1996. The beginning of caprine domestication in south-west Asia. See Harris (1996), 238–62.

Loftus, R.T., E.M. David, D.G. Bradley, P.M. Sharp, P. Cunningham 1994. Evidence for two independent domestications of cattle. *Proceedings of the National Academy of Sciences, USA* **91**, 2757–61.

MacDonald, K.C. 1996. The Windé Koroji Complex: evidence for the peopling of the eastern Inland Niger Delta (2100–500 BC). *Préhistoire Anthropologie Méditerranéennes* **5**, 147–65.

Maréan, C. 1992. Hunter to herder: large mammal remains from the hunter–gatherer occupation at Enkapuna Ya Muto Rock Shelter, central Rift, Kenya. *African Archaeological Review* **10**, 65–128.

Marshall, F. 1994. Archaeological perspectives on East African pastoralism. In *African pastoralist systems*, E. Fratkin, K. Galvin, E. Roth (eds), 17–43. Boulder: Lynne Rienner.

Marshall, F., K. Stewart, J. Bartheleme 1984. Early domestic stock at Bongodien in northern Kenya. *Azania* **19**, 120–27.

Marks, A.E., A. Mohammed-Ali, J. Peters, R. Robertson 1985. The prehistory of the central Nile Valley as seen from its eastern hinterlands: excavations at Shaqadud. *Journal of Field Archaeology* **12**, 262–78.

McIntosh, R.J. 1993. The pulse model: genesis and accommodation of specialization in the Middle Niger. *Journal of African History* **34**, 181–220.

Muzzolini, A. 1989. La "néolithisation" du Nord de l'Afrique et ses causes. In *Néolithisations*, O. Aurenche & J. Chauvin (eds), 145–86. Oxford: BAR.

Muzzolini, A. 1993. The emergence of a food-producing economy in the Sahara. See Shaw et al. (1993), 226–39.

Neumann, K. 1989. Holocene vegetation of the eastern Sahara: charcoal from prehistoric sites. *African Archaeological Review* **7**, 97–116.

Neumann, K. 1993. Holocene vegetation of the eastern Sahara: charcoal from prehistoric sites. See Krzyzaniak, Kobusiewicz, Alexander (1993), 153–69.

Nicholson, S.E. 1994. Recent rainfall fluctuations in Africa and their relationship to past conditions over the continent. *The Holocene* **4**(2), 121–31.

Oren, E.D. 1979. Land bridge between Asia and Africa. In *Sinai: pharaohs, miners, pilgrims, and soldiers*, B. Rothenberg & H. Wayer (eds), 181–91. Berne: Kummerly & Frey.

Pachur, H.J. & G. Braun 1980. Paleoclimatic implications of late Quaternary lacustrine sediments in western Nubia, Sudan. *Quaternary Research* **36**, 257–76.

Peters, J. 1993. Animal exploitation between the fifth and the six cataracts *c.* 8500–7000 BP: a preliminary report on the fauns from El Damer, Abu Darbein and Aneibis. See Krzyzaniak et al. (1993), 413–19.

Petit-Maire, N. 1990. Will greenhouse green the Sahara? *Episodes* **13**(2), 103–7.

Posnansky, M. 1984. Early agricultural societies in Ghana. See Clark & Brandt (1984), 147–51.

Roberts, N., H.F. Lamb, N. El Hamouti, P. Barker 1994. Abrupt Holocene hydro-climatic events: palaeolimnological evidence from North-West Africa. In *Environmental change in drylands: biogeographical and geomorphological perspectives*, A.C. Millington & K. Pye (eds), 163–75. New York: John Wiley.

Robertshaw, P. 1993. The beginnings of food production in south-western Kenya. See Shaw et al. (1993), 358–71.

Rognon, P. & M.A.J. Williams 1977. Late Quaternary climatic changes in Australia and North Africa: a preliminary interpretation. *Palaeogeography, Palaeoclimatology, Palaeoecology* **21**, 285–327.

Rosen, S.A. 1984. The adoption of metallurgy in the Levant: a lithic perspective, *Current Anthropology* **25**(4), 504.

Roset, J.-P. 1987. Néolithisation, Néolithique et post-Néolithique au Niger nord-oriental. *Bulletin de l'Association française d'études Quaternaires* **4**, 203–14.

Roubet, C. 1979. *Économie Pastorale Préagricole en Algérie Orientale: Le Néolithique de Tradition Capsienne*. Paris: CNRS.

Roubet, C. & P.I. Carter 1984. La domestication au Maghreb: état de la question. See Krzyzaniak & Kobusiewicz (1984), 437–51.

Sadr, K. 1993. Environmental change and the development of nomadism in the east-central Sudan. See Krzyzaniak et al. (1993), 421–30.

Sadr, K., A. Castiglioni, G. Negro 1994. Archaeology in the Nubian Desert. *Sahara* **6**, 69–75.

Said, R. 1993. *The River Nile, geology, hydrology, and utilisation*. Oxford: Pergamon Press.

Shaw, T. 1976. Early crops in Africa: a review of the evidence. In *Origins of African plant domestication*, J.R. Harlan, J.M.J. de Wet, A.B. Stemler (eds), 107–53. The Hague: Mouton.

Shaw, T. 1984. Archaeological evidence and effects of food production in Nigeria. See Clark & Brandt (1984), 152–5.

Shaw, T., P. Sinclair, B. Andah, A. Okpoko (eds) 1993. *The archaeology of Africa. Foods, metals and towns*. London: Routledge.

Smith, A.B. 1974. *Adrar Bous and Karkarinchikat: examples of post-Palaeolithic human adaptation in the Saharan and Sahel Zones of West Africa*. Unpublished PhD dissertation, University of California, Berkeley.

Smith, A.B. 1976. A microlithic industry from Adrar Bous. Ténéré Desert, Niger. In *Proceedings of the 7th PanAfrican Congress for Prehistory and Related Studies (Addis Ababa)*, B. Abebe, J. Chavaillon, J.E.G. Sutton (eds), 181–96. Addis Ababa.

Smith, A.B. 1979. Biogeographical considerations of colonisation of the Lower Tilemsi Valley in the second millennium BC. *Journal of Arid Environments* **2**, 255–361.

Smith, A.B. 1984a. Origins of the neolithic in the Sahara. See Clark & Brandt (1984), 84–92.

Smith, A.B. 1984b. The origins of food production in northeast Africa. In *Palaeoecology of Africa and the Surrounding Islands* **16**, 317–24.

Smith, A.B. 1986. Cattle domestication in North Africa. *African Archaeological Review* **4**, 197–203.

Smith, A.B. 1989. The Near eastern connection: early to mid-Holocene relations between North Africa and the Levant. See Krzyzaniak & Kobusiewicz (1989), 69–77.

Smith, A.B. 1992. *Pastoralism in Africa: origins and development ecology*. London: C. Hurst.

Street-Perrott, A. & R.A. Perrott 1990. Abrupt climate fluctuations in the tropics: the influence of Atlantic ocean circulation. *Nature* **343**, 607–12.

Sutton, J.E.G. 1996. The growth of farming and the Bantu settlement on and south of the Equator. *Azania* **29–30**, 2–14.

Thomas, J. 1996. The cultural context of the first use of domesticates in continental central and north-west Europe. In Harris (1996), 310–22.

Uerpmann, H.-P. 1996. Animal domestication in southwest Asia. See Harris (1996), 227–37.

Vermeersch, P., P. Van Peer, J. Moeyersons, W. Van Neer 1996. Sodmein Cave Site, Red Sea Mountains (Egypt). *Sahara* **6**, 31–40.

Wasylikowa, K., J.R. Harlan, J. Evans, F. Wendorf, R. Schild, A.E. Close, H. Krolik, R.A. Housely 1993. Examination of botanical remains from early neolithic houses at Nabta Playa, western Desert, Egypt, with special reference to sorghum grains. See Shaw et al. (1993), 154–64.

Wendorf, F., A.E. Close, R. Schild 1987. Early domestic cattle in the eastern Sahara. *Palaeoecology of Africa* **18**, 441–8.

Wendorf, F., A.E. Close, R. Schild 1992/3. Megaliths in the Egyptian Sahara. *Sahara* **5**, 7–16.

Wendorf, F. & R. Schild 1994. Are the early Holocene cattle in the eastern Sahara domestic or wild? *Evolutionary Anthropology* **3**(4), 118–28.

Wendorf, F. & R. Schild 1995. The Saharan neolithic and the emergence of ranked societies in Egypt. *Dynamics of populations, movements, and responses to climatic changes in Africa, Abstracts*. Assembled by A.A. Longhi and M.C. Gatto, Rome: Forum for African Archaeology and Cultural Heritage (Mimeographed).

Wendorf, F. & R. Schild (assemblers) & A.E. Close (ed.) 1984. *Cattle keepers of the eastern Sahara. The neolithic of Bir Kiseiba*. Dallas: Southern Methodist University Press.

Zvelebil, M. 1996. The agricultural frontier and the transition to farming in the circum-Baltic region. In Harris (1996), 323–45.

CHAPTER SIX

Livestock in Saharan Rock Art

Alfred Muzzolini

1. Introduction

As with rock art everywhere in the world, the tens of thousands of pictures known from the Sahara (Fig. 6.1) largely represent animals. A large part of these are domestic animals. If we are allowed to advance a merely subjective evaluation – an intuitive estimate, just to give a hint about the proportions – we think that animals account for approximately 75 per cent of the Saharan pictures, around a third to a half of them being domestic ones, mainly cattle. Therefore, these representations of domestic animals include an impressive mass of data relating to description and history of fauna. They provide drawings that are sometimes fairly detailed, at least

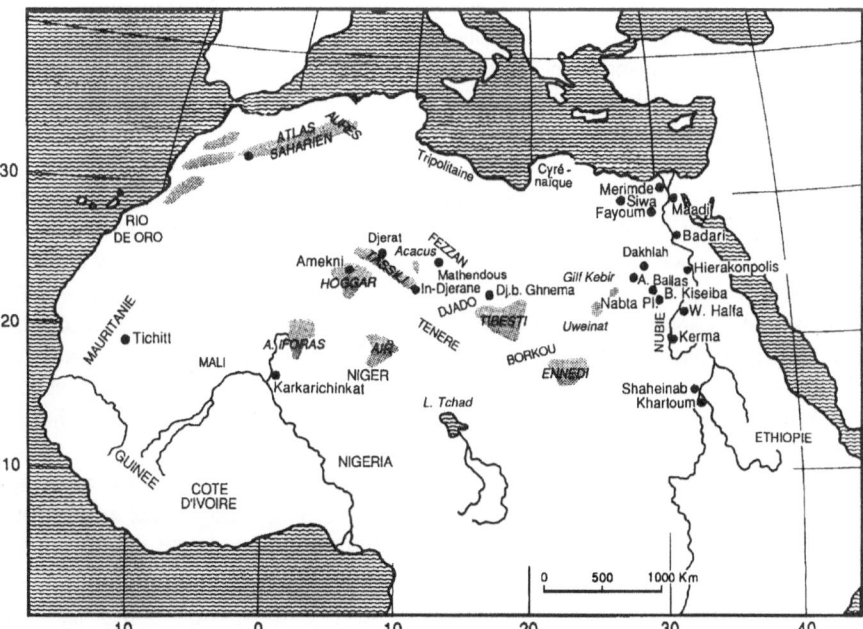

Figure 6.1 Map of the sites mentioned in the text.

within the schools of the naturalistic style. It is amazing that archaeozoologists make so little use of this stock of data.

Why are they so reluctant to use it? Mainly because they think that rock art pictures are not datable and consequently that one does not know the page of faunal history to which these documents relate. However, this is incorrect. In the Sahara, as elsewhere, rock art can be classified into "schools" that can be dated, admittedly in a crude way and inevitably with an important margin of uncertainty. Once this margin is known and accepted, observations and inferences become possible.

Our purpose here is to expose the main pieces of information the Saharan rock art provides on diverse African domestic species. However, we will begin by explaining how the rock art researcher gets his or her data and establishes the classification and chronology. We want to show that the approach proceeds with rigour and that the results, or at least some of them, are compelling.

2. The classification and chronology of Saharan rock art

The rock art researcher begins by grouping thousands of images into classes – i.e. sets of elements joined together by criteria considered to be important. These classes may include: same age (as attested by similar patina or fauna), analogous techniques, characteristic animals (such as horse, buffalo, oryx, etc.), weapons, and above all *style*. The last criterion is defined by Sackett (1977) as "a highly specific and characteristic manner of doing something ... always peculiar to a specific time and place". However it can be used only for figurative (or "iconic") pictures. In the Saharan rock art the differences between the styles of the diverse regional groups are generally clear-cut and immediately recognizable. Spatially limited sets of images usually have a close relationship – through patina, technique, themes or chiefly style – and are obviously different from all others.

The classes or schools constituted in this way are then given a sequence in relative chronology. Criteria peculiar to rock art (superimpositions and patinas) are used, combined with data from other disciplines. For instance, schools that show an extinct species, such as the giant buffalo *Bubalus antiquus*, are very likely anterior to schools which do not. "Ethiopian" fauna, with elephant, rhinoceros, hippopotamus, must be contemporaneous with a wet episode prior to the inception of the Actual Arid Phase when they are absent. The horse and the camel, chariots and writing, are recent everywhere in Africa, and associated images are likewise recent. The bow is replaced in recent millennia by the spear, sword and shield, hence schools showing these weapons may also be sequenced.

This relative chronology is summarily tied to an absolute chronology, by using absolute dates from other disciplines. General archaeology allows rock art chariots and the first domestic horses to be dated after 700 BC.[1] The early period as a whole can be linked to the *post quem* dates provided by archaeozoology for the domestication of animals in Africa (4500–4000 bc, except for the controversial cases of Nabta Playa and Bir Kiseiba in the western Egyptian Desert) and the dating of recent wet/arid oscillations (Muzzolini 1992). However, the latter dates are admittedly rather vague – giving the time of the events only down to a millennium.

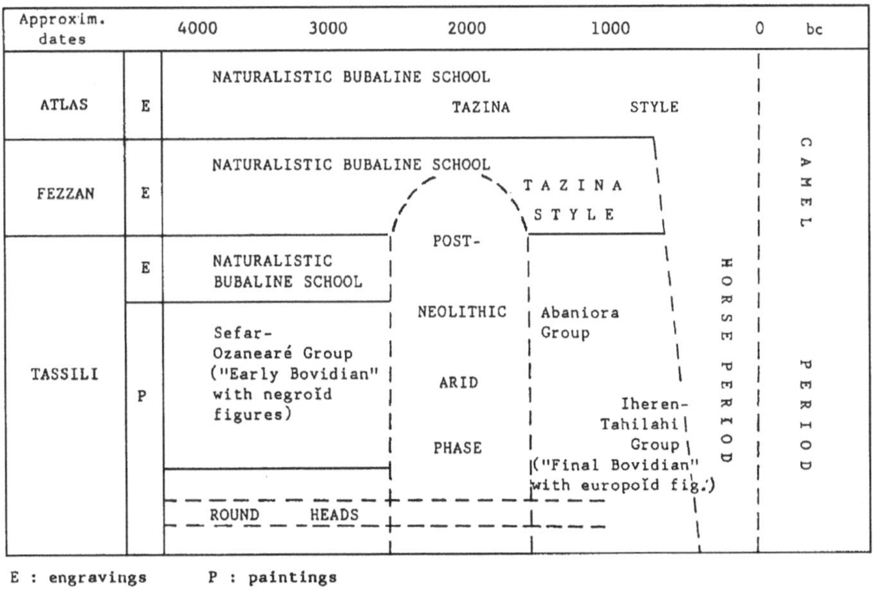
Approxim. dates
4000
3000
2000
1000
0
bc
ATLAS
E
NATURALISTIC BUBALINE SCHOOL
TAZINA
STYLE
FEZZAN
E
NATURALISTIC BUBALINE SCHOOL
TAZINA
STYLE
TASSILI
E
NATURALISTIC BUBALINE SCHOOL
P
Sefar-Ozaneaté Group ("Early Bovidian" with negroïd figures)
POST-NEOLITHIC ARID PHASE
Abaniora Group
Iheren-Tahilahi Group ("Final Bovidian" with europoïd fig.)
ROUND HEADS
HORSE PERIOD
CAMEL PERIOD
E : engravings
P : paintings

Figure 6.2 Classification and chronology of the Saharan rock art.

This is not the place to detail the arguments justifying the chronological sequence to be adopted. Controversies still remain unresolved. The traditional theory (from the 1930s) which appears in all popular accounts maintains that the earliest Saharan rock images, the engravings of the "Bubaline" school (so called because *Bubalus antiquus* is frequently represented), depict only wild fauna. This school would then correspond to a period said to be pre-pastoral, i.e. before animal domestication in Africa. The present writer has questioned this thesis suggesting that it reflects only the limited evidence available in the 1930s (Muzzolini 1983, 1995). Since then, material has accumulated showing numerous and undeniably domestic cattle and sheep within the Bubaline school. If we consider this is defined by patina, technique and style, not by fauna, the argument is not circular. We see in it cattle carrying humans, cattle with collars, pendants or saddles, cattle held with a lunging rein, a milking scene and "ornamented rams" in the Atlas mountains. Figure 6.2 summarizes this position, combining recent data from archaeozoology, climatology and general archaeology. It should be noted that this scheme is not compatible with the traditional thesis, and that this paper will only use the classification/chronological framework of Figure 6.2. Additionally, the absolute chronology and in part also the relative chronology of this scheme were obtained from the absolute dates provided by archaeozoology, climatology and general archaeology; therefore it cannot be inferred from Figure 6.2 that rock art confirms the dates of archaeozoology, climatology or general archaeology.

The evidence presented below relates mainly to two sets of animals; cattle, ovicaprines and donkeys, present from the earliest images (the Bubaline school) of the neolithic Wet Phase (*c*. 5000–3000 BC); and horses and camels, which only appear in recent, protohistoric periods.

Figure 6.3 An ox of the *primigenius* type. Painting, Early Bovidian. In-Itinen (Tassili, Algeria). (L of the ox = *c*. 50 cm.)

3. Cattle

All schools of Saharan rock art, including the earliest, show images of domestic cattle. Except for the figurative Round Heads, representations of cattle are the image most frequently shown. The two types of cattle defined by Rütimeyer in the nineteenth century can be recognized; the long-horned *Bos primigenius* and the short-horned *B. brachyceros*. However, the former comprises the bulk of the depictions.

The Saharan *primigenius* cattle (Figs 6.3, 6.4) appear large-sized, if we judge it according to the most realistic pictures available. These cattle look very much like the cattle of classical Egyptian iconography, with the long horns pictured in front view in spite of the body shown in profile. The literature sometimes gives them the name *B. africanus* (e.g. Grigson 1991, and also Ch. 4 in this volume) but this name means neither an original species nor even a subspecies. Because of the large size and the long horns these cattle must surely be attributed to the *primigenius* type (i.e. the same biological type as that of the wild aurochs), *B. primigenius*, since domestication did not create either an original species or a new type needing a new Latin binomial. During the whole Palaeolithic, and during several millennia around the beginning of domestication, in Africa as in Asia and Europe, this *primigenius* type remained the only type known from either iconography or excavation finds.

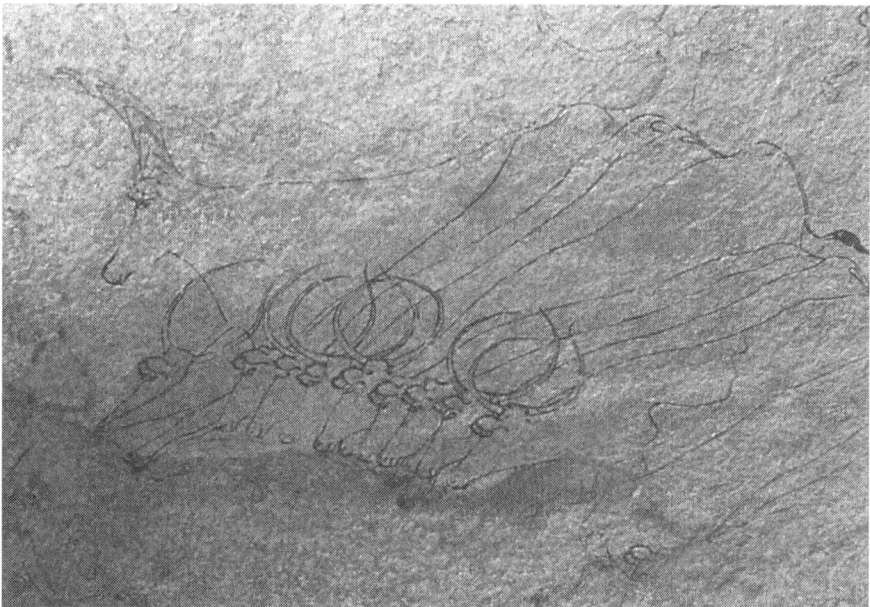

Figure 6.4 Cattle at the watering place. All are of the usual *primigenius* type with thin horns, except for the first ox which is of the forward-pointing horn type. Note on it the two tips carefully shown at the extremity of the "single" very thick horn. It does not water and seems to watch over the herd. Painting, Iheren-Tahilahi school, Final Bovidian. Iheren (Tassili, Algeria). (L of an ox = *c*. 30 cm.)

It seems obvious that African cattle were domesticated from an autochthonous species of aurochs.

When the pictures of this Saharan type of domestic *primigenius* are compared with those of the figures who accompany them, the withers heights can be estimated as being fairly large, towards the upper limit of the *primigenius*, in the range of 130–145 cm. Such are the usual sizes of the Egyptian domestic cattle as well, known either by excavations or by iconography. In the Saharan pictures the horns, the departure of which is almost horizontal, are not only long but also thin. If we accept again to estimate at least crudely their lengths, these would be in the order of 50–70 cm, with base diameters around 6–7 cm. These thin and long horns, sometimes ending in a lyre-shaped form, are very largely predominant on the rock walls of Nubia, Uweinat, Ennedi, Tibesti, Aïr, and also in the Pharaonic representations.

Although in the rock art of the Saharan Atlas, Southern Morocco, Tripolitania and the central Sahara, this aforementioned type also predominates, another type is represented as well: the cattle with forward-pointing horn (Fig. 6.4). They are also large-sized and long-horned, but typically only one horn is pictured, curved forward and very thick at the base. This type appears in those regions in large numbers: up to almost 50 per cent at the Oued Djerat (Tassili), and approximately 30 per cent in the Mathendous area (Libya). It is to be found also in the pictures of the Egyptian pre-Dynastic, but appears very rarely among those from Pharaonic Egypt and those from Nubia or Uweinat rock art. Of course the "forward-pointing horn" means two horns viewed in absolute profile, since one-horned cattle have never existed (!), and because in many carefully drawn pictures a bifid end of the "single" horn correctly shows the two little terminal tips. We also know of some figurations where two perfectly distinct and entire horns are both pointing forward.

What do these forward-pointing horns represent? Prehistoric, Egyptian and Meso-potamian artists, in regions and periods without any relation (e.g. in Franco-cantabrian Palaeolithic art and Saharan art), have nearly always pictured their animals in absolute profile. This, of course, caused a problem when they had to represent the horns. The usual solution consisted of rejecting the actual visual perception, con-ventionally representing horns in front view above bodies drawn in profile. This convention is neither more nor less problematic than the usual Pharaonic conven-tion of figures pictured in profile, except for the shoulders which are shown in front view. Could the forward-pointing horn be no other than an alternative artistic convention, more "true" to the standards of modern Westerners?

In the Sahara we are not so sure about it, because these cattle with "forward-pointing horn" show special features. On the most carefully executed and realistic figurations, this horn is nearly always shorter (about 40–50 cm, approximately) than the usual *primigenius* type with front-viewed horns, and above all it is defin-itely thicker at the base. Moreover, the departure is almost vertical, and not hor-izontal, even on the rare realistic pictures that show both horns pointing forward. All this suggests that these peculiar features might have been genetically fixed. Their originality – without intermediate types – when they are compared with those of the long and thin horns of the usual *primigenius* type, their constancy and their non-random geographical distribution which roughly shows a cline (forward-pointing horns being more frequent in northern regions, rarer to the east and the south) suggest the existence of a true domestic "breed" (i.e. a homozygous stock with fixed genetic features). But how could they be maintained as such, when these cattle appear on the same walls, inter-mixed with the usual *primigenius* showing two thin horns? That both kinds of cattle seem to exist at liberty within the same figured "herds" constitutes a serious difficulty.

Could these cattle with "forward-pointing horn" represent only bulls? In many breeds, bulls have horns thicker and shorter than cows. However, such an inter-pretation does not seem acceptable. Admittedly most of the cattle with "forward-pointing horn" are pictured without an udder, but this is merely a negative argument and some of them do have an udder (Muzzolini 1983:490). Within the set of the usual *primigenius* with front-viewed horns, udders are not frequent either (*c.* 10 per cent of the total), but this set also includes many cattle that are surely males.

Moreover, forward-pointing horns are pictured in nearly all regions from Sumer to the Atlantic, including pre-Dynastic Egypt, and chronologically from the earlier Bubaline school to the Horse and Camel Periods. It appears unlikely that this feature may have spread throughout so many different peoples and diverse cultures of various ages if it was a mere convention of artistic rendering and not a real biological feature.

Hence, the most likely explanation seems to be that the biological feature of forward-pointing thick horns did really exist in the ancient genetic pool of the *primigenius* type. Being a mere variety of horn it did not involve any notable genetic distance within the herd, therefore the special cattle possessing it were bred together with the usual type. However, they were culturally differentiated, exactly in the same way as among Egyptian cattle black spots on the forehead and the neck differentiated an eminent animal, the *Apis bull*. Indeed, these cattle with "forward-pointing horn" are often seen standing in special situations, for example as heads of the herd. (Fig. 6.4). Finally, a few finds from African excavations show thick-based horns that could correspond to this type of forward-pointing horns (including the *opisthonomus* skulls of Pleistocene and Holocene ages described by Pomel 1894, and a Pleistocene skull from the Fayum, Epstein 1971:233).

This trait, and the cultural values possibly tied to it, disappeared in the course of the recent periods. The disappearance occurred suddenly in Egypt at the end of the pre-Dynastic, probably as a result of conscious selection in the Pharaonic estates. Since the Old Kingdom one breed was privileged there, the "*Bos africanus*" with long and thin horns, which we know to have been imported in numbers from Nubia. The phenomenon took place more slowly and later among the Saharan breeders who were not organized so well. On their rock pictures some forward-pointing horns are still observable during the Horse and Camel Periods.

We must stress the point that, notwithstanding their special feature of forward-pointing horns, these cattle too belong to the *primigenius* type. We have fought (Muzzolini 1980, 1983) against an error, frequently found in the Saharan literature, which presents them either as cattle belonging to the short-horned type *Bos brachyceros*, or as *B. ibericus*, or as "*B. brachyceros-ibericus*". Such taxa have no basis. The erroneous attribution to *B. brachyceros* was popularized by H. Lhote who had adopted it from the untenable theories of Pomel (1894) concerning his so-called *B. opisthonomus* (cattle said to graze walking backwards, which Pomel explained as a confirmation of a strange saying from Herodotus). As for the correlation with *B. ibericus*, we are dealing here with another long-cherished though out-of-date idea from the Saharan literature. This name was very often used for an allegedly distinct species of small-sized aurochs, whose provenance in Sanson's theories at the end of the nineteenth century was the Iberian peninsula. As a matter of routine all the numerous finds of small-sized cattle bones dating from Palaeolithic or Epipalaeolithic in the Maghreb and the Sahara were attributed to this "*B. ibericus*". However, this feature, the small size, is not sufficient to differentiate a species. Biologically these "*B. ibericus*" are no other than small common "*B. primigenius*" aurochs. Small size is a frequent feature of varieties among many species in arid or semi-arid areas (Muzzolini 1982).

Another type that appears on rock walls, though in small quantity, is the true short-horned type (formerly termed *brachyceros*) with a size usually smaller than that of *primigenius*. In Africa as in Europe this type emerges only late, and only within domestic breeds. No sure representation of it is perceived in the Sahara before the Final Bovidian or the Horse Period, after *c.* 1000 BC (some older short horns on pictures deteriorated too much or too schematic to be convincing must be abandoned) (Muzzolini 1980).

However, for Grigson (1975–80), Rütimeyer's two types (*primigenius* and *brachyceros*) would only represent the two ends of a continuum of biological types. Indeed, for passing from one type to the other, a few environmental factors favouring genetic drift, or a conscious selection on the part of breeders, are sufficient.[2] A gene can be eliminated or fixed by selection after a few tens of generations, only a few centuries. Therefore an autochthonous origin of the short-horned type in the Sahara is conceivable. However, the hypothesis of imports of short-horned breeds from any region around the Mediterranean cannot be excluded either, for this type had become common everywhere around the end of the second millennium BC. Moreover, we see its appearance in the central Sahara precisely when this region establishes the first certain contacts with the countries of the Mediterranean littoral – such as those revealed by the introduction of the horse and the chariot. The scarcity of short-horned cattle in Saharan rock art, and the schematic nature of the depictions, do not allow us to reach any definite conclusions. In the Sahara the *brachyceros* type can reflect a mutation, a case of local selection, or an import.

We should also mention the many *hornless cattle* in the images. Breeds in which this absence of horns is a genetically fixed character currently exist and could exist. In Egypt, from the Old Kingdom, some breeding sites possessed an important grazing stock consisting of this variety. However, the special word to name it appears only during the Middle Kingdom (Muzzolini 1983:212). Although the scribes count them separately, these hornless cattle could merely reflect the usual domestic *primigenius* cattle whose horns had been sawn or burnt in the first years. This practice, whose aim is to preventing fighting animals from wounding each other, has been widely described. In the Sahara, the hornless cattle in the rock pictures, mixed together with the usual *primigenius*, probably represent such a practice.

Finally, a few depictions show cattle with humps that are generally not very prominent (Fig. 6.5). Some of them are not surprising because they belong to later schools, for instance the recent phase of the "Libyan Warrior" school in Aïr, or the Camel Period of the Ennedi. Indeed, such cattle were present in the current millennium, when the diffusion of the zebu (*Bos indicus*) or intermediate types (mainly Sanga cattle and Fulani cattle) is undeniable. But other cattle, in the central Tassili, in the Oued Djerat, in Libya (Muzzolini 1983, Ch. 8), go back to the Final Bovidian and the Horse Period (i.e. at least to the early first millennium BC). They have long horns and the hump is most frequently in the thoracic position; these features make them clearly distinct from the short-horned and cervico-thoracic humped zebus of Asiatic origin which are known in Egypt, in small numbers, from the second millennium BC onwards. On the other hand these Saharan humped cattle are earlier than the imports of Asiatic zebus which we know to have occurred from Arabia to Ethiopia. The earlier archaeological evidence of these imports in Ethiopia go back only to the late Aksumite period around the fourth century AD and they became important only after the seventh to eighth centuries AD (Epstein 1971, Clark 1977). In Kenya, however, two sites have provided a few fragments of skulls that could

Figure 6.5 Humped cattle. Painting, Iheren-Tahilahi school, Final Bovidian. Weiresen (Tassili, Algeria). (After J. Kunz.)

belong to *B. indicus* (or already an intermediate Sanga type). Their date is around 100 BC (Marshall 1990), which would be the earliest osteological evidence for *B. indicus* in Africa. But, the feature used for this identification, the shape of the orbital rim, seems to be uncertain (cf. Grigson 1991:135).

Could these Saharan humped cattle have been imported from Asia as well, but at an earlier date? This idea appears archaeologically unlikely; such relations would have left other pieces of evidence and intermediate landmarks. Are we dealing in these images not with true humps but merely very high withers? Such is Grigson's (1991) interpretation: she asserts that these cattle have no other feature of zebus except the hump. However, we ought to specify; no other feature of *modern* zebus. Indeed, the features of the modern zebu are those that have been fixed in Asia and eastern Africa after millennia of interbreeding and selection. Such features could be lacking on an ancient breed of humped cattle, in the central Sahara. Moreover we must not expect subtle zoological details from the Saharan pictures, which are often crude.

The hypothesis of an indigenous emergence in the central Sahara of a breed of humped cattle independent of the Asiatic one could explain the zoological differences pointed by Grigson. If either the hump or abnormally protruding withers, which both allow the storage of fat, were long-inscribed in the genotype of *B. primigenius*, and if these features have an adaptive value for cattle in arid areas, then the laws of genetics predict that this feature will episodically appear. Consequently, natural or man-made selection will result in an increasing proportion of these features. The difference would be that, in Asia such a feature underwent an artificial selection and was eventually fixed in a type *B. indicus*, clearly distinct from *B. primigenius*, whereas in the central Sahara there was no strict selection.

However, another hypothesis is suggested by the evidence we have of the early presence of humped cattle in the Sahara, that is, before the historically known imports of Asiatic zebus with cervico-thoracic hump. These Saharan humped cattle could have taken part, through the interbreeding with the indigenous long-horned *B. primigenius*, in the making up of the breeds commonly considered intermediate and called Fulani or Sanga cattle. We must mention that Grigson (1991) sees the Sanga cattle as a very ancient breed, anterior to the crossings with *primigenius* or *indicus*, and so this breed, which was very numerous, would have been the origin of most of the African stock. But even in this case Saharan humped cattle could have been crossed very early with Sanga cattle.

The issue is not clear because archaeozoological data for the decisive millennia – first millennium BC and first millennium AD – are lacking, and because morphological comparison with current breeds is never a simple matter. Moreover, diverse biochemical studies based upon genetic markers of blood groups, polymorphisms of various proteins or DNA result in contradictory indications of origins (see Manwell-Baker 1980, Muzzolini 1983:514; and Bradley & Loftus Ch. 13 in this volume). The idea of an independent Saharan cradle for African humped cattle only constitutes a hypothesis, but it seems more likely than the possibility of very ancient Asiatic imports.

4. Ovicaprines

Domestic sheep and goats are also observable in the earliest or Bubaline period of Saharan rock art. As for sheep we are mainly dealing with the famous "ornamented rams" of the Saharan Atlas. They are hairy and long-tailed. Admittedly, the head decorations (spheroids and feathers), ornamented collars, and pieces of harness that characterize them could also be imagined on wild sheep adorned for the sacrifice. This was certainly Lhote's (1970:180) interpretation. However, Lhote never gave any explanation about the contradiction between his interpretation and the thesis, usually sustained by archaeozoologists, who deny any existence of wild *Ovis* in Africa at any time. But as we hold critical views about this thesis, the reasons why we think that Lhote's interpretation is not tenable are different (Muzzolini 1990). Indeed the domestic character of the "ornamented rams" is attested by biological features which are never found among wild sheep; mainly long tail, lop-ears, convex nose and short downwards-curved horns (the so-called "Amon" horns).[3] In the same way, the goats of the Bubaline school show twisted horns, acquired by selection, whereas the wild species shows only a simple form, the "scimitar" one (Epstein 1971).

In the central Sahara, long-tailed sheep and goats also appear in the engravings of the Bubaline school but most depictions are in the Fezzan, and no pictographs of ovicaprines are known from the early period. We see them only in the Final Bovidian, where they are undeniably domestic. They are usually pictured in herds which often show goats and sheep together, and frequently in camp scenes. These are the favourite themes in the Iheren–Tahilahi school and in a contemporary school of Acacus (Uan Amil Shepherds). We do not know of any trace of weaving activities – neither from excavations (spindle-whorls or loom-weights) nor from rock art. Hence the sheep represented by these schools are very likely still hairy as well. Fat-tailed sheep, which were introduced from the Middle East into Egypt during the Middle Kingdom and into the Maghreb during the historical period (probably by the Phoenicians), do not appear on the Saharan rock pictures.

In Egypt, the type of sheep during the pre-Dynastic and the Old Kingdom is the so-called "*Ovis palaeoaegyptiaca*" with long twisted horizontal horns. It is unknown in the images of the central Sahara, Tibesti, Ennedi, and even Uweinat, but a unique, unmistakable specimen has been found in the Saharan Atlas at El Richa. "*Ovis palaeoaegyptiaca*," which is contemporaneous with the "ornamented rams" of the Atlas mountains, and also the common woolly Egyptian sheep with small "Amon" horns, are long-tailed. However, the Atlas rams, the horns of which are not horizontal, cannot be confused with "*Ovis palaeoaegyptiaca*", and they also cannot be confused with the woolly Egyptian sheep because these rams are still hairy. Moreover, they remain clearly distinguishable from all other types on account of their convex nose.

The sheep of the Iheren–Tahilahi group, which are contemporaneous with the New Kingdom, have nothing to do with the woolly Egyptian sheep or the Atlas "ornamented rams"; since these Tassili sheep have a straight nose, erect ears (some-times), and (most importantly) a short tail, all of which are archaic traits characteriz-ing wild species or recently domesticated varieties. Since an evolutionary throwback is almost inconceivable, these sheep of the Iheren–Tahilahi group can come neither from the Pharaonic sheep stock nor from the Atlas "ornamented ram" stock. So, where do they come from?

Here the traditional story about the origin of African sheep contents itself with repeating the nineteenth century diffusionist account: Saharan sheep would have to be imported from the Middle East. This facile explanation comes up against several archaeological improbabilities:

(1) Before the Ghassulian period in Palestine, around 4000 bc, there hardly existed in the Middle East any organized breeding-site liable to export fixed types. In Egypt, the earliest dates for domestic ovicaprines are those from Badari, Merimde and Fayum, around 4000 bc as well. And even without taking into account linguistic arguments (see below), ^{14}C dates for ovicaprines from Saharan excavations (Nabta Playa in the Egyptian western Desert, Haua Fteah in Cyrenaica, Grotte Capeletti in the Aurès mountains, Achakar near Tangiers, "Shaheinab neolithic" in Sudan, maybe Uan Muhuggiag in Acacus, Dakhlah) (Muzzolini 1990) are at least as early and rather earlier: 4000–4800 bc (the 6000 bc date from Nabta Playa is questionable).

(2) Around 4000 bc Sinai was still inhabited by hunter–gatherers (Dayan et al. 1986) and some sites in Sinai and Negev provide bones of still wild ovicaprines (Davis et al. 1982, Tchernov & Bar-Yosef 1982). How could an "introduction" of domestic ovicaprines from the Middle East avoid this

section, since any spread was likely to be step by step? Moreover, how could a sheep type different from the Egyptian "*palaeoaegyptiaca*" type be "imported" into the Sahara, avoiding Egypt as well?

(3) The archaic Tassili type is unknown in Palestine after 4000 bc and in Egypt in all periods, therefore it remains unexplainable in this diffusionist scheme.

(4) The main argument of the diffusionist thesis is that in the Sahara and in the Nile Valley no bone of wild *Ovis* or *Capra* are reported in pre-neolithic or neolithic fauna records, and that consequently these animals could only have been imported from the Middle East in an already domestic status. However, this argument may be flawed:

   (a) even if importations occurred, other origins are possible.

   (b) even if there were no definite mentions of *Ovis* in pre-neolithic fauna records, this may mean that the animals could have come during the neolithic period from elsewhere. But not necessarily though human agency – they could have taken advantage of the favourable early Holocene climatic episode and spread into the Sahara naturally (Muzzolini 1995).

   (c) indeed there are even a few mentions of wild sheep bones in the faunal record (cf. Vaufrey 1955:391; Bouchud 1975; von den Driesch & Boessneck 1985), but they are systematically attributed to a closely related species, *Ammotragus lervia* (the Barbary Sheep of northern Africa), or the mentioned animals are declared domestic by virtue of the thesis that "wild ovicaprines never existed in Africa" – circular thinking.

(5) Linguists have recently added a weighty argument to the thesis that indigenous wild ovicaprines occurred in Africa before domestication. Within the Afro–Asiatic family, Blench (1995) finds roots for "sheep" and "goat" in proto-Chadic, a language Greenberg dated to around 4000 BC, but which according to Ehret and Blench is now considered to be older (around 6000– 7000 BC). Moreover, Ehret (1993, n.d.) finds two or three terms for "sheep" and "goat" at the divergence between Omotic and Cushitic ("Erythræan" level) – a divergence which glottochronological methods assign to the tenth millennium (?). Some other such words in the proto-Cushitic date between the sixth and seventh millennia BC. Within the Nilo–Saharan family he identifies words for "goat", "sheep", "ram", "lamb", as soon as the proto-Sahelian, a (reconstructed) language spoken as early as the seventh millennium BC. Glottochronology does not claim to provide precise dates. However, if ovicaprines were extant in Africa around the sixth to seventh millennia BC, they could, in this early period, only have been wild.

In conclusion, the existence of pre-neolithic wild ovicaprines in Africa seems plausible. These wild animals may have been very few in number, and would therefore be rare in the faunal record. They might have arrived around the early Holocene by way of a natural migration of the species from southwestern Asia, through Sinai or the Bab-el-Mandeb. In any event, on the basis of current data, this hypothesis seems at least as probable as the traditional diffusionist one.

Figure 6.6 "Flying gallop" chariot, a biga (two horses), one-pole type. Horse Period. Painting, Tasakarot (Tassili, Algeria). (L = *c*. 45 cm.)

5. The horse

The domestic horse has appeared in large numbers on Saharan rock faces since a period called "Horse Period" in the central Sahara. We shall see that this period takes place after 700 BC. The existence of earlier, "neolithic" wild equids in the Sahara has often been evoked in literature. More particularly Lhote (1970:166) sustained this thesis. Since the problem lies in making the distinction between horse and donkey on the pictures, we shall deal with it below in the section about donkeys, where it is concluded that a wild "neolithic" equid is not attested in the Saharan Atlas and remains dubious in the central Sahara.

As for the central Sahara, the domestic horse was obviously introduced from elsewhere, since we find it for the first time exactly when chariots were introduced (Fig. 6.6). Hence it was used at first as a draught animal. Where and when did the chariots and horses come from?

A thesis traditionally put forward since the 1930s, places their origins in Pharaonic Egypt where the horse had been in use since the Middle Kingdom (pre-1700 BC), even before the Hyksos. Unfortunately, this thesis embroiders a story built upon the sayings of the scribes and sculptors who servilely relate the defeat of the coalition of the "Sea Peoples" by the pharaohs Merneptah and Ramses III, around 1200 BC. This coalition was said to be led by "Libyans" – so say the texts, without geographically positioning these "Libyans." On the evening of the last battle, the

Figure 6.7 "Flying-gallop" chariot, a triga (three horses), two-pole type. Horse Period. Painting, Ti-n-Anneuin (Acacus, S.-W. Libya). (L = *c*. 75 cm.)

fleeing "Libyans" were evidently pursued by the Pharaonic army, and this is all we know. The thesis we question adds that the "Libyans" fled westwards with their chariots, to the Tassili, 2,000 km away. There they would have arrived with their chariots still in working order. But the expectation of life for these light and entirely wooden devices, on the stony grounds that are usual throughout the Sahara, is no more than a few kilometres. On the other hand the "Libyans" may have merely represented in the scribe language a stereotype, that of the western enemies. By reconsidering the nineteenth century romantic interpretations about the "Sea Peoples" some Egyptologists (Nibbi 1986) even hold that these "Libyans" and "Sea Peoples" were only rebels from the western part of the Delta.

The chariot and the horse, which were long in use among many countries around the eastern Mediterranean, were certainly introduced from one of these countries into the central Sahara. However, there is no reason for privileging Egypt as the place of their origin. Indeed the written evidence in no way suggests any Tassili chapter. Once more the usual thesis of the Egyptian origin of the chariot and the horse in the Sahara is no more than a reflection of the traditional "*ex Oriente lux*".

In any event, some rock art data prevents a link with the Egyptian episode of the Sea Peoples of the thirteenth century BC. We know of around 130 depictions of the famous Tassili-Acacus and Hoggar painted chariots called "flying-gallop chariots" (on account of the horse legs shown horizontal or at least out-stretched). Most of them are two-horse and one-pole bigas. But half a dozen are quadrigas, a dozen are trigas and some fifteen are two-pole chariots (Fig. 6.7). Now quadrigas and trigas are unknown in Egypt until the late Dynastic. They appeared in the Middle East during the ninth century BC and became common there only in the eighth century BC. As for the two-pole chariots, the first we know are some from Cypriot tombs around 700 BC (Littauer & Crouwel 1979). Therefore, there cannot be a question of Saharan chariots and horses before the date of 700 BC. Even a moderate proposal recently made by Camps (1987) who admitted a lower date for the quadrigas while maintaining a second millennium BC "Egyptian origin" for the other chariots cannot be accepted. This is because quadrigas, trigas and two-pole chariots are mixed without any further distinction with all the other types of chariots on the same walls. It is apparent that the group of the chariots as a whole must be later than 700 BC.

In conclusion, the horse was introduced from some country on the Mediterranean coast at a date later than 700 BC. We can add that the data suggest, only as a probability, that the origin of these Saharan chariots and horses might be Cyrena, a Greek city founded in 631 BC on the African coast by colonists coming from Thera. This Cyrenian origin is based on the following arguments:

(1) An Egyptian origin of the chariot and the horse around 700 BC appears as unlikely as in the thirteenth century, because we do not have any evidence of relations, either peaceful or warlike, between Egypt and Cyrenaica – nor consequently with the central Sahara – before the conquest of Egypt by the Assyrians in the seventh century BC. An origin from Carthage or some Punic factory also appears unlikely because the Phœnicians did not practise chariot-racing, whereas the "flying-gallop" chariots are mainly racing chariots.

(2) Conversely, Cyrena had a high reputation for her chariot races in which the "Libyan" aurigas won a great fame. Quadrigas and two-pole chariots, very similar to the Tassili ones, are known from inscriptions and bas-reliefs of Cyrena (Muzzolini 1995:179).

(3) At least from the "Horse Period" and until today, the central Sahara has been occupied by Berbers. Now among the usual Berber names for the horse, one form, that which refers to the horse *"par excellence" – ayis –* belongs to the Indo–European stock (Greek *hippos*, Latin *equus*) (Chaker 1995). Conversely, we do not find in Berber any trace of the Egyptian name, *susim*, a Semitic term adopted by the Egyptians of the Middle Empire when they introduced the horse from the Middle East (Yoyotte 1970:51).

These remarks provide a strong presumption that the horse was introduced into the central Sahara after 631 BC through some kind of relation with Cyrena and her famed chariot racing. The Saharans, on the basis of the graphic evidence, used the horse mainly for pulling their chariots. The attitudes of horses and drivers on Tassili paintings confirm that chariots were used there for racing (Fig. 6.6). Only a few chariots are shown in hunting scenes, and still more rarely in prestige functions; and on only one picture (at Aboteka, Tassili) is a woman carried beside the driver. There is no graphic evidence that the chariots were used for war. A few riders are painted close to chariots, however this becomes frequent only later, in the Camel Period (Fig. 6.8).

The horses which are painted with the flying-gallop chariots belong to a school that uses a rather schematic style. Identifying racial varieties on such pictures is hardly possible, and seems imprudent. In former decades a thesis was sometimes held, which attributed the horses of the flying-gallop chariots to the barb, the type of the current horses of the Maghreb. Such a thesis has no grounds. On the

Figure 6.8   Horse and rider, in the schematic style of the early Cameline school. Painting, Tizzeine (Tassili, Algeria). (L = *c.* 30 cm.)

contrary, even if we considered the most realistic pictures of this stereotyped school to be reliable what would come out is rather an elongated type, fairly close to the modern horse called "Arabian".

The rock walls of Tibesti, Ennedi and Uweinat show neither horse nor chariot in the schools contemporaneous with the Horse Period of the central Sahara. This fact constitutes further evidence that horses and chariots were really introduced into Tassili-Acacus and Hoggar from Mediterranean countries in the north, and not from the east. Only much later do horse riders appear in the eastern Sahara, all within recent schools dating to the Christian era.

Towards the south, in Djado, Aïr and Adrar des Iforas, we know of a few images of chariots, some of them with the animals. In the latter case we find two cattle more often than two horses. However, very schematic riders frequently appear in the engravings of the middle phase of the "Libyan Warrior" school which is con-temporary with the Tassili Horse Period. It seems very likely that the horse was introduced here also from the north – from Hoggar or Tassili – during the second half of the first millennium BC.

Towards the north, in the Saharan Atlas and more generally in the Maghreb, we have no sure data, either from excavations or from rock art (discussed below) for the existence of an *Equus caballus*, either domestic or wild, anterior to the histor-ical period (the beginning of which is vaguely contemporaneous with the Tassili Horse Period).

6. The donkey

The donkey, another equid, is found first (albeit rarely) within the paintings and engravings of the central Sahara. We find it after *c.* 5000 BC in the earliest schools, that of the Bubaline engravings from the Oued Djerat (Tassili) or Mathendous (Libya) and that of the Round Head paintings, but also in the recent schools. However it does not seem that we are dealing in the Sahara with domestic donkeys, which is surprising. In Egypt the donkey is still lacking in the faunas of the Fayum and Merimde at around 4500–4000 BC, but it appears, already domesticated, in the fauna from Maadi by *c.* 4000–3500 BC (Midant-Reynes 1992). Afterwards it was a very common domestic animal throughout Pharaonic times, being used almost solely as a pack animal. However, the Saharan rock art shows no unambiguous scene of a domesticated donkey. In depictions naturalistic enough to allow the identification of the species we never see donkeys harnessed to a chariot, nor even donkeys mounted by men.[4] In the Maghreb the presence of the donkey is well attested from the faunas of half a dozen neolithic sites but we do not have any evidence certifying that it might have been domesticated there before historical times. In the Sahara the finds of donkey bones in layers from the neolithic or the early Holocene (such as those from Ti-n-Torha East dating from *c.* 7500–6000 BC) (Gautier 1982) are very rare, and on the rock pictures this animal is still hunted, or cut into pieces after killing. It seems particularly incredible that the Iheren–Tahilahi school, a Final Bovidian school which represents many detailed camp-site scenes and herds of cattle, goats and sheep, would be without images of its donkeys if they were domesticated as well. Some paintings of this school illustrate people moving from one camp-site to another and show details of packs attached to cattle horns, but we never see donkeys with their burdens as we do in Egypt. As for the very recent periods some mosaics in the Maghreb still represent wild donkey hunting in the Roman period (Hippone, Djemila, Hadrumeta).

Out of the central Sahara the countless rock pictures of Aïr, Adrar des Iforas, Mauritania, Djado, Tibesti, Ennedi, Uweinat, do not include any unambiguous donkeys. However, the Saharan Atlas represents it from the earliest engravings, those of the Bubaline school. These engravings raise a problem.

The problem is that the Saharan Atlas depictions are of short-eared equids (Fig. 6.9). Lhote's (1970:166) opinion was that we are dealing, at least for some of them, with true horses *Equus caballus*. Moreover, he specified these to be hunted horses, since one of his main theses was that the whole Bubaline school was "pre-pastoral" (i.e. anterior to domestication). But arguments claiming that we are dealing with horses, and hunted horses at that, are thin, circular, and questionable. Camps (1984) after having at first rightly refuted the assertion of a wild horse prior to the modern *E. caballus* in the Atlas, compared ear length in the donkey, the horse and the quagga (quagga ears are almost as short as horse ears). Then comes a conclusion which is difficult to understand; instead of identifying this short-eared equid with the quagga, Camps (1984:380) identifies it with "a variety of donkey with short ears ... which perhaps is only an asinine species more robust than and distinct from *E. asinus africanus*". We cannot grasp the rationale of such a conclusion,

Figure 6.9 "The she-ass and her foals". Note the short ears, the shoulder cross, chevron stripes on the legs. Engraving, Naturalistic Bubaline school, El Richa (Saharan Atlas, Algeria). (L = *c*. 160 cm for the main animal.)

since it is explained earlier in Camps' text that ear lengths cannot be used to discriminate between the quagga and the horse.

We will consider elsewhere the issue of these short-eared equids, mainly of the Atlas, but also from the Sahara (Muzzolini in prep.). The debates in the literature are often based upon rough or inaccurate sketches, or upon images from schools in which schematism is the rule and detail distorted. Therefore it is best to discard the crude pictures of: indeterminate "quadrupeds"; equids of unknown styles or ages (because they could be true *E. caballus*, but modern); equids without sufficient detail to establish species (e.g. those from the Tazina school, which are normally very schematic, making interpretations of anatomical detail dubious); or, finally, definite caballine types from periods too recent to be relevant here. After careful scrutiny we also discarded three Atlas engravings which were often put forward in the debates, because they appear to us useless for our problem; those of El Arouia, El Hadj Mimoun and Gouiret bent Saloul. After such pruning, the file no longer allows a statistical approach, but becomes more reliable. It leads to the following conclusions, *only valid for the figurations from the early period* of Saharan rock art ("early" meaning anterior to the chariot periods of the various regions):

(1) In the Atlas mountains only two engravings seem to show undeniable donkeys; the well known El Richa engraving (Fig. 6.9) of "the she-ass and her foals" (she has short ears), and at a little distance westwards in the same locality an engraving of a lone ass (with mid-long ears). Another engraving,

at Aïn Naga (close to "the shy lovers"), very probably represents two asses, but they are very crudely executed (Lhote 1984:103).

(2) Contrarily, undeniable asses are observable in greater number among the engravings of the Naturalistic Bubaline school at the Oued Djerat (Tassili) and Mathendous (Libya). Their ears are sometimes short, sometimes mid-long, sometimes long. In the Mathendous area they are also to be found on masks worn by humans. On these masks, which reflect some symbolic value, the ears are very long. An ass is also known in a Final Bovidian painting of Jabbaren (Tassili). We must notice that the feature "ear length" is difficult to use, because the variability (biological or artistic) appears too wide. The two asses of Aïn Naga, in file, show mid-long ears for one, long ears for the other. An engraving in the Oued Tilizzaghen (Libya) (Jelinek 1985:142) shows a file of three asses with variable ear lengths; short, mid-long, exaggerated on the last one. The following equation is suggested: long ears = donkey, short ears = undetermined equid.

(3) Hardly half a dozen engravings or paintings can be attributed to the group "probable or possible early *E. caballus*" – none of them appears undeniable and these are exclusively found in Tassili.

The lack of sure *E. caballus* within the images of the Atlas mountains during the early period constitutes a negative argument which alone is not enough to certify that domestic horses were really absent in this northern biotope. However, such a total lack of results both from excavations in the Maghreb[5] and on the walls of the rich area of rock art comprising the Saharan Atlas and its extension into Morocco, constitutes a strong presumption that they were not extant there in this time. We must conclude that the existence of an *E. caballus* during the early period is thus very unlikely in the Saharan Atlas, and dubious in the central Sahara. However, even in the latter region, the images liable to represent it would be very few, and silent about its status – wild or domestic. We know of only one exception, which seems to show a mounted horse. Maybe we are dealing with the first imports of domestic horses?

Inversely, the donkey is attested both from excavations and from rock pictures, since the beginning of the early period, in the central Sahara as in the Maghreb. But here we are dealing with wild donkeys.

These donkeys from the Saharan Atlas as those from rock pictures in the central Sahara and those illustrated in Roman mosaics, seem indeed to represent the same type of wild donkey, with the black shoulder cross often wide and well-marked. In some rare cases (e.g. at El Richa) the legs show chevron stripes. Neither the size, nor the black cross, nor the coat colour, nor other details which are not reliable enough on documents of this kind, allow us to specify with certainty to which of the two subspecies, *E. asinus somaliensis* and *E. asinus africanus* (the "Somali" and the "Nubian"), the specimen must be attributed. The definitive discrimination of these subspecies is difficult, and indeed clinal for most of the features (Groves 1986).

In conclusion, the donkey was not originally domesticated in the Sahara nor in the Maghreb, although wild ancestors were locally available. Domestication occurred either in Egypt or in the Middle East (the last possibility cannot be discarded, cf.

Groves 1986, Blench, Ch. 21 in this volume) and diffused westwards only at very recent dates.

7. The camel

The dromedary appears only in the most recent period of the Saharan rock art, aptly named the "Camel Period". The earliest depictions show dromedaries mounted (Fig. 6.10), or used as pack animals. In the central Sahara we never see them used for agricultural labour, whereas Tripolitania bas-reliefs of the Roman period often show them pulling a plough (Brogan 1954). In rather late Saharan engravings which date from the Christian era camels carry armed men, with some battle scenes mixing horse-riders and camel-riders together in both parties (we are never dealing with horse-riders fighting against camel-riders). These scenes witness the use of the animal for warfare, a use which became common only in historical epochs.

When did the camel arrive in the central Sahara? In Tassili-Acacus, Hoggar, Aïr, Adrar des Iforas, the Camel Period follows the Horse Period. Riders, which were

Figure 6.10 Mounted camel. The rider shows his weapons; shield, spear, sword, dagger. Saddle with a cross, as those of the current Tuaregs. Painting, early Cameline school, Oued Djerat (Tassili, Algeria). (H = c. 20 cm.)

very few during the Horse Period, become more frequent in the Camel Period. But chariots disappear, they are almost never associated with camels. Moreover, in the same massifs the Libyco-Berber inscriptions, which were extremely rare during the Horse Period, suddenly appear in the Early Camel phase. We can thus extrapolate from dates that are historically known for the abandonment of the chariot and of the diffusion of the Phoenician alphabet or its late variants. This allows an approximate positioning of the beginning of the Camel Period; a few centuries before the Christian era (Muzzolini 1995:177). However, ancient texts and coins only mention the camel in the Maghreb during the first century BC. Furthermore, historians are silent about it until the third and fourth centuries AD, when we are sure that it had already come into general use. It had probably diffused a few centuries earlier in the central Sahara, where its usefulness for carrying people and burdens turned out to be obviously greater because of the increasing aridity.

Where did the animal come from? In Aterian times, probably until around 70,000 BP, a camel, *Camelus thomasi*, was extant within the Saharan fauna and even reached the Negev. It was a large-sized animal. Mainly for this reason Gautier (1966) proposed to link it with *C. bactrianus*, the two-humped camel. However, this thesis is questioned because the geographical area of *C. bactrianus* of central Asia did not fluctuate very much and has never spread westwards beyond Iran. At the time of the "post-Aterian Arid Phase" (*c.* 70,000–12,000 BP) *C. thomasi* disappeared from the Sahara. Did a few isolates subsist in Africa, for instance on the Ethiopian plateau? We do not know of any sure trace of them.[6] Around 2500 BC a camel reappears, but towards the southeast of the Arabian peninsula, in the Abu Dhabi emirate (Firefly 1975). This was a one-humped *C. dromedarius* – which then cannot reflect any western extension of *C. bactrianus* – and it is believed that it might already be a domestic animal (Compagnoni & Tosi 1978). Could it be a descendant of the ancient African *C. thomasi*, or of some other camels, also very ancient, which are known from the Mousterian of Palestine (Gilead & Grigson 1984)? We do not know. We are also in the dark about its possible relation with camel bones found at Ele Bor, near Lake Turkana (Kenya) and dating from *c.* 1000 BC (Phillipson 1985:143).

Around 1100 BC the Arabian dromedary is perceived as a common animal in Syria, and the Assyrians adopted it as a pack animal for their armies. Assarhadon introduced it into Egypt at the time of the Assyrian conquest in 670 BC. The origin of the animal in the central Sahara is probably linked with this Egyptian camel through a northern route – the southern way through Ennedi and Tibesti is less likely, because no picture of camel is known there before the Christian era. Moreover, words for "camel" in Nigeria and Cameroon come from Berber, not from Nilo–Saharan languages (Blench 1995). An introduction from the Punic territories must be excluded, because Carthage did not breed this animal.

During Antiquity the dromedary represented a domestic animal specifically linked to Africa and the Near East, as it does still today. However, some corps of the Roman army occasionally used it in Europe as a pack animal, and it also participated in circus games (Morales et al. 1995).

8. Conclusions

As far as animal domestication is concerned we can verify that the vast Sahara sometimes preceded and sometimes followed the advances of surrounding regions. Rock art faithfully reflects the introductions of the basic animals into the domestic livestock and their importance in the economic sphere. Within the symbolic field the first domestications, those of cattle and sheep, found expression as major events generating appropriated myths.

We can observe that in these neolithic millennia the distribution of breeds is not homogeneous, the breeds are few, and not the same at any time in each region. This, as a general rule, is as valid for population genetics as for cultural sets. When we are dealing with isolates – a typical feature in arid zones – the variability within the isolate is small, although the differences between isolates are conspicuous.

Notes

1. We refer to ^{14}C dates according to the usual convention; bc = raw (uncalibrated) dates, BC and AD = calibrated or historical dates.
2. In the latter case the selection need not necessarily be for horn length, it can be only for a genetic feature linked by some means or other to this length, so the *brachyceros* type is selected unconsciously.
3. The problem of distinguishing *Ammotragus lervia* and *Ovis ammon* in pictures does not exist: *Ammotragus lervia* always has gigantic horns (bent backwards and divergent). This feature does not exist in *Ovis*.
4. We must mention the picture of a donkey mounted by a man, reported by Lhote (1976:160) from Iheren (Tassili). However, this image has never been published. We could not relocate it at Iheren and do not know from which school it dates (maybe cameline?).
5. A so-called "caballine" *Equus* was identified by Vera Eisenmann from the Allobroges site (Algiers; see Bagtache et al. 1984) and has often been quoted in discussions about the present issue. However, it has nothing to do with it. It dates from the Aterian and the word "caballine" is used here in reference to a zoological group within the division of the genus *Equus* made by Vera Eisenmann, and not to the modern species *E. caballus*. This Aterian "caballine" *Equus* had disappeared from the Sahara, as the whole fauna also did, during the long Post-Aterian Hyper-arid Phase (probably 70,000 BP to 12,000 BP). The equid which replaced it during the neolithic cannot have any genetic relation to it.
6. The literature mentions some strange finds; a broken molar coming from the Capsian layers of Medjez II (Algeria) (Bouchud 1975), a tooth from Gobedra (Ethiopia) (Phillipson 1985) previously claimed to be 5,200 years old, etc. Such finds, isolated and obviously odd within their contexts, are not reliable data. However, a camel rib from Sayala going back to the C-Group of Nubia (c. 2000 BC) seems to hold out against criticism (Midant-Reynes & Braunstein-Silvestre 1977). Thus, the hypothesis of some relict African population cannot be excluded.

References

Bagtache, B., D. Hadjouis, V. Eisenmann 1984. Présence d'un *Equus* caballin (*E. algericus*, n.sp.) et d'une autre espèce d'*Equus* (*E. melkiensis*, n.sp.) dans l'Atérien des Allobroges, Algérie. *Comptes Rendus de L'Academie des Sciences* **298**[II,14], 609–12.

Blench, R. 1995. A history of domestic animals in northeastern Nigeria. *Cahiers de Sciences Humaine, ORSTOM* **31**, 181–237.

Brogan, O. 1954. The camel in Roman Tripolitania. *Papers of the British School at Rome* **22**, 126–31.

Bouchud, J. 1975. La faune de Medjez II. In *Un gisement Capsien de facies Sétifien, Medjez II, El Eulma (Algérie)*. H. Camps-Fabrer (ed.), 377–91, Paris: CNRS.

Camps, G. 1984. Quelques réflexions sur la représentation des Equidés dans l'art rupestre Nord-Africain et Saharien. *Bulletin de la Societé Préhistorique Française* **81** (10–12), 371–81.

Camps, G. 1987. Les chars sahariens. *Travaux du LAPMO 1987*, 107–24.

Chaker, S. 1995. Linguistique et préhistoire; autour de quelques noms d'animaux domestiques en Berbère. In *L'Homme méditerranéen* (Mélanges, G. Camps), R. Chenorkian (ed.), 259–64. Aix-en-Provence: LAPMO.

Clark, J.D. 1977. The origins of domestication in Ethiopia. *Proceedings of the 8th Congress of the PanAfrican Association for Prehistory and Related Studies*, R.E. Leakey & B.A. Ogot (eds), 268–70. Nairobi.

Compagnoni, B. & M. Tosi. 1978. The camel; its distribution and state of domestication in the Middle East during the 3d millennium BC in the light of finds from Shar-i Sokhta. In *Approaches to faunal analysis in the Middle East*, R.H. Meadow & M.A. Zeder (eds), 91–103. Cambridge, Mass.: Peabody Museum.

Davis, S., N. Goring-Morris, A. Gopher 1982. Sheep bones from the Negev Epipalæolithic. *Paléorient* **8**, 87–93.

Dayan, T., E. Tchernov, O. Bar-Yosef, Y. Yom-Tov, 1986. Animal exploitation in Ujrat el-Mehed, a neolithic site in southern Sinaï. *Paléorient* **12**, 105–16.

Ehret C. 1993. Nilo-Saharans and the Saharo-Sudanese neolithic. In *The archaeology of Africa, foods, metals and towns*, T. Shaw, P. Sinclair, B. Andah, A. Okpoko (eds), 104–25. London: Routledge.

Ehret C. n.d. The African roots of Egyptian culture; linguistic and comparative ethnographic indicators. Communic. *9ème Semaine d'Etudes Africaines*, Centre d'Estudis Africans, mars 1996, Barcelona.

Epstein, H. 1971. *The origin of the domestic animals of Africa* [2 volumes]. New York: Africana Publishing.

Firefly, K. 1975. Archäologische Forschungen am Persischen Golf. *Antike Welt* **5**, 15–24.

Gautier, A. 1966. *Camelus thomasi* from the northern Sudan and its bearing on the relationship *C. thomasi – C. bactrianus. Journal of Palaeontology* **40**, 1368–72.

Gautier, A. 1982. Prehistoric fauna from Ti-n-Torha (Tadrart Acacus, Libya). *Origini* **11**[1977–1982], 87–127.

Gilead, I. & C. Grigson 1984. Farah II; a Middle Palæolithic open-air site in the Northern Negev, Israel. *Proceedings of the Prehistoric Society* **50**, 71–98.

Grigson, C. 1975–80. The craniology and relationships of four species of *Bos. Journal of Archaeological Science* **1**, 353–79; **2**, 109–28; **3**, 115–36; **5**, 123–52; **7**, 3–32.

Grigson, C. 1991. An African origin for African cattle? – some archaeological evidence. *African Archaeological Review* **9**, 119–44.

Groves, C.P. 1986. The taxonomy, distribution and adaptations of recent equids. In *Equids in the Ancient World*, R.H. Meadow & H.P. Uerpmann (eds), 11–51. Wiesbaden: Ludwig Reichert.

Jelinek, J. 1985. Tilizahren, the key site of Fezzanese rock art. Part I. Tilizahren West Galleries. Part II. Tilizahren East, analyses, discussion, conclusions. *Anthropologie* **23/2**, 125–65; **23/3**, 223–75.

Lhote, H. 1970. *Les gravures rupestres du Sud-Oranais*. CRAPE, Mém. 16, Paris: CRAPE.

Lhote, H. 1984. *Les gravures rupestres de l'Atlas Saharien. Monts des Ouled-Naïl et region de Djelfa*. Alger: Office Parc Nationale Tassili.

Littauer, M.A. & J.H. Crouwel 1979. *Wheeled vehicles and ridden animals in the ancient Near East*. Leiden: Brill.

Manwell, C. & C.M.A. Baker 1980. Chemical classification of cattle, 2; phylogenetic tree and specific status of the Zebu. *Animal Blood Groups and Biochemical Genetics* **11**, 151–62.

Marshall, F. 1990. Cattle herds and caprine flocks. In *Early pastoralists of southwestern Kenya*, P. Robertshaw (ed.), 205–60. British Institute in eastern Africa, Memoir 11. Nairobi: BIEA.

Midant-Reynes, B. 1992. *Préhistoire de l'Egypte. Des premiers hommes aux premiers pharaons*. Paris: A. Colin.

Midant-Reynes, B. & F. Braunstein-Silvestre 1977. Le chameau en Egypte. *Orientalia* **46**, 337–62.

Muzzolini, A.1980. L'âge des peintures et gravures du Djebel Ouenat et le problème du *Bos brachyceros* au Sahara. *Travaux de l'Institut d'Art Préhistorique* [Univ. Toulouse-Mirail] **22**, 347–71.

Muzzolini, A. 1982. Une "relecture" de la littérature archéologique relative au *Bos ibericus*. *Bulletin Société Méridionale. Spéléologie Préhistoire* **22**, 11–29.

Muzzolini, A. 1983. *L'art rupestre du Sahara central; classification et chronologie. Le boeuf dans la préhistoire Africaine*. Thèse 3e cycle Université de Provence, Aix-en-Provence.

Muzzolini, A. 1990. The sheep in Saharan rock art. *Rock Art Research* **7**, 93–109 (with comments by A. Gautier, J. Jelinek, F. Soleilhavoup, A. von den Driesch).

Muzzolini, A. 1992. Dating the earliest central Saharan rock art; archaeological and linguistic data. In *The followers of horus*, R. Friedman & B. Adams (eds), 147–54. Oxford: Oxbow.

Muzzolini, A. 1995. *Les images rupestres du Sahara*. Toulouse: Ed. par l'auteur.

Muzzolini, A. in prep. Les équidés dans les figurations rupestres sahariennes.

Nibbi, A. 1986. *Lapwings and Libyans in ancient Egypt*. Oxford: DE Publications.

Phillipson, D.W. 1985. *African archaeology*. Cambridge: Cambridge University Press.

Pomel A. 1894. *Les Boeufs-Taureaux*. Carte géologique Algérie, Paléontol., Monogr., 1–106.

Riquelme, J.A., C. Liesau von Lettow-Vorbeck, A. Morales Muniz 1995. Dromedaries in Antiquity; Iberia and beyond. *Antiquity* **69**, 368–75.

Sackett, J.R. 1977. The meaning of style in archaeology; a general model. *American Antiquity* **42**, 369–80.

Tchernov, E. & O. Bar-Yosef 1982. Animal exploitation in the pre-pottery neolithic B period at Wadi Tbeik, Southern Sinai. *Paléorient* **8**, 17–37.

Vaufrey, R. 1955. *Préhistoire de l'Afrique. Tome 1. Maghreb*. Paris: Masson

von den Driesch, A. & J. Boessneck 1985. *Die Tierknochenfunde aus der neolitischen Siedlung von Merimde-Benisalame am westlichen Nildelta*. Munich: Institut für Palæoanatomie Domestikation Forschung und Geschichte Tiermedizin Universität München.

Yoyotte, J. 1970. Article "Cheval". In *Dictionnaire de la civilisation égyptienne*, G. Posener, S. Sauneron, J. Yoyotte (eds), 51–2. Paris: Hazan.

CHAPTER SEVEN

African livestock remains from Saharan mortuary contexts

François Paris

1. Introduction

In the Sahara, livestock remains or representations found in mortuary contexts are very rare, unlike the Nile Valley where they are frequent: in Egypt during the pre-Dynastic period (Nagada 1, Maadien) or in North Sudan with the Nubian A-Group or in the late neolithic of Khartoum (El Kidded). Indeed, the sites of the eastern Sahara area (Egyptian or Sudanese) have yielded only a few early livestock burials, although it should be remembered that all funerary monuments appear to be rare in this region. Thus, the appearance of domestic species, replacing offerings of wild species (which may persist), is unfortunately poorly known during the neolithic periods of the Sahara.

Hypotheses concerning the advent of pastoralism often indicate an important ideological or ritual change reflecting this new economic adaptation. The rarity of such evidence could contradict hypotheses that have been put forward concerning possible centres of cattle domestication in the the Egyptian Western Desert (Gautier 1987) or even the central Sahara (Leclant & Huard 1980). However, while the presence of livestock in burials *is* evidence of their domestication, the burying of an animal, alone or associated with human corpses, is not necessarily directly connected with the first appearance of domestication, although it does indicate religious behaviour. The disassociation between domestication and the use of animals in funerary rituals is well known at Kadero in Sudan, where no sacrifices of domestic species have been noted despite their frequent identification in midden contexts (Gautier 1984). Here it is argued that the presence of livestock in a mortuary context can be explained in two ways: either animal graves *sensu stricto* or the association of animals with human tombs.

2. Animal tombs

The diagnosis of burial, for an animal in the Saharan context, is not as clear-cut as it might seem. First, the possibility of natural burial must be eliminated. Then, among the intentional inhumations, ritual burials must still be distinguished from common or utilitarian burials. At the archaeological level it is sometimes difficult to discriminate between these three possibilities. Usually, simple burials are identified by the good conservation of the anatomical connections of the skeleton, but sometimes burials are found containing dismembered remains which may be the manifestation of a ritual. It is sometimes difficult to distinguish such burials from simple waste pits (Clark et al. 1973). Natural burials would have been rare enough during the humid periods of the Saharan neolithic, but the covering of an animal carcass by aeolian sands could have been more frequent. However, if articulations are to be more or less preserved, this burying by wind-blown sand must take place shortly after the death of the animal. If not, the carcass would almost immediately be eaten by hyenas, jackals or vultures which effectively dislocate and scatter a skeleton within a short period. In any case, the hypothesis of natural burial cannot be completely rejected. This is certainly Lhote's (1976, 1979) interpretation of the articulated cattle found in the Talak area and Carter & Clark's (1976) for the *Bos* skeleton excavated at Agoras in Tast (Adrar Bous).

Due caution should be exercised with Lhote's position, since he considers that the great number of articulated skeletons found in the Talak–Timersoï area is the result of an epizootic followed by strong sand storms. During research in the Ighazer wan Agadez area (south of the Talak–Timersoï) we have excavated several *Bos* skeletons, buried in neolithic dwelling sites, and in each case we have concluded them to be an inhumation, ritual or not, performed by people. The case of the *Bos* skeleton of Agoras in Tast (Adrar Bous) is less certain. Additionally, excavations by J.P. Roset and myself in the same region at the Tenerean site of Adrar n'Kiffi, have revealed *Bos* inhumations in pits dug by people, but we cannot say whether they were ritual inhumations or not.

3. Animal inhumations of the western Aïr (Talak–Timersoï and Ighazer wan Agadez)

Lhote was the first to point out the existence of articulated *Bos* skeletons found usually at the edge of dwelling sites on which there are sometimes also human cemeteries. Those are the sites of Tagudalt, Awkare, Akarao, Alabakat and Tin Kulna (Fig. 7.1) (Lhote 1976). Alabakat, in particular, is a very large site which yielded not only *Bos* skeletons but also articulated sheep (or goats) and dogs, more to the south (see below). Several human skeletons also lie at the edge of this important site. All of these places are south of Talak–Timersoï, and none are north of the town of Arlit. Lhote believed that the carcasses of these animals had been abandoned at their death, victims of an epizootic or a catastrophic drought and then naturally covered by sand. This view does not seem to square with the archaeology, since it would be curious if all these animals had all died in the same orientation: the head towards the east.

In the Anu Zegeren and Ikawaten area (northern Ighazer wan Agadez), Grébénart (in Poncet 1983) mentions several neolithic sites, at least one of which has human

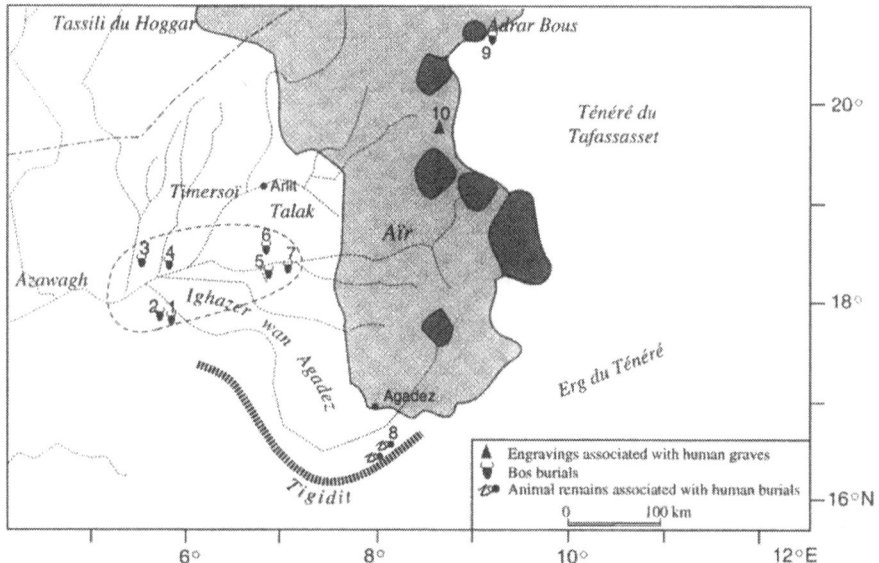

1: Chin Tafidet; 2: In Tuduf; 3: Tin Kulna; 4: Alabakat; 5: Tagudalt; 6: Akarao; 7: Awkare;
8: Afunfun (TAG 9, TAG 12); 9: Adrar Bous; 10: Iwelen

Figure 7.1 Animal remains from funerary contexts in the Nigerian Sahara.

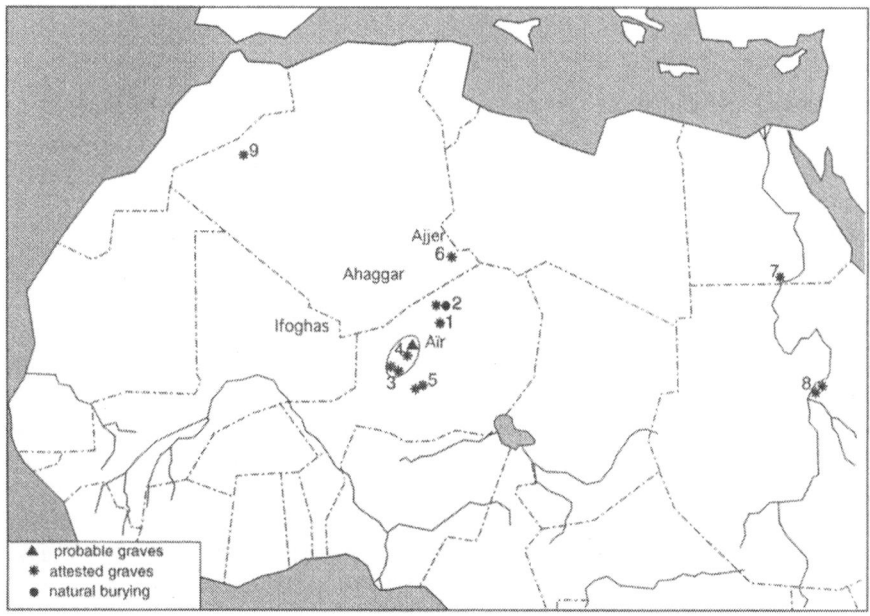

1: sites of Adrar Bous; 2: Iwelen; 3: In Tuduf and Chin Tafidet; 4: region of Talak-Timersoï;
5: Afunfun (TAG 9; TAG 12); 6: Mankhor; 7: Nabta; 8: el Kadada and el Ghada; 9: Torba.

Figure 7.2 Main sites with livestock remains from funerary contexts in the Sahara.

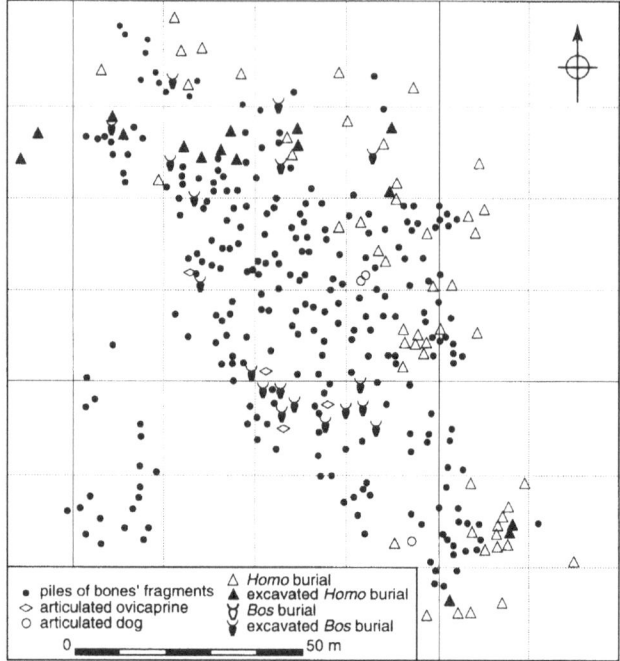

Figure 7.3   Chin Tafidet, location of animal remains and human graves.

burials and articulated *Bos* skeletons, although there is apparently no association between the human and *Bos* sepultures.

The study of the neolithic village of Chin Tafidet has permitted us to have a better understanding of the question of livestock burials (Paris 1984, 1992). The maps of the site (Figs 7.1, 7.2) show an area with animal inhumations relatively separate from a human cemetery although there is some overlapping owing to the long occupation of the place (3900–3300 bp).

We have plotted the animal remains and classified them according to their conservation and degree of anatomical connection (Figs 7.3, 7.4). The location of the human and animal skeletons shows two human cemeteries (C1 & C2) and two zones with animal burials (F1 & F2). F1 is clearly separated from the areas of human graves but F2 occupies partially the north part of C1. In a first analysis, all the articulated skeletons and the piles of partially articulated bones coming from the same animal have been interpreted as possible inhumations. The piles of partially articulated bones belong to animals cut up into pieces and then buried. We have one example of this type (burial B4). Eighteen *Bos*, seven goats or sheep and three dogs' burials have been located.

The *Bos* inhumations are the ones most likely to be linked to a ritual but it is not possible to affirm whether this ritual concerns the *Bos* itself, like the Egyptian bull tombs, or whether they were an offering to a separate individual or entity. Four *Bos*

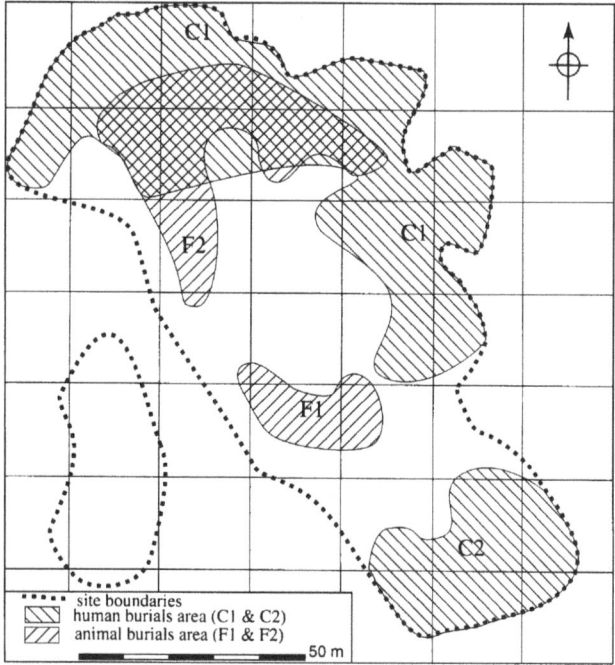

Figure 7.4 Chin Tafidet, the human and animal cemeteries.

burials have been excavated, three contained more or less complete articulated skeletons, the fourth was simply a pile of *Bos* bones (Figs 7.5, 7.6). The three first (B1, B2, B3) are small animals whose stature is 1–1.10 m at the withers.[1] The poor conservation of the skulls did not allow examination of the horn cores. The incisors of B3 indicate an age of 4–5 years old (Faye 1984), and the long bones indicate that all the excavated animals were adult. These three skeletons lie on their right side, the forequarter facing east and the hindquarter to the west. Cut marks have been observed on the 5th cervical vertebra of B2, a probable indication of cutting the throat. All the 18 articulated *Bos* skeletons located lay on their right side. The skeletal orientation could be estimated for only 11 animals: they lay in an east-west direction, 9 with the forequarter to the east, 2 to the west. The excavation of the B4 bone pile revealed a pit 1.20 m in diameter and 0.50 m deep, filled up with partially articulated bones of a *Bos* placed upside down. This *Bos* is complete but cut up into several pieces: haunches, pieces of thorax, the four legs, the neck and the head. The skull is very damaged but a well preserved strong and short horn core shows a low implantation of the horns. All the characteristics of the skull, which confirm those of B1, seem to indicate that this *Bos* belongs to the *B. taurus* species.

The other more or less articulated skeletons are seven ovicaprines and three dogs. We have no evidence for the ritual burial of these animals (regular orientation, for example). They are probably burials of animals performed without religious or

Figure 7.5 Chin Tafidet (B. Faye & F. Paris excavation, 1979); *Bos* burial B1, example of articulated skeleton. There is no association with the human grave which is earlier.

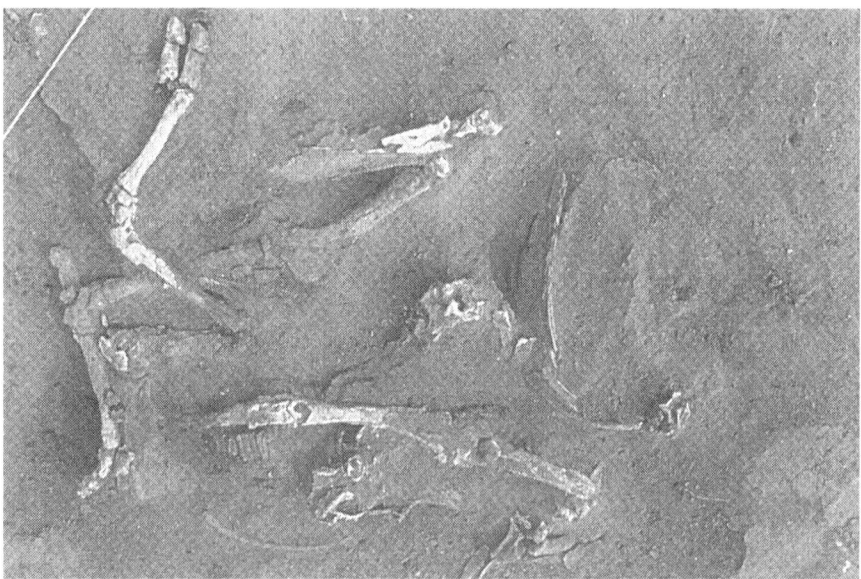

Figure 7.6 Chin Tafidet (B. Faye excavation, 1979); *Bos* burial B4, example of partially articulated bones buried in a pit.

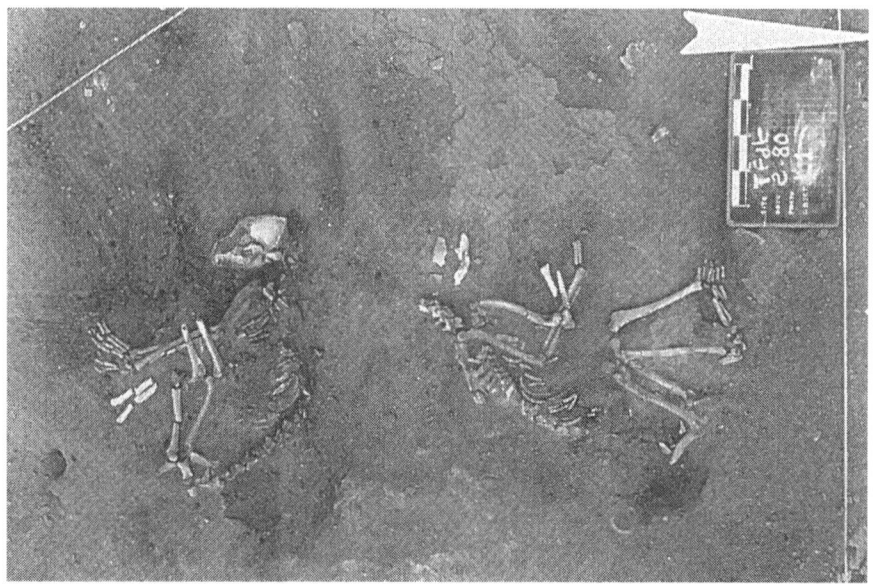

Figure 7.7 Chin Tafidet (B. Faye excavation, 1979), burials of two *Canis* (L1 & L2, articulated skeletons).

ritual signification, perhaps so that their corpses would not decompose on the village rubbish dump. That is how we explain the three excavated domestic dog burials (Fig. 7.7). The well preserved skull of one dog (ratio length occipital-maxillaries/ bicondylian width = 2.4) and the morphological characters of the limbs and the spinal column (limbs length and ratio between legs and spinal column) of these dogs indicate that they belong to a greyhound group. However, they are smaller than the actual *sloughi* of the Azawagh (Faye 1984).

Our knowledge of *Bos* inhumations has been widened by the excavations of Columeau (pers.comm.) during a survey in 1987 at In Tuduf, 10 km northwest of Chin Tafidet. The livestock are principally cattle, goat and sheep, the isolated broken bones of which are scattered upon the whole site. Among the cattle *Bos taurus* is prevalent but, after an initial examination in the field, Columeau believes that some *Bos* could belong to the *B. indicus* family. This hypothesis must be confirmed by more advanced study but if verified these will be the most ancient *B. indicus* remains in Africa (3500 bp). Unfortunately, it is now impossible to return to this area owing to the Tuareg revolt.

The excavation of an apparently articulated *Bos* skeleton (*B. indicus* ?) has also been conducted by Columeau (Figs 7.8, 7.9). The skeleton, relatively well preserved, lay on its right side, oriented towards the east-southeast. The skull is almost completely broken, and only a few splinters from the maxilla and mandible are still preserved. The vertebral column, like the rib cage, are in anatomical connection. For the forequarter, metacarpals, radius and femur are dislocated and placed near

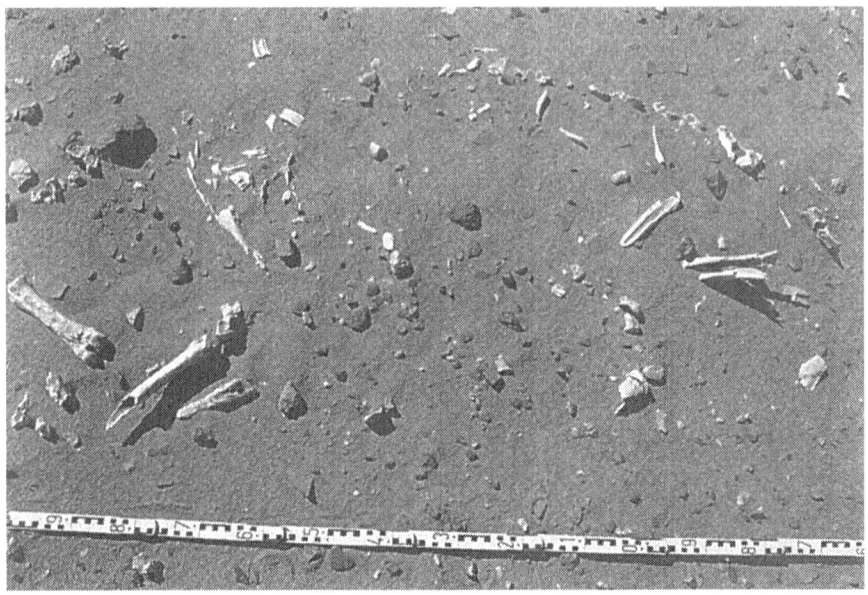

Figure 7.8   In Tuduf (TDF1), *Bos* skeleton, before excavation.

Figure 7.9   In Tuduf (TDF1, P. Columeau excavation, 1985), *Bos* burial B1, example of a partially disarticulated and "re-built" skeleton.

the skull. For the hindquarter, the right back member is partially connected but its position, in hyperextension with the pelvis, is not natural.

The skeleton is lying on a 5 cm layer of ashes, particularly under the right back member and the vertebral column. No butchery marks have been observed on the ribs, and the poor condition of the cervical vertebra prevents us from observing cut marks.

The remains' layout resembles an animal which, after slaughter, has been cut up, excluding the trunk. In a first stage, the two back members, the forelegs and perhaps the head would have been separated. Then, all these pieces would have been replaced and buried, the forequarter laid towards the east-southeast, the hindquarter to the west. A short distance from there another *Bos* skeleton is lying in a comparable arrangement: more or less articulated members separated from the body, spread on the right side, forequarter towards the east. These observations may be connected with those made on the *Bos* B4 of the Chin Tafidet site.

All of these findings show that these *Bos* have been slaughtered, cut up and then buried by people, as part of a ritual. All the pieces are buried, often with the intention of reconstructing the articulation of the animal's body. Moreover the body has always been placed on its right side, with an east–west direction, almost always with the forequarter to the east. This orientation is also that of the human skeletons in the area (Paris 1996). However, nothing permits us to link these cattle sacrifices with human burials and cattle burials are less numerous than human tombs: 5/38 at In Tuduf 1, 1/30 at In Tuduf 3, 18/70 at Chin Tafidet. Therefore, if a relation does exist this is only in specific cases. For all these sites, cattle inhumation zones are separate from human cemeteries.

With due caution, and without indicating any direct link with the neolithic period, this ritual cattle butchery recalls present-day sacrifices by the Wodaabe in Niger, at the occasion of the clan (lineage) meeting before the beginning of the *Gerewol* ceremony. This sacrifice could be called the alliance bull sacrifice. When the clans gather, the host family sacrifices a bull, cutting it up, except for the head, tail and hocks which are preserved, left attached to the skin. Then all the pieces of meat and the entrails are grilled. When the meat seems ready, the pieces are placed on the spread-out skin, reconstructing the bull's anatomy. This is the signal for a ritual sequence (P. Paris, in press). In contrast to the neolithic, where the cattle were buried, the Wodaabe of Niger eat all the meat at the end of the rituals. Indeed, the skin is cut up and distributed to make prized sandals or amulets.

4. Adrar Bous burials

At Agoras in Tast the almost complete and articulated skeleton of a young *Bos* (between 2–5 years old) was excavated by the British Expedition to the Aïr Mountains (Carter & Clark 1976). It has been dated on bone collagen to 5760±500 bp (UCLA 1658). For the authors it is not an intentional burial but a natural interment probably owing to wind-blown sand. They have also excavated some concentrations

Figure 7.10   Adrar Bous (AB S1); tumulus T1 before excavation.

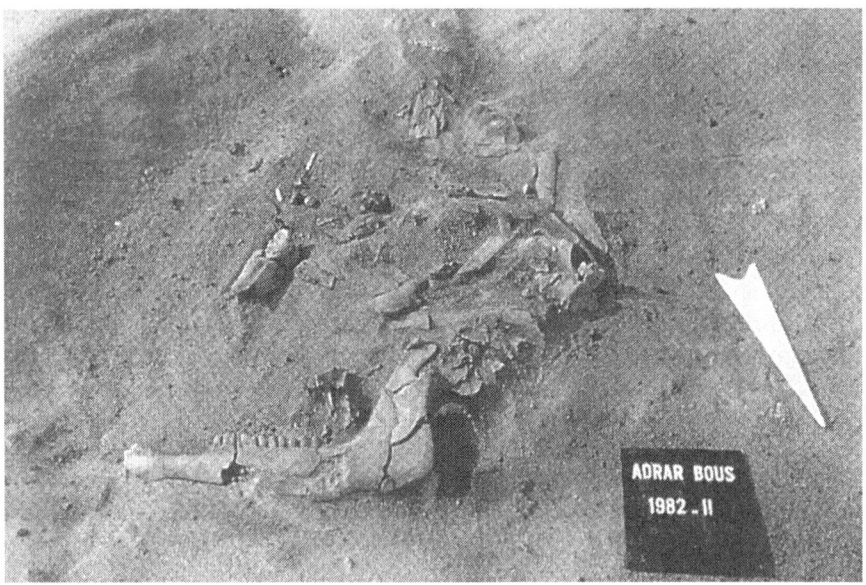

Figure 7.11   Adrar Bous (AB S1, F. Paris & J.-P. Roset excavation, 1981): articulated remains of a *Bos'* right forequarter covered by the tumulus T1.

Figure 7.12   Adrar Bous (AB S1); tumulus T3 before excavation.

of charred bone remains of one or more animals piled together in a small pit, some-times covered by stones. They have been interpreted as remains of meals left in the hearth where they had been cooked except for one case where the bones lay in a pit 75 cm deep (Clark et al. 1973).

North of Adrar n Kifi, at the northern limit of the Tenerian site AB S1, in 1981 we excavated a small stone tumulus (1.40 m diameter and 0.40 m high), believing that it was a human tomb (Fig. 7.10). However, it was found to cover the articu-lated remains of a *Bos* right forequarter, i.e. half right mandible, skull with horn cores, cervical and dorsal vertebrae with the first ribs and the right foreleg (Fig. 7.11). These bones had been burnt. A second structure, on the same site, was excavated in 1989. This tumulus (Fig. 7.12) is comparable but larger (2 m diameter and 0.50 m high). It also contained charred *Bos* bones which, in spite of their very bad preser-vation, seem to belong to an entire skeleton (Fig. 7.13). These bones have been dated on bone collagen to 6200±250 bp (Pa 753). A similar structure excavated by Roset in 1985 has also been dated to a comparable age, 6325±300 bp (Pa 330) (Roset 1987:207). It is not possible to say that they are ritual structures but they are intentional as the bones are placed in a pit closed by stones. The articulated half forequarter of the first tomb could be the remains of a ritual butchery. On the other hand they can be compared to those excavated by the British Expedition, which are interpreted as pit waste. Two human burials have been located on the middle of the site. One has been excavated and dated, on bone apatite, to 4350±250 bp (Pa 1112), a more recent age than the cattle (Paris 1996).

Figure 7.13   Adrar Bous (AB S1, F. Paris excavation, 1989); disarticulated and charred *Bos* bones covered by the tumulus T3. Bones gave a radiocarbon date of 6200±250 bp (Pa 753), on collagen.

5. Human burials with livestock associated

For the neolithic burials (without stone tumuli) there are only two sites with livestock associated with human tombs. They belong to the late neolithic (4000–3500 bp) and they are located in the Afunfun area, beneath the Tigidit cliff 40 km southeast of Agadez. They differ from the other sites of the same period by the generalized existence of mortuary offerings: pottery (broken or not), shells of freshwater bivalves (*Unio sp.*) and sometimes a small livestock skeleton. We know of two examples of small livestock remains at TAG 9 (H1 et H22) and TAG 12 (H3 et H18). But these sites are very damaged by erosion and other examples may exist. In four cases there was a more or less complete and articulated skeleton of a very young animal (a kid or lamb) placed near a human skeleton (Fig. 7.14). These deposits recall the burial traditions described at el-Ghaba and el-Kadada, in the Sudan, during the fifth and the fourth millennia BC (Gautier 1986, Geus 1986). In particular the same specific type of concave-opening pottery is found in these Sudanese sites and the Afunfun graves.

The other examples of associated animals are in tumuli. Two cases may be distinguished: drawings (paintings or engravings) and bone remains. Drawings of domestic livestock in funerary contexts are exceptional and only two examples are known today. The first case consists of paintings on the northern edge of the Sahara at Torba (30 km south of Bechar in oued Guir) where Colonel Lihoreau has

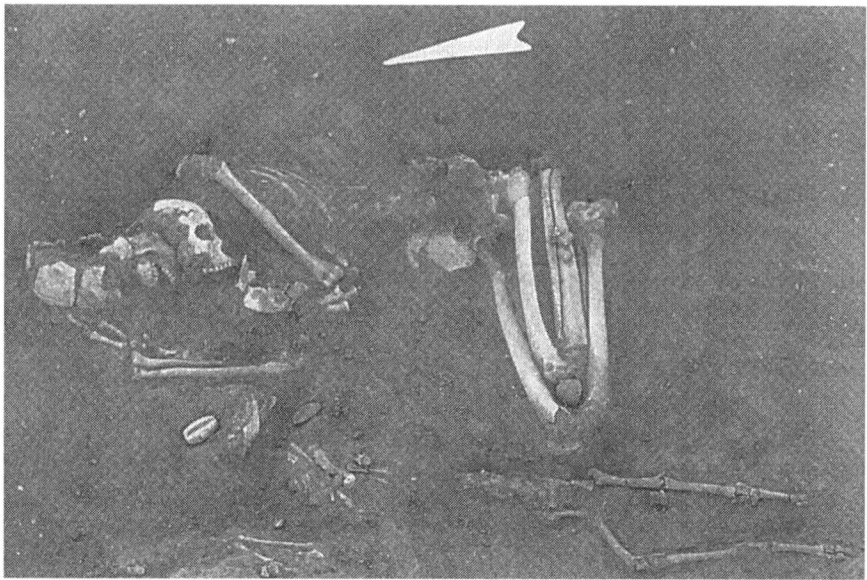

Figure 7.14 Afunfun (TAG 12, F. Paris excavation, 1990). Human burial H18 with deposit of one young ovicaprine (articulated skeleton). Bones gave a radiocarbon date of 3710±80 bp (Pa 1103), on apatite.

conducted the excavation of several funerary monuments (Lihoreau 1994). Some contain horses painted on flagstones. Three tombs have been dated, on the bone apatite of human skeletons: 2230±60 bp (Pa 1252) for T 19, 1760±60 bp (Pa 1257) for T 28 and 1700±60 bp (Pa 1260) for T 27. Another example consists of engravings located in the Aïr mountains at Iwelen where there are engraved blocks of stone around the funerary room of a crater tumulus (wln 7) dated, on the bone apatite of human skeleton, to 2675±200 bp (Pa 525). One of these engravings represents a rhinoceros.

The necropolis of Iwelen and particularly the crater type tumuli belong to the pastoral populations of the Libyco–Berber Period (3500–2500 bp) who, despite having left many cattle engravings, appear not to have placed such drawings in their tombs. Additionally, the populations who painted or engraved cattle during the important Pastoral Phase (5000–4000 bp) did not place such artwork in their tombs. Indeed, we have excavated many funerary monuments dating to this period (Paris 1990, 1996), at Adrar Bous and Iwelen (platform cairns) and in the Emi Lulu mountains of the northeast Tenere (keyhole tumuli), and have found no associated cattle representations.

There are other signs of livestock, including traces of leather shrouds in which corpses were placed. These traces are relatively frequent in the platform cairns. Analyses of two tombs at Iwelen (wln 68) and Mammanet (mmnt 3) also indicate the presence of cattle leather. There is only one example of cattle bones associated with a tumulus. This is at Iwelen where a fragment of cattle mandible was found near a human skeleton in a crater type tumulus (wln 24) (dated on the remains of the leather shroud, to 2550±350 bp).

6. Conclusion

As Saharan tombs associated with livestock remains are rather rare, this overview examined all possible forms of animal burials. In spite of this broadened approach our information remains sparse. Therefore this poverty must be considered from a cultural point of view. It is a well known fact, corroborated by the many cattle paintings or engravings, that livestock was largely integrated within Saharan neolithic cultural life. The lack of livestock remains in tombs may thus be explained only by a particular conception of funerary ritual and, at another level, by a special idea of the hereafter.

Until recently, it seemed that livestock burials strictly speaking, i.e. animals naturally dead and buried with a funerary ritual, did not exist; we had found only the buried remains of ritually sacrificed cattle.[2] Some animals seem to have died naturally, domestic dogs for example, but their burials cannot be automatically considered as ritual or ceremonial as they may simply be pits dug by a sentimental owner or, more prosaically, for hygienic reasons. The only possible case of real animal tombs is related by a traditional Tuareg legend which tells that horse tumuli exist in the Aïr mountains, specifically the Bagezans area. Unfortunately we have never been able to verify this traditional story.

Associations between livestock remains and human skeletons are exceptional in funerary monuments: there is one case at Iwelen for the Libyco–Berber period, and none for the funerary monuments of the neolithic period. Only the neolithic graves of the Afunfun area have lamb or kid remains buried with human skeletons. The presence of bivalve shells (*Unio sp.*) and systematic deposits of pottery distinguish the Afunfun sites from others in the Ighazer region, but recall the funerary customs of some Sudanese neolithic tombs (el-Ghaga and el-Kadada). Therefore it is possible that the Afunfun people were influenced by Sudanese culture, with the form and the decor of their ellipsoidal pottery seeming to confirm this hypothesis. It would indicate a migration of Nilotic Sudanese people, or at least of the diffusion of their cultural influence, along the southern edge of the Sahara at this time.

In spite of the small number of cases, livestock remains from mortuary contexts bring new and important information to the archaeological picture of the southern Saharan neolithic period:

- the parallels between the ritual cattle butchery in the Ighazer area and those still observed among modern pastoral populations (Wodaabe of Niger),
- the specific funerary customs at Afunfun suggesting a relation, if only cultural, with the Sudanese Nile.

The initial information provided here on livestock burials in the Sahara will no doubt be added to in the coming years. The early age and good preservation of these remains should provoke further and more detailed hypotheses concerning the relations of people and livestock during the Saharan neolithic.

Notes

1. The stature was estimated from measures taken in the field during the excavation on the articulated skeleton.
2. Recently, other *Bos* tombs have been discovered, in the Egyptian western Desert and in southeast Algeria. In the western Desert, at Nabta, the Combined Prehistoric Expedition, led by Professor F. Wendorf, excavated some mounds containing *Bos* remains. In one of them, a *Bos* was fully articulated, buried in an oval chamber covered by a wooden roof. Wood from the roof gave a radiocarbon date of 6470 bp (F. Wendorf and A. Gautier, pers. comm. 1996). In southeast Algeria, at Mankhor, a very large site (about fifty hectares) has been discovered (Ferhat et al. 1996 and pers. comm.). About thirty pits containing *Bos* bones have been identified. One of them was dated on bone apatite to 4870±120 bp (Pa 1455).

References

Aumassip, G., M. Tauveron, R. Vernet 1995. L'élevage au Sahara. *Milieux Hommes et Techniques*, CNRS, Paris, 137–57.

Carter, P.L. & J.D. Clark 1976. Adrar Bous and African cattle. In *Proceedings of the 7th PanAfrican Congress of Prehistory, Addis Ababa 1971*, A. Berhanou, J. Chevaillon, J.E.G. Sutton (eds), 487–93. Addis Ababa: Provisional Military Government of Socialist Ethiopia, Ministry of Culture.

Clark, J.D., M.A.J. Williams, A.B. Smith 1973. The geomorphology and archeology of Adrar Bous, central Sahara: a Preliminary Report. *Quaternary* 17, 245–97.

Faye, B. 1984. Etude ethnozoologique du site de Chin Tafidet. In *Les sépultures du Néolithique final à l'Islam*, F. Paris, 72–5. Études Nigériennes No. 50. Niamey: I.R.S.H.

Ferhat N., K.H. Striedter, M. Tauveron 1996. Un cimetière de bœuf dans le Sahara central: la nécropole de Mankhor. In *La Préhistoire de l'Afrique de l'ouest, nouvelles données sur la période récente*, 102–7, Saint-Maur: Édition Sépia.

Gautier, A. 1984. The fauna of the neolithic site of Kadero (central Sudan). In *Origin and early development of food-producing cultures in northeastern Africa*, L. Krzyzaniak & M. Kobusewicz (eds), 317–20. Poznan: Muzeum Archeologiczne w Poznaniu.

Gautier, A. 1984. Archaeozoology of the Bir Kiseiba region, eastern Sahara. In *Cattle-Keepers of the eastern Sahara. The neolithic of Bir Kiseiba*, F. Wendorf, R. Schild, A.E. Close (eds), 49–72. Dallas: SMU Press.

Gautier, A. 1986. La faune de l'occupation néolithique d'el Kadada (secteurs 12–22–32) au Soudan central. *Archéologie du Nil Moyen* 1, 59–112.

Gautier, A. 1987. Prehistoric men and cattle in North Africa: a dearth of data and a surfeit of models. In *Prehistory of arid North Africa*, A.E. Close (ed.), 163–87. Dallas: SMU Press.

Geus, F. 1986. La section française de la direction des antiquités du Soudan. Travaux de terrain et de laboratoire en 1982, 1983. *Archéologie du Nil Moyen* 1, 13–58

Leclant, J. & P. Huard 1980. *La culture des chasseurs du Nil et du Sahara*. Mémoire du CRAPE XXIX, Alger: S.N.E.D.

Lhote, H. 1976. *Vers d'autres Tassili. Nouvelles découvertes au Sahara*. Paris: Ed. Arthaud.

Lhote, H. 1979. Au Sahara, il y a 5,000 ans, une civilisation de pasteurs de boeufs. *Archéologia* 129, 51–7.

Lihoreau, M. 1995. *Djorf Torba, nécropole saharienne antéislamique*. Paris: Kartala.

Paris, F. 1984. *Les sépultures du Néolithique final à l'Islam*. Études Nigériennes No. 50. Niamey: I.R.S.H.

Paris, F. 1990. Les sépultures monumentales d'Iwelen (Niger). *Journal de la Société des Africanistes* **60**, 47–74.

Paris, F. 1992. Chin Tafidet village néolithique. *Journal de la Société des Africanistes* **62**, 33–53.

Paris, F. 1996. *Les sépultures du Sahara nigérien du Néolithique à l'Islamisation: coutumes funéraires, chronologie, civilisation*. Paris: ORSTOM.

Paris, P. in press. Les taureaux de l'alliance, *Journal de la Société des Africanistes*.

Poncet, Y. (ed.) 1983. *Atlas de la région d'In Gall, Tegidda-n-Tessemt, Programme Archéologique d'Urgence (PAU)*, Études Nigériennes No. 47. Niamey: I.R.S.H.

Roset, J.-P. 1987. Néolithisation, Néolithique et post-Néolithique au Niger nord-oriental. *Bulletin de l'Association Française pour l'Étude du Quaternaire* **4**, 203–14.

CHAPTER EIGHT

The origins and development of domesticated animals in arid West Africa

Kevin C. MacDonald &
Rachel Hutton MacDonald

1. Introduction

For a number of years we have been involved in the study of animal bones from archaeological sites in arid West Africa (MacDonald 1989, 1992, 1993, 1995a,b, 1996a,b; MacDonald & MacDonald 1996, n.d.; MacDonald & Van Neer 1994). Our work has benefited from large, carefully recovered assemblages, and improvements in archaeozoological methodology not available to most previous workers in the region. Thus, this review is heavily reliant on our own primary research on the period between 2000 BC and AD 1400 in the West African Sahel, although findings from other researchers and from adjoining regions will be discussed where appropriate. The location of sites mentioned in the text is shown in Fig. 8.1.

The first evidence of domesticated animals west of Lake Chad has long been those associated with the so-called "Ténérian neolithic" in the vicinity of the Aïr Massif of modern Niger. These remains, including the well known Adrar Bous cow, were recovered in 1970 by Desmond Clark and Andrew Smith. A direct date of 5760±500 bp was run on bone collagen from the cow itself, with further "Ténérian" dates, associated with cattle bones, ranging from 5400 bp to 4910 bp (Clark et al. 1973, Carter & Clark 1976, Smith 1980, Gautier 1987). However, recent research from Saharan Niger Republic adds almost a thousand years to the known age of domesticated cattle in West Africa, with cattle inhumations from the vicinity of Adrar n'Kiffi being dated to 6325±300 bp and 6200±250 bp, respectively (Paris, Ch. 7 in this volume).

On the basis of current archaeological evidence, it would seem that neither cattle nor ovicaprines entered West Africa's main river basins until after 2500 BC. This may be attributable, as Smith (1979, 1992) suggests, to constraints imposed by humidity-related diseases (e.g. trypanosomiasis) in the pre-2500 BC Sahel, that were eased by subsequent droughts which opened areas of low tsetse challenge for

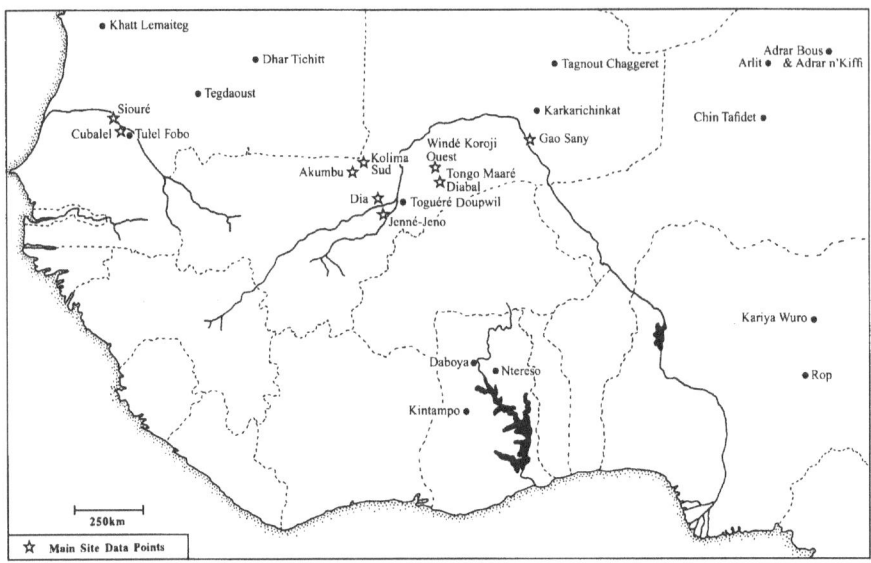

Figure 8.1   Location of archaeological sites mentioned in the text.

colonization southwards. However, this traditional explanation does not account for, or attempt to explore, the origins of the modern trypanotolerant cattle of West Africa (e.g. West African dwarf shorthorn and Ndama). An alternative hypothesis might suggest that this expansion was the result of the successful development of these trypanotolerant breeds around 2000 BC. This proposition *might* conceivably be tested with more metric information from archaeological populations, with dwarfing suggesting an acquisition of trypanotolerance. Dwarfing in African animals has been attributed to selective pressures operative in humid, forested environments, with a more poorly understood linkage to the resistance of fly-borne diseases (cf. Epstein 1971, Blench 1993:91, 1994). The development of resistance to humidity-related diseases in African livestock is not restricted to dwarf populations (cf. the Kuri cattle of Lake Chad). Thus, while dwarf cattle, sheep and goat breeds today are normally trypanotolerant and resistant to other humidity-related diseases, it is not impossible that dwarf prehistoric breeds may have lacked this tolerance, or that non-dwarf breeds may have possessed it. With these caveats in mind, it is interesting to note that, on the basis of the limited metric data, size reduction in breeds had occurred by 2000 BC (see below).

By 2000 BC cattle and ovicaprine remains are documented not only from Sahelian sites such as Chin Tafidet (Paris 1992), Karkarichinkat (Smith 1975), Khatt Lemaiteg (Vernet 1992), and Dhar Tichitt (Munson 1971), but also from more southerly localities, such as: Windé Koroji Ouest south of the Niger bend (MacDonald et al. 1994, MacDonald 1996b), Gajiganna in the Nigerian savanna (Breunig et al. 1996), and Ntereso in northern Ghana (Carter & Flight 1972, Gautier 1987; but cf. Van Neer, Ch. 9 in this volume). Elsewhere in the savanna, remains for early cattle and ovicaprines are more equivocal. Daboya in northern Ghana has no confirmed cattle remains in its earliest "Kintampo" layers (Shinnie & Kense 1989), and the identification of cattle from Kintampo itself has been questioned (Stahl 1985, Van Neer, Ch. 9 in this volume).

The first domestic animals present in West Africa south of the Sahara for which there is osteological evidence appear to have been simply cattle, sheep and goats. Although domestic dogs dating to at least the early second millennium BC have been recovered by Paris (1984, 1992) from Chin Tafidet in Niger, they have not yet been definitely documented from any other West African sites dating to before 200 BC. Other domestic animals now common in West Africa, including the donkey, the horse, the camel, the pig and the chicken are without tangible evidence until the early first millennium AD when a period of widening economic interaction brought them to the Niger and Senegal basins from North and East Africa. The domestic guinea-fowl is the sole potential indigenous domesticate considered for this region, although a definite means of osteologically differentiating wild and domestic varieties has not yet been found (MacDonald 1992, 1995b).

We will now present the evidence for these domestic animals on a species by species basis, paying particular consideration, where possible, to the metric characteristics of the remains.

2. Domestic cattle

The earliest cattle remains yet known from either the Senegal or Niger river basins are those from Windé Koroji which date from the late third or early second millennium BC (MacDonald 1996b). All earlier remains are from within the modern Sahara (see above and Table 8.1). The osteological nature of these early Saharan cattle remains, which probably supplied the genetic core of diverse modern West African cattle populations, is poorly understood. With the exception of the Adrar Bous cow, for whom a stature estimate of 100–125 cm is available (close to that of the modern Ndama breed), little is known about the morphological or metrical characteristics of the early cattle of West Africa (Carter & Clark 1976, Gautier 1987). However, from periods subsequent to the spread and diversification of this "core group" more information is available.

Although measurable elements from West African sites are usually scarce, some metrical divergence seems to be visible in cattle remains from the assemblages we have analyzed. From these there would appear to be at least three broad cattle size classes present in West Africa between 2000 BC and AD 1000 (Fig. 8.2). The middle group, those from Kolima Sud (1400–800 BC) and the Middle Senegal Valley (Siouré, Cubalel and Tulel Fobo, c. AD 0–950) are comparable in size to modern Ndama breeds (MacDonald & MacDonald n.d., Van Neer & Bocoum 1991). The smaller type, as documented at Jenné-Jeno, is close to that of the West African dwarf shorthorn, although it should be noted that a "middle-sized" breed is also in evidence in Phases I through III at Jenné-Jeno – a situation probably stemming from seasonal trade with nomadic pastoralists (MacDonald 1995a). Limited metric data

Table 8.1 Summary of domestic taxa in arid West Africa by site and date.

Sites	Dates (1 Sigma limits of series)	Cattle	Ovicap (dwarf)	Ovicap (savanna)	Dog	Ass	Horse	Camel	Guinea-fowl	Chicken
Adrar n'Kifi (AB S1)	4680–4000 bc	X	–	–	–	–	–	–	–	–
Adrar Bous (Agoras-in-Tast)	5000–3350 BC	X	–	X	–	–	–	–	–	–
Arlit	4300–3700 BC	X	–	X	?	–	–	–	–	–
Tagnout Chaggeret MK42	3650–3350 BC	X	–	–	–	–	–	–	–	–
Karkarichinkat Nord & Sud	2900–1450 BC	X	–	X	–	–	–	–	–	–
Chin Tafidet	2600–1300 BC	X	–	X	X	–	–	–	–	–
Windé Koroji Ouest	2200–950 BC	X	?	X	–	–	–	–	–	–
Dhar Tichitt (Khimiya–Chebka)	2000–800 BC	X	–	X	?	–	–	–	–	–
Khatt Lemaiteg	2000–850 BC	X	–	?	–	–	–	–	–	–
Kolima Sud	1400–800 BC	X	X	–	–	–	–	–	–	–
Kobadi	1360–200 BC	X	–	–	–	–	–	–	–	–
Jenné-jeno Ph. I/II	200 BC–AD 400	X	X	–	X	–	–	–	–	–
Cubalel/Siouré Ph. IA	AD 0–250	X	–	X	X	X	–	–	X	–
Cubalel/Siouré Ph. IB	AD 250–400	X	–	X	X	X	–	X	X	–
Tulel Fobo	AD 300–900	X	–	X	X	–	–	–	X	–
Jenné-jeno Ph. III	AD 400–800	X	X	X	X	–	–	–	?	X
Cubalel/Siouré Ph. II	AD 400–600	X	–	X	X	X	–	–	X	X
Akumbu (Early)	AD 400–600	X	–	X	X	–	–	–	–	–
Tongo Maaré Diabal (Hz's 2–3)	AD 450–850	X	?	X	–	–	–	–	X	?
Tegdaoust (Occup. I)	AD 600–900	X	–	X	–	–	–	X	–	?
Cubalel/Siouré Ph. III	AD 600–950	X	–	X	X	–	–	–	X	–
Akumbu (early middle)	AD 600–1000	X	–	X	X	X	–	–	–	–
Gao Ancien (Period IV)	AD 800–1000	X	–	X	–	–	–	–	?	?
Jenne-jeno Ph. IV	AD 800–1400	X	X	X	X	–	X	–	?	X
Gao SA93	AD 900–1200	X	–	X	?	–	–	–	?	X
Tegdaoust (Occup. II-V)	AD 900–1200	X	–	X	X	X	–	X	–	?
Siouré Ph. IV-V	AD 950–1400	X	–	X	X	X	–	–	X	–
Akumbu (middle & late middle)	AD 1000–1400	X	–	X	X	X	X	–	?	X
Dia (D6)	AD 1400–1600	X	–	X	–	–	–	–	?	X

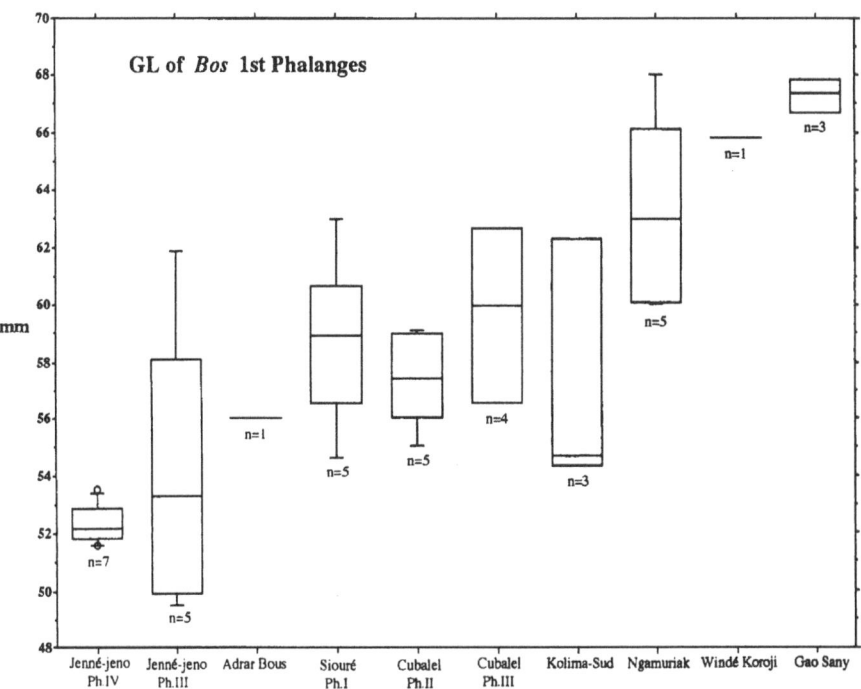

Figure 8.2   Glpe (after von den Driesch 1976) of cattle 1st phalanges.

available from elsewhere (although unfortunately more often on the breadth of phalanges than their length), might also place some or all of the cattle from Dhar Tichitt, Kintampo, Chin Tafidet, and Ntereso within this smaller size category (Carter & Flight 1972, Holl 1986, Paris, Ch. 7 in this volume). The large size class, coming from Windé Koroji (2200–950 BC), Gao Sany (AD 900–1200) and Siouré Phase IV–V (AD 950–1400) is comparable in height to the large cattle of Ngamuriak, Kenya (Marshall 1990, and Ch. 10 in this volume).

One must pose the question – what relation do these ancient cattle bear to those in West Africa today? In other words, were they humped or humpless? There exist in West Africa today large (Kuri), medium (Ndama) and small (West African dwarf shorthorn) humpless cattle breeds (Blench 1993). Do these breeds in some way equate with our three archaeological size categories, or are some of these ancient remains also ancestral to modern humped cattle? This issue has been clouded by the re-appearance of *Bos africanus* in the literature (Grigson 1991, and Ch. 4 in this volume). As early as 1983 Muzzolini noted the presence of stylistically early cattle images with humps in the Saharan rock art (Muzzolini 1983, Blench 1993:77). Subsequently, Grigson (1991) suggested that the indigenous African cattle (cf. *B. africanus*) may have had a low cervico-thoracic hump (as opposed to the high thoracic hump of *B. indicus*). If this is true, then many questions arise: were the progenitors of modern Ndama and West African dwarf shorthorn breeds a type of

African cattle with a low cervico-thoracic hump? What of the Fulani cattle breeds, conventionally thought to be a mixture of zebu *(B. indicus)* and humpless race? While such questions are at present unanswerable, we will consider such facts as are available.

With the exception of the Daima III (northern Nigeria) terracotta cattle figurine dating to *c.* AD 1150 (Connah 1981), we know of no other humped representations of cattle recovered archaeologically in West Africa. In this particular instance the hump would appear to be cervico-thoracic (Connah 1981:Fig. 8.9). From Kolima Sud in Mali, in deposits dating to *c.* 1400–800 BC two intact cattle figurines were recovered, neither of which had any trace of a hump, although one of them features an udder (MacDonald 1994, 1996a; MacDonald & Van Neer 1994). Two possible representations of cattle recovered from Kursakata (Borno State, Nigeria; 800 BC– AD 400) also appear to be humpless (Gronenborn 1996:Fig. 4, 1–2). Likewise, all of the clay cattle figurines from the Jenné-Jeno sequence (200 BC–AD 1400) were humpless (McIntosh 1995).

Bifurcated vertebral spines (indicative of some types of humped cattle – being all Zebu breeds, but only some Sanga breeds, including the Africander; cf. Epstein 1955, 1971i:388,477) have not been recovered from any of the assemblages we have worked with. However, jugals with flattened rims, a trait sometimes ascribed to *B. indicus* (or Zebu), have been identified from Phase IV contexts at Jenné-Jeno and Phase IV/V contexts at Siouré (both *c.* AD 800–1400) (Marshall 1989, Grigson, Ch. 4 in this volume).

Metrically, the large cattle type of Windé Koroji, Gao-Sany, and Siouré may simply represent a large humpless longhorn breed; just as the other two smaller varieties of cattle may simply represent long-term adaptations of such large cattle to wetter and more sedentary living conditions. The osteology of the putative *B. africanus* must in itself be clarified before the origins of these types of early cattle may be more fully addressed.

The study of livestock must go beyond mere osteomorphology and osteometrics, as we are ultimately trying to learn about ancient pastoral economies. However, the analysis of cattle herd structure is made difficult by our small cattle sample size at most sites. One possible exception is the assemblage of cattle remains from the Middle Senegal Valley. Here, osteological evidence indicates a consistent cattle herding practice over space and time, geared for maintaining maximum herd size with little culling of younger animals (based on dental eruption and wear data (after Grant 1982), see Fig. 8.3). This profile is also supported by epiphyseal data, which, although scanty, also shows a preponderance of adult animals. In the case of the first phalanx, for example (which fuses at between 18 mo and 24 mo (Grigson 1982)), of 16 first phalanges recovered from Phase I/II contexts, 15 were fully fused. A practice of maintaining maximum herd size by avoiding culling is in keeping with many modern African cattle herd management strategies, particularly those of nomadic populations (Amanor 1995). Such a practice is ideal for the accumulation of "wealth on the hoof" or to preserve herds in difficult climatic conditions, but it is hardly the most efficient model for meat production.

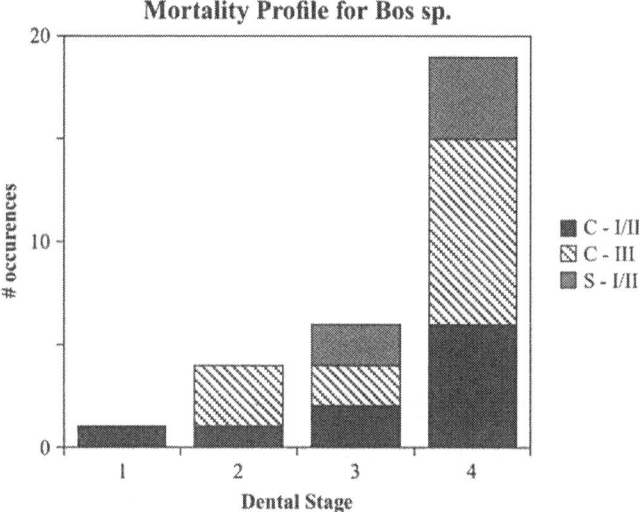

Figure 8.3 Cattle mortality profile for Cubalel and Siouré phase I/II (AD 0–600), and Cubalel phase III (AD 600–950). Dental stages after Grant 1982.

3. Domestic ovicaprines

Accompanying these cattle at virtually every site, except the early (pre-3000 BC) localities of Adrar n'Kiffi and Tagnout Chaggeret (Table 8.1), are sheep and goats (Gautier 1991, Paris, Ch. 7 in this volume). The differentiation of sheep and goats in the archaeological record is sometimes difficult, but may be facilitated by the rigorous application of diagnostic criteria to key elements. For all of the assemblages worked on by the authors, the criteria of Boessneck (1969) and Prummel & Frisch (1986) were used. With other researchers, when direct attributions have been made, they seem to have been on the basis of intact crania or horn cores. In most instances, both sheep and goat seem to be equally numerous throughout site sequences, but at some sites, such as Tegdaoust, one type (sheep) seems to dominate (Bouchud 1983).

Like cattle, the major osteological variable in these ovicaprine populations is size – with goats often being of a dwarf variety and sheep tending to be consistently larger (see Appendix). True dwarf goats are already present in small numbers at Kolima-Sud by the early first millennium BC (MacDonald 1994, 1996a), with goats from the initial (200 BC) occupation of the urban site of Jenné-Jeno also being dwarf (MacDonald 1995a). As with cattle, this size reduction is *usually* linked with the development of trypanotolerance (Epstein 1971) – an essential adaptation for breeds living year round in floodplain conditions.

By AD 400, larger breeds of both sheep and goat appear at Jenné-Jeno, but in smaller numbers relative to the dwarf breeds. This pattern then continues for almost

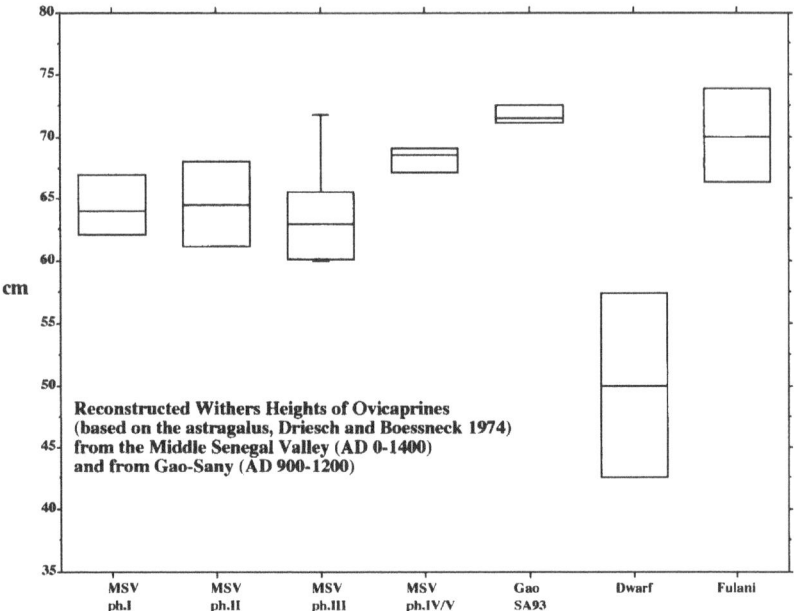

Figure 8.4 Reconstructed withers heights of ovicaprines (based on the astragalus, after von den Driesch & Boessneck 1974). Modern dwarf and Fulani breed ranges are from Epstein (1971).

a thousand years (MacDonald 1995a). Such large, "savanna" breeds could represent animals traded in by nomadic pastoral groups on seasonal visits to the Delta. Today, such trade between transhumant Fulbe pastoralists and sedentary farmers and fisherfolk of the Delta is a common phenomenon (Gallais 1967, 1984).

Unlike the ovicaprines of Jenné-Jeno, those of Cubalel, Siouré, and Gao appear to be non-dwarf savanna varieties (cf. MacDonald 1995a and Epstein 1971). Owing to excessive carnivore scavenging at these sites no intact metapodials or other long bones were recovered. In spite of this, phalanges and astragali were present in quantity and were often intact. Estimates for the stature of living populations can be derived from astragali by the use of factors in von den Driesch & Boessneck (1974). Combined Phase withers height percentile plots, presented for the Middle Senegal Valley and Gao-Sany ovicaprines, demonstrate a marked homogeneity in stature over time, always falling within the size range of modern "savanna" ovicaprine populations rather than dwarf ones (Fig. 8.4; breed statures from Epstein 1971) First phalanx length (measured by the authors as Glpe = greatest length of the peripheral [abaxial] half) shows a similar pattern: the Cubalel and Siouré samples are of a larger "breed" than the dwarf ovicaprines from Jenné-Jeno or modern African comparatives (Fig. 8.5).

Although livestock herd structure is an important issue, small assemblage size usually prevents it from being addressed in West Africa. However, the Middle

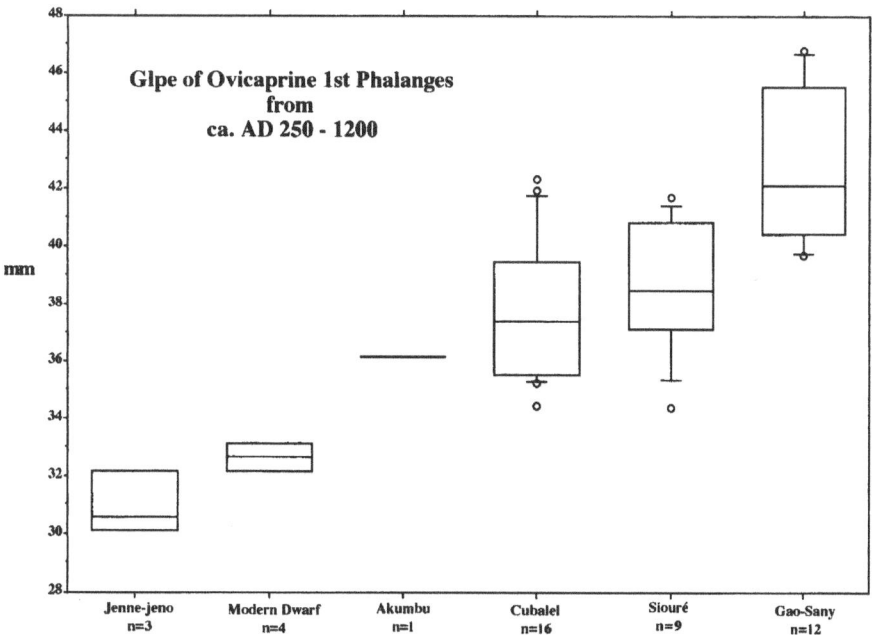

Figure 8.5   Glpe (after von den Driesch 1976) of ovicaprine 1st phalanges from *c*. AD 250–1200 modern comparative African dwarf goat measurements courtesy of Dr Richard Meadow, Harvard University.

Senegal Valley ovicaprine remains are sufficiently numerous to make a first essay. Ovicaprine age at death is determined following the methodology of Grant (1982), and Payne (1973). While we do not have enough data to discuss mortality profiles in Phase IV, interesting trends can be discerned in the temporal aggregations of Phases I/II and III(a+b) (Figs 8.6, 8.7). Our combined sample from Phases I and II (Fig. 8.6), shows a pattern that may best be equated with a pattern of natural attrition. In age-structure, if broken down into immature (0–6 mo), young (6–24 mo) and adult animals (24+ mo), this pattern would most resemble Cribb's (1985:100–101) "Saxon" profile. A "Saxon" mortality structure is a herd management strategy geared to keeping animals alive as long as possible. Explanations for this may include wealth accumulation or, if in a time of economic stress, a policy of maximizing herd viability (Dahl & Hjort 1976). The aggregated Phase III sample shows a strikingly different trend (Fig. 8.7). Here the structure indicates the introduction of a culling practice to maximize meat and milk production (Cribb's (1985) "Roman" profile). Thus, around AD 600 a decision may have been made to change local ovicaprine herding practice, abandoning a strategy of keeping animals alive as long as possible. It is hoped that future large-scale excavation campaigns in Sahelian West Africa will yield samples sufficiently extensive to address herd management strategies with greater certainty and more detail.

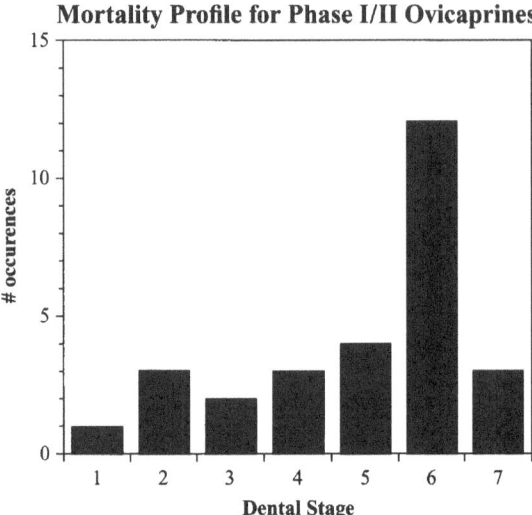

Figure 8.6   Ovicaprine mortality profile for Cubalel and Siouré phases I and II (AD 0–600) after Payne (1973).

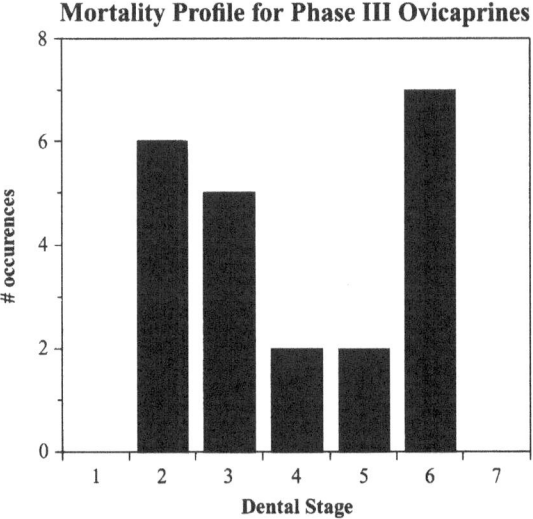

Figure 8.7   Ovicaprine mortality profile for Cubalel and Siouré phase III (AD 600–950) after Payne (1973).

4. Domestic dog

The antiquity of the domestic dog in West African contexts, seems to be only slightly less than that of cattle. The difficulty lies in separating the remains of *Canis familiaris* from African jackals and hunting dogs. At some early sites, such as at

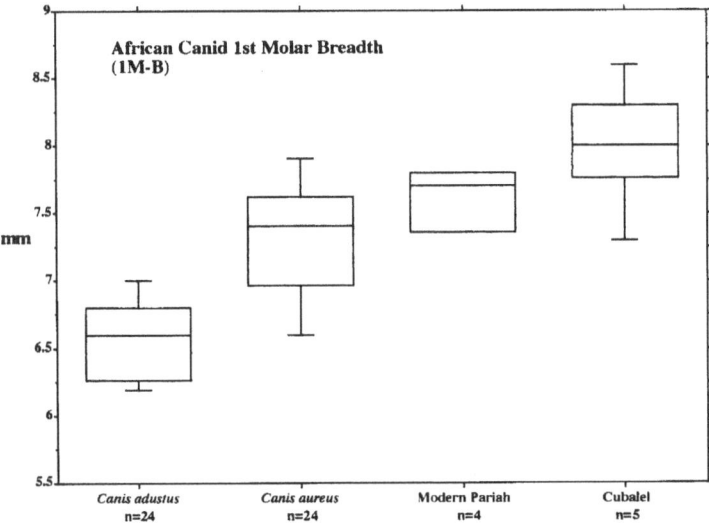

Figure 8.8   African Canid 1st molar breadth (after von den Driesch 1976).

Chin Tafidet (Paris 1984, 1992, Ch. 7 in this volume), inhumations of entire dog skeletons allow a relatively straightforward identification of their remains. From cranial morphology, the Chin Tafidet animals most resemble modern greyhounds (Paris, this volume). However other early occurrences, when canid remains occur within comminuted "kitchen middens", are less easily determined. Thus, occurrences such as "jackal" at Khatt Lemaiteg during the second millennium BC (Vernet 1992), and "possible dog" remains from Arlit in the fourth millennium BC (Smith 1980:457) cannot be properly judged without definite diagnostic criteria.

We have identified domestic dog (*Canis familiaris*) from numerous first millennium AD contexts at Jenné-jeno, Akumbu, Tongo Maaré Diabal and the Middle Senegal Valley sites (MacDonald 1993, 1995a, MacDonald & MacDonald n.d., MacDonald et al. 1994). The identifications are based upon metric criteria developed by the authors with the aid of extensive comparative ranges housed at the US National Museum of Natural History, the Museum of Comparative Zoology (Harvard) and the Natural History Museum (London). Domestic dogs, wild hunting dogs (*Lycaon pictus*) and jackals can be separated by the use of numerous multi-dimensional metric criteria – particularly those reflecting relative mandibular robusticity (MacDonald & MacDonald, in prep.). The figures show a relatively clear-cut example, where differences in mandibular depth and molar row length differentiate the Iron Age canids of Cubalel and Siouré from indigenous wild canids (Figs 8.8, 8.9). Van Neer has also identified domestic dog from the contemporary Middle Senegal Valley site of Tulel Fobo (Van Neer & Bocoum 1991). We have had the opportunity to measure these specimens ourselves and our criteria confirm this identification.

Secondary evidence for the extensive presence of domestic dogs at both the Middle Senegal Valley Sites and at Gao-Sany (from which we have as yet no

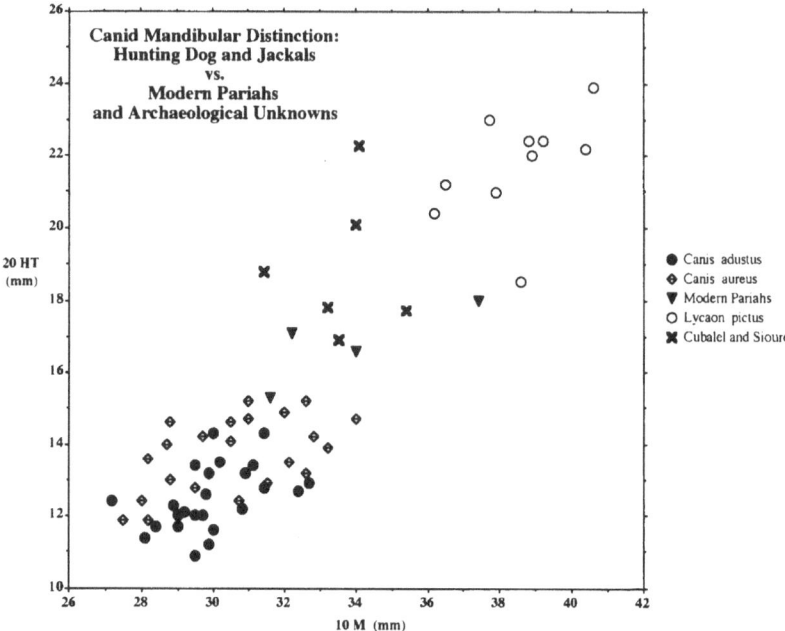

Figure 8.9   Scattergram of Cubalel and Siouré Canid mandibular height (20 HT) compared with mandibular molar row length (10 M). These samples are compared with modern jackal (*Canis adustus* and *C. aureus*), hunting dog (*Lycaon pictus*) and Indian pariah dog (*Canis familiaris*) specimens.

diagnostic canid elements) includes a striking absence of ungulate long-bone epiphyses as well as carnivore gnaw marks on numerous ungulate bones. Given the midden contexts of the faunal assemblages recovered from these sites, it is most likely that this taphonomic pattern was the result of a carnivore whose presence within the settlement was tolerated by the inhabitants. Furthermore, the size and pattern of puncture and gnaw marks on the bones indicate a carnivore the size of a dog (Binford 1981:51).

Regarding the function of dogs recovered from these sites we have but little information. Modern non-Islamic West African peoples have used dogs for hunting, guarding and for a source of meat, sometimes keeping two different breeds for these purposes (e.g. the Bakosi of Cameroon; Epstein 1971i:44–5). The Middle Senegal Valley dogs form the most substantial canid assemblage with which we have worked. However, they were not complete inhumations, but rather, from their frequent charring and pre-depositional fragmentation, appear to have been food remains. Van Neer & Bocoum (1991) notes that some of the Tulel-Fobo dog remains are charred, agreeing with the evidence for dog-eating at Cubalel and Siouré. Similar evidence is available from our initial investigations of first and early second millennium AD contexts at Tongo Maaré Diabal. Even in the ninth to thirteenth century contexts at the Saharan trading entrepôt of Tegdaoust, Bouchud

(1983:362) observed "*La présence parmi des vestiges «alimentaires» de restes non douteux de chien ... Il ne s'agit jamais de squelettes entiers d'animaux éventuellement inhumés, mais des fragments isolés*". The presence of cynophagy at this ancient Berber trading town is not so surprising when one considers that Al-Bakri noted the frequent eating of dogs by the Berbers of southeastern Morocco and Tunisia during the tenth century (Lewicki 1974:89). The ritualistic continuation of cynophagy by Saharan Berbers in recent times has been recorded by Briggs (1960:24–5). Lewicki (1974) even goes so far as to suggest that dog-eating may have been introduced into sub-Saharan Africa by (non-Islamic) Berbers. Indeed, the distribution and endurance of cynophagy in Africa remains an interesting and ideologically meaningful topic for future archaeozoological investigation, being until now only touched upon in the anthropological literature (e.g. Mauny 1954, You 1955, Blench, Ch. 2 in this volume).

5. Domestic equids

Numerous rock art depictions and allusions within classical histories to Saharan "horse-drivers" have placed the horse in the Sahara and its margins by the first millennium BC (Law 1980:1–2). It has been presumed that the horse and the ass became the dominant means of transport in the Sahara and the Sahel at this time and remained so until the introduction of the camel in the first millennium AD. It is thus remarkable that equid remains have only been recovered in small numbers from the tell sites of the Niger and Senegal river basins, with no finds from the southern Sahara to the coast pre-dating the birth of Christ (Table 8.1). True, we do have the controversial equid tooth from Rop rockshelter in Nigeria, supposedly dating to the first millennium BC (Sutton 1985). However, in the total absence of other contemporary evidence, we are sceptical of this single find from a site of shallow stratigraphy which also features a recent occupational layer, and will remain so until it is directly dated (e.g. ^{14}C bio-apatite) (see also Blench, Ch. 21 in this volume).

It is important to maintain a healthy scepticism regarding the actual penetration of "Saharan charioteers" to the southern margin of the desert – particularly since depictions seem to become progressively more schematic the farther to the south and the west that one goes. Indeed, in his synthesis on the subject, Lhote (1982:152–4) noted that naturalistic depictions of horse-drawn chariots are only common in the Tassili n Ajjer (central Sahara). Elsewhere, they are often represented without horses or, as is commonly the case for the petroglyphs of Mauritania, they are represented with cattle in the harness (ibid.).

It should also be remembered that there are no zebra or wild ass remains to confound determinations – the Holocene evidence for both of these animals in the western third of Africa outside of the Maghreb (whether osteological or artistic) remains dubious or nonexistent. So, the absence of early equid remains in the Sahel may be seen, for the moment, to be convincingly total. We thus believe the lack of verified domestic equid remains in West Africa before the time of Christ *might* truly represent their extreme rarity or total absence at that time (but see discussion in relation to linguistic evidence in Blench 1995, and Ch. 21 in this volume).

Our earliest and best osteological evidence for domestic equids in West Africa comes from the Middle Senegal Valley tell sites and Akumbu during the first millennium AD (MacDonald 1993, MacDonald & MacDonald n.d., Togola 1996). These sites are conspicuous by their placement at the northern edge of the recent Holocene savanna. Remains from floodplain and savanna forest environments seem to be later, with finds from Jenné-Jeno, Daboya (in Ghana) and Kariya Wuro rockshelter (in Nigeria) all dating to *c.* AD 1000–1400 (Allsworth-Jones 1982, Shinnie & Kense 1989, MacDonald 1995a).

It is often difficult to separate horse (*Equus caballus*) from ass/donkey (*E. asinus*) with only fragmentary remains. However, our earliest equid remains from the Middle Senegal River Valley (*c.* AD 0–400) include a small, complete metatarsal from Cubalel which would reconstruct to a withers height of 116 cm (MacDonald & MacDonald n.d.). This is approximately the size of a mature male modern West African domestic ass (which ranges between 90–115 cm), and smaller than the pony breeds of Mali and Senegal which stand between 120–145 cm at the withers (cf. Doutresoulle 1952, Epstein 1971, von den Driesch & Boessneck 1974). Indeed, when compared directly against a range of reference specimens, the remaining elements from the Middle Senegal would also appear to fall more comfortably within the size range of ass (*E. asinus*) than that of even the smallest horse breeds (*E. caballus*). Thus, all of the recovered remains from Cubalel and Siouré have been identified provisionally as *Equus* cf. *asinus*.

The Akumbu remains from AD 600–1000 are a different matter – falling in size comfortably within the range of modern ponies and horses, based on the criteria of Eisenmann (1986), and with dentition that is identifiable morphologically as *E. caballus* (MacDonald 1994). During this same time a few smaller remains would indicate that *E. asinus* is also present, although they could represent a second, smaller breed of horse.

Excavations in 1996 carried out by the authors and T. Togola at the tell site of Tongo Maaré Diabal in Mali have produced one of the earliest known equestrian representations in West Africa. This is a terracotta figurine of a horse and rider, recovered within an ash pit outside the entrance to a house and dated on the basis of associated charcoal to AD 820–1020 (1105±80 bp; GX-21728). The rather schematically represented beast would appear to be a horse rather than a camel on the basis of the presence of a mane (Fig. 8.10). The date of this find accords well with the slightly earlier date for the Akumbu horse remains from the left bank of the Niger. Interestingly, there have as yet been no equid remains recovered from the Tongo Maaré Diabal excavations.

Blench (1993, and Ch. 21 in this volume), on historical, artistic, and linguistic evidence, has argued for the presence of small horse breeds in West Africa during the first millennia BC and AD. The osteological data presently available can neither confirm nor refute this hypothesis. Our earliest osteological evidence for donkeys precedes that of horses, but our excavated sample is small. Moreover, most remains from within settlements only include consumed animals, or carcasses scavenged by

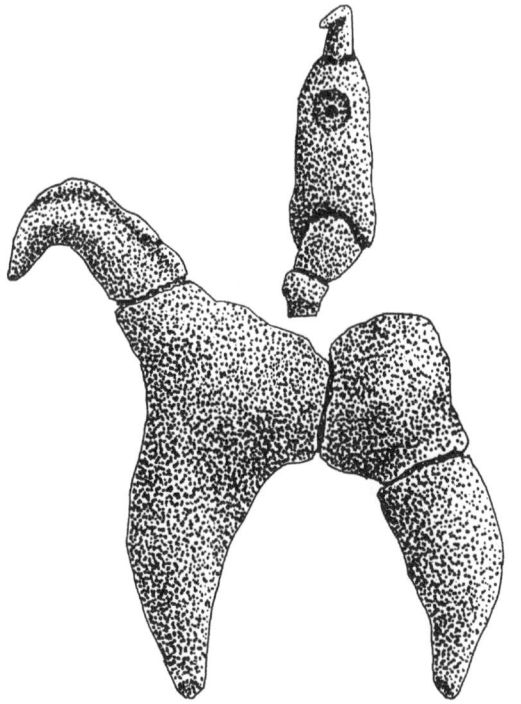

Figure 8.10   Equestrian terracotta figurine recovered
from recent excavations at Tongo Maaré Diabal (Mali).

local dogs, with large unconsumed equid carcasses probably being left to decay (as
today) beyond the limits of the town. However, the complete absence of reliable
osteological evidence for equids from before the first millennium AD, does point to
their absence or scarcity before that time.

6. Domestic camel

Camel (*Camelus dromedarius*) has been identified in West Africa south of the
Sahara in only two archaeological localities. The earliest of these finds is a char-
red first phalanx, recovered from a Phase 1B context at Siouré (MacDonald &
MacDonald n.d., McIntosh et al. 1992). The presence of camel along the Middle
Senegal Valley between AD 250 and AD 400 is worthy of special comment, as this is
the earliest known sub-Saharan osteological identification of the animal.

Bulliet (1990:113) notes that camel was not mentioned by Roman historians as
being in the Maghreb until AD 300 when they were reported to be abundant. The
earliest evidence from northeast Africa comes from the Egyptian site of Qasr Ibrim,
where camel remains and dung have been dated to the early first millennium BC by
Rowley-Conwy (1988). This accords with the view of some researchers (e.g. Bulliet

1990), who favour an early trans-Sahelian over a trans-Saharan route for the introduction of camels to the Berbers in West Africa. Our new evidence is still too recent to confirm this view. But the presence of camel, presumably as a pack animal, between AD 250 and AD 400 along the Sahara's southern edge, might support numismatic arguments by Garrard (1982) for a more substantial (pre-Arab) trans-Saharan trade than has been generally recognized.

The other camel identification comes from Tegdaoust (Mauritania), and dates to the advent of the trans-Saharan trade at this major entrepôt. Camel remains have been found throughout the eighth to thirteenth centuries AD occupation layers (Bouchod 1983).

7. Domestic chickens

The osteological evidence for domestic chickens (*Gallus gallus*) in West Africa is already described in some detail by MacDonald (1992, 1995a,b). Recent historical linguistic considerations of the history of the domestic fowl in Africa, which suggest a more ancient introduction in sub-Saharan Africa than has been previously supposed, are presented elsewhere by Williamson (Ch. 23 in this volume).

Before AD 900 chickens are known osteologically in West Africa only from the urban site of Jenné-Jeno. MacDonald (1995b) has suggested elsewhere that this is probably due to the status of Jenné-Jeno as an early West African trade entrepôt. It is also possible that at the chicken's first introduction to West Africa during the early first millennium it was confined to elite households, and only after several hundred years became a widely available comestible. Thus, it is only during the early second millennium AD that chicken appears to reach settlements throughout modern Mali, with numerous confirmed finds from Gao-Sany (deposits dated to AD 900–1200; Insoll 1994), Akumbu (Mound A, AD 1000–1400; Togola 1993), Toguéré Gailia (located near Jenné-Jeno, from mixed period III contexts, AD 1000–1600; Bedaux et al. 1978), Toguéré Doupwil (from mixed period in contexts, AD 1225–1400; Bedaux et al. 1978), and Dia (from Unit D6, AD 1400–1600; MacDonald 1992). Interestingly, no chicken has yet been identified along the Middle Senegal.

Chickens have also been identified from pre-AD 900 contexts at Tegdaoust (perhaps dating back to as early as the seventh to eighth century). However, as the only criteria Bouchud (1983) lists for their identification was the length of their longbones, these remains may just as easily be guinea-fowl. A re-examination of this material would be most useful.

The small size of the early West African chicken has been commented upon previously (MacDonald 1992, MacDonald & Edwards 1993). Findings from Jennéjeno have been borne out in the Gao-Sany assemblage where small chicken remains are numerous, as well as from more isolated West African finds (Fig. 8.11). As Thesing (1977) has noted for Europe, the size explosion of chicken seems only to have occurred in Roman and modern times – with European Iron Age and Medieval chickens all being of a similarly diminutive size (Blench & MacDonald, in press).

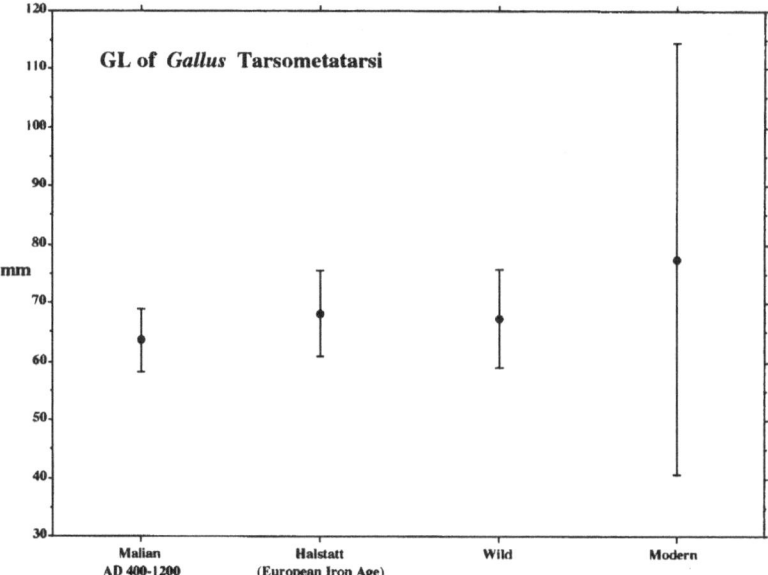

Figure 8.11   GL (after von den Driesch 1976) of *Gallus* Tarsometatarsi.

We have argued elsewhere that this is most probably owing to both poor nutrition (leaving chickens to forage for themselves) and an inattention to breeding for size (MacDonald 1992, 1995a,b). Development work has reached similar conclusions regarding the small size of modern Malian fowls (cf. Wilson et al. 1987).

8. Domestic guinea-fowl

Despite the status of guinea-fowl (*Numida meleagris*) as the only putative indigenous West African domesticate, our database regarding it in West Africa is sparse. Gautier (1990) indicates an initial domestication for *Numida* in the Maghreb, with a subsequent spread across the Sahara or perhaps down the Nile Valley. It was certainly known in Greece by the mid first millennium BC, and from much of the rest of Europe during the time of the Roman Empire, so an assertion of North African domestication – where it also occurs in its wild state – is not out of the question (Mongin & Plouzeau 1984).

Using morphological criteria, guinea-fowl (wild or domestic), has only been positively identified from the first millennium AD Middle Senegal Valley sites (MacDonald & MacDonald n.d., Van Neer & Bocoum 1991). Curiously, contexts in Mali, including those possessing chicken, have produced no definite guinea-fowl remains. Unfortunately, the guinea-fowl remains from Senegal have suffered greatly from both human and scavenger gnawing, so although they are numerous only five elements were measurable. These few measurements lie only at the lower margin of size ranges we have documented for modern wild specimens (see Appendix). This may represent some form of human intervention, but the samples remain too few to form the basis for any arguments. Perhaps the best secondary evidence for guinea-fowl domestication in Senegal are large quantities of eggshell recovered from the Cubalel and Siouré excavations. We hope to undertake positive taxonomic identification of this eggshell – although even if it is *Numida* it could simply represent the raiding of wild guinea-fowl nests.

9. Conclusions

In outline, the evidence available indicates that cattle reached West Africa by at least *c.* 6300 bp, perhaps as a result of an indigenous domestication in northeastern Africa, and perhaps before the arrival of sheep and goats. Domestic dogs, probably to aid in hunting and herding, may have accompanied some of these early cattle-keepers. However, our earliest definite evidence comes from the third to second millennia dog inhumations from Chin Tafidet (Paris 1992, Ch. 7 in this volume). Of the equids, asses (or very small horses) seem to have preceded the arrival of large horse breeds in the West African Sahel. Small equids are documented from the beginning of the first millennium AD whereas large horses are scarce until post *c.* AD 800. Camel remains are rare, but are now known to date to at least as early as AD 400 on the southern shore of the Sahara. There is not yet any early osteological evidence for the intensive exploitation or domestication of guinea-fowl in West Africa. It becomes frequent, like the imported chicken, only from AD 400 onwards in early urban or town sites.

Acknowledgements

Our work would not have been possible without both the prodigious efforts of our own field team members and the kindness of other researchers who have invited us to work on their bone material. These latter individuals include: Professors Roderick and Susan McIntosh (Jenné-Jeno, Dia, Cubalel, and Siouré), Dr Timothy Insoll (Gao Sany and Gao Ancien), and Dr Téréba Togola (Akumbu).

We would also like to thank the following individuals for access to and assistance with comparative collections: Dr Juliet Clutton-Brock (Natural History Museum, London), Jo Bailey (Ornithological Division of the Natural History Museum, Tring), Dr Maria Rutzmoser (Museum of Comparative Zoology, Harvard University) Dr Rosie Luff (Cambridge Faunal Remains Unit) and Jessica Hale (the "Bone Room", University of Cambridge).

Thanks also to Roger Blench for our many discussions and debates concerning the origins of domesticated animals in West Africa.

Appendix (Tables 8.2–8.6)

Listing of all measurements taken by the authors of West African domestic mammal remains, with the exception of those from Jenné-Jeno, already published in MacDonald 1995a. All measurements are after von den Driesch 1976.

Table 8.2 Cattle measurements.

Bos sp.

Site	Phase	Element	Measurements					
Kolima Sud	1400–800 BC	Mandibular Third Molar	L = 35.0	B = 13.8				
Akumbu	AD 1000–1200	"	L = 31.6	B = 11.1				
Cubalel	AD 0–250	"	L = 35.5	B = 14.1				
Cubalel	AD 400–600	"	L = 34.3	B = 12.1				
Cubalel	AD 400–600	"	L = 37.4	B = 15.1				
Cubalel	AD 600–950	"	L = 39.1	B = 13.5				
Cubalel	AD 600–950	"	L = 33.5	B = 11.8				
Cubalel	AD 600–950	"	L = 36.4	B = 12.1				
Kolima Sud	1400–800 BC	Radius	Bp = 63.5	BFp = 58.6				
Kolima Sud	1400–800 BC	Metacarpal			Bd = 52.9	Dd = 28.0		
Kolima Sud	1400–800 BC	"			Bd = 51.1	Dd = 27.0		
Kolima Sud	1400–800 BC	"			Bd = 49.4	Dd = 25.3		
Dia	AD 1400–1600	"	Bp = 61.6					
Cubalel	AD 600–950	"			Bd = 59.1	Dd = 23.0		
Gao Sany	AD 900–1200	"	Bp = 57.9	Dp = 35.1				
Gao Sany	AD 900–1200	"			Bd = 61.1	Dd = 32.6		
Cubalel	AD 600–950	Femur	DC = 41.1					
Kolima Sud	1400–800 BC	Tibia	Bd = 63.0	Dd = 46.2				
Cubalel	AD 400–600	Metatarsus	Bp = 47.9	Dp = 47.3				
Cubalel	AD 400–600	"			Bd = 50.9		Dm = 22.2	
Kolima Sud	1400–800 BC	Astragalus	GLl = 53.1	GLm = 48		Dl = 29.3	Dm = 28.2	
Kolima Sud	1400–800 BC	"	GLl = 57.4	GLm = 50.8		Dl = 31.9	Dm = 31.0	Bd = 36.0
Kolima Sud	1400–800 BC	"	GLl = 59.1	GLm = 53.7		Dl = 32.2	Dm = 33.2	Bd = 36.8
Akumbu	AD 600–1000	"	GLl = 68.7	GLm = 61.8		Dl = 36.7	Dm = 32.4	Bd = 43.4
Cubalel	AD 400–600	"	GLl = 68.8	GLm = 61.3		Dl = 37.7	Dm = 37.4	Bd = 43.5
Cubalel	AD 400–600	"	GLl = 57.8	GLm = 53.2		Dl = 32.1	Dm = 30.9	Bd = 35.2

Table 8.2 (cont'd)

Bos sp.

Site	Phase	Element	Measurements				
Gao Sany	AD 900–1200	Astragalus	GLl = 71.5	GLm = 64.4	Dl = 37.9	Dm = 39.6	Bd = 42.6
Gao Sany	AD 900–1200	"	GLl = 66.0	GLm = 63.5	Dl = 38.0		Bd = 41.0
Kolima Sud	1400–800 BC	First Phalanx	GLpe = 54.7	Bp = 29.6		SD = 22.9	Bd = 27.1
Windé Koroji Ouest	1700–950 BC	"	GLpe = 65.8	Bp = 36.9			Bd = 32.8
Akumbu	AD 600–1000	"	GLpe = 68.7	Bp = 29.1			
Akumbu	AD 1200–1400	"	GLpe = 62.0			SD = 25.3	
Siouré	AD 250–400	"	GLpe = 63.0	Bp = 30.0	Dp = 32.0		Bd = 29.2
Siouré	AD 250–400	"	GLpe = 57.1	Bp = 27.4			Bd = 28.1
Siouré	AD 250–400	"	GLpe = 54.6	Bp = 25.7	Dp = 27.7		Bd = 25.0
Siouré	AD 250–400	"	GLpe = 59.9	Bp = 31.0	Dp = 31.7		Bd = 27.7
Cubalel	AD 400–600	"	GLpe = 55.0	Bp = 28.2	Dp = 28.8		Bd = 24.4
Cubalel	AD 400–600	"	GLpe = 59.1	Bp = 27.1	Dp = 28.4		Bd = 25.3
Cubalel	AD 400–600	"	GLpe = 58.9				Bd = 26.0
Cubalel	AD 400–600	"	GLpe = 57.4	Bp = 27.1	Dp = 30.5		
Cubalel	AD 400–600	"	GLpe = 56.3	Bp = 27.5			
Siouré	AD 400–600	"	GLpe = 63.5	Bp = 27.3			Bd = 25.9
Siouré	AD 400–600	"	GLpe = 59.3	Bp = 26.7			Bd = 25.1
Siouré	AD 400–600	"	GLpe = 59.5	Bp = 26.6			Bd = 26.4
Siouré	AD 400–600	"	GLpe = 57.6	Bp = 25.3	Dp = 25.6		Bd = 24.2
Siouré	AD 400–600	"	GLpe = 59.2				Bd = 25.0
Cubalel	AD 600–950	"	GLpe = 64.5	Bp = 27.6	Dp = 31.8		Bd = 26.0
Cubalel	AD 600–950	"	GLpe = 53.8	Bp = 26.1	Dp = 27.0		Bd = 23.0
Cubalel	AD 600–950	"	GLpe = 60.7	Bp = 26.5			Bd = 27.1
Gao Sany	AD 900–1200	"	GLpe = 67.4	Bp = 30.8		SD = 25.3	Bd = 30.3

Gao Sany	AD 900–1200	"	GLpe = 68.0	Bp = 33.0		Bd = 31.8
Gao Sany	AD 900–1200	"	GLpe = 66.4	Bp = 29.9	SD = 22.4	Bd = 26.0
Kolima Sud	1400–800 BC	Second Phalanx	GLpe = 34.0			Bd = 25.0
Akumbu	AD 600–1000	"	GLpe = 37.5			Bd = 23.7
Dia	AD 1400–1600	"	GLpe = 35.4			
Siouré	AD 250–400	"	GLpe = 39.9	Bp = 27.2		Bd = 23.4
Siouré	AD 250–400	"	GLpe = 41.1	Bp = 25.4	Dp = 26.1	Bd = 23.0
Siouré	AD 250–400	"	GLpe = 36.9	Bp = 28.5	Dp = 28.3	Bd = 23.8
Cubalel	AD 400–600	"	GLpe = 39.9	Bp = 29.1		Bd = 23.2
Cubalel	AD 400–600	"		Bp = 28.6		Bd = 23.3
Cubalel	AD 400–600	"		Bp = 30.0		Bd = 25.7
Cubalel	AD 400–600	"	GLpe = 39.6	Bp = 26.1	Dp = 27.5	Bd = 22.2
Cubalel	AD 400–600	"	GLpe = 36.9	Bp = 25.1	Dp = 26.2	Bd = 21.2
Cubalel	AD 400–600	"	GLpe = 34.0	Bp = 27.1	Dp = 26.3	Bd = 23.8
Siouré	AD 400–600	"	GLpe = 38.4	Bp = 26.7	Dp = 29.4	
Siouré	AD 400–600	"	GLpe = 39.2	Bp = 27.3		Bd = 23.4
Cubalel	AD 600–950	"	GLpe = 36.0	Bp = 25.3	Dp = 25.5	
Cubalel	AD 600–950	"		Bp = 26.0		Bd = 22.2
Cubalel	AD 600–950	"	GLpe = 37.6	Bp = 29.7		Bd = 25.9
Cubalel	AD 600–950	"	GLpe = 38.5	Bp = 28.3		Bd = 22.6
Cubalel	AD 600–950	"	GLpe = 40.1	Bp = 26.7	Dp = 28.6	
Gao Sany	AD 900–1200	"	GLpe = 45.8	Bp = 30.7		Bd = 29.1
Gao Sany	AD 900–1200	"	GLpe = 44.4	Bp = 30.8		
Gao Sany	AD 900–1200	"	GLpe = 41.4	Bp = 29.3		Bd = 26.7
Kolima Sud	1400–800 BC	Third Phalanx	DLS = 84.4	Ld = 61.1	MBS = 26.7 Bp = 24.6	
Kolima Sud	1400–800 BC	"	DLS = 63.4		MBS = 22.7 Bp = 23.9	
Dia	AD 1400–1600	"	DLS = 74.4	Ld = 51.3		

measurements after von den Driesch (1976).

Table 8.3 Sheep and goat measurements.

Ovis / Capra

Site	Phase	Element	Measurements	
Cubalel	AD 250–400	Mandibular Third Molar	L = 23.1	B = 8.6
Siouré	AD 250–400	"	L = 21.0	B = 7.2
Cubalel	AD 400–600	"	L = 24.7	B = 8.6
Cubalel	AD 400–600	"	L = 21.3	B = 7.4
Cubalel	AD 400–600	"	L = 23.4	B = 9.0
Cubalel	AD 400–600	"	L = 21.7	B = 7.8
Cubalel	AD 400–600	"	L = 22.5	B = 8.1
Cubalel	AD 400–600	"	L = 21.5	B = 7.5
Cubalel	AD 600–950	"	L = 24.5	B = 10.7
Cubalel	AD 600–950	"	L = 22.1	B = 8.7
Cubalel	AD 600–950	"	L = 22.8	B = 8.6
Akumbu	AD 600–1000	"	L = 24.5	B = 9.1
Akumbu	AD 600–1000	"	L = 24.4	B = 9.9
Akumbu	AD 600–1000	"	L = 24.2	B = 9.4
Kolima Sud	1400–800 BC	Humerus	BT = 25.9	
Akumbu	AD 600–1000	"	Bd = 28.8	BT = 28.7
Akumbu	AD 600–1000	Radius	Bp = 29.0	BFp = 28.2
Dia (D6)	AD 1400–1600	"	Bp = 31.1	
Cubalel	AD 400–600	Metacarpal	Bp = 21.6	Dp = 14.7
Siouré	AD 400–600	"	Bp = 21.0	Dp = 14.0
Cubalel	AD 600–950	"	Bp = 22.8	Dp = 16.4
Cubalel	AD 600–950	"	Bp = 24.5	Dp = 17.4
Cubalel	AD 600–950	"	Bp = 22.9	Dp = 16.7
Akumbu	AD 600–1000	"	Bp = 23.6	Dp = 17.1
Akumbu	AD 600–1000	"	Bp = 23.0	Dp = 17.5
Siouré	AD 950–1400	"	Bp = 21.5	Dp = 15.5

Site	Date	Element	Measurements
Akumbu	AD 1200–1400	"	Bp = 26.7; Dp = 17.1
Cubalel	AD 400–600	Femur	DC = 19.5
Cubalel	AD 400–600	"	DC = 19.4
Cubalel	AD 400–600	"	DC = 19.5
Cubalel	AD 600–950	"	DC = 19.4
Siouré	AD 250–400	Metatarsal	Bp = 21.9; Dp = 19.3
Cubalel	AD 400–600	"	Bp = 19.2
Cubalel	AD 400–600	"	Bp = 19.8; Dp = 20.5
Cubalel	AD 400–600	"	Bp = 18.9
Cubalel	AD 400–600	"	Bp = 21.3; Dp = 19.4
Cubalel	AD 600–950	"	Bp = 19.2; Dp = 18.0
Cubalel	AD 600–950	"	Bp = 18.4; Dp = 16.4
Siouré	AD 950–1400	"	Bp = 22.3; Dp = 21.1
Akumbu	AD 1000–1200	"	Bp = 20.5; Dp = 19.1
Dia (D6)	AD 1400–1600	"	Bp = 20.4
Akumbu	AD 400–600	Astragalus	GLl = 27.3; GLm = 26.6; Dl = 15.4; Bd = 19.1
Cubalel	AD 400–600	"	GLl = 29.5; GLm = 26.3; Dl = 16.7; Bd = 16.6
Cubalel	AD 600–950	"	GLm = 26.9; Dm = 14.0; Bd = 17.2
Gao Sany	AD 900–1200	"	GLl = 34.0; GLm = 31.4; Dl = 18.6; Dm = 19.7; Bd = 21.2
Gao Sany	AD 900–1200	"	GLl = 34.4; GLm = 33.0; Dl = 18.7; Dm = 20.0; Bd = 22.2
Gao Sany	AD 900–1200	"	GLl = 34.7; Bd = 22.5
Cubalel	AD 400–600	First Phalanx	Bp = 12.9; Dp = 13.7
Cubalel	AD 400–600	"	Bp = 12.1; Dp = 14.3
Cubalel	AD 600–950	"	Bp = 12.7; Dp = 13.8
Cubalel	AD 600–950	"	Bp = 10.5; Dp = 12.6
Akumbu	AD 600–1000	"	GLpe = 44.5; Bp = 14.5; Bd = 11.5
Gao Sany	AD 900–1200	"	GLpe = 39.7; Bd = 12.8
Gao Sany	AD 900–1200	"	GLpe = 44.0; Bp = 13.9
Siouré	AD 950–1400	"	GLpe = 36.9; Bp = 12.1

Table 8.3. (cont'd)

Ovis / Capra

Site	Phase	Element	Measurements			
Siouré	AD 250–400	Second Phalanx	GLpe = 27.7	Bp = 12.8	Dp = 13.9	Bd = 8.4
Siouré	AD 250–400	"	GLpe = 24.5	Bp = 11.7	Dp = 12.1	Bd = 11.1
Siouré	AD 250–400	"	GLpe = 25.9	Bp = 14.2	Dp = 14.5	Bd = 10.2
Siouré	AD 250–400	"	GLpe = 22.4	Bp = 13.2	Dp = 13.1	Bd = 8.2
Cubalel	AD 600–950	"	GLpe = 22.1	Bp = 10.3	Dp = 11.1	Bd = 8.7
Cubalel	AD 600–950	"	GLpe = 21.2	Bp = 10.6	Dp = 11.4	
Cubalel	AD 600–950	"		Bp = 10.0	Dp = 10.9	
Cubalel	AD 600–950	"	GLpe = 21.5	Bp = 12.9		Bd = 10.0
Cubalel	AD 600–950	"	GLpe = 23.9	Bp = 11.3	Dp = 12.8	Bd = 10.3
Cubalel	AD 600–950	"		Bp = 11.4		Bd = 9.1
Cubalel	AD 600–950	"	GLpe = 20.0	Bp = 11.6		Bd = 9.3
Cubalel	AD 600–950	"	GLpe = 21.0	Bp = 10.2	Dp = 10.9	Bd = 8.2
Cubalel	AD 600–950	"	GLpe = 21.7	Bp = 12.0	Dp = 12.4	Bd = 9.4
Cubalel	AD 600–950	"	GLpe = 21.0	Bp = 10.8	Dp = 11.7	
Cubalel	AD 600–950	"	GLpe = 20.5	Bp = 10.4	Dp = 11.5	
Cubalel	AD 600–950	"		Bp = 12.0		Bd = 8.4
Gao Sany	AD 900–1200	"	GLpe = 26.6	Bp = 12.8	Bd = 10.2	
Siouré	AD 950–1400	"	GLpe = 24.6	Bp = 14.1	Dp = 14.7	Bd = 10.9
Akumbu	AD 1000–1200	"	GLpe = 21.1	Bp = 10.3		Bd = 7.3
Akumbu	AD 1200–1400	"	GLpe = 27.2			Bd = 8.7
Akumbu	AD 1200–1400	"	GLpe = 25.3			
Akumbu	AD 1200–1400	"		Bp = 12.0		

measurements after von den Driesch (1976).

Table 8.4 Sheep measurements.

Ovis aries

Site	Phase	Element					
Gao Sany	AD 900–1200	Humerus	Bd = 34.5	BT = 33.3			
Gao Sany	AD 900–1200	"	Bd = 34.9	BT = 32.2			
Cubalel	AD 250–400	Metacarpal	Bp = 21.9	Dp = 16.1			
Cubalel	AD 250–400	"	Bp = 21.6	Dp = 15.6			
Cubalel	AD 400–600	"	Bp = 23.8				
Cubalel	AD 600–950	"	Bp = 23.2	Dp = 16.8			
Cubalel	AD 600–950	"	Bp = 21.6	Dp = 16.8			
Cubalel	AD 600–950	"			Bd = 24.5	Dd = 18.2	DC = 15.7
Gao Sany	AD 900–1200	"			Bd = 27.7	Dd = 18.5	
Siouré	AD 950–1400	"			Bd = 25.3	Dd = 10.6	DC = 16.6
Siouré	AD 950–1400	"			Bd = 24.7	Dd = 8.8	DC = 16.5
Cubalel	AD 0–250	Tibia	Bd = 26.0	Dd = 21.9			
Siouré	AD 250–400	"	Bd = 25.6	Dd = 20.0			
Cubalel	AD 400–600	"	Bd = 26.8	Dd = 20.9			
Cubalel	AD 400–600	"	Bd = 30.3	Dd = 23.4			
Cubalel	AD 600–950	"	Bd = 27.3				
Cubalel	AD 600–950	"	Bd = 27.0	Dd = 22.6			
Cubalel	AD 600–950	"	Bd = 26.8	Dd = 20.4			
Cubalel	AD 600–950	"	Bd = 27.3	Dd = 20.5			
Siouré	AD 950–1400	"	Bd = 26.3	Dd = 19.5			
Siouré	AD 950–1400	"	Bd = 24.3	Dd = 19.6			
Akumbu	AD 1000–1200	"	Bd = 24.1	Dd = 19.0			
Akumbu	AD 1000–1200	"	Bd = 26.1	Dd = 21.0			
Akumbu	AD 1200–1400	"					

Table 8.4 (cont'd)

Ovis aries

Site	Phase	Element					
Siouré	AD 950–1400	Metatarsus		GLm = 25.6	Bd = 23.0	Dd = 8.6	DC = 16.0
Siouré	AD 250–400	Astragalus	GL1 = 27.9		Dl = 15.2		Bd = 16.0
Siouré	AD 250–400	"	GL1 = 30.3		Dl = 16.7		Bd = 19.2
Cubalel	AD 400–600	"	GL1 = 30.6		Dl = 16.4		Bd = 20.1
Cubalel	AD 400–600	"		GLm = 29.7		Dm = 15.5	Bd = 16.7
Cubalel	AD 600–950	"		GLm = 26.0		Dm = 16.5	
Cubalel	AD 600–950	"	GL1 = 28.2	GLm = 26.8			Bd = 18.7
Cubalel	AD 600–950	"		GLm = 25.3			
Cubalel	AD 600–950	"		GLm = 26.2		Dm = 16.5	
Cubalel	AD 600–950	"	GL1 = 31.0	GLm = 29.6	Dl = 16.8	Dm = 17.5	Bd = 19.4
Cubalel	AD 600–950	"	GL1 = 28.2	GLm = 26.2		Dm = 15.8	Bd = 17.3
Cubalel	AD 600–950	"	GL1 = 32.1	GLm = 29.7	Dl = 17.4	Dm = 18.3	Bd = 20.1
Cubalel	AD 600–950	"	GL1 = 30.5	GLm = 28.4	Dl = 16.2	Dm = 16.9	Bd = 20.2
Gao Sany	AD 900–1200	"	GL1 = 34.0		Dl = 18.5		Bd = 20.7
Gao Sany	AD 900–1200	"		GLm = 31.1			Bd = 20.1
Siouré	AD 950–1400	"	GL1 = 30.2	GLm = 28.5	Dl = 16.2	Dm = 18.8	Bd = 19.0
Akumbu	AD 1000–1200	"		GLm = 26.8	Dl = 16.0	Dm = 15.2	
Cubalel	AD 250–400	Calcaneus	GL = 57.0	GB = 20.7			
Cubalel	AD 250–400	"	GL = 52.4	GB = 16.4			
Cubalel	AD 600–950	"	GL = 53.0	GB = 18.9			

Site	Date	Element				
Gao Sany	AD 900–1200	"	GL = 64.7	GB = 20.9		Bd = 11.4
Gao Sany	AD 900–1200	"	GL = 66.2	GB = 22.6		Bd = 12.9
Siouré	AD 950–1400	"	GL = 60.4	GB = 19.4		Bd = 13.2
Cubalel	AD 0–250	First Phalanx	GLpe = 36.6	Bp = 12.1		Bd = 11.6
Siouré	AD 250–400	"	GLpe = 34.4	Bp = 15.2	Dp = 14.2	Bd = 10.9
Siouré	AD 400–600	"	GLpe = 38.6	Bp = 12.3	Dp = 14.3	Bd = 11.3
Cubalel	AD 600–950	"	GLpe = 35.5	Bp = 12.3	Dp = 13.6	Bd = 12.7
Cubalel	AD 600–950	"	GLpe = 35.3	Bp = 11.8	Dp = 12.9	Bd = 11.3
Cubalel	AD 600–950	"	GLpe = 39.4	Bp = 13.0	Dp = 15.5	Bd = 12.3
Cubalel	AD 600–950	"	GLpe = 39.5	Bp = 12.6	Dp = 14.7	Bd = 10.8
Cubalel	AD 600–950	"	GLpe = 38.2	Bp = 11.8	Dp = 15.0	Bd = 9.3
Cubalel	AD 600–950	"	GLpe = 39.3	Bp = 13.3	Dp = 16.0	Bd = 12.4
Cubalel	AD 600–950	"	GLpe = 35.5	Bp = 12.0	Dp = 14.0	
Akumbu	AD 600–1000	"	GLpe = 36.2	Bp = 11.4		Bd = 12.9
Gao Sany	AD 900–1200	"	GLpe = 41.6	Bp = 12.5		Bd = 12.0
Gao Sany	AD 900–1200	"	GLpe = 40.3	Bp = 11.9		Bd = 12.6
Gao Sany	AD 900–1200	"	GLpe = 46.8			Bd = 11.8
Gao Sany	AD 900–1200	"	GLpe = 46.2	Bp = 13.3		
Gao Sany	AD 900–1200	"	GLpe = 46.6	Bp = 14.0		
Gao Sany	AD 900–1200	"	GLpe = 44.9			
Siouré	AD 950–1400	"	GLpe = 41.0	Bp = 13.3		Bd = 13.5

measurements after von den Driesch (1976).

Table 8.5 Goat measurements.

Capra hircus

Site	Phase	Element			
Gao Sany	AD 900–1200	Humerus	Bd = 31.3	BT = 30.5	
Gao Sany	AD 900–1200	"	Bd = 32.7	BT = 29.0	
Gao Sany	AD 900–1200	"	Bd = 34.9	BT = 32.2	
Cubalel	AD 400–600	Metacarpal	Bd = 23.6	Dd = 10.6	
Cubalel	AD 400–600	"	Bd = 25.5	Dd = 10.4	DC = 15.6
Cubalel	AD 400–600	"	Bp = 22.4	Dp = 15.8	
Cubalel	AD 400–600	"	Bp = 21.6	Dp = 15.6	
Cubalel	AD 600–950	"	Bp = 22.0	Dp = 15.9	
Cubalel	AD 600–950	"	Bp = 19.9	Dp = 14.4	
Cubalel	AD 600–950	"	Bd = 21.5		DC = 15.2
Cubalel	AD 600–950	"	Bd = 22.2		DC = 14.0
Gao Sany	AD 900–1200	"	Bd = 27.2	Dd = 16.7	
Gao Sany	AD 900–1200	"	Bd = 24.4	Dd = 16.5	
Cubalel	AD 0–250	Tibia	Bd = 26.0	Dd = 19.5	
Siouré	AD 250–400	"	Bd = 26.3	Dd = 19.7	
Cubalel	AD 400–600	"	Bd = 27.1	Dd = 19.9	
Cubalel	AD 400–600	"	Bd = 28.6	Dd = 20.6	
Cubalel	AD 400–600	"	Bd = 29.2	Dd = 22.9	
Cubalel	AD 600–950	"	Bd = 23.3	Dd = 19.7	
Cubalel	AD 600–950	"	Bd = 27.3	Dd = 20.0	
Siouré	AD 950–1400	"	Bd = 23.7	Dd = 18.9	
Siouré	AD 950–1400	"	Bd = 23.8	Dd = 18.4	

Site	Date	Element					
Cubalel	AD 400–600	Metatarsal		Bp = 18.8	Bd = 24.5	Dd = 8.9	DC = 16.5
Cubalel	AD 400–600	"			Bd = 24.7	Dd = 10.0	DC = 17.4
Cubalel	AD 400–600	"					
Windé Koroji Ouest	1700–950 BC	Astragalus	GLl = 21.4	GLm = 23.3			Bd = 16.9
Siouré	AD 0–250	"	GLl = 27.9	GLm = 25.6	Dl = 14.5	Dm = 14.8	Bd = 17.9
Cubalel	AD 250–400	"	GLl = 27.0	GLm = 25.7	Dl = 15.6	Dm = 16.9	Bd = 19.3
Siouré	AD 250–400	"		GLm = 27.5			Bd = 18.1
Siouré	AD 400–600	"	GLl = 27.5	GLm = 24.6			Bd = 17.4
Cubalel	AD 600–950	"	GLl = 26.6	GLm = 25.1	Dl = 14.2	Dm = 14.7	Bd = 16.4
Cubalel	AD 600–950	"	GLl = 27.0	GLm = 26.4	Dl = 13.8	Dm = 15.1	
Cubalel	AD 600–950	"	GLl = 26.6	GLm = 25.0	Dl = 14.5	Dm = 16.0	Bd = 16.7
Cubalel	AD 600–950	"		GLm = 25.0			Bd = 16.8
Akumbu	AD 600–1000	"	GLl = 26.8	GLm = 25.0	Dl = 15.2	Dm = 14.9	Bd = 18.5
Siouré	AD 950–1400	"	GLl = 30.1	GLm = 27.5	Dl = 15.3	Dm = 16.7	Bd = 17.2
Siouré	AD 950–1400	"	GLl = 28.9	GLm = 26.9	Dl = 15.5		
Gao Sany	AD 900–1200	"		GLm = 30.2		Dm = 19.3	Bd = 20.8
Cubalel	AD 400–600	Calcaneus	GL = 55.0	GB = 18.3			
Cubalel	AD 600–950	"					
Siouré	AD 250–400	First Phalanx		Bp = 12.1			
Cubalel	AD 400–600	"	GLpe = 40.7	Bp = 12.6	Dp = 15.5	Bd = 13.2	
Cubalel	AD 400–600	"	GLpe = 42.3	Bp = 13.4	Dp = 16.5	Bd = 12.1	
Cubalel	AD 400–600	"	GLpe = 39.3	Bp = 13.0	Dp = 13.9	Bd = 12.3	
Cubalel	AD 600–950	"	GLpe = 35.4	Bp = 12.6	Dp = 13.8	Bd = 12.4	
Cubalel	AD 600–950	"	GLpe = 36.5	Bp = 11.8			
Cubalel	AD 600–950	"	GLpe = 35.9	Bp = 12.3	Dp = 14.5	Bd = 10.8	
Cubalel	AD 600–950	"	GLpe = 34.5	Bp = 11.0	Dp = 14.0	Bd = 9.9	

Table 8.5 (cont'd)

Capra hircus

Site	Phase	Element				
Akumbu	AD 600–1000	First Phalanx				Bd = 11.0
Gao Sany	AD 900–1200	"	GLpe = 40.8	Bp = 12.9		Bd = 12.7
Gao Sany	AD 900–1200	"	GLpe = 42.6	Bp = 12.9		Bd = 12.0
Gao Sany	AD 900–1200	"	GLpe = 40.4	Bp = 12.1		Bd = 12.0
Gao Sany	AD 900–1200	"	GLpe = 39.8	Bp = 12.5		Bd = 13.2
Siouré	AD 950–1400	"	GLpe = 40.8	Bp = 13.2	Dp = 15.6	Bd = 12.5
Siouré	AD 950–1400	"	GLpe = 38.4	Bp = 13.0	Dp = 14.6	Bd = 10.9
Siouré	AD 950–1400	"	GLpe = 38.5	Bp = 12.2	Dp = 15.1	Bd = 12.3
Siouré	AD 950–1400	"	GLpe = 37.1	Bp = 12.7	Dp = 15.0	Bd = 12.1
Siouré	AD 950–1400	"	GLpe = 41.7	Bp = 13.2	Dp = 15.6	
Kolima Sud	1400–800 BC	Third Phalanx	DLS = 23.7	LD = 19.2	H = 12.5	BPf = 7.5
Gao Sany	AD 900–1200	"	DLS = 32.2			
Gao Sany	AD 900–1200	"	DLS = 31.7			

measurements after von den Driesch (1976).

Table 8.6 Horse/donkey and dog measurements.

Equus

Site	Phase	Element					
Akumbu	AK1-64	Scapula	BG = 40.6	GLP = 78.0	LG = 43.0		
Cubalel	AD 250–400	Metacarpal	GL = 185.4	GL1 = 183.6	L1 = 181	Bp = 38.4	Dp-Tp = 27.3
			Bd = 36.3	Dd-Td = 27.1			
Akumbu	AK1-16	Pelvis	LA = 60.0				
Akumbu	AK1-6	Tibia	Dd = 31.0				
Akumbu	AK1-15	Astragalus	GH = 51.4	LmT = 53.5			
Akumbu	AK1-63	First Phalanx	BFp = 48.2	Bp = 52.1	Dp = 33.7		
Akumbu	Ak1-37	Second Phalanx	GL = 44.0				

Canis

Site	Phase	Element					
Cubalel	AD 600–950	Maxilla	#15 = 65.1	#16 = 18.1	#17 = 49.6		
Cubalel	AD 600–950	Maxillary Fourth Premolar	L = 18.3	GB = 8.9	B = 6.6		
Cubalel	AD 600–950	Maxillary First Molar	L = 14.2	B = 14.6			
Siouré	AD 250–400	Maxillary Second Molar	L = 7.3	B = 9.4			
Cubalel	AD 0–250	Mandible			#12 = 29.0		#20 = 20.7
Cubalel	AD 400–600	"				#19 = 20.4	
Cubalel	AD 400–600	"				#19 = 21.0	
Cubalel	AD 400–600	"		#10 = 34.1		#19 = 25.5	#20 = 22.3
Cubalel	AD 400–600	"	#8 = 72.2	#10 = 34.0	#12 = 33.1	#19 = 24.3	#20 = 20.1
Cubalel	AD 400–600	"		#10 = 35.2		#19 = 26.0	
Cubalel	AD 400–600	"	#8 = 66.4	#10 = 31.4	#12 = 29.4	#19 = 23.1	#20 = 18.8
Siouré	AD 400–600	"	#8 = 73.0	#10 = 35.4	#12 = 35.2	#19 = 22.3	#20 = 17.7

Table 8.6 (cont'd)

Canis Site	Phase	Element						
Siouré	AD 400–600	Mandible	#8 = 68.4	#10 = 33.2	#12 = 31.8	#19 = 19.9	#20 = 17.8
Siouré	AD 400–600	"					#20 = 18.0
Cubalel	AD 600–950	"	#8 = 70.8	#10 = 33.5	#12 = 32.7	#19 = 21.7	#20 = 16.9
Cubalel	AD 0–250	Mandibular First Molar		B = 7.9			
Cubalel	AD 400–600	"	L = 20.0	B = 8.0	#14 = 19.7		
Cubalel	AD 400–600	"	L = 20.5	B = 8.6	#14 = 19.9		
Cubalel	AD 400–600	"	L = 19.8	B = 7.3	#14 = 17.9		
Cubalel	AD 400–600	"		B = 8.2			
Akumbu	AK1–10	"	L = 19.7	B = 7.7			
Cubalel	AD 400–600	Mandibular Second Molar	L = 9.0	B = 7.1			
Cubalel	AD 400–600	"	L = 8.9	B = 6.6			
Cubalel	AD 400–600	"	L = 8.4	B = 6.4			
Cubalel	AD 400–600	"	L = 8.5	B = 6.1			
Akumbu	AK1–67	Pelvis	LA = 18.5				
Akumbu	AK3–29	Tibia	Bp = 30.2				
Akumbu	AK1–55	Astragalus	GL = 23.5				
Akumbu	AK1–2	Calcaneus	GL = 36.2	GB = 13.3			
Gao Sany	AD 900–1200	"	GL = 38.8	GB = 12.2			

measurements after von den Driesch (1976).

References

Allsworth-Jones, P. 1982. Kariya Wuro faunal report. *Zaria Archaeological Papers* **4**, 6–9.

Amanor, K.S. 1995. Dynamics of herd structures and herding strategies in West Africa: a study of market integration and ecological adaptation. *Africa* **65**, 351–94.

Bedaux, R.M.A., T.S. Constandse-Westermann, L. Hacquebord, A.G. Lange, J.D. van der Waals 1978. Recherche archéologiques dans le Delta Interieur du Niger (Mali). *Palaeohistoria* **20**, 91–220.

Binford, L. 1981. *Bones: Ancient men and modern myths*. New York: Academic Press.

Blench, R.M. 1993. Ethnographic and linguistic evidence for the prehistory of African ruminant livestock. In *The archaeology of africa: foods, metals and towns*, T. Shaw, P. Sinclair, B. Andah, A. Okpoko (eds), 71–103. London: Routledge.

Blench, R.M. 1994. The expansion and adaptation of Fulbe pastoralism to subhumid and humid conditions in Nigeria. *Cahiers d'études Africaines* **133–5**, 197–212.

Blench, R.M. 1995. A history of domestic animals in northeastern Nigeria. *Cahiers de Sciences Humaine, ORSTOM*, **31**(1), 181–238.

Blench, R.M. & K.C. MacDonald, in press. Domestic fowl. In *Cambridge Encyclopaedia of Nutrition*, Vol. I. Cambridge: Cambridge University Press.

Boessneck, J. 1969. Osteological differences between sheep (*Ovis aries*, Linné) and goat (*Capra hircus*, Linné). In *Science and archaeology*, 2nd edn, D.R. Brothwell & E. Higgs (eds), 331–58. London: Thames & Hudson.

Bouchud, J. 1983. Paléofaune de Tegdaoust. In *Tegdaoust III Recherches sur Aoudaghost: Campagnes 1960–1965*, 355–63. Paris: Éditions Recherche sur les Civilisations.

Breunig, P., K. Neumann, W. Van Neer 1996. New research on the Holocene settlement and environment of the Chad Basin in Nigeria. *African Archaeological Review* **13**, 111–45.

Briggs, L.C. 1960. *Tribes of the Sahara*. Cambridge, Mass.: Harvard University Press.

Bulliet, R.W. 1990. *The camel and the wheel*, 2nd edn. New York: Columbia University Press.

Carter, P.L. & J.D. Clark, 1976. Adrar Bous and African cattle. In *Proceedings of the PanAfrican Congress of Prehistory, Addis Ababa 1971*, A. Berhanou, J. Chevaillon, J.E.G. Sutton (eds), 487–93. Addis Ababa: Provisional Military Government of Socialist Ethiopia, Ministry of Culture.

Carter, P.L. & C. Flight 1972. A report on the fauna from the sites of Ntereso and Kintampo rock shelter six in Ghana, with evidence for the practice of animal husbandry in the second millennium BC. *Man* **7**, 277–82.

Clark, J.D., M.A.J. Williams, A.B. Smith 1973. The geomorphology and archaeology of Adrar Bous, central Sahara: a Preliminary Report. *Quaternaria* **17**, 245–67.

Connah, G. 1981. *Three thousand years in Africa*. Cambridge: Cambridge University Press.

Cribb, R. 1985. The analysis of ancient herding systems: an application of computer simulation in faunal studies. In *Beyond domestication in prehistoric Europe: investigations in subsistence archaeology and social complexity*, G. Barker & C. Gamble (eds), 75–106. New York: Academic Press.

Dahl, G. & A. Hjort 1976. *Having herds: pastoral herd growth and household economy*. Stockholm Studies in Social Anthropology, 2. Stockholm: University of Stockholm.

Doutresoulle, G. 1952. *L'elevage au Soudan Français: son économie (deuxieme édition)*. Algiers: E. Imbert.

Eisenmann, V. 1986. Comparative osteology of modern and fossil horses, asses and half asses. In *Equids in the Ancient World*, R. Meadow & H.-P. Uerpman (eds), 67–116. Wiesbaden: Ludwig Reichert.

Epstein, H. 1955. Phylogenetic significance of *spina bifida* in Zebu cattle. *Indian Journal of Veterinary Science* **25**, 313–6.

Epstein, H. 1971. *The origin of the domestic animals of Africa*. New York: Africana Publishing.

Gallais, J. 1967. *Le delta intérieur du Niger et ses bordures: étude de géographie régionale*. [2 volumes]. Dakar: IFAN.

Gallais, J. 1984. *Hommes du Sahel*. Paris: Flammarion.

Garrard, T. 1982. Myth and metrology. The early Transaharan gold trade. *Journal of African History* **23**, 443–61.

Gautier, A. 1987. Prehistoric men and cattle in North Africa: a dearth of data and a surfeit of models. In *Prehistory of arid North Africa*, A. Close (ed.), 163–87. Dallas: SMU Press.

Gautier, A. 1990. *La Domestication: et l'homme créa l'animal.* Paris: Editions Errance.
Gautier, A. 1991. Mammifères fossiles Holocénes du Sahara Malien. In *Paléoenvironments du Sahara: Lacs holocénes à Taoudenni (Mali)*, N. Petit-Maire (ed.), 173–6. Paris: Editions du CNRS.
Grant, A. 1982. The use of tooth-wear as a guide to the age of domestic animals. In *Ageing and sexing animal bones from archaeological sites*, B. Wilson, C. Grigson, S. Payne (eds), 91–108. Oxford: BAR.
Gronenborn, D. 1996. Kundiye: archaeology and ethnoarchaeology in the Kala-Balge area of Borno State Nigeria. In *Aspects of African archaeology: Papers from the 10th Congress of the PanAfrican Association for Prehistory and Related Studies*, G. Pwiti & R. Soper (eds), 449–59. Harare: University of Zimbabwe Publications.
Grigson, C. 1982. Sex and age determination of some bones and teeth of domestic cattle: a review of the literature. In *Ageing and sexing animal bones from archaeological sites*, B. Wilson, C. Grigson, S. Payne (eds), 7–23. Oxford: BAR.
Grigson, C. 1991. An African origin for African cattle? – some archaeological evidence. *African Archaeological Review* **9**, 119–44.
Holl, A. 1986. *Economie et Société Néolithique du Dhar Tichitt (Mauritanie).* Paris: Editions Recherche sur les Civilisations, Mémoire No. 69.
Insoll, T. 1994. Preliminary results of excavations at Gao, September and October 1993. *Nyame Akuma* **41**, 45–8.
Law, R. 1980. *The horse in West African history.* Oxford: Oxford University Press.
Lewicki, T. 1974. *West African food in the Middle Ages.* Cambridge: Cambridge University Press.
Lhote, H. 1982. *Les Chars Rupestres Sahariens: des Syrtes au Niger, par le pays des Garamantes et des Atlantes.* Toulouse: Editions des Hespérides.
MacDonald, K.C. 1989. *The identification and analysis of animal bones from West African archaeological sites.* BA Honors thesis, Department of Anthropology, Rice University.
MacDonald, K.C. 1992. The domestic chicken (*Gallus gallus*) in Sub-Saharan Africa: a background to its introduction and its osteological differentiation from indigenous fowls (Numidinae and Francolinus sp.). *Journal of Archaeological Science* **19**, 303–18.
MacDonald, K.C. 1993. *An initial report on the faunal remains of Akumbu (Mali).* Appendix to the PhD thesis of Téréba Togola, Department of Anthropology, Rice University.
MacDonald, K.C. 1994. *Socio-economic diversity and the origins of cultural complexity along the Middle Niger (2000 BC to AD 300).* PhD thesis, Department of Archaeology, University of Cambridge.
MacDonald, K.C. 1995a. The faunal remains (mammals, birds, and reptiles). In *Excavations at Jenné-Jeno, Hambarketolo, and Kaniana (Inland Niger Delta, Mali), the 1981 season*, S.K. McIntosh (ed.), 291–318. Berkeley: University of California Press.
MacDonald, K.C. 1995b. Why chickens? The centrality of the domestic fowl in West African ritual and magic. In *The symbolic role of animals in archaeology*, K. Ryan & P.J. Crabtree (eds), 50–56. Philadelphia: MASCA (MASCA Research Papers in Science and Archaeology vol. 12).
MacDonald, K.C. 1996a. Tichitt-Walata and the Middle Niger: evidence for cultural contact in the second millenium BC. In *Aspects of African archaeology: Papers from the 10th Congress of the PanAfrican Association for Prehistory and Related Studies*, G. Pwiti & R. Soper (eds), 429–40. Harare: University of Zimbabwe Publications.
MacDonald, K.C. 1996b. The Windé Koroji complex: evidence for the peopling of the eastern Inland Niger Delta (2100–500 BC). *Préhistoire Anthropologie Méditerranéennes* **5**, 147–65.
MacDonald, K.C. & D.N. Edwards 1993. Chickens in Africa: the importance of Qasr Ibrim. *Antiquity* **67**, 584–90.
MacDonald, K.C. & R.H. MacDonald, n.d. *Report on the mammalian, avian and reptilian remains from the 1990 season at the sites of Cubalel and Siouré (Sénégal).* Unpubl. ms., Department of Anthropology, Rice University, Houston.
MacDonald, R.H. & K.C. MacDonald 1996. A preliminary report on the faunal remains recovered from Gao Ancien and Gao Saney (1993 season). In *Islam, Archaeology, and History: Gao Region (Mali) c. AD 900–1250*, T. Insoll, 124–6. Oxford: BAR.
MacDonald, R.H. & K.C. MacDonald, in prep. The domestic dog (*Canis familiaris*) in Sub-Saharan Africa: a background to its introduction and osteometric differentiation from African jackals and the hunting dog (*Lycaon pictus*).

MacDonald, K.C., T. Togola, R.H. MacDonald, C. Capezza 1994. Douentza, Mali. *Past* 17, 12–14.

MacDonald, K.C. & W. Van Neer 1994. Specialised fishing peoples in the later Holocene of the Méma region (Mali). In *Fish exploitation in the past*, W. Van Neer (ed.), 243–51. Tervuren: Musée Royal de l'Afrique centrale (Annales du Musée Royal de l'Afrique centrale, Sciences Zoologiques, No. 274).

McIntosh, S.K. (ed.) 1995. *Excavations at Jenné-Jeno, Hambarketolo, and Kaniana (Inland Niger Delta, Mali), The 1981 Season.* Berkeley: University of California Press.

McIntosh, S.K., R.J. McIntosh, H. Bocoum 1992. The Middle Senegal Valley Project: preliminary results from the 1990–91 field season. *Nyame Akuma* 38, 47–61.

Marshall, F. 1989. Rethinking the role of *Bos indicus* in Sub-Saharan Africa. *Current Anthropology* 30, 235–9.

Marshall, F. 1990. Cattle herds and caprine flocks. In *Early pastoralists of south-western Kenya. Memoir 11, the British Institute in Eastern Africa*, P. Robertshaw (ed.), 205–60. Nairobi: British Institute in eastern Africa.

Mauny, R. 1954. La cynophagie en afrique occidentale. *Notes Africaines* 64, 114 only.

Mongin, P. & M. Plouzeau 1984. Guinea-fowl. In *Evolution of domesticated animals*, I.L. Mason (ed.), 322–5. London: Longman.

Munson, P.J. 1971. *The Tichitt Tradition: a late prehistoric occupation of the southwestern Sahara.* PhD thesis, Department of Anthropology, University of Illinois at Urbana-Champaign.

Muzzolini, A. 1983. *L'Art rupestre du Sahara central, classification et chronologie; le bouef dans le préhistoire saharienne.* Unpublished PhD dissertation (3me cycle), Université d'Aix-en-Provence.

Paris, F. 1984. *La Région d'In Gall – Tegidda N Tesemt (Niger): Programme Archéologique d'urgence 1977–1981. III. Les Sépultures du Néolithique Final 'a l'Islam. Etudes Nigériennes No. 50.* Niamey: Institut de Recherches en Sciences Humaines.

Paris, F. 1992. Chin Tafidet, village néolithique. *Journal des Africanistes* 62, 33–54.

Payne, S. 1973. Kill-off patterns in sheep and goats: the mandibles from Asvan Kale. *Anatolian Studies* 23, 281–303.

Prummel, W. & H.-J. Frisch 1986. A guide for the distinction of species, sex and body side in bones of sheep and goat. *Journal of Archaeological Science* 13, 567–77.

Rowley-Conwy, P. 1988. The camel in the Nile valley: new radiocarbon accelerator (AMS) dates from Qasr Ibrim. *Journal of Egyptian Archaeology* 74, 245–8.

Shinnie, P.L. & F.J. Kense 1989. *Archaeology of Gonja, Ghana: excavations at Daboya.* Canada: University of Calgary Press.

Smith, A.B. 1975. A note on the flora and fauna from the post-Palaeolithic sites of Karkarichinkat Nord and Sud. *West African Journal of Archaeology* 5, 201–4.

Smith, A.B. 1979. Biogeographical considerations of colonization of the Lower Tilemsi Valley in the second millenium BC. *Journal of Arid Environments* 2, 355–61.

Smith, A.B. 1980. The neolithic tradition in the Sahara. In *The Sahara and the Nile*, M.A.J. Williams & H. Faure (eds), 451–65. Rotterdam: Balkema.

Smith, A.B. 1992. *Pastoralism in Africa: origins and development ecology.* London: Hurst.

Stahl, A. 1985. Reinvestigation of Kintampo 6 Rock shelter, Ghana: implications for the nature of culture change. *African Archaeological Review* 3, 117–50.

Sutton, J.E.G. 1985. The antiquity of horses and asses in West Africa: a correction. *Oxford Journal of Archaeology* 4, 117–18.

Thesing, R. 1977. *Die Großentwicklung des Haushuhns in vor-und Fruhgeschichtlicher Zeit.* Dissertation der Universitat Munchen.

Togola, T. 1993. *Archaeological investigations of Iron Age sites in the Méma region.* PhD thesis, Department of Anthropology, Rice University.

Togola, T. 1996. Iron Age occupation in the Méma region, Mali. *African Archaeological Review* 13, 91–110.

Van Neer, W. & H. Bocoum 1991. Etude archéozoologique de Tulel-Fobo, site protohistorique (IVe–Xe siècle) de la moyenne vallée du Fleuve Sénégal (République du Sénégal). *Archaeozoologia* 4, 93–114.

Vernet, R. 1992. *Les Sites Neolithiques de Khatt Lemaiteg (Amatlich) en Mauritanie Occidentale.* Paris: Copy Top Voltaire.

von den Driesch, A. 1976. *A guide to the measurement of animal bones from archaeological sites.* Cambridge, Mass.: Peabody Museum of Archaeology and Ethnology (Peabody Museum Bulletin No. 1).

von den Driesch, A. & J. Boessneck 1974. Kritische Anmerkungen zue Widerristhohenberechnung aus Langenmassen vor-und frühgeschichtlicher Tierknochen. *Saugetierkundliche Mitteilungen* 22, 325–48.

Wilson, R.T., A. Traore, H.G. Kuit, M. Slingerland 1987. Livestock production in central Mali: reproduction growth and mortality of domestic fowl under traditional management. *Tropical Animal Health and Production* 19, 229–36.

You, R. 1955. Cynophagie. *Notes Africaines* 66, 40 only.

CHAPTER NINE

Domestic animals from archaeological sites in Central and west–central Africa

Wim Van Neer

1. Introduction

Information on the presence and propagation of domestic animals in Central Africa is scanty. This is a combined result of the few excavations carried out thus far, and the poor chances for preservation of bone in this region. The acidity of the soils, particularly in forested regions, is responsible for the low survival chances of faunal remains. The paucity of sites with fauna is clearly illustrated on Figure 9.1 which indicates both Iron Age and Stone Age sites in the region under consideration.

Archaeozoological material occurs mainly in open air sites of very recent date and in caves or rock shelters. The only sites in the equatorial rainforest where faunal remains have thus far been found are Matupi cave and the seventeenth to nineteenth century AD open air site of Nkile. Other cave sites such as Dimba, Ngovo and Ntadi Yomba lie presently in a heavily wooded environment (the forest–savanna mosaic). In the Grassfields of Cameroon, faunal remains were also found in a series of caves. Rapid, and by preference, deep burial of bone is necessary to make good preservation possible in open air sites. Alluvial and lacustrine deposits can, therefore, sometimes yield abundant faunal remains. This is the case of certain sites found in the Ishango region where lake deposits yielded faunal remains in a sequence studied in the 1950s (de Heinzelin 1957, Hopwood & Misonne 1959). Research was resumed in the late 1980s north of Ishango in the alluvial Semliki Beds, and also yielded sites with good faunal preservation (Brooks & Smith 1987, Peters 1990, Brooks et al. 1995). Another type of deposit where preservation conditions can be favourable are human burials such as those in the Upemba Depression region of Zaïre where animal remains were found associated with human skeletons. The filling of pits is another type of deposit where preservation can be quite good when infilling is rapid. This is certainly true for the pits found at Nkang in Cameroon for instance. Finally, there are some faunal assemblages where bone is well-preserved in shell middens owing to the alkaline environment created by the shells. This is the case for Oveng in Gabon and several sites in Angola.

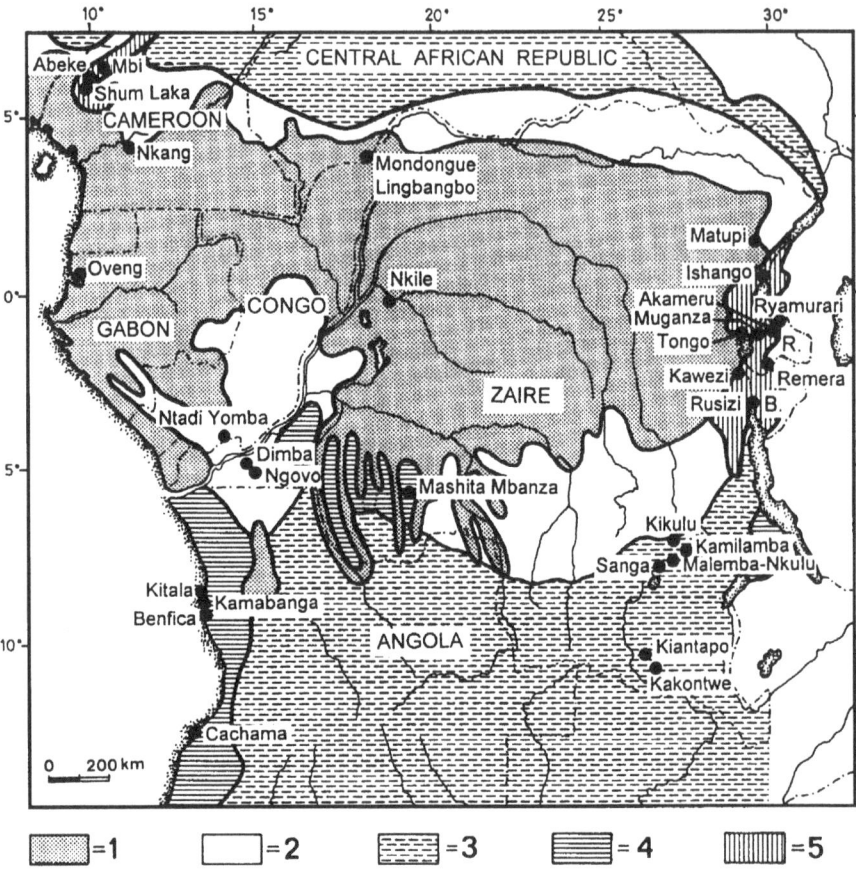

Figure 9.1 Map showing Central African Stone Age and Iron Age sites with faunal remains. The main vegetation belts are also indicated. 1 = evergreen forest; 2 = forest–savanna mosaic; 3 = woodland savanna; 4 = grass savanna; 5 = montane habitats.

2. Overview of sites

The region considered in this overview comprises the area from about 5°N to 13°S and from the Atlantic coast to the interlacustrine region. Included are all sites with domestic animals as well as some additional settlements where such animals were lacking despite the relatively recent age of their deposits. Well-preserved bones allowing measurements occur only sporadically, with these data mostly given in the text. The measurements are in millimetres and have been taken according to von den Driesch (1976). Shoulder heights of domestic cattle, sheep and goat were calculated whenever complete long bones were available. The indices used are those summarized by von den Driesch & Boessneck (1974). A regional overview of the faunal data available in the literature and some unpublished data are given below.[1]

A list of radiometric dates from the sites with domestic animals is given in Table 9.9. This separate list indicates the laboratory dates as well as the corresponding calibrated minimum and maximum dates (One Sigma) calculated with the OxCal v2.15-programme. The calibrated values are used in the text.

2.1. Cameroon

2.1.1. Nkang The site of Nkang lies on the top of a hill that gradually slopes towards the Sanaga river. It belongs to the so-called Obobogo tradition, which refers to the earliest village communities in Cameroon, typically installed on hill tops near Yaoundé since the end of the second millennium BC. The precise extension of the Obobogo tradition still remains to be established but Nkang is the most southerly indication for it. The archaeological material from these sites derives mainly from large pit structures and comprises, in general, pottery, grinding stones, polished stone axes, metal slag, and remains of *Elaeis guineensis* and *Canarium schweinfurthii*. The latter finds have sometimes been interpreted as a possible indication for some form of agriculture (de Maret 1991:45), but indications for animal husbandry was at that time lacking from sites of the Obobogo tradition. This is owing to the fact that faunal remains were poorly preserved, if at all, in previous excavations. Obobogo is the only open air site in southern Cameroon, besides Nkang, to have yielded fauna. The only preserved bone fragment at Obobogo is from a brush-tailed porcupine *Atherurus africanus* (de Maret 1991:45) and was found in a pit dated to the end of the first millennium BC. The archaeological material comprises iron slag and ceramics of the beginning of the early Iron Age. One potsherd showed an impression of a *Pennisetum* grain (Claes 1985).

The site of Nkang comprises a number of refuse pits distributed over an area of about one hectare. Thirteen of these pits were excavated in 1990 (Mbida 1996) and yielded pottery, a polished stone axe fragment and other lithic material, iron slag, burnt remains of *Canarium* and *Elaeis*, and a total of about 100 bone fragments. Radiocarbon dates indicate that Nkang was inhabited between the eighth and second centuries BC. Table 9.1 gives an overview of the animals identified (Van Neer, unpubl. data). The majority of the fauna represents human food waste, with only a few remains being considered as intrusive elements. This is the case for the *Limicolaria* shells and for the small rodents. The former typically colonize human habitation areas and may be more or less contemporaneous with the occupation phase. The small rodents, on the other hand, are less fossilized than the rest of the faunal material indicating that they represent animals that burrowed and died in the archaeological layers at a more recent date.

The fauna discovered at Nkang allows a reconstruction of the former environment. The *Achatina* shells, the bushbuck, the forest duikers and the forest buffalo indicate a rather dense forest, whereas the waterbuck and kob are typical of more open habitats. The former environment may thus have resembled the modern one, i.e. a forest–savanna mosaic.

The subsistence of the inhabitants of Nkang relied, as far as animals are concerned, on the collecting of molluscs, fishing, hunting and animal husbandry. Both terrestrial (*Achatina*) and aquatic molluscs (*Aspatharia*) were harvested. The latter,

Table 9.1 Faunal remains found in different pits at the site of Nkang, Cameroon. Figures indicate number of individual specimens (NISP).

	surf.	F1	F3	F5	F6	F7	F7bis	F7NF	F9	F13	Total
Lanistes libycus	–	–	–	–	2	–	–	–	–	–	2
Potadoma cf. *freethii*	–	–	–	–	1	–	–	1	3	–	5
Achatina sp.	–	1	–	–	–	2	–	–	7	–	10
Limicolaria sp.	1	–	–	–	–	–	2	1	2	–	6
unidentified gastropods	–	–	–	–	–	–	1	1	3	–	5
Aspatharia sp.	–	–	–	–	1	–	–	–	2	–	3
freshwater crab	–	–	–	–	–	–	1	–	–	–	1
catfish 1 (*Chrysichthys* sp.)	–	–	–	–	–	–	1	–	–	–	1
catfish 2 (Clariidae)	–	–	–	–	1	–	–	–	–	–	1
Nile perch (*Lates niloticus*)	–	–	–	–	–	–	1	–	–	–	1
small rodents	–	–	1	–	–	–	–	–	–	1	2
cane rat (*Thryonomys* sp.)	–	–	–	–	1	1	–	–	1	–	3
hippopotamus (*Hippopotamus amphibius*)	–	–	–	–	–	–	1	–	–	–	1
bushbuck (*Tragelaphus scriptus*)	–	–	–	–	–	1	–	–	–	–	1
waterbuck (*Kobus ellipsiprymnus*)	–	–	–	–	–	–	7	1	–	–	8
kob (*Kobus kob*)	–	–	–	–	–	–	5	–	–	–	5
medium-sized duikers (*Cephalophus* spp.)	–	–	–	1	2	–	7	–	–	–	10
forest buffalo (*Syncerus caffer nanus*)	–	–	–	–	1	–	2	–	1	–	4
goat (*Capra aegagrus* f. *hircus*)	–	–	–	–	1	–	–	–	–	–	1
sheep (*Ovis ammon* f. *aries*)	–	–	2	–	–	–	–	–	–	–	2
goat/sheep	–	–	1	–	–	–	1	–	–	–	2
unidentified mammals	–	–	–	1	8	–	15	–	1	–	25
total identified	1	1	4	1	10	4	28	3	16	1	69

as well as freshwater crabs, were probably taken from the nearby tributary of the Sanaga river. Fishing may have played an important role as well and relied on various techniques judging from the habitat preferences of the species. Various parts of the river were exploited. The fish captured comprise animals typical of open, deep water (*Lates niloticus*), bottom-dwellers (*Chrysichthys* and Clariidae), as well as small schooling fish occurring inshore near the surface (*Alestes/Brycinus*). Remains of the latter group of fish, as well as some small cyprinids were found in a sieve sample not included in Table 9.1. Hunting concentrated mainly on antelopes of medium to large size but included the forest buffalo and smaller animals such as the cane rat. The wide variety of exploited animals illustrated by only a small faunal sample may be considered as typical of relatively sedentary communities cf. Flannery 1972, Tchernov 1993). The presence of small stock is attested to by five bones only. They were found in the following pit structures: F3 dated between 760 BC and 400 BC, F6 dated between 830 BC and 540 BC, and F7bis dated to 760–260 BC. A goat humerus with its proximal end unfused was found in F6 and had the following dimensions: Bd 24.2 mm; BT 23.2 mm. These measurements, as well as the overall size of the bone, indicate a rather small breed, comparable in size to the dwarf goats found in equatorial Africa today (Epstein 1971:211). The presence of sheep is attested in pit F3 by a distal fragment of a metatarsal belonging to a sub-adult specimen and by a complete third phalanx (DLS 21.6 mm; Ld 18.0 mm). Both specimens belong to a small breed comparable to the dwarf breed of thin-tailed hair sheep that is widely distributed today in the tropical forest of Africa (Epstein 1971:48). This is the oldest evidence for ovicaprines in Central Africa.

2.1.2. Cave sites   The excavation since the 1980s of a number of caves and rock shelters in the Grassfields has yielded large faunal assemblages. The initial purpose of the research conducted in this area was to document the period of transition between the Late Stone Age and the Metal Ages and studying the possible relationship with the first stages of Bantu expansion (de Maret 1980). Major faunal assemblages are available from the sites of Abeke, Mbi Crater and Shum Laka I and II.

Abeke, explored in 1980 by Pierre de Maret, yielded a Late Stone Age industry in layers dated to the fifth millennium BC. The overlying surface layers contained a few pottery sherds but have not been dated. These upper levels yielded the same forest fauna as the deeper ones; there is no evidence for domestic species at all (de Maret et al. 1987).

Excavations at Mbi Crater yielded a Late Stone Age industry in two phases. A shift is seen through time from a savanna fauna, in the earliest phase dated between about 25,000 bp and 18,000 bp, to an assemblage typical of a more densely wooded vegetation, from about 9000 bp onwards. The earliest occurrence of pottery is seen in a level dated to 3050–2500 BC; the upper layer is dated to 1090–810 BC (Warnier & Asombang 1982, Asombang 1988). Some *Canarium schweinfurthii* and *Elaeis guineensis* were found but domestic animals are lacking throughout (Asombang 1988).

The site of Shum Laka I has yielded thousands of faunal remains. A test excavation was carried out in 1980 (de Maret et al. 1987) and extensive excavations followed in the 1990s (Asombang & de Maret 1992, de Maret et al. 1993, 1995, Cornelissen et al. 1995). A detailed stratigraphic sequence of 3 m is now available. Iron Age pottery was found in the upper grey ash layers dated from the fourth century BC to the thirteenth century AD, with dates for the deeper layers ranging between 6000 bp and 30,000 bp. I have analyzed material from all of these layers, but was unable to see any shifts in the fauna through time. Only forest animals occur throughout the sequence, although botanical analyses (pollen, phytoliths and charcoal) have only yielded evidence for savanna species. This seems to suggest that Shum Laka cave was situated at the fringe of the savanna and forest. The giant forest hog (*Hylochoerus meinertzhageni*) and forest buffalo (*Syncerus caffer nanus*) predominate in all levels, whereas cercopithecids and duiker antelopes are also very common. No savanna animals appear in the younger levels and there is no trace of domestic animals either. A similar situation was found at the smaller Shum Laka II, a nearby rockshelter with shallower archaeological deposits.

2.2. Central African Republic

Mondongué and Lingbangbo are two open air sites in the tropical forest south of Bangui. They yielded fauna, pottery, metal slag, metal objects and few lithic remains (Koté 1992). Radiometric dates place the occupation of Mondongué between the sixteenth and twentieth centuries AD, whereas Lingbangbo was inhabited some time between the fourteenth and seventeenth centuries AD. The present-day inhabitants of the area keep chickens, goats and dogs. The fauna from the site comprises mainly wild animals but domestic species occur in small numbers (Table 9.2). Seven fragments of ovicaprines were found at Lingbangbo but it was impossible to identify these to a species level. However, a distal humerus (BT 26.5 mm) found at the site of Mondongué could be definitely identified as goat. The dimensions of this piece indicate a small breed. Domestic dog is represented by at least six remains of a medium-sized breed at Lingbangbo and may also have been present at Mondongué. The shoulder height of the dogs from Lingbangbo is estimated at about 45 cm on the basis of direct comparison of a distal humerus (Bd 23.0 mm) with modern skeletons.

2.3. Gabon

2.3.1. Oveng   This site is a shell midden at 12 km northeast of Libreville. It consists of a 30 m high hilltop surrounded by mangroves and equatorial forest. A small tributary of the Bombié river, running to the Mondah bay, lies close to the site. Excavations yielded pottery, iron slag, faunal remains, carbonized nuts of *Elaeis guineensis* and *Coula edulis*, and charcoal (Clist 1987, 1989). Analysis of the archaeological material and the stratigraphy shows that this is a single-component site dated between the first century AD and the sixth century AD. The shells comprise *Anadara senilis* (78 per cent), *Tympanotus fuscus* and *T. radula* (13.5 per cent), *Ostrea tulipa* (5 per cent), and *Semifusus morio* (3.5 per cent) (Van Neer & Clist 1991). About 3,000 vertebrate remains were found of which only a small portion

Table 9.2 Faunal remains from the sites of Mondongué and Lingbango, Central African Republic. Figures indicate number of individual specimens (NISP).

	LINGBANGBO					MONDONGUE	
	main trench	test trench					
	layer	layer				layer	
	1	1	3	5	6	1	2
Achatina sp.	–	–	–	–	–	–	1
Polypterus sp.	–	–	–	3	–	–	–
Mormyridae indet.	–	–	–	1	–	–	–
Cyprinidae indet.	–	–	–	1	–	–	–
catfish 1 (Clariidae)	–	–	–	10	–	–	–
catfish 2 (*Auchenoglanis* sp.)	–	–	–	1	–	–	–
unidentified catfish	1	–	–	10	–	–	–
Parachanna sp.	1	–	–	3	–	–	–
unidentified perciform	–	–	–	1	–	–	–
Kinixys sp.	1	–	–	–	–	–	–
Pelomedusidae	–	–	–	2	1	–	–
Cyclanorbis sp.	–	–	–	1	–	–	–
Python sp.	–	–	–	2	–	–	–
Crocodylidae	–	–	–	2	–	–	–
guenon (*Cercopithecus* sp.)	1	–	–	–	–	–	–
Cercopithecidae	–	–	–	–	–	–	1
giant rat (*Cricetomys* sp.)	–	–	–	6	–	–	–
porcupine (*Atherurus africanus*)	–	–	–	–	–	–	1
elephant (*Loxodonta africana)*	–	–	1	–	–	–	–
bushpig (*Potamochoerus porcus*)	5	–	1	2	1	–	2
blue duiker (*Cephalophus monticola*)	–	–	–	9	5	–	–
medium-sized duikers (*Cephalophus* spp.)	7	2	–	9	2	1	–
sitatunga (*Tragelaphus spekei*)	–	–	–	–	–	1	1
waterbuck (*Kobus ellipsiprymnus*)	–	–	–	–	–	–	1
African buffalo (*Syncerus caffer*)	–	–	–	1	–	–	–
dog (*Canis lupus* f. familiaris)	1	–	–	4	1	–	1?
goat (*Capra aegagrus* f. hircus)	–	–	–	–	–	–	1
goat/sheep	–	–	–	7	–	–	–
unidentified mammals	15	2	4	25	7	10	8
total identified	17	2	2	75	10	5	11

was identifiable owing to the extreme fragmentation. More than 97 per cent of the complete bone collection belongs to fish. The species present, and their small reconstructed size, indicate that fishing was practised in the mangrove rivers near the site. Mammal bones were heavily under-represented. The only identifiable mammalian remains were from small, intrusive rodents. The remaining mammal bone fragments were unidentifiable but belonged to medium-sized animals (size of duiker antelope or ovicaprines). The faunal data hence indicate that the inhabitants of Oveng relied mainly on shell-collecting and fishing.

2.4. Congo (Zaïre)

2.4.1. Matupi   This cave, which is situated in the rainforest at about 70 km west of Lake Mobutu, yielded a microlithic LSA industry dated between about 3000 BP and more than 40,000 BP (Van Noten 1977). The upper layers comprise Iron Age material and have been dated to the thirteenth and fourteenth centuries AD. The only possible indication for domestic animals is a lower second premolar of a canid found in the Iron Age levels (Van Neer 1989). It was, however, impossible to identify this piece with certainty as domestic dog. The tooth is also very similar to the corresponding one from jackals (*Canis adustus* and *Canis mesomelas*) living in the region.

2.4.2. Ngovo and Dimba   The excavation of these two caves, and six other sites, in Lower Zaïre was aimed at detailing the archaeological context of polished stone tools frequently found at the surface in the savannas just south of the equatorial forest. These polished stone tools are systematically found in association with pottery of the so-called "Group VI" which de Maret (1986) proposed to designate as the "Ngovo Group". The industry is placed within the last few centuries BC by a series of dates in good agreement. The Ngovo Group has been considered by several authors as possible evidence for the colonization of this part of the continent by Bantu-speaking farmers. The systematic absence of metal in the Ngovo Group sites seems to indicate that it was pre-metallurgical. It is only in the second century AD that the first iron appears in the region. A fragment of a *Canarium schweinfurthii* nut was found at Ngovo which de Maret (1986) considers as a possible indication of a form of agriculture or arboriculture. No domestic animals occur among the faunal assemblages from Ngovo (Van Neer 1986) and Dimba (Van Neer 1990). The samples are both very small and comprise only 38 and 98 identifiable fragments, respectively. The faunal remains are all from wild animals, with in both cases a clear preponderance of species typical of closed environments. The fauna seems to indicate that the immediate surroundings of both sites were more heavily wooded than they are today, a hypothesis that is also supported by botanical evidence for Ngovo (Deschamps 1986).

2.4.3. Kawezi   Kawezi is an open air site in the Ruzizi valley that has yielded early Iron Age pottery, charcoal and animal bones (Maquet & Hiernaux 1969). No radiometric dates are available. The pottery differs from the types usually encountered in the region and resembles most closely the channelled ware of Zambia. The terrestrial fauna has been partially identified by S. Frechkop (Brussels) and comprises the remains of hyena, buffalo, a reduncine antelope (*Kobus ellipsiprymnus?*), *Tragelaphus*, and some antelopes of larger size (eland antelope?). No domestic animals have been reported but it would be interesting to re-investigate this material, especially the specimens identified as buffalo. Using Peters' (1988) diagnostic criteria for separating the post-cranial elements of buffalo and domestic cattle it should be possible to see if domestic cattle were present at Kawezi.

2.4.4. Upemba   Extensive fieldwork has been carried out in a series of Iron Age necropolises in the Upemba Depression (de Maret 1985a). The excavations took place in four areas: Kamilamba and Kikulu at the shore of Lake Kabamba, Malemba Nkulu at the right bank of the Lualaba, and Sanga along Lake Kisale. A chronological framework has been established for the human occupation in the region on the basis of more than 40 radiocarbon dates (not mentioned in Table 9.9; see de Maret 1982, Geyh & de Maret 1982). The Kamilambian phase starts in the sixth century AD and is replaced by the Ancient Kisalian during the eighth century. The Classic Kisalian starts around the tenth century and is followed by the Kabambian A in the thirteenth to fourteenth centuries. The latter is replaced by the Kabambian B in the sixteenth century and lasts until the eighteenth century.

The faunal remains were found in close association with the human skeletons or in the pottery that accompanied them (Van Neer 1990). Except for some intrusive rodent skeletons, all the faunal remains are considered as funerary gifts. Fish bones are dominant throughout the sequence and only at Sanga were substantial numbers of other vertebrates (mammals and birds) found. The graves from Sanga belong to the Classic Kisalian and date between the tenth and thirteenth/fourteenth centuries. They yielded a large number of metapodials from bovids. One of them was a goat metacarpal of which the distal ends were fusing. This specimen has the following dimensions: GL 89.0 mm; Bp 21.1 mm; SD 14.4 mm; Bd 25.4 mm. The corresponding shoulder height of this animal was about 51 cm, which is more or less the upper limit for dwarf breeds (Epstein 1971:211). Another grave from Sanga yielded a complete skeleton of a galliform bird identified as a domestic chicken.

2.4.5. Mashita Mbanza   Mashita Mbanza is situated at about 50 km southeast of Kikwit in a grassland environment. The site comprises a series of about ten small mounds in a semicircle. The genesis and the function of these mounds is unclear. They contain earth, charcoal, metal slag, pottery, metal objects, kitchen debris and, in some cases, cattle dung. Oral traditions of the Pende refer to Mashita Mbanza as the area where they lived before migrating to their present-day homeland. The site was discovered by Maes (1935) and re-excavated by de Maret in 1984. Radiocarbon dates place the occupation between the sixteenth and twentieth centuries AD (Pierot 1987). About 100 mammal remains have been found of which only 16 were identifiable (Van Neer 1990). They comprise hunted animals as well as domestic species. Wild mammals identified from this site are cane rat (*Thryonomys swinderianus*), banded mongoose (*Mungos mungo*), warthog (*Phacochoerus aethiopicus*) and bush duiker (*Sylvicapra grimmia*). The latter two species indicate that the environment was also open at the time of the occupation. Three bones of ovicaprines were found, one of which is identifiable as sheep. Firm osteological evidence for cattle is not available. There is one incisor of a large bovid which cannot be attributed with certainty to either domestic cattle or African buffalo. The excavators reported, however, that cattle dung was present in some of the mounds.

2.4.6. Nkile   An extensive archaeological survey carried out along the main rivers of the equatorial forest in Zaïre (Eggert 1983, 1984, 1987) showed that faunal preservation is very poor in this area. Animal remains were only discovered at Nkile, a village situated along the Ruki river at about 3 km downstream Bokuma. Most of the faunal remains are from levels belonging to the Botendo horizon. The corresponding occupation covers the nineteenth century but might go back to the eighteenth or seventeenth centuries (Eggert pers.comm.). The fauna comprises molluscs, most of which are probably intrusive, as well as fish and mammal bones representing food remains (Table 9.3). Although the assemblage is small, it seems that the inhabitants relied mainly on fishing. Animal husbandry and hunting were apparently practised to a lesser degree. The wild mammals indicate a densely wooded environment (guenon *Cercopithecus* sp. and blue duiker *Cephalophus monticola*). Domestic animals comprise pig, ovicaprines and dog. The lower jaw of a suid could be assigned to domestic pig on the basis of the dental characters distinguishing *Sus scrofa* from *Potamochoerus porcus* (Cooke & Wilkinson 1978). Ovicaprines are represented by a molar and a fragment of a first phalanx. An ulna of a domestic dog shows cutmarks.

2.4.7. Kiantapo   Kiantapo cave, situated on the western slope of the Biano plateau in Shaba, is famous for its rock engravings (Breuil 1952, Mortelmans 1952). Excavations carried out in front of the cave by Cabu (1938) and in 1955 by Mortelmans have not allowed a precise dating of the engravings. The archaeological material of the 1955 fieldwork belongs mainly to the late Iron Age and is not more than two centuries old (Mortelmans & de Maret 1989). The pottery is of the type still locally produced by the Bena Mitumba. Oral traditions stipulate that the Bena Mitumba are Luba coming from the region of Kanda Kanda. After they settled in the Biano area, conflicts occurred with other Luba groups and with Msiri people. It is likely that the Kiantapo cave, and many other caves in the area, had been used as a refuge during these periods of instability. The numerous faunal remains found in Kiantapo cave indicate that the inhabitants relied almost exclusively on hunting (Van Neer 1989). This was practised both on the grassy plains on top of the Biano plateau, and on its wooded slopes. The only domestic animal that was found is chicken. The absence of other domestic species is surprising for a site of such a recent date. This may be related to the fact that the deposits were formed by people seeking refuge in periods of insecurity. They may have been able to only bring along the smallest domestic stock they had.

2.4.8. Tongo   This site, which lies at about 50 km north of Lake Kivu, was discovered during road construction works between Masisi and Rutshuru. The

Table 9.3 Faunal remains from the site of Nkile, Zaïre. Figures indicate number of individual specimens (NISP).

	NISP
Homorus sp.	2
Limicolaria sp.	1
Subulina cf. *avakubiensis*	1
unidentified gastropods	2
Protopterus sp.	1
Polypterus sp.	8
Mormyrops sp.	5
Cyprinidae	1
Clariidae	20
Synodontis sp.	3
Bagridae	4
Tilapiini	7
Ctenopoma sp.	1
Parachanna sp.	49
unidentified fish	80
crocodile	2
Cercopithecidae	1
Genetta sp.	1
Cephalophus monticola	2
Cephalophus sp.	3
small bovid	2
pig	1
sheep/goat	2
dog	1
unidentified mammals	22

vegetation in the area is mainly grassland with patches of forest. Because of its high elevation the region is free of trypanosomiasis which renders it very suitable for herding. Excavations yielded pottery with twisted cord roulette decoration and sherds of the Urewe tradition. Aside from the surface layers, three main entities have been distinguished (Kanimba pers. comm.): the levels between 70 cm and 150 cm comprise roulette-decorated pottery; between 150 cm and 210 cm Urewe pottery was found, whereas the levels between 210 cm and 300 cm yielded a mixture of Urewe pottery and finely decorated sherds with a polished surface. Associated with the Urewe pottery were also worked shells, some iron objects and iron slag (Kanimba & Gatare 1992). Radiocarbon dates place the Urewe occupation between the third and sixth centuries AD. Human skeletons were found as well, but their age is uncertain. Faunal remains occur throughout the sequence (Table 9.4). Domestic cattle and ovicaprines are present in all levels and make up about 40 per cent of the total

Table 9.4 Faunal remains from the different levels at the site of Tongo, Zaïre. Figures indicate number of individual specimens (NISP).

	Late Iron Age		Early Iron Age		
	surface	70–150 cm	150–210 cm	210–300 cm	total
Achatina sp.	–	1	–	–	1
large Gastropoda	1	–	3	1	5
Clariidae	1	1	–	4	6
Bagrus sp.	–	–	1	1	2
Tilapiini	–	20	14	5	39
fish identified	1	21	15	10	47
fish indet.	–	14	3	1	18
Francolinus sp.	–	–	1	–	1
Galliformes indet.	–	1	1	–	2
bird identified	–	1	2	–	3
bird indet.	–	3	2	–	5
small rodents	–	4	8	6	18
giant rat (*Cricetomys* sp.)	1	5	4	3	13
baboon (*Papio anubis*)	–	–	–	1	1
colobus monkey (*Colobus* sp.)	–	–	–	1	1
cercopithecids (Cercopithecidae)	–	–	2	2	4
serval (*Felis serval*)	–	–	–	1	1
small carnivores	–	2	2	–	4
hyrax (Procaviidae)	–	1	–	–	1
elephant (*Loxodonta africana*)	–	1	–	–	1
giant forest hog (*Hylochoerus meinertzhageni*)	1	2	2	3	8
cattle (*Bos primigenius* f. taurus)	6	16	7	5	34
sheep (*Ovis ammon* f. aries)	–	–	1	–	1
goat (*Capra aegagrus* f. hircus)	1	2	1	2	6
sheep/goat	6	17	8	9	40
duiker (*Cephalophus* sp.)	–	1	1	–	2
bushbuck (*Tragelaphus scriptus*)	3	6	3	–	12
small bovid	2	20	8	15	45
roan (*Hippotragus equinus*)	–	1	–	–	1
mammal identified	20	78	47	48	193
mammal indet.	40	165	98	79	382

number of identified specimens. Worth noting is that among the galliform remains no domestic fowl were recognized; only *Francolinus* could be identified. Fishing was concentrated on shallow water taxa (clariid catfish and tilapia). Hunting seems to have been practised to a greater extent in densely forested areas (giant forest hog, cercopithecids, duikers) than in more open habitats (baboon and roan).

Table 9.5 summarizes the most important measurements that could be taken on the sheep and goat bones. The complete ovicaprine metapodials allowed the

Table 9.5 Measurements of some selected ovicaprine bones from Tongo, Zaïre. The remains come from the surface levels, late Iron Age, and early Iron Age deposits.

metacarpus		Early Iron Age Ovis		
	GL	96.3		
	Bp	19.1		
	SD	12.4		
	Bd	22.6		
metatarsus		Late Iron Age Capra	Late Iron Age Capra	surface Capra
	GL	97.2	–	–
	Bp	17.2	17.7	18.0
	SD	10.7	–	–
	Bd	–	–	–
phalanx 1		Early Iron Age Capra		
	Glpe	32.3		
	Bp	11.6		
	SD	9.9		
	Bd	10.8		

calculation of shoulder heights for a sheep (47 cm) and for a goat (52 cm), showing that we are dealing with dwarf forms. No complete long bones of cattle were present for the calculation of height at withers but the available measurements suggest a medium to large-sized breed (Table 9.6).

2.5. Rwanda

2.5.1. Muganza I Fieldwork carried out in Rwanda and Burundi by Van Grunderbeek et al. (1982, 1983) concentrated on the early Iron Age period and aimed also at reconstructing the evolution of the former environment. Cores had been taken, among others, in the peat deposits along the western shore of Lake Ruhondo in northern Rwanda. Palynological analysis of sediments from Muganza I, dated between 400 BC and AD 250, revealed that wooded savanna with gallery forests existed near the site. Animal bones were found as well but there was no archaeological material associated with them. The rather complete state of preservation of the skeletal elements and the kind of deposit may indicate that the bones represent animals that have been trapped accidentally in the peat deposits. They may, therefore, be of a more recent date than the surrounding sediment dated to 2300 BP. An attempt to directly date the bones failed, however, owing to the poor collagen content. The faunal remains, which must be 2,300 years or younger, did not comprise any domestic species. Warthog *Phacochoerus aethiopicus*, a medium-sized antelope (size of bushbuck *Tragelaphus scriptus*), and African buffalo *Syncerus caffer* were the only animals recognized. The latter species is represented by a

Table 9.6 Selected cattle measurements from various sites in the interlacustrine region. Most of the data are from the literature (for references see description of sites), those from Tongo are new. Surface finds, late Iron Age and early Iron Age material from Tongo is indicated separately.

scapula		Akameru	Ryamurari		
	BG	±43	±48		
humerus		Tongo surf	late		Akameru
	BT	61.5	68	Bd	±70
radius		Tongo early			
	Bp	74			
metatarsus		Tongo late		Akameru	
	Bp	50.7		48	
phalanx 1		Akameru and Cyinkomane			Ryamurari
	GLpe	63	65	66	65
phalanx 1		Gisagara VII	Ruzizi		
	GLpe	59.0	59.3	62.6	65.0
	Bp	29.8	25.7	27.7	31.1
	SD	24.0	21.8	21.3	26.2
	Bd	28.6	24.4	26.9	31.1
phalanx 2		Tongo early	Akameru	Ruzizi	
	GLpe	55.5	41.5	40.9	
	Bp	–	–	27.8	
	SD	–	–	20.7	
	Bd	30.5	–	24.0	

pelvis fragment and a distal tibia which could be discriminated from domestic cattle on the basis of the large size and the diagnostic morphological criteria established by Peters (1988).

2.5.2. Akameru and Cyinkomane   These two caves are situated in the Musanza region a few kilometres west of Ruhengeri. Excavations, which were limited to a square metre test trench in each cave, yielded the same pottery throughout. It comprises mainly sherds with rouletted decoration, belonging to the late Iron Age (Van Noten 1983). Two radiocarbon dates place the corresponding occupation at Akameru between the ninth and thirteenth centuries AD. About 200 identifiable vertebrate remains were found (Gautier 1983a). Hunted animals comprise mainly antelopes (sitatunga *Tragelaphus spekei* and duiker *Cephalophus nigrifrons*) and large rodents (giant rat *Cricetomys* and cane rat *Thryonomys*). Domestic cattle, sheep, goat and chicken were found as well. They represent about 35 per cent of the consumed animals at Akameru, and 56 per cent at Cyinkomane. The measurements of a few post-cranial elements of domestic fowl indicate that it was a small race. Sheep predominate among the ovicaprines. The heavy fragmentation of the cattle remains and the absence of horn cores precluded any attempt of racial attribution. The measurements, however, would indicate a breed with a withers height between 105 cm and 130 cm (Gautier 1983a).

2.5.3. Remera I, Kabuye XIV and Gisagara   These Iron Age sites lying in southern Rwanda have yielded faunal remains that have only been partially published. Excavations at Remera I yielded a furnace dated between AD 250 and AD 340. Associated with this structure were two cattle teeth identified by Gautier (Van Grunderbeek et al. 1983:41). They comprise a fragment of an upper molar, and a lower third molar of 34.8 mm long which was only slightly worn. This tooth was attributed to a young adult of which the shoulder height was estimated at 110 cm. Similar cattle teeth were also found at the early Iron Age site Kabuye XIV in association with pottery. A somewhat larger assemblage of animal remains was found at Gisagara, 15 km east of Butare. This site comprises circular structures with roulette decorated pottery typical of the late Iron Age. Animal remains were found at Gisagara II, III, V and VII, showing that ovicaprines and cattle were kept. Twenty remains of cattle and one ovicaprine tooth were identified. A radiocarbon date for Gisagara II places the occupation in the eleventh to twelfth centuries AD. Comparison of the measurements (see also Table 9.6) with those from Manching would indicate a small breed of cattle with a shoulder height between 110 cm and 120 cm (Gautier, n.d.).

2.5.4. Ryamurari   This proto-historic site is the capital of the ancient Ndorwa kingdom. It is situated at the top of the Mukana hill in northeast Rwanda. Two types of structure have been discovered near the flat hilltop. Circular-shaped, artificial depressions of 6–8 m diameter and 1–2.5 m deep occur, at the base of which sometimes granitic blocks were found. The large, flat stones bear cavities that fill with water during heavy rains. Cattle would have been watered here during the rainy season when the marshy riverbanks made access to the nearby rivers difficult. Small enclosures were found at the southwest of the hilltop which, according to oral traditions, were inhabited by the servants. The royal enclosures occur at the southeastern end of the settlement. They are much larger and are arranged around a great, central circular structure 48 m in diameter. The latter structure was surrounded by ramparts much higher than those of the other enclosures. The enclosures served for both stock-keeping and human habitation. The majority of the pottery found during the excavations was with rouletted decoration. Radiocarbon dates situate the beginning of the occupation in the mid seventeenth century. The site was inhabited until the early twentieth century (Tshilema Tshiluka 1983). The faunal remains collected during the first field season have been studied by Gautier (1983b). The royal enclosures yielded only three identifiable bones, all belonging to domestic cattle. In the servants' enclosures 20 cattle remains were found as well as three bones of hare. More than half of the cattle remains are from juvenile specimens. The proportion of juveniles may, in reality, have been much higher taking into account the relatively low probability of preservation for their bones. The remains of these young individuals probably represent male animals that were eliminated early from the herds. Meat from these young bulls and from older cows may have been of less importance in the economy of Ryamurare. It is likely that cattle were kept mainly for the production of milk as is still done today by Cushitic and Nilotic peoples. Additional faunal remains have been found during later excavations. The preliminary study of this material (Tshilema Tshiluka 1983) shows that ovicaprines were also kept. Some well-preserved horn cores of cattle indicate that they belonged to a race with long, lyrate horns.

2.6. Burundi

The only faunal remains known thus far from this country come from a mound in the Ruzizi plain (De Meulemeester & Waleffe 1973). This site, which has not been named, yielded archaeological material comprising roulette decorated pottery, some pipe fragments, iron objects and glass beads. Radiocarbon dating places this late Iron Age assemblage between the seventeenth and twentieth centuries AD. The shape and the dimensions of the structure indicate an enclosure for livestock. Fauna is well-preserved as a result of the fast accumulation of cattle dung, kitchen refuse, and sand that was either transported by man or windblown. The vertebrate fauna is predominantly composed of fish (Van Neer 1990). The mammal remains comprise a few specimens of wild species (bushbuck *Tragelaphus scriptus* and zorilla *Ictonyx striatus*). The commensal black rat (*Rattus rattus*) is present as well. The majority of the identified mammal remains belong, however, to domestic cattle. A few measurable phalanges indicate animals of medium size (Table 9.6).

2.7. Angola

2.7.1. Kamabanga I and Kitala   The excavation of Kamabanga, just south of Luanda, yielded a hearth in a layer of *Anadara senilis* shells. It dates to the ninth to tenth centuries AD (de Maret 1985b). Some sherds of finely decorated pottery and a few faunal remains were associated with the hearth. The bones comprise about ten fish remains, an astragalus of warthog (*Phacochoerus aethiopicus*), tibia fragments of Burchell's zebra (*Equus burchelli*) and a first phalanx of domestic cattle with the following measurements: Glpe 62.6 mm; Bp 30.0 mm; SD 25.8 mm; Bd 28.6 mm.

Kitala lies closely to the preceding site and has been dated to the thirteenth to fourteenth centuries AD (de Maret 1985b). The only faunal remains that were found are a molar of Burchell's zebra and a carpal bone (*carpale magnum*) of domestic cattle.

2.7.2. Benfica   This site lies at the seashore about 17 km south of Luanda. It yielded some surface lithic material comparable to the Lupembo–Tshitolian complex of Central Africa, which was, however, never found in stratigraphical context. A thick layer of shells was found between 15 cm and 25 cm below the surface. It comprises mainly *Anadara senilis*, but also some *Ostrea* sp., *Fucus* sp., *Murex* sp. and *Conus*. This level has been excavated over a surface of 40 m² (Dos Santos & Ervedosa 1970) and yielded substantial quantities of pottery with a wide variety of incised decorations. Radiometric dates place the occupation for this shell midden between the second and fourth centuries AD (Lanfranchi & Clist 1987). The fauna included numerous fish vertebrae, mammals and, possibly, also birds. No domestic animals are mentioned from this site. Among the mammal remains, J.W. Kitching identified a phalanx and an upper incisor of Burchell's zebra (Dos Santos & Ervedosa 1970:49).

2.7.3. Cachama   Cachama is a coastal site that lies close to Baïa Farta and has been excavated in different loci (Pinto 1988). Cachama 3, analyzed over a surface of 120 m², yielded an undated Iron Age level. Cachama 1 was excavated over a surface of 165 m² and yielded late Iron Age and Iron Age horizons. The upper horizon contained pottery and was rich in faunal remains but is thus far undated. The fauna comprises mainly marine fish, some mammal bone and few remains of birds. A preliminary analysis showed that no domestic animals were present (Van Neer, unpubl. data).

3. Discussion and conclusions

3.1. Presence of the different domestic species

The faunal record for Central Africa is very limited as a result of the poor preservation chances and the relatively low number of sites excavated in this vast region. The foregoing overview of sites with fauna, dating roughly to the last three millennia, shows the presence of some relatively early finds of domesticates on certain sites, whereas, at the same time, such animals are lacking on several settlements where their presence could not be excluded *a priori* on the basis of the archaeological context and the age of the site.

The keeping and propagation of domestic stock was certainly hampered by environmental factors, as it is today. Trypanosomiasis limits the distribution of zeboid cattle in humid and forested areas although certain breeds such as the Ndama and Muturu cattle have adapted to the tsetse challenge of this environment. Since the larger part of the area considered in this overview is forested, it is unsurprising that few indications for the presence of domestic cattle have been found. Archaeozoological evidence for the presence of cattle is available in the interlacustrine area at Tongo (Zaïre: third to sixth centuries AD), Remera I (third to fourth centuries AD), Kabuye XIV and Gisagara (Rwanda: eleventh to twelfth centuries AD), Akameru and Cyinkomane (Rwanda: ninth to thirteenth centuries AD), Ryamurari (Rwanda: seventeenth to twentieth centuries AD) and at a mound on the Ruzizi plain (Burundi: seventeenth to twentieth centuries AD). The only other finds are from Kamabanga (ninth to tenth centuries AD) and Kitala (thirteenth to fourteenth centuries AD) in Angola and from Mashita Mbanza (sixteenth to twentieth centuries AD) in Zaïre. In all these areas cattle are also kept today.

The sheep and goats that occur in forested areas of Africa today are dwarf breeds. Faunal remains from Nkang, where both dwarf sheep and goat were found, show that these adaptations go back in this region to at least the first millennium BC. Dwarf goat has also been found at Sanga (Zaïre: tenth to thirteenth centuries AD) in a lakeside site; the goat and sheep from Tongo (third to sixth centuries AD) are also of rather small stature (52 cm and 47 cm withers height, respectively). Evidence for small goats was also found at Mondongué (Central African Republic: sixteenth to twentieth centuries AD). Unidentified ovicaprines are present at Gisagara (Rwanda: eleventh to twelfth AD), whereas both sheep and goat were found at Akameru and Cyinkomane (Rwanda: ninth to thirteenth centuries AD). No indications on the size of those ovicaprines is available, however.

Domestic chicken has been found at several second millennium AD sites of which the tenth to thirteenth century AD finds from a Kisalian grave in the Upemba region and the ninth to thirteenth century finds of Akameru and Cyinkomane (Rwanda) are the earliest. The eighteenth to nineteenth century levels of Kiantapo in Shaba, Zaïre yielded also chicken remains of a relatively small breed. None of the galliform bird remains found at the third to sixth centuries AD site of Tongo belong to domestic fowl.

Evidence for other domestic species is scanty. Domestic pig has only been found at the site of Nkile, dated to the seventeenth to nineteenth centuries AD. Given the relatively young age of this material it is likely that the pigs kept at Nkile resembled those still found in the region, i.e. with slender legs and a long snout. Nkile and Lingbangbo (Central African Republic: fourteenth to seventeenth centuries AD) are the only Central African sites where the presence of domestic dog has been attested with certainty. It was impossible to decide whether the few canid remains from the Iron Age levels of Matupi belong to a jackal or to domestic dog.

3.2. Origin and propagation of Central African domesticates

The southward migration of pastoralists from the Sahara, roughly between 3000 BC and 2000 BC, is well documented (Gautier 1987). Several sites illustrate the further southward propagation of ovicaprines and domestic cattle in the savanna belt. South of Lake Chad small and large livestock dated to the second half of the first millennium BC have been found at Maroua in northern Cameroon (Quéchon 1974) and at Daima in Nigeria, close to the Cameroonian border (Connah 1976, 1981). The more extensive and more securely dated material from Gajiganna, Nigeria is much older (Breunig 1995, Breunig et al. 1996). The oldest cattle, sheep and goat remains occur in levels dated between about 2300 BC and 1900 BC.

Domestic stock has also been reported from Ntereso and Kintampo Rockshelter 6 in Ghana (Carter & Flight 1972). The former site is located in dry savanna woodland, whereas the latter is situated at the fringe of the high forest. At Ntereso, dated between about 2100 BC and 1300 BC, goat bones are rare and have been attributed to a dwarf breed. A bovine second phalanx is present as well but the authors hesitate in attributing it to domestic cattle. Kintampo Rockshelter 6 compares closely in date to Ntereso (Stahl 1985). The bovine remains found in a level dated between 2130 BC and 1740 BC were attributed to domestic cattle by Carter &

Flight (1972) on the basis of a comparative metrical analysis of modern cattle and African buffalo. These identifications have been questioned in the literature (Stahl 1985), but apparently the material has never been re-analyzed using the morphological criteria established by Peters (1988). Kintampo K6 also yielded bones of goat which were identified as a dwarf breed. The length of the mandibular row of a Kintampo specimen compares closely with measurements of dwarf goats from Malakal, Sudan (Carter & Flight 1972). The values observed for this measurement in normal goats is, however, similar to those found in the Kintampo specimen (Gautier 1987), despite the fact that dwarf goats have a skull which is, in comparison to normal goats, reduced in length, width and height (Chang & Landauer 1950). Given the location of Kintampo close to the rainforest, it is very likely that a dwarf breed of goat was indeed present, but a re-study of the material using other measurements and considering more modern material would be preferable to confirm this. The site of Nkang near Yaoundé, Cameroon, is located in a savanna–forest mosaic and yielded sheep and goat bones from a small breed. This shows that a process of dwarfing has gone along with the introduction of ovicaprines in the forested areas of Africa. No other west–central African sites are available to illustrate the further early spread of domestic animals to the south.

It has been suggested on the basis of linguistic evidence that demographic pressure forced Bantu-speaking populations to emigrate from their initial homeland in the Grassfields of Cameroon. These people, whose lifestyle was not adapted to the forest, would have migrated from west to east through the savanna to finally reach the interlacustrine region (Bastin et al. 1979, 1982, Meussen 1980). After a relatively long stay in this area, a farther southward and then westward migration would have brought them to the savanna south of the equatorial forest. Certain groups, however, would have been able to cross the rainforest belt by taking a more westerly route along the coastline or by travelling south along the waterways farther east (Bastin 1978, David 1980). Archaeological data to support these hypothetical itineraries are scanty (Maret 1986) but botanical evidence shows that, as a result of the last Holocene dry phase, savanna corridors existed in the forest (Schwartz et al. 1990, Roche 1991) which would have made it easier for Bantu populations to reach the southern savannas. In that area both the eastern and the western migratory waves would have met. Thus far no evidence is available for domestic animals in the sites of the Grassfields of Cameroon. The same is true for Oveng, which is the only early site with fauna in west–central Africa. People relied mainly on fishing and shell collecting. Mammals were heavily under-represented at this site which yielded pottery, iron slag, *Elaeis* and *Coula*. In the region just south of the Central African rainforest, sites with domesticates dating to the end of the first millennium BC or the first half of the first millennium AD are also absent. The Ngovo Group in Lower Zaïre is the only known archaeological assemblage, dated shortly before our era and located at the southern edge of the forest, that comprises pottery and polished stone tools. It has been suggested that the Ngovo Group represents the first, or in any case very ancient, Bantu-speaking populations that would have crossed the rainforest and settled in the savannas to the south (de Maret 1986). Only two sites of the Ngovo Group, Ngovo and Dimba, have yielded fauna. Both assemblages are small and lack domestic animals. Hence the archaeozoological data available thus far do not support the hypothesis that early Bantu-speaking populations were responsible for the propagation of food production. It is not clear to what extent the present faunal record can be generalized. It should be underlined that the assemblages from Ngovo and Dimba caves were small and that the excavated deposits may have been partly related to ritual practices. The cave sites of Cameroon lie in the Grassfields but their immediate surroundings were heavily wooded. Sites like Shum Laka may not represent living sites, but could instead have been places where hunters halted temporarily and where their dead were buried. Firm botanical evidence for food production is not available either. Methodological problems do not allow confirmation of the domestic status of the sporadic remains of *Elaeis* and *Canarium* (Eggert 1993).

The interlacustrine region probably played a more important role in the propagation of domestic animals. It is very suitable for herding (Epstein 1971i:410) as a result of the extensive grassland vegetation in highly elevated areas where tsetse fly and trypanosomiasis are lacking. It is not clear whether the first inhabitants of the interlacustrine region were Bantu-speaking populations coming from the west or if they were Nilotic pastoralists coming from the east, or a mixture of both populations. Both groups may have started their migrations towards this area during the second millennium BC. Palynological evidence shows that early Iron Age groups, belonging to the Urewe culture, had a profound impact on the environment, especially during the first centuries of our era. The cumulative effect of the overexploitation of the environment and a dry climatic phase negatively influenced human habitation in the area. Especially in Rwanda, this resulted in a reduction of the exploited territory during the first few centuries. This in turn allowed a regeneration of the environment. The following late Iron Age period was characterized by the use of new metallurgical techniques and of roulette-decorated pottery. From this period onwards the need for more pasture and new agricultural land increased pressure on the environment (Roche 1991).

It has been suggested in the literature that a small breed of zebu cattle was first introduced in the interlacustrine region during the first millennium BC. A larger breed of humpless longhorn cattle would have been brought in during the late Iron Age by a new wave of Nilotic pastoralists coming from the east. The archaeozoological data to support this model are very scanty, however, and new evidence from Tongo seems to contradict it. The third to fourth centuries AD cattle finds from Remera I, Rwanda comprise only two teeth. On the basis of the length of the lower third molar it is supposed that a small breed of about 110 cm shoulder height was present (Van Grunderbeek et al. 1983). This could correspond to the small zebu cattle that was supposedly introduced from the northeast in the first millennium BC of which the present-day *inkuku* cattle, a small breed of shorthorn zebu, might be a descendant (Gotanègre et al. 1974, Roche 1991:201).

According to oral tradition, the large longhorn Batutsi cattle would have been introduced in the region later during the twelfth (Kagame 1972) or the fourteenth centuries AD (Vansina 1962). The seventeenth to twentieth century faunal material from Ryamurari comprises two almost complete skulls with the horn cores

Table 9.7 Contribution of hunting, fishing and stock herding at some Central African sites and at Gajiganna, Nigeria. The calculations were based on the relative abundance of the remains. Sample size indicates NISP. (* dog remains have not been included; ** dog remains have been included because cut marks were present.)

	sample size	fishing	hunting	herding
Gajiganna A	1610	24.2	7.4	68.4
Gajiganna B	1336	60.7	4.6	34.7
Nkang	40	7.5	80.0	12.5
Lingbangbo & Mondongué	109	29.4	63.3	7.3*
Mashita Mbanza	14	–	78.6	21.4
Nkile	112	88.4	8.0	3.6**
Tongo late Iron Age	75	28.0	25.4	44.6
Tongo early Iron Age	83	32.6	31.5	36.9
Akameru	51	–	64.7	35.3
Cyinkomane	118	–	44.1	55.9
Ruzizi	753	96.3	0.4	3.3

preserved. They are very long and lyre-shaped and correspond well to the Batutsi longhorn (Tshilema Tshiluka 1983). Gautier (1983b) estimates the shoulder height of the Ryamurari cattle at 110–130 cm, which would not exclude an attribution to the large breed. At the ninth to thirteenth centuries sites of Akameru and Cyinkomane, shoulder height was estimated at 105–130 cm (Gautier 1983a). A closer look at the available measurements and comparison with the material from Tongo shows, however, that the observed variation is not very large (Table 9.6). With the exception of a large second phalanx from the early Iron Age levels of Tongo, the measurements do not seem to exceed the expected variation in a single breed. The large size of a single early Iron Age second phalanx from Tongo, furthermore, contradicts the suggestion that only a small breed would have preceded the larger Batutsi type. Thus, the raw data are insufficient to draw any reasonable conclusions concerning the breeds of ancient cattle in the interlacustrine region.

3.3. The importance of domestic animals in subsistence

There seems to be considerable variation in the degree of reliance on livestock herding in Africa. Typical pastoralist communities relying almost exclusively on domestic stock have been described from several sites outside Central Africa. Examples are the 2600–1000 BC sites from Wadi Howar in Sudan (Van Neer & Uerpmann 1989) and several Pastoral Neolithic sites in East Africa (Marshall & Stewart 1994). Other sites such as Gogo Falls in Kenya (ibid.) and the Gajiganna series in Nigeria (Breunig et al. 1996) show that, in addition to keeping of domestic stock, hunting and fishing could also play a major role in the subsistence of pastoralists. Table 9.7 indicates the relative importance of hunting, fishing and herding, expressed as percentages of the number of identified specimens (NISP) of the different taxa for a number of sites. The data from Gogo Falls cannot be presented in the

Table 9.8 Relative abundance of domestic and wild animals (excluding fish) of Central African sites compared to that of Gajiganna, Nigeria and Gogo Falls, Kenya. The sites have been arranged in descending order of the contribution of domestic animals.

	sample size	hunting	herding
Gajiganna A	1221	9.8	90.2
Ruzizi	28	10.7	89.3
Gajiganna B	525	11.6	88.4
Tongo late Iron Age	54	35.2	64.8
Tongo early Iron Age	58	43.1	56.9
Cyinkomane	118	44.1	55.9
Gogo Falls	612	46.4	53.6
Akameru	51	64.7	35.3
Nkile	13	69.2	30.8
Mashita Mbanza	14	78.6	21.4
Nkang	37	86.5	13.5
Lingbangbo & Mondongué	77	89.6	10.4

same way since the analysis of the mammal fauna considered only teeth, thereby differing from the approach at the other sites where all the mammal remains were included. The importance of hunting versus stock herding has been presented in Table 9.8 by omitting the fish remains. This also allows the inclusion of the data from Gogo Falls if we assume that the ratio given by the dental remains is comparable to the one that would be obtained if post-cranial remains were also included. The ratio of domestic versus wild animals (excluding fish) is about 9:1 at Gajiganna and at the mound in the Ruzizi plain. In the interlacustrine region, Ryamurari, Gisagara, Kabuye and Remera have also yielded a fauna dominated by domestic stock, although it should be emphasized that the available samples are small. More extensive faunal assemblages will be needed in order to confirm that specialized, livestock-oriented groups occurred since the early Iron Age.

Other sites from the interlacustrine region (Tongo, Cyinkomane and Akameru) have higher proportions of wild animals. Their contribution is comparable to that seen at Gogo Falls. The relatively high proportion of wild animals at Gogo Falls is seen as a possible result of environmental constraints (Marshall & Stewart 1994). Drought and disease on that site may have resulted in a diversification of subsistence strategies. The palaeoenvironmental reconstruction based on the wild fauna indicates a mosaic habitat including grassland, bush and woodland in which tsetse fly and trypanosomiasis probably occurred. The faunas from Tongo, Cyinkomane and Akameru comprise substantial numbers of remains from woodland species indicating that at the time of the considered Iron Age occupation the environment was more forested than today. As a result of the high elevation, between 1700 m and 1800 m a.s.l., the environment of both sites in Rwanda was free of trypanosomiasis. The surroundings of Tongo lie at an altitude between 1400 m and 1500 m and are suitable for herding as well. Cyinkomane and Akameru being cave sites, it

Table 9.9 Summary of radiometric dates for Central and west–central African archaeological sites with domestic animal remains.

Country	Site	Dates (bp)	Dates (calibrated, 1 sigma)	Sheep/Goat	Cattle	Dog	Chicken
Nigeria	Gajiganna A&B		RANGE: 1520–810 BC	X	X	X	X
		3150±70 bp (UtC-2330)	1490–1318 BC				
		3140±110 bp (UtC-2332)	1520–1310 BC				
		2960±50 bp (UtC-2795)	1286–1022 BC				
		2930±60 bp (UtC-2329)	1236–1009 BC				
		2740±50 bp (UtC-2331)	967–813 BC				
Cameroon	Nkang		RANGE: 830–120 BC	X			
		2580±70 bp (LV-1940)	830–540 BC				
		2490±80 bp (Lv-1943)	790–520 BC				
		2490±110 bp (Lv-1944D)	800–510 BC				
		2420±70 bp (Lv-1939)	760–400 BC				
		2400±60 bp (Lv-1942)	760–390 BC				
		2340±70 bp (Lv-1941)	760–260 BC				
		2310±90 bp (Lv-1945)	520–200 BC				
		2170±80 bp (Lv-1946)	370–120 BC				
C.A.R.	Lingbangbo		RANGE: AD 1300–1660	X		X	
		559±77 bp (Bdy-582)	AD 1300–1430				
		430±180 bp (Bdy-255)	AD 1300–1660				
C.A.R.	Mondongué	140±240 bp (BM-2425)	AD 1529–…	X		?	
Congo (Zaïre)	Tongo I, 1		RANGE: AD 240–550	X	X		
		1690±80 bp (Gif-9010)	AD 240–430				
		1620±90 bp (Gif-9006)	AD 260–550				
Congo (Zaïre)	Sanga		RANGE: AD 900–1300	X			
Congo (Zaïre)	Matupi (tr.E., 15–20 cm)	720±45 bp (GrN-7244)	AD 1240–1370			?	X

Table 9.9 (cont'd)

Country	Site	Dates (bp)	Dates (calibrated, 1 sigma)	Sheep/Goat	Cattle	Dog	Chicken
Congo (Zaïre)	Mashita Mbanza	265±55 bp (Hv-13451)	RANGE: AD 1510–1800	X	?		
		260±55 bp (Hv-13453)	AD 1510–1800				
		205±55 bp (Hv-13452)	AD 1640–...				
		140±55 bp (Hv-13454)	AD 1670–...				
Congo (Zaïre)	Nkile		RANGE: AD 1600–1900	X		X	
Rwanda	Remera I	1730±30 bp (GrN-9663)	AD 250–340		X		
Rwanda	Akameru		RANGE: AD 810–1260	X	X		X
		1075±95 bp (GrN-7671)	AD 810–1040				
		845±75 bp (GrN-7672)	AD 1050–1260				
Rwanda	Gisagara II	925±30 bp (GrN-9661)	AD 1030–1160	X	X		
Rwanda	Ryamurari		RANGE: post AD 1650	X	X		
		185±45 bp (GrN-7589)	AD 1650–...				
		155±45 bp (GrN-7588)	AD 1660–...				
Burundi	Ruzizi Plain Mound		RANGE: post AD 1650		X		
		190±45 bp (GrN-6109)	AD 1650–...				
		100±45 BP (GRN-6108)	AD 1690–1930				
Angola	Kamabanga I	1120±60 bp (Gif-6182)	AD 860–990		X		
Angola	Kitala II	720±60 bp (Gif-6011)	AD 1220–1380		X		

had been suggested by Gautier (1983a) that the high ratio of wild animals, and also the observed variation in their contribution, could be related to the function of the caves as refuge places during times of instability. Since Tongo is an open air site the frequent occurrence of wild animals cannot be explained as a result of troubles in the area due to migrations or other political factors. Moreover, it is striking that at Tongo the ratio of wild animals is equally high during both early and late Iron Age. A mixed economy of herding and hunting may have been a deliberate choice in this area, differing from the present-day lifestyle of stock-oriented pastoralists, agricultural groups or hunter–gatherer societies.

South of the equatorial forest, at Mashita Mbanza, the contribution of domestic stock (ovicaprines and cattle) is low despite the location of the site in an open environment suitable for herding. It is not clear whether this is an artefact related to the low number of identifiable remains collected at that site. The remaining sites, Nkile, Nkang, Lingbangbo and Mondongué, are located in the rainforest or in forest–savanna mosaic and yielded low numbers of domestic animals (mainly dwarf ovicaprines). Although these data should be used with caution because of the small sample sizes, it seems that the keeping of large herds of domestic stock was never practised, possibly because the demographic pressure in these areas was not so high as in savanna environment.

Note

1. This text presents research results of the Belgian programme on Inter-University Poles of Attraction initiated by the Belgian State, Prime Minister's Office, Federal Services. I also thank Prof Achilles Gautier (University of Ghent) for putting at my disposal the unpublished faunal data from some sites in Rwanda. Mr Alain Reygel (Tervuren) prepared the drawing for this paper.

References

Asombang, R. 1988. *Bamenda in prehistory: the evidence from Fiye Nkwi, Mbi crater and Shum Laka rockshelters*. PhD thesis, Institute of Archaeology, University of London.

Asombang, R. & P. de Maret 1992. Re-investigating Shum Laka: the December 1991–March 1992 campaign. *Nsi* **10/11**, 13–16.

Bastin, Y. 1978. Statistique grammaticale et classification des langues bantoues. *Linguistics in Belgium* **2**, 17–37.

Bastin, Y., A. Coupez, B. de Halleux 1979. Statistique lexicale et grammaticale pour la classification des langues bantoues. *Bulletin de l'Académie royale des Sciences d'Outre-Mer* **3**, 375–87.

Bastin, Y., A. Coupez, B. de Halleux 1982. Classification lexicostatistique des langues bantoues (214 relevés). *Bulletin de l'Académie royale des Sciences d'Outre-Mer* **27**(2), 173–99.

Breuil, H. 1952. Les figures incisées et ponctuées de la grotte de Kiantapo (Katanga). *Annales du Musée du Congo Belge, sér. in 8°, Sciences de l'Homme. Préhistoire* **1**, 1–33.

Breunig, P. 1995. Gajiganna und Koduga. Zur frühen Besiedlung des Tschadbeckens in Nigeria. *Beiträge zur Allgemeinen und Vergleichenden Archäologie* **15**, 3–48.

Breunig, P., K. Neumann, W. Van Neer 1996. New research on the holocene settlement and environment of the Chad Basin in Nigeria. *The African Archaeological Review* **13**, 111–45.

Brooks, A.S. & C.C. Smith 1987. Ishango revisited: new age determinations and cultural interpretations. *African Archaeological Review* **5**, 65–78.

Brooks, A.S., D.M. Helgren, J.S. Cramer, A. Franklin, W. Hornyak, J.M. Keating, R.G. Klein, W.J. Rink, H. Schwarcz, J.N.L. Smith, K. Stewart, N.E. Todd, J. Verniers, J.E. Yellen 1995. Dating and context of three Middle Stone Age sites with bone points in the Upper Semliki Valley, Zaïre. *Science* **268**, 548–53.

Cabu, F. 1938. Premières notes d'ensemble de la mission de recherches préhistoriques au Katanga (Congo Belge). *Bulletin de la Société Préhistorique Française* **25**, 172–86.

Carter, P.L. & C. Flight 1972. A report on the fauna from the sites of Ntereso and Kintampo Rock Shelter 6 in Ghana: with evidence for the practice of animal husbandry during the second millennium B.C. *Man* **7**, 277–82.

Chang, T.K. & W. Landauer 1950. Observations on the skeleton of African dwarf goats. *Journal of Morphology* **86**, 367–76.

Claes, P. 1985. *Contribution à l'étude de céramiques anciennes des environs de Yaoundé*. Mémoire de licence, Université Libre de Bruxelles.

Clist, B. 1987. Early Bantu settlements in west Central Africa: a review of recent research. *Current Anthropology* **28**, 380–82.

Clist, B. 1989. La campagne de fouilles 1989 du site Age du Fer ancien d'Oveng: province de l'Estuaire (Gabon). *Nsi* **5**, 15–18.

Connah, G. 1976. The Daima sequence and the prehistoric chronology of the Lake Chad region of Nigeria. *Journal of African History* **17**, 321–52.

Connah, G. 1981. Man and a lake. In *Le Sol, la Parole et l'Ecrit. Mélanges en Hommage à Mauny*. Société Française d'Histoire d'Outre-Mer (ed.), 161–78. Paris: Société Française d'Histoire d'Outre-Mer.

Cooke, H.B.S. & A.F. Wilkinson 1978. Suidae and Tayassuidae. In *Evolution of African mammals*, V.J. Maglio & H.B.S. Cooke (eds), 435–82. Cambridge, Mass.: Harvard University Press.

Cornelissen, E., J. Moeyersons, P. de Maret 1995. Fouilles archéologiques à Shum Laka (Cameroun). *Nouvelles de la Science et des Technologies* **13**, 319–22.

David, N. 1980. Early Bantu expansion in the context of Central African prehistory: 4000–1 BC. Colloque du C.N.R.S. *L'expansion bantoue*, n.s. **2**: 265–78.

de Heinzelin de Braucourt, J. 1957. Les fossiles d'Ishango. *Exploration du Parc National Albert*, 2e série, fasc. 1. Brussels: Institut des Parcs Nationaux du Congo Belge.

de Maret, P. 1980. Preliminary report on 1980 fieldwork in the Grassfields and Yaounde, Cameroon. *Nyame Akuma* **17**, 10–12.

de Maret, P. 1982. New survey of archaeological research and dates for West Central and North Central Africa. *Journal of African History* **23**, 1–15.

de Maret, P. 1985a. Fouilles archéologiques dans la vallée du Haut-Lualaba, Zaïre. II, Sanga et Katongo, 1974. *Annales du Musée Royal de l'Afrique Centrale, Sciences Humaines* No. 120. Tervuren: Musée Royal de l'Afrique Centrale.

de Maret, P. 1985b. Recent archaeological research and dates from Central Africa. *Journal of African History* **26**, 129–48.

de Maret, P. 1986. The Ngovo Group: an industry with polished stone tools and pottery in Lower Zaïre. *African Archaeological Review* **4**, 103–33.

de Maret, P. 1991. La recherche archéologique au Cameroun. In *Actes de la Journée d'Etude "La Recherche en Sciences humaines au Cameroun" (Bruxelles, 20 juin 1989)*, P. Salmon & J.-J. Symoens (eds), 37–51. Brussels: Académie Royale des Sciences d'Outre-Mer.

de Maret, P., B. Clist, W. Van Neer 1987. Résultats des premières fouilles dans les abris sous roche de Shum Laka et Abeke au nord-ouest du Cameroun. *L'Anthropologie* **91**, 559–84.

de Maret, P., R. Asombang, E. Cornelissen, P. Lavachery, J. Moeyersons, W. Van Neer 1993. Preliminary results of the 1991–1992 field season at Shum Laka, Northwestern Province, Cameroon. *Nyame Akuma* **39**, 13–15.

de Maret, P., R. Asombang, E. Cornelissen, P. Lavachery, J. Moeyersons 1995. Continuing research at Shum Laka rock shelter, Cameroun (1993–1994 field season). *Nyame Akuma* **43**, 2–3.

De Meulemeester, J. & A. Waleffe 1973. Résultats des travaux de fouilles dans une butte de la plaine de la Ruzizi (Burundi). *Africa-Tervuren* **19**, 16–24.

Deschamps, R. 1986. Remains of flora. *African Archaeological Review* **4**, 113 only.

Dos Santos Jnr, R. & C. Ervedosa 1970. A estaçao arqueologica de Benfica. *Ciencias Biologicas (Luanda)* **1**, 33–51.

Eggert, M.K.H. 1983. Remarks on exploring archaeologically unknown rain forest territory: the case of Central Africa. *Beiträge zur Allgemeinen und Vergleichenden Archäologie* 5, 283–322.

Eggert, M.K.H. 1984. Imbonga und Lingonda: zur frühesten Besiedlung des zentralafrikanischen Regenwaldes. *Beiträge zur Allgemeinen und Vergleichenden Archäologie* 6, 247–88.

Eggert, M.K.H. 1987. Imbonga and Batalimo: ceramic evidence for early settlement of the equatorial `rainforest. *African Archaeological Review* 5, 129–45.

Eggert, M.K.H. 1993. Central Africa and the archaeology of the equatorial rainforest: reflections on some major topics. In *The archaeology of Africa. Food, metals and towns*, T. Shaw, P. Sinclair, B. Andah, A. Okpoko (eds), 289–329. London: Routledge.

Epstein, H. 1971. *The origin of the domestic animals of Africa* [2 volumes]. New York: Africana Publishing.

Flannery, K.V. 1972. The cultural evolution of civilizations. *Annual Review of Ecology and Systematics* 3, 399–426.

Gautier, A. 1983a. Les restes osseux des sites d'Akameru et de Cyinkomane (Ruhengeri, Rwanda). In *L'histoire archéologique du Rwanda*, F. Van Noten (ed.), *Annales du Musée Royal de l'Afrique Centrale, Sciences Humaines* No. 112, 104–20. Tervuren: Musée Royal de l'Afrique Centrale.

Gautier, A. 1983b. Les restes de mammifères du gisement protohistorique Hima à Ryamurare (Rwanda, 17e–18e siècle). In *L'histoire archéologique du Rwanda*, F. Van Noten (ed.), *Annales du Musée Royal de l'Afrique Centrale, Sciences Humaines* No. 112, 121–6. Tervuren: Musée Royal de l'Afrique Centrale.

Gautier, A. 1987. Prehistoric men and cattle in North Africa: a dearth of data and a surfeit of models. In *Prehistory of arid North Africa*, A.E. Close (ed.), 163–87. Dallas: SMU Press.

Gautier, A. n.d. Les restes de mammifères de Gisagara près de Butare. Unpublished report.

Geyh, M.A. & P. de Maret 1982. Histogram evaluation of ^{14}C dates applied to the first complete Iron Age sequence from West Central Africa. *Archaeometry* 24, 158–63.

Gotanègre, J.F., C. Prioul, P. Sirven 1974. *Géographie du Rwanda*. Brussels: De Boeck.

Hopwood, A.T. & X. Misonne 1959. Mammifères fossiles. *Exploration du Parc National Albert. Mission de Heinzelin de Braucourt, 1950*, fasc. 4, 111–19. Brussels: Institut des Parcs Nationaux du Congo Belge.

Kagame, A. 1972. *Un abrégé de l'ethno-histoire du Rwanda*. Université nationale du Rwanda, collection Muntu, 3(1).

Kanimba, M. & S. Gatare 1992. Archaeological and ethnoarchaeological research in the zones of Rutshuru and Masisi in Northern Kivu. *Nyame Akuma* 38, 66–71.

Koté, L. 1992. *Naissance et développement des économies de production en Afrique centrale*. Unpublished Doctoral thesis, Paris X.

Lanfranchi, R. & B. Clist 1987. Mission de recherches et de formation en R.P. d'Angola, October 1987. *Nsi* 2, 4–8.

Maes, J. 1935. Le camp de Mashita Mbansa et les migrations des Bapende. *Congo* 2(5), 713–24.

Maquet, E. & J. Hiernaux 1969. Un site à poterie cannelée en République Démocratique du Congo: Kawezi (vallée de la Ruzizi). *Journal de la Société des Africanistes* 39, 159–71.

Marshall, F. & K. Stewart 1994. Hunting, fishing and herding pastoralists of western Kenya: the fauna from Gogo Falls. *Archaeozoologia* 7, 7–27.

Mbida, M.C. 1996. *L'émergence de communautés villageoises au Cameroun méridional. Etude archéologique des sites de Nkang et de Ndindam*. Doctoral thesis, Université Libre de Bruxelles.

Meussen, A. 1980. Apports nouveaux en matière de classification et du degré d'archaïsme des langues bantoues. Colloque du C.N.R.S. *L'expansion bantoue*, n.s. 2, 457–72.

Mortelmans, G. 1952. Les dessins rupestres gravés, ponctués et peints du Katanga. Essai de synthèse. *Annales du Musée du Congo Belge, sér. in 8°, Sciences de l'Homme. Préhistoire* 1, 33–55.

Mortelmans, G. & P. de Maret 1989. Résultats des fouilles de 1955 devant la grotte de Kiantapo au Shaba. In *Contribution to the archaeozoology of Central Africa*, W. Van Neer (ed.), *Annales du Musée Royal de l'Afrique Centrale, Sciences Zoologiques* No. 259, 137–40. Tervuren: Musée Royal de l'Afrique Centrale.

Peters, J. 1988. Osteomorphological features of the appendicular skeleton of African buffalo, *Syncerus caffer* (Sparrman 1779) and of domestic cattle, *Bos* primigenius f. taurus Bojanus, 1827. *Zeitschrift für Säugetierkunde* 53, 108–23.

Peters, J. 1990. Late Pleistocene hunter–gatherers at Ishango (eastern Zaïre): the faunal evidence. *Revue de Paléobiologie* **9**, 73–112.

Pierot, F. 1987. *Etude ethnoarchéologique du site de Mashita Mbanza (Zaïre)*. Mémoire de licence, Université Libre de Bruxelles.

Pinto, L.P. 1988. Le Musée National d'Archéologie de Benguela (Angola): bilan des premiers travaux: 1979–1987. *Nsi* **3**, 5–14.

Quéchon, G. 1974. Un site protohistorique de Maroua (Nord-Cameroun). *Cahiers de Sciences Humaines, ORSTOM* **11**, 3–46.

Roche, E. 1991. Evolution des paléoenvironnements en Afrique centrale et orientale au Pléistocène supérieur et à l'Holocène. Influences climatiques et anthropiques. *Bulletin de la Société géographique de Liège* **27**, 187–208.

Schwartz, D., H. de Foresta, R. Deschamps, R. Lanfranchi 1990. Découverte d'un premier site de l'Age du Fer ancien (2.110 B.P.) dans le Mayumbe congolais. Implications paléobotaniques et pédologiques. *Comptes Rendus hebdomadaires des Séances de l' Académie des Sciences Paris* **310**(2), 1293–8.

Stahl, A.B. 1985. Reinvestigation of Kintampo 6 rock shelter, Ghana: implications for the nature of culture change. *The African Archaeological Review* **3**, 117–50.

Tchernov, E. 1993. The impact of sedentism on animal exploitation in the southern Levant. In *Archaeozoology of the Near East. Proceedings of the first international symposium on the archaeozoology of southwestern Asia and adjacent areas*, H. Buitenhuis & A.T. Clason (eds), 10–26. Leiden: Universal Book Services.

Tshilema Tshiluka 1983. Ryamurare, capitale de l'ancien Royaume du Ndorwa. In *L'histoire archéologique du Rwanda*, F. Van Noten (ed.), *Annales du Musée Royal de l'Afrique Centrale, Sciences Humaines* No. 112, 149–53. Tervuren: Musée Royal de l'Afrique Centrale.

Van Grunderbeek, M.C., E. Roche, H. Doutrelepont 1982. L'âge du fer ancien au Rwanda et au Burundi. Archéologie et environnement. *Journal de la Société des Africanistes* **52**, 5–88.

Van Grunderbeek, M.C., E. Roche, H. Doutrelepont 1983. *Le premier Age du Fer au Rwanda et au Burundi. Archéologie et environnement*. Institut National de Recherche Scientifique, Publication 23. Butare: Institut National de Recherche Scientifique.

Van Neer, W. 1986. Faunal remains. *African Archaeological Review* **4**, 109 and 113.

Van Neer, W. 1989. Contribution to the archaeozoology of Central Africa. *Annales du Musée Royal de l'Afrique Centrale, Sciences Zoologiques* 259. Tervuren: Musée Royal de l'Afrique Centrale.

Van Neer, W. 1990. Les faunes de vertébrés quaternaires en Afrique centrale. In *Paysages quaternaires de l'Afrique centrale atlantique*, R. Lanfranchi & D. Schwartz (eds), 195–220. Paris: ORSTOM.

Van Neer, W. & B. Clist 1991. Le site de l'Age du fer Ancien d'Oveng (Province de l'Estuaire, Gabon), analyse de sa faune et de son importance pour la problématique de l'expansion des locuteurs bantu en Afrique Centrale. *Comptes Rendus des Séances de l' Académie des Sciences Paris* **312**(II), 105–10.

Van Neer, W. & H.-P. Uerpmann 1989. Palaeoecological significance of the Holocene faunal remains of the B.O.S.-missions. In *Forschungen zur Umweltgeschichte der Ostsahara*, R. Kuper (ed.), *Africa Praehistorica* **2**, 307–41.

Van Noten, F. 1977. Excavations at Matupi cave. *Antiquity* **51**, 35–40.

Van Noten, F. (ed.) 1983. Histoire archéologique du Rwanda. *Annales du Musée Royal de l'Afrique Centrale, Sciences Humaines* No. 112. Tervuren: Musée Royal de l'Afrique Centrale.

Vansina, J. 1962. *L'évolution du royaume Rwanda des origines à 1900*. Mémoire de l'Académie royale des Sciences d'Outre-Mer, Classe des Sciences morales et politiques, 26(2). Brussels: Académie Royale des Sciences d'Outre-Mer.

von den Driesch, A. 1976. *A guide to the measurement of animal bones from archaeological sites*. Peabody Museum Bulletin 1. Cambridge, Mass.: Peabody Museum.

von den Driesch, A. & J. Boessneck 1974. Kritische Anmerkungen zur Widerristhöhenberechnung aus Längenmassen vor- und frühgeschichtlicher Tierknochen. *Säugetierkundliche Mitteilungen* **22**, 325–48.

Warnier, J.P. & R. Asombang 1982. Archaeological research in the Bamenda Grassfield, Cameroun. *Nyame Akuma* **21**, 3–4.

CHAPTER TEN

The origins and spread of domestic animals in East Africa

Fiona Marshall

1. Introduction

It is possible on the basis of the distribution of wild progenitors of domestic cats (African wild cat, *Felis sylvestris* (Schreber 1777)), donkeys (The Nubian wild ass, *Equus africanus* (Fitzinger 1857)), or the domestic guinea-fowl (Helmeted guinea-fowl, *Numida meleagris* (Linnaeus 1758)), that these animals could have been domesticated in eastern Africa. In addition, classical sources, and depictions of elephants at Meroitic sites, suggest that by 2,000 years ago elephants (*Loxodonta africana*, Blumenbach 1797) from the Sudan and northern Ethiopian regions were systematically captured and tamed (Shinnie 1967, Scullard 1974, Burstein 1989). But there is currently no zooarchaeological evidence that any of the diverse and abundant mammalian fauna of eastern Africa were domesticated. Instead, there seems to be a pattern of long-term successful use of wild mammals, with a gradual spread of Near eastern and North African domesticates, especially cattle, sheep and goat, through eastern Africa from the Sudan and more arid areas to the north 6,300–2,000 years ago (Ambrose 1984a, Gautier 1984a,b, Barthelme 1985, Marshall 1989, Marean 1992).

The cattle (*Bos taurus cf africanus*) were probably domesticated in the Sahara from North African populations of *B. primigenius*, perhaps as early as 9000 bp (Gautier 1984c, Grigson 1991, Ch. 4 in this volume; Bradley et al. 1996, Bradley & Loftus Ch. 13 in this volume; cf. Clutton Brock 1989, Smith 1992, MacDonald Ch. 1 in this volume). The sheep and goats are thought to be ultimately of Near eastern origin, occurring in North Africa by the seventh millennium bp in North Africa (Gautier 1987, Smith 1992, Clutton Brock 1993). Contacts with the Sahara, and movement of people into the Sudanese and East African regions from the north, relate to the appearance of domestic animals such as taurine cattle, sheep and goat which are found in the Sudan by 6000 bp and in East Africa by 4000–3500 bp. By contrast, trade in the Nile Valley, Red Sea and Indian Ocean is likely to have influenced the appearance of animals such as camels (Bulliet 1990), and humped cattle (*B. indicus*) in East Africa (Payne 1964, Epstein 1971, Marshall 1989).

The early pattern of adoption of food production in eastern Africa appears to be slow and patchy, and often associated with pastoralism, rather than with cultivation (Marshall 1990b). Mechanisms envisioned for this spread include limited movement of pastoralists southwards, and the adoption of herding by local hunter–gatherers. However, many prehistoric hunter–gatherer groups in eastern Africa remained focused on wild resources. This appears to contrast with patterns of adoption of food production in the Near East (Bar Yosef & Meadow 1995) and historic patterns of rapid and widespread adoption of new livestock such as zebu cattle in East Africa (Epstein 1957, 1971). Why is this pattern of slow and sporadic adoption of food production so characteristic of much of Africa, and especially marked in East Africa, prior to the beginnings of metallurgy?

Field studies of adoption of food production by contemporary African hunter–gatherers (Brooks et al. 1984, ten Raa 1986, Marshall et al., in prep.) raise interesting questions for zooarchaeologists regarding relationships between the degree of social, cultural and economic change associated with adoption of domestic livestock, and larger patterns of livestock spread. I hypothesize here that herding is more challenging for hunter–gatherers than cultivation. And that slow and patchy spread of livestock is to be expected when adoption of new animals, such as cattle, small stock, or camels, is associated with major socio-economic or ideological shifts within a group. By contrast rapid and widespread patterns of adoption of domesticates are more likely to be associated with conditions of minimal internal social change, as for instance, when new cattle breeds are adopted by established herders. In following sections of this paper, I review evidence for the appearance of domestic animals in the archaeological record, discuss regional patterns of spread of food production, and explore more fully factors affecting the adoption and rate of spread of domestic taxa in East Africa.

2. The evidence

This review uses a broad definition of East Africa, concentrating on zooarchaeological material from the horn of Africa: Eritrea, Ethiopia, Somalia, and then from Kenya, and Tanzania. Although Uganda is important, and falls within my defined area, there is so little faunal data from the area that I do not incorporate it in the current review (research is currently in progress by Robertshaw, Reid, and others). Because the patterns of spread of domesticates in East Africa are integrally tied to the Sudanese data, I will start with this region.

2.1. The Sudan

Gautier (1984a,b, 1986) and Peters (1986, 1989, 1992) have recently conducted extensive analyses of faunal assemblages from the Sudan and discussed much of the material from this region. Most prehistoric faunal data comes from sites in the Sudanese Middle Nile around Khartoum. These are the datasets closest to East African sites considered here, lying only about 400 km from western Eritrea, but still more than 1,500 km from northern Kenya (Fig. 10.1).

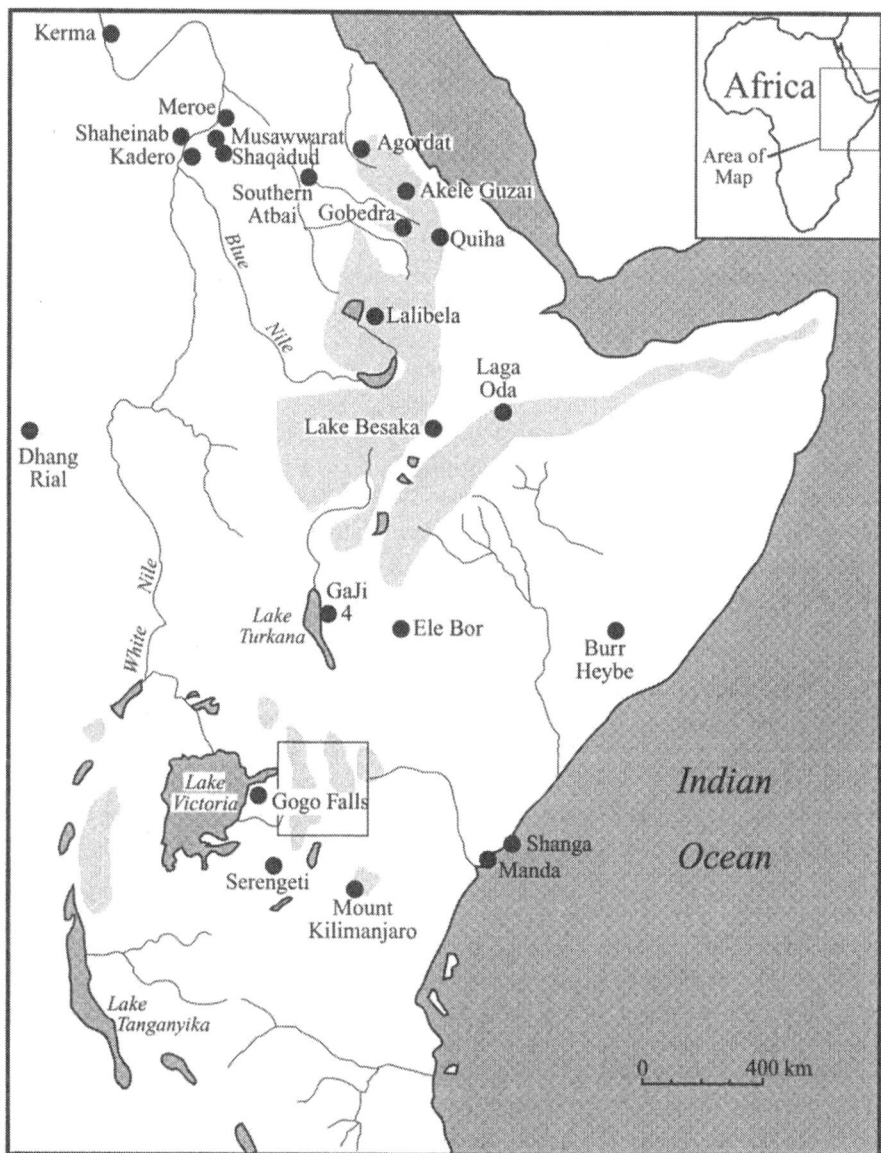

Figure 10.1 Key sites and localities mentioned in the text (see also Fig. 10.7).

By the sixth millennium bp small numbers of domestic cattle and small stock were present at the Khartoum neolithic site of Esh Shaheinab (Gautier 1984a, Peters 1986). However at Kadero, during the sixth millennium bp, domestic animals are well represented including cattle, sheep, and probably domestic dogs (Gautier 1984b, Kryzaniak 1978, 1984, 1991). The cattle from Kadero are thought to have been quite large, 110–130 cm at the withers (Gautier 1984b). Cattle bucrania in graves at the later neolithic (fifth millennium bp, *c.* 4850–4600 bp) site of Kadada near Shaqadud suggest that animals were long-horned (Gautier 1986, 1987).

In fourth millennium bp contexts at Shaqadud (*c.* 4046–3615 bp) the fauna is predominantly wild. Guinea-fowl remains from Shaqadud do not differ from modern wild specimens, and they are the only guinea-fowl reported from East African Holocene sites (Peters 1986, 1991; for West Africa, see MacDonald 1992, MacDonald & MacDonald Ch. 8 in this volume). The few domestic specimens from the fourth millennium bp at Shaqadud include sheep/goat, cattle, and one metatarsus of a domestic donkey (Peters 1986, 1991). In the eastern Sudan, during the fifth millennium bp, cattle and small stock are found at Butana phase sites in the southern Attbai. In the fourth millennium bp, cattle, small stock and donkey are documented from the Gash phase at Mahal Teglinos (Geraads 1983, Peters in Sadr 1991:138,142, Fattovich 1993).

It appears that cattle were present near the Egyptian border, at A-Group sites in Lower Nubia between the sixth to fifth millennia bp, although the faunal evidence is very limited (Reisner 1910, Nordstrom 1972). Nevertheless, it is possible that domesticated cattle were on the Egypt/Sudan frontier as early as any in the Khartoum area. Long-horned cattle used by later C-group pastoralists in Lower Nubia date to approximately 4300–3500 bp, using evidence from rock art (Adams 1977, Gautier 1984a), depictions on pottery (Adams 1977:153), and cattle skulls from C-group graves at Faras (Grigson 1991). Gautier notes that at C-group burials at Wadi Halfa, sheep are of the ammon type, which replaces the earlier long-legged breed with twisted horns known in Egypt from the twelfth dynasty (Gautier 1984a:50).

At Kerma in Lower Nubia (*c.* 4500–3700 bp) there are cattle, small stock, donkey and dog burials, sometimes in very large numbers, and cattle and sheep seem to have played a significant role in ritual (Reisner 1923, Chaix & Grant 1987, 1992, 1993). These sheep are of a long-legged hairy breed (Ryder 1984). They also appear to have been thin rather than fat-tailed, and desiccated skins show mostly black or white coats (Chaix & Grant 1987).

In Upper Nubia and the Middle Nile during the Napatan and Meroitic periods, cattle are still depicted as humpless longhorns (Shinnie 1967:18,129). Some horses may also have been used during this period, although identifications are not secure (Shinnie 1967). Horses are known from the Nile north at Buhen by *c.* 1675 BC (Clutton-Brock 1974, 1993). At the site of Musarwwarat es Sofra (first century AD), there are pens believed to have been for keeping elephants, as well as depictions, including one showing a king riding an elephant (Shinnie 1967:95). Scullard (1974) suggests that these animals may have been domesticated.

Camel dung from Napatan levels at Qasr Ibrim in Egyptian Lower Nubia has been dated to as early as 2690+90 bp, or the early first millenium BC (calibrated) (Rowley-Conwy 1988), but camels are not thought to have been common on the Middle Nile until the Christian era, after the sixth century AD (Shinnie 1967). Domestic chickens are also known from Qasr Ibrim, by the late fifth or early sixth century AD (MacDonald & Edwards 1993). Humped cattle do not appear until the fourteenth or fifteenth century AD, when humped cattle figurines are found at Dang Rhial in the southern Sudan (David et al. 1981, David 1982).

Table 10.1 Domestic animals from Djibouti, Eritrea, Ethiopia, and Somalia: the faunal evidence.

Approx. Date	Site	Cattle	Sheep/Goat	Donkeys	Camels
~3500 bp	Lake Besaka	X	X		
~3500 bp	FeJx 3				
~3500 bp	Laga Oda	X			
~3500 bp	Asa Koma	X			
~2000 bp	Gobedra	X			
~2000–4000 bp	Quiha	X		?	
< 3500 bp	Gogoshiis Qabe	X	X		
14th C AD	Lalibela	X	X		
14th–17th C AD	Laga Oda				X

Information from: Brandt & Carder 1978, Clark 1980, 1988, Clark & Prince 1978, Clark & Williams 1978, Dombroswski 1970, Guerin & Faure 1996, Phillipson 1977b, 1990, 1993.

2.2. The Horn of Africa

There has been little study of later prehistoric sites in this region, with only three major Holocene sequences excavated in Ethiopia, and others excavated but with fauna as yet unpublished in Somalia (Brandt & Carder 1987). Substantially more archaeological research has been done on Aksumite sites dating from approximately the second to ninth centuries AD, although thus far with very little emphasis on animal remains. This region does provide interesting insights into the nature of early livestock through its well-preserved rock art and Aksumite figurines.

2.2.1. Animal bones Cattle are the earliest domesticates in the area (Table 10.1). They are present at *c.* 3500 bp at the sites of Asa Koma in Djibouti, Lake Besaka in the Rift Valley and at Laga Oda in the eastern highlands, although at the latter two localities the faunal samples are very small and identification mostly based on dental fragments (Clark & Prince 1978, Clark & Williams 1978, Brandt 1980, 1984, Brandt & Carder 1987, Guerin & Faure 1996). At Asa Koma, cattle form a part of an ecomically broad subsistence regime including substantial hunting, fishing and grain gathering components (Guerin & Faure 1996). At Lake Besaka, cattle are reported from the site of FeJx3, associated with low lake levels when fishing was not practiced. A shift in the lithic industry, including adoption of scrapers similar to those used by contemporary tanners, may suggest increased hide processing at this time (Brandt 1980, 1984, Brandt & Carder 1987). A stone bowl similar to those found associated with early pastoralists in Tanzania and Kenya was also found. Brandt (1980, 1984) has suggested that these *c.* 3500 bp contexts at Lake Besaka were created by early pastoral peoples. At Laga Oda, rock paintings of cattle are thought to be of the same age as the animal remains (Clark 1980) (Table 10.2).

At Gobedra in the northern highlands of Ethiopia, cattle are present in levels dating to 2000 bp (Phillipson 1977b, 1993). Domestic cattle and perhaps donkey

Table 10.2 Domestic animals from Djibouti, Eritrea, Ethiopia, and Somalia: figurines and rock art.

Approx. Date	Site	Cattle, cf. *Bos t.*	Cattle, cf. *Bos i.*	Sheep/Goat	Donkeys	Camels
	PAINTINGS					
?	Zeban Ona Libanos	x				
?	Zeban Cabessa	x				
?	Ba'atti Sollum	x				
	& many other sites					
?	Zeban Ona Libanus	x-milking				
?	Serekama			x-fat tailed		
?	Adi Quaza		x			
?	Laga Oda		x			
?	Addi Gelemo		x			
?	Karin Heegan					x
Christian period	Addi Alauti					x
?	Ba'ati Facada	x-plough				
	FIGURINES					
>2C AD	Zeban Kutur		x			
Early Axumite	Matara		x			

Information from: Anfray 1967, 1968, Brandt & Carder 1987, Cervicek & Braukamper 1975, Clark 1976, 1980, Clark & Williams 1978, Drew 1954, Graziozi 1964a, 1964b, Joussaume 1981, Ricci 1955–58.

(Marshall in Clark 1988) are also present at Quiha near Mecalle in northern Ethiopia, a site excavated in the 1940s for which there are no radiocarbon dates. On the basis of the ceramics, the animal remains are thought to be contemporary with the Gash phase in the eastern Sudan, dating to about 4000 bp (Clark 1988). However, there is a considerable discrepancy between this estimation and the much later one of 71+107 bp based on obsidian hydration (Clark 1988), so the chronological position of this material remains unclear.

Cattle and sheep/goat are found at Gogoshiis Qabe in Somalia in levels dating to earlier than 3500 bp (Brandt & Carder 1987). At Lalibela in the Ethiopian highlands they are present in levels dating to the fourteenth century AD (Dombrowski 1970). No more specific information is available on cattle or small stock breeds from any of these sites. Evidence for domestic camel is fairly late in the Horn of Africa, although given their early presence in the Nile Valley, this may be an artefact of sampling. At Laga Oda, camels are first found in levels dating to between c. AD 1300–1600 (Clark & Prince 1978, Clark & Williams 1978). Direct dates are currently awaited on a single camel tooth from Gobedra found in levels dating to between c. 7000–3000 bp. (Phillipson 1993). Since direct dating has shown early finger millet reported from the site to be intrusive (Phillipson 1990), the presence of the camel tooth in these levels may also be the result of bioturbation.

2.2.2. Rock art Rock paintings are known from many areas of the Horn of Africa, and especially from the Akelle Guzai province of Eritrea, the eastern highlands of Ethiopia around Harar, and parts of Somalia (Clark 1954, 1976, 1980, Drew 1954, Franchini 1959, Graziosi 1964a,b, Brandt & Carder 1987, Phillipson 1993). The depiction of animal forms contributes interesting information on the possible distribution and timing of longhorn versus shorthorn, and humped versus humpless cattle in East Africa, and the appearance of camels, fat-tailed sheep and draft animals. But it is not nearly as reliable as direct zooarchaeological information for analyzing these issues because of the difficulties of interpreting and dating rock art.

Many paintings depict long-horned humpless cattle (Table 10.2). Those paintings without zebu or camel have been interpreted as earlier, predating both the pre-Aksumite period (about 2500 bp) and close contacts with southern Arabian populations (Graziosi 1964b, Cervicek 1979, Clark 1980, Joussaume 1981, Brandt & Carder 1987). Paintings showing camels, humped cattle, milking and ploughing scenes have been interpreted as later (Graziosi 1964a,b, Cervicek & Braukamper 1975, Clark 1980, Joussaume 1981, Brandt & Carder 1987). Some time-related distinctions have also been made between groups of paintings on stylistic grounds, but for the purposes of a relative ordering of the appearance of domestic animals in East Africa current systems of rock art dating are not useful, since their reasoning is largely circular. Unfortunately, there are no direct dates for any of the sites.

Examples of paintings showing humpless cattle with long horns include those from Ba'atti Sollum (Graziosi 1964b:190; see Fig. 10.2, Table 10.2), Zeban Ona Libanos and Zeban Cabessa. Many cattle are depicted with distinctive patchy markings, usually in brown and reddish tones, possibly also representing cattle brands.

Figure 10.2   Humpless longhorn cattle from a rock-painting at Ba'atti Sollum, Akele Guzai, Eritrea. After Graziosi (1964a, Fig. 4).

Figure 10.3   A milking scene from a rock-painting at Zeban Ona Libanos, Akele Guzai, Eritrea. After Graziosi (1964a).

Milking scenes are present at sites such as Zeban Ona Libanos in Eritrea (Fig. 10.3). Examples of paintings of humped cattle include those from Adi Quanza and Laga Oda near Harar (Graziosi 1964a:97) and Addi Gelemo in Eritrea (Ricci 1959). Fat-tailed sheep are depicted at Serekama in the eastern highlands of Ethiopia (Clark

Figure 10.4   Fat-tailed sheep from a rock-painting at Serekama in the eastern Highlands of Ethiopia. Detail from Clark & Williams (1978:Fig. 13), with the kind permission of J.D. Clark.

Figure 10.5   A cattle-drawn plough from a rock-painting at Ba'ati Facada in Tigre, northern Ethiopia. After Drew (1954).

& Williams 1978, Fig. 10.4). Humped cattle and camels at Addi Alauti in Eritrea are associated with figures bearing a cross, probably showing Christian influence (Graziosi 1964b:187) and dating to the fourth century AD or later. Camels are shown at Karin Heegan in Somalia (Brandt & Carder 1987:203) while at Ba'ati Facada in Tigre, northern Ethiopia a cattle-plough is depicted (Drew 1954) (Fig. 10.5).

2.2.3. Humped cattle figurines   Currently, the earliest date for humped cattle in the Horn is a figurine dating to about the second century AD from an early Axumite

Figure 10.6   Small (7.5 × 9.3 cm) bronze cow figurine from early Axumite context at Zeban Kutur, Akele Guzai, Eritrea. Drawing based on photo from Ricci (1955–58:Fig. 1).

context at Zeban Kutur (Ricci 1955–58) in the Akele Guzai region of Eritrea. A humped cattle figurine is also known from early Aksumite context at the nearby site of Matara (Anfray 1967, 1968). The small bronze figurine at Zeban Kutur was found by Franchini in early Aksumite levels. It is clearly Ethiopian rather than south Arabian because of the inscription and is thought to represent a cow rather than a bull (Ricci 1955–58) (Fig. 10.6).

2.3.   East African data

2.3.1. Neolithic sites   The earliest large sample of securely dated domesticates is from northern Kenya, at the site of Dongodien (GaJi 4) east of Lake Turkana dated to *c*. 4000 bp and associated with an Nderit pottery tradition (Marshall et al. 1984) (Fig. 10.7). This assemblage is dominated by domestic caprini, (sheep/goat) but a few cattle bones are also present. Large quantities of fish and some wild animal bones are also present. In Kenya, the term Pastoral Neolithic is applied to this site, and others where a Later Stone Age lithic technology is present, along with ceramic vessels, and an economy heavily reliant on domestic stock (Bower et al. 1977).

At Ele Bor, also in northern Kenya, one sheep/goat incisor and a camel molar and cuneiform were found in levels with grindstones and pottery, and predominantly wild fauna, dated to between 6000–4000 bp (Phillipson 1984). These dates are interesting and suggestive, but because of the very small sample size and potential problems with bioturbation in rock shelters, direct dating of these specimens is needed before the presence of camel in Kenya at this early date can be accepted.

In the south central Rift Valley, near Lake Navishiva, there are similar problems related to sample size and rockshelter stratigraphy, associated with the dating of the earliest appearance of domesticates at the long-sequence rockshelter site of Enkapune Ya Muto. In discussing the fauna from this site I will refer to several different levels or recognizable stratigraphic entities Many of the levels at Enkapune can also

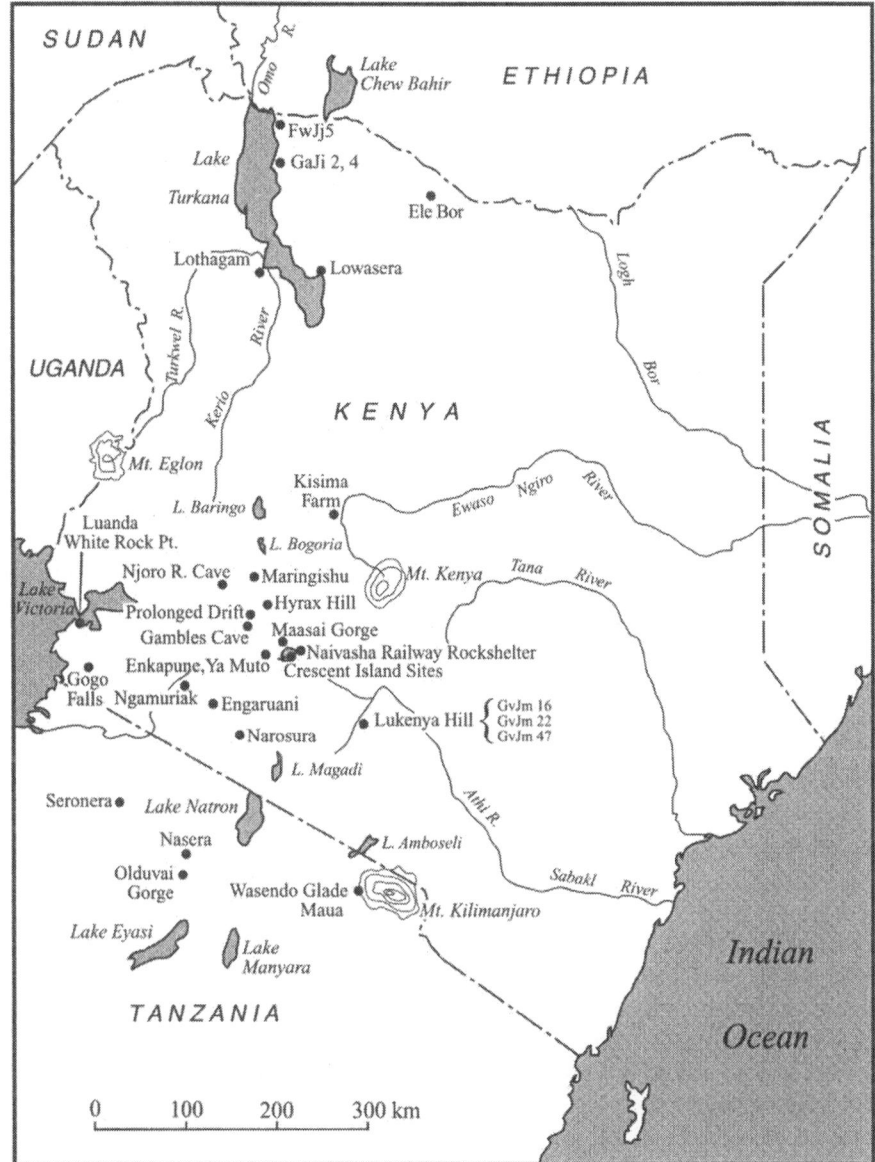

Figure 10.7 Kenyan and northern Tanzanian sites.

be differentiated by their associated material culture, especially lithic assemblages (Ambrose 1984b). The early fauna is found in Eburran levels with a lithic technology associated with a long tradition of late Stone Age occupation at the site. A pastoral Elmenteitan material culture with characteristic lithics and pottery is not found until the uppermost levels at Enkapune Ya Muto, after *c.* 3000 bp (Ambrose 1984b, Marean 1992).

It is possible that the dates for the appearance of domestic stock at Enkapune, in Eburran level RBL2.1, dating to *c*. 4900–3000 bp, are the earliest domestic stock in Kenya (Marean 1992). It should be stressed that the samples are very small, with a NISP (number of identifiable specimens) of only eight for domestic caprini and two for cattle. These are associated with a wild fauna, and a varied ceramic assemblage, including Nderit ware (Marean 1992). This is interesting, since Nderit ware is associated with early pastoralists at sites such as GaJi 4 in northern Kenya. However, the pattern of representation of domesticates does not persist or strengthen in the more recent layers at Enkapune, since the higher Eburran level RBL1, dating to *c*. 3390 bp, preserves only two cattle teeth, with wild animals dominating the rest of the assemblage. There were no ceramics associated with this level.

It is not until *c*. 3000 bp, level BS1, that domestic animals (mostly caprini) first dominate Enkapune's assemblage. Even at this time samples are relatively small (NISP = 474). Ceramics are present but rare, and level BS1 is attributed by Ambrose (1984b) to the Eburran 5, which he associated with hunter–gatherers living near to and interacting with, pastoral groups. Ambrose attributes the varied pottery associated with all later Eburran levels to such later interactions.

Faunal assemblages from upper levels at Enkapune, younger than *c*. 3000 bp, are associated with an Elmenteitan, with lithics and ceramics characteristic of a pastoral group. The faunas from upper Elmenteitan levels are still under study (Mbae, pers. comm.). The very small samples make the presence of early domestic fauna (dating to *c*. 4900–3000 bp) at the site difficult to interpret. But domestic animals are certainly present at the site probably in hunter–gatherer contexts by no later than 3000 bp.

New radiocarbon dates from Gogo Falls also tentatively indicate the presence of domestic stock prior to 3,000 years ago in western Kenya (at *c*. 3480 bp). Here livestock are associated with a different ceramic ware than present at other sites discussed, Kansyore ware (Munene in Hedges et al. 1993). This is interesting because it has been argued that there are resemblances between Kansyore pottery found on later Stone Age sites in East Africa and neolithic sites from the southern Sudan (Chapman 1967), suggesting influences from the north on East African hunter–gatherer groups prior to the adoption of food production in the area (Ambrose 1990). The Gogo Falls evidence is the first to associate Kansyore ware with domestic livestock. But as at other sites mentioned, this evidence should be taken with caution because Gogo Falls is a multicomponent site, with known mixing between levels in some areas (Robertshaw 1991).

Thus in summary, the only site with a substantial sample of domestic animals securely dated is GaJi 4, with cattle and small stock at 4000 bp. Future excavation and direct dates will test the hypothesis of early domesticates at Ele Bor, Enkapune Ya Muto, and Gogo Falls.

By contrast, large faunal samples are well known from the period dating to 3,000–2,000 years ago in central, southern and western Kenya, and northern Tanzania. Assemblages such as those associated with Elmenteitan material culture at Ngamuriak (Marshall 1990a,b), Sambo Ngige (Marshall 1990a,b), and Maasai Gorge (Ambrose 1984b), demonstrate the intensive use of domestic livestock, cattle, sheep and goat, and some donkey. Dogs are not known from these sites, probably owing to problems of sampling and identification (see Table 10.3).

A similar pattern is seen among generally contemporary, and geographically adjacent sites in southern Kenya associated with a different constellation of material culture, and known as the Savanna Pastoral Neolithic (Ambrose 1984a). These include Narosera (Gramly 1972, Gifford-Gonzalez & Kimengich 1984), Crescent Island Main (Onyango-Abuje 1977, Gifford-Gonzalez & Kimengich 1984), Lemek North East (Marshall 1990a,b), and probably West Kilimanjaro sites such as Maua Farm in northern Tanzania (Mturi 1986).

At the same time there are a number of sites attributed to hunter–gatherers on the basis of continuity of aspects of site structure and material culture, especially lithics, and known as Eburran 5 sites (Ambrose 1984a). These sites characteristically contain faunal assemblages where domestic stock is present, associated with large quantities of wild fauna. The presence of domestic animals in sites thought to have been occupied by hunter–gatherers has been attributed to factors such as gifts from pastoral neighbours, raiding, or limited/temporary herding by hunter–gatherers (Marean 1992). Fauna of this type include Enkapune Ya Muto (levels RBL2.1 and RBL1) Crescent Island Causeway (Gifford-Gonzalez & Kimengich 1984) and Naivasha Railway Rockshelter (Ambrose 1984b).

This pattern of mixed wild and domestic assemblages is not however confined to Eburran, hunter–gatherer sites. Two very large faunal samples, from Gogo Falls in western Kenya (Marshall 1990b, Marshall & Stewart 1995), and Prolonged Drift in the central Rift Valley (Gifford et al. 1980), do not fit the categories described here. At both these sites, large quantities of domestic animals, cattle, sheep and goat are present, together with large numbers of large wild ungulates. I have argued that Gogo Falls is an Elmenteitan pastoral site, with an atypical focus on wild resources, caused by the effect of tsetse fly on stock in the region (Marshall 1990b, Marshall & Stewart 1995). Both earlier and later Savanna Pastoral Neolithic sites in the Serengeti, also a tsetse region, are dominated by wild fauna (Bower 1988, 1991).

Prolonged Drift in the central Rift Valley is more difficult to interpret, as the material culture does not fit any recognized grouping. It could be understood in terms of loss of stock by pastoralists for circumstantial reasons, such as raiding, drought or disease, or alternatively as a non-Eburran hunter–gatherer site. Slightly later in time a more regional pattern of diversified animal production may be present at Maringuishu and Akira SPN sites (Bower 1988, 1991). However, not many sites of this type are known and faunal samples are small.

Patterns of variation in material culture and subsistence in the Kenyan neolithic do not fit into neat categories (Marshall 1990b, 1994). This probably indicates fluidity in social and economic systems as a result of interactions within and between hunter–gatherer and pastoralist groups, adoption of food production by hunter–gatherers, and periodic loss of stock by pastoralists. The context, and probably the nature, of interactions between hunter–gatherer's and pastoralists in East Africa has changed greatly in the last 2,000 years. During this period, Iron Age agricultural groups settled fertile agricultural lands, and pastoral societies shifted their axes

Table 10.3 Domestic animals from Kenya and northern Tanzania: key faunal evidence.

Approx. Date	Site	Cattle	Sheep/Goat	Donkeys	Camels	Dogs	Cats	Chickens
	NEOLITHIC							
~6000–4000 bp	Ele Bor		x?		x?			
~4000 bp	GaJi 4	x	x					
~3000–4900 bp	Enkapune Ya Muto:							
	RBL2.1	x?	x?					
~3480 bp	Gogo Falls-Kansyore	x?	x?					
~3000–2000 bp	Ngamuriak	x	x	x				
	Maasai Gorge	x	x					
	Narosura			x				
	Crescent Island Main	x	x					
	Lemek North East	x	x					
	Enkapune Ya Muto:							
	RBL1	x?	x?					
	BS1	x	x					
	Crescent Island Causeway	x	x					
	Naivasha Railway Rockshelter	x	x					
	Gogo Falls-Elmenteitan	x	x					
	Prolonged Drift	x	x					
	Serengeti sites	x	x					
	Wasendo-Kilimanjaro	x	x					
~2000–500 bp	IRON AGE							
	INLAND SITES							
	Deloraine	x	x					
	Hyrax Hill	*	x			x		
	Lanet	*	x					
	COASTAL SITES							
	Shanga	x	x				x	x
	Manda	x	x		x		x	x

Information from: Leakey 1945, Posnansky 1967, Gramly in Odner 1972, Onyango-Abuje 1977, Gifford & Kimengich 1984, Ambrose 1984b, Chittick 1984, Gifford-Gonzalez 1985, Phillipson 1984, Marshall 1989, 1990b, Mudida in Ambrose 1984c, Kyule 1991, pers. comm. 1996, Marean 1992, Horton & Mudida 1993.
Key to Symbols: ? = questionable dates, * = c.f. *Bos indicus*.

of interaction, no longer operating in exclusively hunter–gatherer socio-economic and political contexts. Nevertheless, nonlinear shifts between pastoral and hunter–gatherer social and subsistence systems have continued until recent times, where the adoption of a hunting and gathering way of life by pastoralists, is especially well documented in response to the disasters of the nineteenth century (Spencer 1973, Kenny 1981, Lamprey & Waller 1990).

2.3.1.1. Humped cattle   On the basis of the conformation of orbital rims from cattle of this period and measurements of metapodials, Marshall (1989, 1990a,b) suggested that the cattle found on Kenyan sites between 3000–2000 bp were *Bos indicus*. However I now withdraw this interpretation of the material, following Grigson's recent (1991) discussion of possible differences between Eurasian and African *B. taurus*, and the difficulty of finding morphological criteria for distinguishing between African *B. taurus* and *B. indicus*. Her re-analysis of measurements of cattle from African and Near eastern sites in this volume suggests that the Kenyan material that I describe may conform to the recently defined African *B. taurus cf. africanus*. Grigson's arguments have been reinforced by the molecular support for the hypothesis of a separate African domestication of *B. primigenius* (Bradley et al. 1996, Bradley & Loftus Ch. 13 in this volume). All measurements for complete specimens for cattle and sheep and goat from the Ngamuriak assemblage, and other Kenyan assemblages that I have measured are included in Table 10.4.

It is important to note that only 0.4 per cent of the more than 60,000 bones studied at Ngamuriak were measurable. This is typical of the assemblages that I, and others (also Gifford-Gonzalez 1985, pers. comm.) have studied in East Africa. In East African pastoral sites, bone is heavily processed as evidenced by cut and chop marks, and highly fragmented, preserving few morphological features (Marshall 1990b). It seems likely that the long-term future of the study of cattle "breeds" from these assemblages may be best served by attempting to isolate genetic material from archaeological remains.

2.3.2. Iron Age sites   Unlike the neolithic, there is very little information on Iron Age faunal assemblages from sites in eastern Africa. This is partly because iron-using pastoralists are less archaeologically visible than stone-using pastoralists, and partly because the focus of archaeological research in the Iron Age is on Bantu origins and ceramics, rather than subsistence data.

At Gogo Falls, Elmenteitan contexts are overlain by early Iron Age levels dating to the middle of the first millennium AD, suggesting cultural discontinuity in the region. However, little evidence is available on this or other early Iron Age fauna from East Africa (Robertshaw 1991). At the somewhat later site of Deloraine (*c.* 1100 bp), on the eastern flanks of the Mau Escarpment, there is possibly evidence for continuity and cultural change within the Elmenteitan, including: stylistic changes to Elmenteitan pottery, the introduction of the use of iron, and probable cultivation of domestic finger millet (Collett & Robertshaw 1983, Ambrose 1984c, Sutton 1987). The large faunal assemblage from Deloraine preserved a high

Table 10.4 Measurements of complete cattle long bones and long bone ends from Kenyan sites (in mm to the neasest 0.5 mm) following conventions established by von den Driesch (1976).

Site	Humerus Bd	Scapula GLP	Metacarpal LG	Metacarpal Bp	Metacarpal Bd	Lt. Cuneiform GB	Tibia Bd	Fibula GD	Metatarsal Bp	Metatarsal Bd	Metatarsal GL	Astragalus Bd	Astragalus GIl	Astragalus GLm	Calcaneum GL	Calcaneum GB	Phalanx 1 GL	Phalanx 2 Bp	Phalanx 2 Bd	Phalanx 2 Gl
Crescent Island	72.00	69.00	64.00						50.50	52.00	226.3	42.00	67.50	61.50						
GvJm 44	83.00		60.00									45.00	72.00	65.50						
GvJm 44 Akira level										55.00		41.00	69.00	62.00						
										65.50		42.00	66.00	61.50						
												51.00	69.00	64.00						
GvJm 44 Narosura level												50.00	70.00	64.00						
Narosura	70.00									52.00			64.00							
	77.00									54.00		40.00	69.00	63.00						
	78.00									55.00		40.50	69.00	63.00						
	78.00									56.00		41.00	64.00	58.00						
	86.00									56.50		41.00	69.00	63.00						
										61.00		42.00	67.50	61.00						
										65.00		42.00	70.00	65.00						
									52.00			42.50	66.00	59.00						
									52.50			42.50	66.00	61.50						
												42.50	68.50	62.00						
												43.00	67.50	62.00						
												44.00	65.00	59.00						
												44.00	68.00	61.50						
												46.00	71.00	65.50						
												46.00	71.00	65.50						
Ngamuriak	74.00				58.50	34.00	62.00	27.00		53.50		34.00	60.00	63.50		42.00	60.00	24.50	21.00	38.00
				50.50		35.00	62.00	30.00		55.00		34.00	67.00	61.50		42.00	60.00	28.00	23.00	40.00
						35.00	65.00	33.00		57.00		39.00	64.00	57.00		44.00	63.00	29.00	23.00	42.00

35.00	66.00	33.00	62.00			39.50	65.00	60.00	126.00	41.00	65.50	29.00	23.50	39.00
35.00		34.00	50.50			40.00	64.00	59.00	126.00	47.00	68.00	30.00	25.00	45.00
35.50		34.00	51.50	57.20	247.7	40.00	64.00	59.00	134.00	42.00		30.00	38.00	40.00
35.50		34.00				40.50			140.00	44.00		31.00	25.00	42.00
36.00		34.00				41.00		63.00				31.00	25.00	46.00
36.00		34.00				41.00	67.00	60.00				32.00	24.50	45.00
36.00		34.50				42.00	64.50	56.00				36.00	28.00	40.00
36.00		35.00				43.00	67.50	60.00						
36.00		35.00				43.00	68.00	61.50						
36.00		35.00				43.50	66.50	61.00						
37.00		35.00				45.00		63.00						
37.00		36.00				45.00	67.00	60.00						
37.50		36.00				45.00	67.00	61.50						
38.00		36.00				46.50	66.00	62.50						
38.00		36.00				46.50	68.00	63.00						
38.00		36.00				47.00	72.00	63.00						
38.00		37.00				47.00	72.00	67.00						
38.00		37.00				47.50		66.00						
40.00		37.50				47.50	73.00	65.50						
40.00		38.00				48.50		63.00						
40.00		38.50				48.50	72.00	64.50						
40.00		39.00				49.00		65.50						
40.50		40.00				50.00	74.00	67.00						
41.00		41.00				55.50	70.50	65.50						
41.00		43.50						57.00						
41.00		45.00												
42.00														
42.50														
44.00														

Prolonged Drift

	56.50	
	41.00	
		43.00	76.00	62.00
		43.00	68.00	63.00
		38.00	62.50

Deloraine

55.50	55.50		38.00	68.00	61.00
56.00	53.00	55.50	43.00		67.00
61.00	53.50		43.00	71.00	64.50
63.50			44.00	63.50	58.00
		56.00	55.50
		61.00
		63.50	53.50

proportion of domestic cattle, small stock and a few wild animals. The cattle from this site were especially large (Mudida in Ambrose 1984c, Marshall 1989, 1990a). I had previously suggested that they fit within the *Bos indicus* size range, Grigson (Ch. 4 in this volume) argues that they fit within the *B. taurus cf africanus* range.

A different pattern of iron-using pastoralism occurs in Kenya slightly later (800 bp) associated with Lanet, twisted-cord rouletted pottery, and "Siriqua" sites distributed throughout the Mau and western highlands (Posnansky 1967, Bower et al. 1977). Sutton (1973, 1987) argues that these sites relate to early Kalenjin populations in Kenya, while others have argued for Maasai connections (Blackburn 1974). Bifid thoracic vertebrae from both were found at both Hyrax Hill (Leakey 1945) and Lanet (Posnansky 1967), dating to the sixteenth century AD. In addition, new specimens were recovered from recent excavations at Hyrax Hill (Kyule 1992, in press). Thus the cattle from these sites now probably stand as the earliest morphological evidence for humped cattle in the region.

By contrast with inland late Iron Age sites, coastal Swahili sites are well known. However, just as inland, research has not been focused on retrieval and analysis of faunal samples, and despite the fact that some faunal lists have been published (e.g. Chittick 1974), very few faunal assemblages have been studied in detail. One exception to this is the site of Shanga in the Lamu archipelago. This assemblage is dominated by fish, with sea turtle and dugong also being important while the proportion of domesticates is relatively small (Horton & Mudida 1993). In the late eighth century AD, the only domesticate represented is the domestic chicken. Ovicaprines and cattle appear in the late ninth century AD and become more common later.

Nina Mudida also studied fauna from the nearby Swahili site of Manda, and a brief species list is included by Chittick in his monograph on the site (Chittick 1984). In contrast to Shanga, at Manda numbers of aquatic fauna and other wild animals is relatively low. Throughout the site's sequence, from the ninth to the fourteenth century AD, the fauna is dominated by sheep/goat and cattle. Domestic cats are also present, and a small number of camels and domestic chickens. Relative proportions rather than sample sizes were published, so it is difficult to know what these small numbers mean. In addition, no measurements or further details on cranial or vertebral morphology and breeds are currently available. This is very tantalizing given hypotheses on the role of Indian Ocean trade on the spread of zebu cattle (cf. Epstein 1955, Payne 1964), and the possible influence of the Indian Ocean trade on the spread of other animals such as chickens or camels.

Taken together, the sample of faunal remains from East African sites is uneven. A great deal of information is published from neolithic sites in Kenya and northern Tanzania. Little information is available for Eritrea, Ethiopia, southern Tanzania, Uganda, and for Iron Age sites in general. While the available information indicates general patterns of appearance and spread of domestic livestock in East Africa, more samples, especially those preserving complete crania, long bones, or vertebrae, and published measurements (following von den Driesch 1976) are badly needed to further understanding of patterns of variation in the distribution of East African livestock breeds over the last 5,000 years.

3. Regional patterns in the spread of food production

3.1. Early patterns of animal acquisition: pastoral migration and adoption of food production by hunter–gatherers

Taurine cattle, sheep and goats, the earliest domestic animals to appear in East Africa, were not domesticated in the region and are found as early as c. 9000–7000 bp in the Sahara. It is clear from the data reviewed in the previous section that domesticates are found earlier in the Sudan, c. 6300–5000 bp, than in East Africa, where they occur c. 5000–3500 bp in Kenya, and >3500 bp in the Horn of Africa. There is a loose gradient in the dates from earlier in the north to later in the south, although the data are much influenced by the varying amount of research conducted in different areas, and particularly by the relative lack of research in Eritrea, Ethiopia and Uganda.

Several lines of evidence suggest the possible spread of domestic livestock into Eritrea and Ethiopia from the eastern Sudan, as a result of cultural connections, and perhaps movement of people, between these areas. Clark (1980, 1988) sees similarities between ceramics from Gobedra and Quiha in the Ethiopian highlands, Lake Besaka in the Rift Valley and the site of Agordat in the Beraka Valley, western Eritrea (see also Brandt & Carder 1987). Furthermore, both Fattovich (1975, 1990; see also Sadr 1991) and Clark (1988) have noted general similarities between this material and Kassala phase, Gash Group, ceramics from the Butana area of the eastern Sudan. They note the similarity of other material from Agordat, such as stone axes and maceheads, with those from the complex societies of Kerma and the Nubian C Group of the Nile Valley (Arkell 1954, Clark 1976, Fattovich et al. 1984). Clark envisions connections between Ethiopia, Eritrea, and the Sudan, in terms of movement of pastoral populations into the Horn of Africa from the Nile Valley, perhaps in response to increasing aridity in the north (Clark 1976, 1988). He also suggests the existence of a seasonal pattern of pastoralism between the plains and highlands of northern Ethiopia and Eritrea and discusses the possibility of contact resulting in the adoption of domesticates by local hunter–gatherer groups (Clark 1976, 1988). It is worth noting, that the rather widespread similarities in ceramic wares that Clark and others describe could be the result of similar but independent long-term traditions of pottery manufacture in Sudan and Ethiopia (Arkell 1953, Fattovich 1975). Excavation of prehistoric sites in Eritrea and northern Ethiopia is badly needed to better understand the nature of the Holocene archaeological sequence in the region.

Fattovich et al. (1984) and Kitchen (1971, 1993) believe that the northeastern Sudan and lowland, northwestern Eritrea is the land of Punt referred to in ancient texts, which traded incense, ebony and short-horned cattle with Queen Hatchepsut's emissaries as early as 1493 BC (Kemp 1982, O'Connor 1982). There are also indications of early connections between the Arabian peninsula and the Eritrean region. Cervicek (1979) notes African influences in South Arabian rock art. Early domesticated African grains, both tef (at c. 2000 bp; van Bleek, 1969) and sorghum (at c. 4500 bp; Cleuziou & Constantini 1980) are known from the Arabian Peninsula.

Finger millet and sorghum are known from Harappan sites by c. 4000 bp (Kajale 1991), with finger millet and pearl millet being known from southern India by c. 3000 bp (Rao 1963, Vishnu-Mittre 1968, 1974). The existence of domestic African grains in Asia by the third millennium bp is especially interesting since they are not known from East Africa at this time. Given these early indications of contacts with the Egyptian and South Arabian trade in the Red Sea, trade presumably played a role in the introduction of food production to parts of the northernmost East Africa. However, it is likely that most systematic trade affecting domestic animals in the region developed later, associated with pre-Aksumite and Aksumite societies. It is at this time (c. 2500 bp; Taka Phase), that Karim Sadr (1991) argues that specialized nomadism arose in Butana, eastern Sudan, in response to opportunities for trade and symbiosis created by the existence of hierarchical agricultural societies in the region.

In Kenya, as in Eritrea and Ethiopia, archaeologists have argued that domestic livestock (cattle, sheep and goat), first appeared as a result of movement of pastoral peoples from the Sudanese region (Phillipson 1977a, Ambrose 1984a, Barthelme 1985, Bower 1988, 1991). It is also likely that pastoralists moved south into Kenya from Ethiopia and possibly Somalia (Bower 1988).

This is supported by the gradient of dates for the Pastoral Neolithic, which are earlier in the north than the south. Despite this, demonstrated similarities between material culture from the Sudan, Ethiopia, Somalia, and Kenya or Tanzania are very limited. The stone bowl found at Lake Besaka has been referred to the Kenyan Savanna Pastoral Neolithic (Brandt 1980), and the obsidian-based blade industry of Quiha has been compared to those of the Elmenteitan lithic industry (Clark 1988). It has also been suggested that there are similarities between ceramics from Quiha and Savanna Pastoral Neolithic Narosera Ware (Bower 1988). Comparisons have also been drawn between Kansyore ceramics and wares from the Khartoum neolithic (Chapman 1967, Ambrose 1990). However, samples are small, and systematic comparisons have not been made between material culture from northern and southern regions of East Africa. Strong continuities in lithic industries between later Stone Age and neolithic sites in southern East Africa suggest, if anything, small-scale population movements into Kenya from the north (Ambrose 1984a,b, Barthelme 1985, Bower 1988, 1991).

In Kenya and northern Tanzania, early pastoral sites have generally been found on the productive rangelands occupied by contemporary pastoralists, and not in the wooded upland areas subsequently occupied by Iron Age agriculturalists. This pattern of site distribution (Robertshaw & Collett 1983), in conjunction with a lack of evidence for a focus on cultivation at the sites, has suggested that these earliest food producers were broad-based pastoralists, probably cultivating, and hunting, but emphasizsing domestic herd production (Marshall 1990b, 1994).

Besides optimal grazing conditions, it has been suggested that the distribution of disease was a factor in the distribution of early pastoralists in East Africa (Marshall 1990b, 1994, Robertshaw 1990, 1991). Tsetse are commonly found in bushy country today in western Kenya and southern Tanzania (Lamprey & Waller 1990). Since East Africa was outside the normal range of wild *Bos primigenius*, it may have posed more disease-related problems for domestic African *B. taurus* than did northern Africa.

Hunter–gatherer sites are found alongside early pastoral sites in the region, showing that adoption of use of cattle and sheep and goat by local pastoralists was by no means a general phenomenon. By 3000 bp specialized pastoralists are found in parts of southwestern. Kenya, making little use of wild animal resources, and probably not cultivating. This specialization on production from domestic herds may have been encouraged by opportunities presented by the development of the bimodal pattern of rainfall characteristic of the region today, and relations with hunter–gatherers (Marshall 1990b).

The pattern of early adoption of domestic livestock in Kenya and northern Tanzania seems to be small scale and slow, halting in northern Tanzania. It is thought that the distribution of tsetse may have played a role in creating this apparent boundary to substantial southwards movement of stone tool using pastoralists (Marshall 1990b, Smith 1992, and Ch. 11 in this volume).

3.1.1. Trade and later patterns of livestock acquisition   Animals adopted after 3000 bp in East Africa probably include: camels, *Bos indicus*, draft animals, fat-tailed sheep, chickens and horses. In contrast with the earliest domesticates, patterns of distribution, together with ancient texts, suggest that trade, rather than migration, played an important role in their appearance in eastern Africa.

There are suggestions that camels were present in northern Ethiopia at Gobedra as early as 7000–3000 bp and in northern Kenya at Ele Bor between 6000–4000 bp. Given the very small number of specimens, and potential for bioturbation, I have reservations about the dating of specimens. There is secure evidence for camels in Napatan levels at Qasr Ibrim in Egyptian Lower Nubia by the early first millennium BC. At Manda on the East African coast, camel bones date from the ninth to fourteenth centuries AD, and at Laga Oda (in the eastern highlands of Ethiopia) camel specimens date to between the fourteenth and seventeenth centuries AD. Camels are represented in rock art at Addi Alauti where they are associated with Christian motifs and are also thought to have become common on the Middle Nile during the Christian era.

The earliest faunal evidence for the presence of (presumably) domestic camels in nearby Arabia is from Abu Dhabi during the later part of the 5th millennium bp (Uerpmann 1987, Grigson et al. 1989). Until more securely dated camel specimens are found, or direct dates are obtained on existing specimens, the earliest date of the appearance of camels in East Africa is unclear. However, the Lower Nubian date, in combination with the dates for the Arabian Peninsula suggest that we may expect to find camels in East Africa by 3000 bp. Its first occurrence may very well be tied to the Red Sea trade (cf. Bulliet 1990). Its further distribution, as suggested by the specimen at Manda, may be related to Swahili coastal trade.

At present the earliest morphological evidence for humped cattle (*Bos indicus* or *B. indicus/taurus* crossbreeds) in East Africa is probably that from Hyrax Hill (Leakey 1945, Kyule 1992, in press) dating to the sixteenth century AD. Humped cattle figurines are known earlier from early Axumite contexts at Zeban Kutur (*c.* second century AD) and Matara in Eritrea. Humped cattle figurines are also found from the fourteenth to fifteenth centuries AD at Dhang Rhial in the southern Sudan (David et al. 1981, David 1982).

This has been a time of flux for understanding the timing and distribution of *B. indicus* in East Africa. The evidence once again points to a relatively late date for its appearance, consistent with hypotheses that tie the earliest occurrence of *B. indicus* in East Africa to the pre-Aksumite or Aksumite trade with southern Arabia (Payne 1964, Clark 1976), and emphasize the role of the Indian Ocean trade in the distribution of humped cattle (Payne 1964, Epstein 1971). To address these hypotheses, well studied faunal samples, and published measurements, are badly needed from Swahili sites, inland Iron Age sites, and from pre-Aksumite and Aksumite sites.

Domestic chickens are known only from the eighth century AD at Shanga, and the ninth to fourteenth centuries AD at Manda on the East African coast. Given their relatively early date at Qasr Ibrim in Sudanese Nubia (fifth to sixth centuries AD) it has been suggested that the chicken's distribution in Africa relates either to trade down the Nile corridor or from the East African coast (MacDonald & Edwards 1993). Manda and Shanga support a Swahili connection. However, it is also possible that Aksumite or pre-Aksumite trade relations could have played a role in the distribution of chickens in East Africa.

Fat-tailed sheep, and horses, are domesticates that appear late in East Africa, and for which there is very little morphological evidence. Fat-tailed sheep in East Africa appear in late rock paintings, but there are no dates for their appearance in East Africa. There is also no osteological evidence for horses in East Africa, although their presence at Buhen, in Lower Nubia, *c.* 1675 BC, and historic evidence for their distribution in Ethiopia and Somalia, suggests that it should be forthcoming.

The role of the Swahili coastal trade of East Africa has been emphasized in discussions of early African–Asian connections. Camels, humped cattle and chickens could all have come directly from Asia to East Africa via the Red Sea trade from 1500 BC (but see Payne 1964). In this regard, Eritrea is a key region for contacts between Africa and Asia (Anfray 1968, Kobasichanov 1979, Munro Hay 1982, 1993, Marshall 1989). In addition to indications that it formed a trade link between Egypt and south Asia by 1400 BC, Eritrea also contained part of the Aksumite civilization (Anfray 1968, Kobasichanov 1979, Munro Hay 1982, 1993, Phillipson 1990). This Ethiopian society had ties with south Arabia and it is argued that it was visited by ships from India and the Far East before the second century BC (Kobasichanov 1979). At its height, Aksum was very active in the Red Sea trade and may have played an important, although heretofore unexplored, role in the distribution of imported livestock.

3.2. Cultural change, and acquisition of domestic animals

I believe that the reason for the slow and patchy distribution of early food producers and domestic stock in East Africa is in part because pastoral in-migration was limited and occurred over a long period. In addition, as has been shown in Kenya, the subsistence of the earliest food producers stressed herding of domestic livestock, rather than horticulture or agriculture. The influence on local hunter–gatherers would therefore have been towards adoption of pastoral ideals, and sheep, goat and cattle herding, rather than cultivation. This may have presented particular problems for hunter–gatherers.

There is mounting ethnographic and ethnoarchaeological evidence that neither the production of domestic plants nor animals is easy for "immediate-return" hunter–gatherers to incorporate into their social and subsistence systems (Ingold 1980, Brooks et al. 1984, ten Raa 1986, Marshall et al., in prep.).

Some of the problems faced by hunter–gatherers in the process of adopting food production include: increased labour costs, constraints on mobility, and the effect of egalitarian social structures with levelling mechanisms such as extensive sharing (Meillassoux 1972, Ingold 1980, North 1981, Brooks et al. 1984). Of these, the latter may be the biggest obstacle to successful farming. Brooks et al. (1984) have documented ways in which adoption of food production into a hunting and gathering subsistence system by the !Kung ultimately resulted in major social changes, and reconfiguration of society.

It is clear, however, that some domesticates, like dogs, are much more easily integrated into hunter–gatherer socio-economic systems than others. Some studies suggest that acquisition of domestic livestock, and successful herding, is especially difficult for hunter–gatherers (ten Raa 1986, Marshall, pers. obs.)

Problems for hunter–gatherers associated with keeping sheep and goat, and especially cattle in any numbers include: the capital necessary for stock acquisition, the constant labour required to pasture, water, milk, detick, and defend animals from predators, as well as long-term limits on mobility, and increased danger of raiding from pastoral groups.

The livestock to start a herd is generally much less easily available, and correspondingly more valuable than plant seeds. Among the *Piik aap Oom* Okiek, seeds for millet, maize or weedy greens were easily obtained as gifts from friends in farming communities, but obtaining the considerable social or monetary capital necessary for a cow was difficult (Marshall et al., in prep.). Sheep and goats were easier to acquire than cattle, but still more difficult than plant seeds. In addition, several studies suggest that keeping animal herds constrains mobility more than casual horticulture. While this may appear counter-intuitive, it was true among the Okiek (Marshall et al., in prep.) and among the Sandawe (ten Raa 1986), because animals can never be safely abandoned for weeks or months in the way that gardens can in some stages of growth or once crops are harvested. Looked at from the perspective of the constant labour requirements associated with livestock versus the more sporadic labour requirements associated with cultivation, the logistics of herd vesus crop management is a generalizable pattern.

In addition, ten Raa (1986) argues, based on the Sandawe, that once you have livestock you have something of value and become much more vulnerable to raiding from neighbouring pastoral and agrricultural groups already possessing efficient social mechanisms for warfare. At the same time, he notes that it is much more difficult for the Sandawe to employ their traditional strategy of avoiding attacks by concealment.

Taken together, these factors suggest that it is easier for hunter–gatherers to take up cultivation than herding. Given the prominence of stock-oriented, early pastoral groups, in East African prehistory, and the difficulties posed for hunter–gatherers of obtaining herd animals, this may have contributed to the slow and sporadic patterns of adoption of cattle, sheep, and goat, by hunter–gatherers in the region.

By contrast, the later adoption of animals such as *Bos indicus* cattle or fat-tailed sheep by established cattle and small stock herders involved very little cultural change and was probably rapid. Later adoption of other herd animals, especially camels, may have involved considerable social and cultural adjustments. Studies of the relationship between contemporary camel herders and social structure in East Africa, suggest that the slow growth rate of camel herds, and differences in ways of organizing economic production for camels compared with cattle, contribute to markedly different social and political organization between cattle and camel herders (Spencer 1973, Fratkin 1986, Fratkin & Smith 1994).

There are considerable advantages to keeping camel herds in arid areas of East Africa. Camels give more milk than cattle (four times as much), for a longer period, and do not dry up in the dry season. They do not need frequent watering, travel quickly, and can take advantage of the saline browse of desert regions (Dahl & Hjort 1976, Fratkin 1986, Fratkin & Smith 1994). However, they also have very low patterns of herd growth (1.5 per cent), compared with cattle (4 per cent), and cannot easily be herded with cattle because of different food and water requirements and different susceptibilities to disease. Spencer (1973) compared Rendille camel-keepers with Samburu cattle-keepers in northern Kenya and argued that low herd growth rates and the management requirements of camels resulted in less polygyny, delayed marriage, inheritance by first son of the first wife only, and larger settlements.

Thus, I hypothesize that there is a significant relationship between the amount of cultural change associated with adoption of a new domesticate, and the speed with which it is adopted. I argue that adoption of domestic livestock by hunter–gatherers in East Africa was slow. But that new livestock breeds with needs and life histories similar to those of the earliest domesticates in East Africa, would have spread quickly among established pastoralists. By contrast, adoption of camel herding by cattle and sheep and goat herders in East Africa, must have been associated with considerable cultural change, and as a result was a slower and more sporadic process. At present sufficient data on the timing and distribution of early camels, humped cattle and fat-tailed sheep in East Africa is not available to test this hypothesis.

4. Conclusions

Domestic animals precede domestic plants in East Africa, and spread into the region from the north and west, starting as early as the fifth to fourth millennia bp associated with limited movements of pastoralists from the Sudan, and adoption of domestic animals by local hunter–gatherer groups. The earliest domestic animals to appear in East Africa include: humpless, long-horned, and possibly short-horned cattle (*Bos taurus cf. africanus*), sheep and goat, and probably donkeys and dogs. Camels may be a part of this early complex, but on present evidence they appear later. Hunter–gatherer subsistence systems continue to be found alongside food producers in East Africa until the present day. Select domestic animals such as dogs, and bees, have been successfully incorporated into such stable hunter–gatherer subsistence systems. We do not yet know how early this occurred.

There is a general trend in East Africa for earlier adoption of domesticates in the north than in the south, but the pattern appears slow and patchy in all regions. This is probably the result of pastoral migration patterns, affected by the distribution of productive range lands and livestock diseases, and the process of adoption of livestock by hunter–gatherer groups. Such a pattern may be related to the particular difficulties of hunter–gatherers attempting to integrate herding, compared with those of incorporating limited cultivation, into a subsistence system based on wild resources.

In the south, small-scale subsistence oriented hunter–gatherer and pastoral societies provide the socio-economic context for the spread of food production. Evidence for trade or exchange is minimal. But in Eritrea, and possibly northern Ethiopia, relations with early complex societies of the Sudan, such as the Nubian A-group, C-group, and Kerma, as well as ties to long distance trade via the Nile and Red Sea, could have influenced patterns of spread of early domestic animals.

Through time there is a gradual trend towards decreased use of wild animals by pastoral groups across East Africa. Specialized pastoralism focused on production from cattle, sheep and goat, appears in the Loita-Mara region of southern East Africa by 3000 bp (Marshall 1990b). Pastoral nomadism appears in the Taka Phase in the southern Atbai region of northern East Africa by 2500 bp (Sadr 1991). But the kind of specialization, and processes leading to it, may have been very different in the two regions. In the Loita-Mara specialization may relate to opportunities for year round production from pastoral herds created by the advent of a modern bimodal rainfall system in East Africa, together with the nature of socio-economic relations with local hunter–gatherer groups. In the southern Atbai, Sadr (1991) argues that the development of specialized nomadism results from long-term relations between increasingly specialized pastoral groups and neighbouring state level societies.

Humped cattle, possibly camels, and chickens, and fat-tailed sheep appear later in East Africa. The spread of these animals seems to have been more influenced by trade (pre-Aksumite, Aksumite, Red Sea and Indian Ocean), than by the processes of migration and exchange that affected the distribution of the earliest domesticates in East Africa. Humped cattle were probably more quickly adopted by later food producing groups in East Africa, and were more widespread than camels because their use necessitated minimal cultural change.

East Africa as a region has an especially long and sustained tradition of specialized cattle, sheep and goat use. Animal use thus has an especially complex history in this region, which has been characterized for the past 5,000 years by highly diverse cultural and subsistence systems, ranging from hunting and gathering, to pastoral, settled agricultural, and hierarchical state level societies. While much work remains to be done, the increasingly fine-grained zooarchaeological data available on the range of variation in animal types, and patterns of animal use in this region, has much to offer in thinking about both past and future human–animal relations in East Africa.

Acknowledgments

I am grateful to Kevin MacDonald and Roger Blench for inviting me to write this paper, and providing such a stimulating multidisciplinary environment in which to talk about early African livestock. I have benefited from discussions of these issues with many colleagues over the years, especially Diane Gifford-Gonzalez, Stanley Ambrose, and Peter Robertshaw. I thank the Government and National Museums of Kenya for Research clearance and am grateful for support from the Leakey Foundation and the National Science Foundation.

References

Adams, W. 1977. *Nubia. Corridor to Africa.* London: Allen Lane.

Ambrose, S.H. 1984a. The introduction of pastoral adaptations to the highlands of East Africa. In *From hunters to farmers*, J.D. Clark & Steven Brandt (eds), 212–39. Berkeley: University of California Press.

Ambrose, S.H. 1984b. Holocene environments and human adaptation in the central Rift Valley, Kenya. Unpublished PhD dissertation, University of California, Berkeley.

Ambrose, S.H. 1984c. Excavations at Deloraine, Rongai, 1978. *Azania* **20**, 29–67.

Ambrose, S.H. 1990. Hunter–gatherer/herder interactions in highland East Africa. Paper presented at the Society of Africanist Archaeologist Meetings, Center for African Studies, University of Florida, Gainesville, 22 March 1990.

Anfray, F. 1967. Matara. *Annales d'Ethiopie* **7**, 33–53.

Anfray, F. 1968. Aspects de l'archfiologie fithiopienne *Journal of African History* **9**, 345–66.

Arkell, A.J. 1953. *Shaheinab*. London: Oxford University Press.

Arkell, A.J. 1954. Four occupation sites at Agordat. *Kush* **2**, 33–62.

Bar Yosef, O. & R. Meadow 1995. The origins of agriculture in the Near East. In *Last hunters–first farmers*, T.G. Price & A.B. Gebauer (eds), 39–94. Santa Fe: School of American Research Press.

Barthelme, J.W. 1985. *Fisher–hunters and neolithic pastoralists in East Turkana, Kenya.* BAR International Series 254. Oxford: British Archaeological Reports.

Blackburn, R. 1974. The Okiek and their history. *Azania* **9**, 139–58.

Bower, J. 1988. Evolution of Stone Age food-producing cultures in East Africa. In *Prehistoric cultures and environments in the late Quaternary of Africa*, J. Bower & D. Lubell (eds), 91–114. BAR International Series 405, Oxford: British Archaeological Reports.

Bower, J. 1991. The Pastoral Neolithic of East Africa. *Journal of World Prehistory* **5**, 49–82.

Bower, J., et al. 1977. The University of Massachusetts later Stone Age/Pastoral Neolithic comparative study in central Kenya: an overview. *Azania* **12**, 119–46.

Bradley, D., D.E. MacHugh, E.P. Cunningham, R.T. Loftus 1996. Mitochondrial diversity and the origins of African and European cattle. *Proceedings of the National Academy of Sciences* **93**, 5131–35.

Brandt, S.A. 1980 Investigation of late Stone Age occurrences at Lake Besaka, Ethiopia. In *Proceedings of the 8th PanAfrican Congress on Prehistory and Quaternary Studies 1977*, R. Leakey & B. Ogot (eds), 239–43. Nairobi: TILLMIAP.

Brandt, S.A. 1984. New perspectives on the origins of food production in Ethiopia. In *From hunters to farmers*, J.D. Clark & S.A. Brandt (eds), 173–90. Berkeley: University of California Press.

Brandt, S.A. & N. Carder 1987. Pastoral rock art in the Horn of Africa: making sense of udder chaos. *World Archaeology* **19**, 195–213.

Brooks, A., D.E. Gelburd, J. Yellen 1984. Food production and culture change among the !Kung San: implications for prehistoric research. See Clark & Brandt (1984), 293–310.

Bulliet, W.R. 1990. *The camel and the wheel*, Morningside edn. New York: Columbia University Press. (1st publ. 1975).

Burstein, S.M. 1989. *Agartharchides of Cnidus*. S.M. Burstein (trans. & ed.). London: Hakluyt society.

Cervicek, P. 1979. Some African affinities of Arabian rock art. *Rassegna di studi etiopici* **27**, 5–12.

Cervicek, P. & U. Braukamper 1975. Rock paintings of Laga Gafra (Ethiopia). *Paideuma* **21**, 47–60.

Chaix, L. & A. Grant 1987. A study of a prehistoric population of sheep (Ovis aries L.) from Kerma (Sudan). *Archaeozoologia* **1**, 93–107.

Chaix, L. & A. Grant 1992. Cattle in ancient Nubia. *Anthropozoologica* **16**, 61–6.

Chaix, L. & A. Grant 1993. Paleoenvironment and economy at Kerma, northern Sudan, during the third millennium BC: archaeozoological and botanical evidence. See Krzyzaniak et al. (1993), 399–404.

Chapman, S. 1967. Kantsyore Island. *Azania* **2**, 165–91.

Chittick, N. 1974. *Kilwa, an Islamic trading city on the East African Coast* [2 volumes]. BIEA Memoir 5. Nairobi: The British Institute in eastern Africa.

Chittick, N. 1984. *Manda*. Nairobi: The British Institute in eastern Africa.

Clark, J.D. 1954. *The prehistoric cultures of the Horn of Africa*. Cambridge: Cambridge University Press.

Clark, J.D. 1976. The domestication process in sub-Saharan Africa with special reference to Ethiopia. In *Origine de l'élevage et de la domestication*, 56–115. Nice: Union International des Sciences préhistoriques et Protohistoriques.

Clark, J.D. 1980. The origins of domestication in Ethiopia. In *Proceedings of the 8th PanAfrican Congress on Prehistory and Quaternary Studies 1977*, R. Leakey & B. Ogot (eds), 268–70. Nairobi: TILLMAIP.

Clark, J.D. 1988. A review of the archaeological evidence for the origins of food production in Ethiopia. In *Proceedings of the 8th International Conference of Ethiopian Studies, Addis Ababa 1984*, Taddese Beyene (ed.) vol. 1, 55–69. Addis Ababa.

Clark, J.D. & S. Brandt (eds) 1984. *From hunters to farmers*. Berkeley: University of California Press.

Clark, J.D. & G.R. Prince 1978. Use-wear on later Stone Age microliths from Laga Oda, Harraghi, Ethiopia and possible functional interpretations. *Azania* **13**, 101–10.

Clark, J.D. & M.A.J. Williams 1978. Recent archaeological research in southeastern Ethiopia (1974–1975): some preliminary results. *Annales d'Ethiopie* **11**, 19–44.

Cleuziou, S. & L. Costantini 1980. Premiers filfiments sur l'agriculture protohistorique de l'Arabie Orientale. *Paléorient* **6**, 245–51.

Clutton-Brock, J. 1974. The Buhen horse. *Journal of Archaeological Science* **1**, 89–100.

Clutton-Brock, J. 1989. Cattle in ancient North Africa. In *The walking larder: patterns of domestication, pastoralism and predation*, J. Clutton-Brock (ed.), 200–206. London: Unwin Hyman.

Clutton-Brock, J. 1993. The spread of domestic animals in Africa. See Shaw et al. (1993), 61–70.

Collett, D. & P. Robertshaw 1983. Pottery traditions of rarly pastoral communities in Kenya. *Azania* **18**, 107–25.

Dahl, G. & A. Hjort 1976: *Having herds. Pastoral herd Growth and household economy*. Stockholm Studies in Social Anthropology 2. Stockholm: University of Stockholm.

David, N. 1982. Prehistory and historical linguistics in Central Africa: points of contact, In *The archaeological and linguistic reconstruction of African history*, C. Ehret & M. Posnansky (eds), 78–95. Berkeley: University of California Press.

David, N., P. Harvey, & C.J. Goudie 1981. Excavations in the southern Sudan. *Azania* **16**, 7–54.

Dombrowski, J. 1970. Preliminary report on excavations in Lalibela and Natchabiet caves, Begemder. *Annales d'Ethiopie* **8**, 21–9.

Drew, S. 1954. Notes from the Red Sea Hills. *South African Archaeological Bulletin* **9**, 101–2.

Epstein, H. 1955. The zebu cattle of East Africa. *East African Agricultural Journal* **2**, 83–95.

Epstein H. 1957. The Sanga cattle of East Africa. *East African Agricultural Journal* **22**, 149–64.

Epstein, H. 1971. *The origin of the domestic animals of Africa*. Vol. 1. London: Holmes & Meier.

Fattovich, R. 1975. The contribution of the Nile Valley's cultures to the rising of the Ethiopian civilization: elements for a hypothesis of work. *Meroitic Newsletter* **16**, 2–8.

Fattovich, R. 1990. Remarks on the pre-Aksumite period in northern Ethiopia. *Journal of Ethiopian Studies* **23**, 1–33.

Fattovich, R. 1993. The Gash Group of the eastern Sudan: an outline. See Krzyzaniak et al. (1993), 439–44.

Fattovich, R., A.E. Marks, A. Mohammed-Ali 1984. The archaeology of the eastern Sahel, Sudan: preliminary results. *The African Archaeological Review* **2**, 173–88.

Franchini, V. 1959. Notizie sa alcune pitture ed incisioni repestre recentemente ritrovate in Eritrea. *Atti del Convegno Internazionale di Studi Etiopici*, Roma, 285–9.

Fratkin, E. 1986. Stability and resilience in East African pastoralism: the Rendille and the Ariaal of northern Kenya. *Human Ecology* **14**, 269–86.

Fratkin, E. & K. Smith 1994. Labor, livestock and land: the organization of pastoral production. In *African Pastoralist Systems*, E. Fratkin, K. Galvin, E. Roth (eds), 91–112. Boulder: Lynne Rienner.

Gautier, A. 1984a. Quaternary mammals and archaeozoology of Egypt and the Sudan: a survey. See Krzyzaniak & Kobusiewicz (1984), 43–56.

Gautier, A. 1984b. The fauna of the neolithic site of Kadero (central Sudan). See Krzyzaniak & Kobusiewicz (1984), 317–19.

Gautier, A. 1984c. Archaeozoology of the Bir Kiseiba region, eastern Sahara. In *Cattle-keepers of the eastern Sahara: the neolithic of Bir Kiseiba*, F. Wendorf & R. Schild (assemblers), A.E. Close (ed.), 49–72. Dallas: Southern Methodist University Press.

Gautier 1986. La faune de l'occupation néolithique d'el Kadada (secteurs KDD-12,22,32) au Soudan Central. *Archaeologie du Nil Moyen* **1**, 59–111.

Gautier, A. 1987. Prehistoric men and cattle in North Africa: a dearth of data and a surfeit of models. In *Prehistory of arid North Africa*, A. Close (ed.), 163–87. Dallas: Southern Methodist University Press.

Geraads, D. 1983 Faunal remains from the Gash Delta. *Nyame Akuma* **23**, 22–3.

Gifford-Gonzalez, D.P. 1985. Faunal assemblages from Masai Gorge Rockshelter and Marula Rockshelter. *Azania* **20**, 69–88.

Gifford-Gonzalez, D.P. & J. Kimengich 1984. Faunal evidence for early stock-keeping in the central Rift of Kenya: preliminary findings. See Krzyzaniak & Kobusiewicz (1984), 357–471.

Gifford, D.P., G.Ll. Isaac, C.M. Nelson 1980. Evidence for predation and pastoralism from a Pastoral Neolithic site in Kenya. *Azania* **15**, 57–108.

Gramly, R. 1972. Appendix B: Report on the teeth from Narosura. *Azania* **7**, 87–91.

Graziosi, P. 1964a. New discoveries of rock paintings in Ethiopia. Part I. *Antiquity* **38**, 91–8.

Graziosi, P. 1964b. New discoveries of rock paintings in Ethiopia. Part II. *Antiquity* **38**, 187–94.

Grigson, C., J.A.J. Gowlett, J. Zarins 1989. The camel in Arabia-A direct radiocarbon date, calibrated to about 7000 BC. *Journal of Archaeological Science* **16**, 355–62.

Grigson, C. 1991. An African origin for African cattle: some archaeological evidence. *African Archaeological Review* **9**, 119–44.

Guerin, C. & M. Faure 1996. Chasse au chacal et domestication du boeuf dans le site Néolithique d'Asa Koma (République de Djibouti). *Journal des Africanistes* **66**, 299–311.

Hedges, R.E.M., R.A. Housely, C. Bronk-Ramsey, G.J. van Klinken 1993. Radiocarbon dates from the Oxford AMS system. *Archaeometry* **35**, 147–67.

Horton, M. & N. Mudida 1993. Exploitation of marine resources: evidence for the origin of the Swahili communities of East Africa. See Shaw et al. (1993), 673–83.

Ingold, T. 1980. *Hunters, pastoralists and ranchers*. Cambridge: Cambridge University Press.

Joussaume, R. 1981. L'Art rupestre de l'Ethiopie. In *Préhistoire Africaine: Mélanges Offerts au Doyen Lionel Balout*, C. Roubet, H.J. Hugot, G. Souville, (eds), 159–75. Paris: A.D.P.F.

Kajale, M.D. 1991. Current status of Indian palaeoethnobotany: introduced and indigenous food plants with a discussion of the historical and evolutionary development of Indian agriculture and agricultural systems in general. In *New light on early farming*, J. Renfrew (ed.), 155–89. Edinburgh: Edinburgh University Press.

Kemp, B. 1982. Old Kingdom, Middle Kingdom, and Second Intermediate period in Egypt. In *The Cambridge History of Africa, vol. I*, J.D. Clark (ed.), 658–761. Cambridge: Cambridge University Press.

Kenny, M.G. 1981. Mirror in the forest: the Dorobo hunter–gatherers as an image of the other. *Africa* **51**, 476–95.

Kitchen, K.A. 1971. Punt and how to get there. *Orientalia* **40**, 184–207.

Kitchen, K.A. 1993. The Land of Punt. See Shaw et al. (1993), 587–608.

Kobishchanov, Y.M. 1979. *Axum*, J.W. Michels (ed.), L.T. Kapitanoff (trans.). University Park: Pennsylvania State University Press.

Krzyzaniak, L. 1978. New light on early food production in the central Sudan. *Journal of African History* **19**, 159–72.

Krzyzaniak, L. 1984. The neolithic habitation at Kadero (central Sudan). See Krzyzaniak & Kobusiewicz (1984), 309–15.

Krzyzaniak, L. 1991. Early farming in the Middle Nile Basin; recent discoveries at Kadero (central Sudan). *Antiquity* **65**, 515–32.

Krzyaniak, L. & M. Kobusiewicz (eds) 1984. *Origin and early development of food-producing cultures in North Eastern Africa.* Poznan: Polish Academy of Sciences and Poznan Archaeological Museum.

Krzyzaniak, L., M. Kobusiewicz, J. Alexander (eds) 1993. *Environmental change and human culture in the Nile Basin and Northern Africa until the second millennium BC.* Poznan: Poznan Archaeological Museum.

Kyule, D. 1992. Economy and subsistence of Iron Age Sirikwa culture at Hyrax Hill, Nakuru: a zooarchaeological approach. Unpublished MA thesis, University of Nairobi.

Kyule, D. in press. The Sirikwa economy: further work at Site II on Hyrax Hill. *Azania* **31**.

Lamprey, R. & R. Waller 1990. The Loita-Mara area in historical times: patterns of subsistence, settlement and ecological change. In *Early pastoralists of south-western Kenya*, P. Robertshaw (ed.), 16–35. Nairobi: British Institute in eastern Africa.

Leakey, M.D. 1945. Report on the excavations of Hyrax Hill, Nakuru, Kenya Colony 1937–1938. *Transactions of the Royal Society of South Africa* **30**, 271–409.

MacDonald, K.C. 1992. The domestic chicken (*Gallus gallus*) in sub-Saharan Africa: a background to its introduction and its osteological differentiation from indigenous fowls (Numidinae and *Francolinus sp.*). *Journal of Archaeological Science* **19**, 303–18.

MacDonald, K. & D. Edwards 1993. Chickens in Africa: the importance of Qasr Ibrim. *Antiquity* **67**, 584–90.

Marean, C. 1992. Hunter to herder: large mammal remains from the hunter–gatherer occupation at Enkapune ya Muto rockshelter, central Rift, Kenya. *The African Archaeological Review* **10**, 65–127.

Marshall, F. 1989. Rethinking the role of *Bos indicus* in sub-Saharan Africa. *Current Anthropology* **30**, 235–39.

Marshall, F. 1990a. Cattle herds and caprine flocks; early pastoral strategies in southwestern Kenya. In *Early pastoralists of south-western Kenya*, P. Robertshaw (ed.), 205–60. Nairobi: British Institute in eastern Africa.

Marshall, F. 1990b. Origins of specialized pastoral production in East Africa. *American Anthropologist* **92**, 873–94.

Marshall, F. 1994. Archaeological perspectives on East African pastoralism. *African Pastoralist Systems*, E. Fratkin, J. Galvin, E. Roth (eds), 17–44. Boulder, Col.: Lynn Rienner.

Marshall, F. & K. Stewart 1995. Hunting, fishing, and herding pastoralists of western Kenya: the fauna from Gogo Falls. *Zooarchaeologia* **7**, 7–27.

Marshall, F., K. Stewart, J. Barthelme 1984. Early domestic stock at Dongodien in northern Kenya. *Azania* **19**, 120–27.

Marshall, F., T. Pilgram, C. Kratz, in prep. The *Piik aap Oom* Okiek, cultivation and livestock.

Meillassoux, C. 1972. On the mode of production of the hunting band. In *French perspectives in African studies*, P. Alexandre (ed.), 39–58. Oxford: Oxford University Press.

Mturi, A.A. 1986. The Pastoral Neolithic of West Kilimanjaro. *Azania* **21**, 53–63.

Munro-Hay, S. 1982. The foreign trade of the Aksumite port of Adulis. *Azania* **17**, 107–25.

Munro-Hay, S.C. 1993. State development and urbanism in northern Ethiopia. See Shaw et al. (1993), 609–21.

Nordström, H.Å. 1972. Neolithic and A-Group sites. *The Scandinavian Joint Expedition to Sudanese Nubia*, vol. 3. Uppsala.

North, D. 1981. *Structure and change in economic history*. New York: W.W. Norton.

O'Connor, D. 1982. Egypt 1552–664 BC. In *The Cambridge History of Africa* Vol. I, J.D. Clark (ed.), 830–925. Cambridge: Cambridge University Press.

Odner, K. 1972. Excavations at Narosura: a stone bowl site in the southern Kenyan highlands. *Azania* 7, 25–92.

Onyango-Abuje, J.C. 1977. A contribution to the study of the neolithic in East Africa with particular reference to the Naivasha Nakuru basin. Unpublished PhD dissertation, University of California, Berkeley, California.

Payne W.J.A. 1964. The origin of domestic cattle in Africa. *Empire Journal of Experimental Agriculture* 32, 97–113.

Peters, J. 1986. A revision of the faunal remains from two central Sudanese sites: Khartoum Hospital and Esh Shaheinab. *Archaeozoologia*, **Mélanges**, 11–35.

Peters, J. 1989. The faunal remains from several sites at Jebel Shaqadud (central Sudan): a preliminary report. In *Late prehistory of the Nile Basin and the Sahara*, L. Kryzaniak & M Kobusiewicz (eds), 469–72. Poznan: Poznan Archaeological Museum.

Peters, J. 1991. The faunal remains from Shaqadud. In *The late prehistory of the eastern Sahel*, A.E. Marks & A. Mohammed-Ali (eds), 197–235. Dallas: Southern Methodist University Press.

Peters, J. 1992. Late Quaternary mammalian remains from central and eastern Sudan and their palaeo-environmental significance. *Palaeoecology of Africa* 23, 91–115.

Phillipson, D.W. 1977a. *The later prehistory of eastern and southern Africa*. London: Heinemann.

Phillipson, D.W. 1977b. The excavation of Gobedra Rock Shelter, Axum: an early occurrence of cultivated finger-millet in northern Ethiopia. *Azania* 12, 53–82.

Phillipson, D.W. 1984. Aspects of early food production in northern Kenya. See Krzyzaniak & Kobusiewicz (1984), 489–95.

Phillipson, D.W. 1990. Aksum in Africa. *Journal of Ethiopian Studies* 23, 55–65.

Phillipson, D.W. 1993. The antiquity of cultivation and herding in Ethiopia. See Shaw et al. (1993), 344–57.

Posnansky, M. 1967. Excavations at Lanet, Kenya, 1957. *Azania* 2, 89–114.

Rao, S.R. 1963. Excavation at Rangpur and other explorations in Gujarat. *Ancient India* 18, 5–208.

Ricci, L. 1955–58. Ritrovamenti archaeologici in Eritrea, II. *Rassegna di studi etiopici* 15, 48–68.

Ricci, L. 1959. Rassegna di studi etiopici. *Ritrrovamenti archeologici in Eritrea* 14, 48–68.

Reisner, G.A. 1910. *The archaeological survey of Nubia, Report for 1907–1908*. Cairo: National Printing Department.

Reisner, G.A. 1923. *Excavations at Kerma*. Cambridge, Mass.: Harvard African Studies.

Robertshaw, P. 1990. *Early pastoralists of south-western Kenya*. Nairobi: British Institute in eastern Africa.

Robertshaw, P. 1991. Gogo Falls: a complex site east of Lake Victoria. *Azania* 26, 63–195.

Robertshaw, P. & D. Collett 1983. The identification of pastoral peoples in the archaelogical record: an example from East Africa. *World Archaeology* 15, 67–78.

Rowley-Conwy, P. 1988. The camel in the Nile valley: new radiocarbon accelorator (AMS) dates from Qasr Ibrim. *Journal of Egyptian Archaeology* 74, 245–48.

Ryder, M.L. 1984. Skin, hair and cloth remains from the ancient Kerma civilization of northern Sudan. *Journal of Archaeological Science* 11, 477–82.

Sadr, K. 1991. *The development of nomadism in ancient northeast Africa*. Philadelphia: University of Pennsylvania Press.

Scullard, H.H. 1974. *The elephant in the Greek and Roman world*. Ithaca: Cornell University Press.

Shaw, T., P. Sinclair, B. Andah, A. Okpoko (eds) 1993. *The archaeology of Africa. Food, metals and towns*. London: Routledge.

Shinnie, P.L. 1967. *Meroe: a civilization of the Sudan*. New York: Praeger.

Smith, A.B. 1992. *Pastoralism in Africa*. London: Hurst.

Spencer, P. 1973. *Nomads in alliance*. Oxford: Oxford University Press.

Sutton, J.E.G. 1973. *The archaeology of the western highlands of Kenya*. Nairobi: British Institute in eastern Africa.

Sutton, J.E.G. 1987. Hyrax Hill and the Sirikwa: new excavations on site II. *Azania* 22, 1–36.

ten Raa, E. 1986. The acquisition of cattle by hunter–gatherers: a traumatic experience in cultural change. *Sprache und Geschichte in Afrika* **7**, 361–74.

Uerpmann, H.-P. 1987. *The ancient distribution of ungulate mammals in the Middle East*. Beihefte zum Tubinger Atlas der Vorderen Orients, Reihe A, Nr.27. Wiesbaden: Ludwig Reichert.

Vishnu Mittre, 1968. Protohistoric record of agriculture in India. *Transactions of the Bose Research Institute* **31**, 87–106.

Vishnu Mittre, 1974. Paleobotanical evidence in India. In *Evolutionary studies in world crops: diversity and change in the Indian subcontinent*, J.B. Hutchinson (ed.), 3–30. Cambridge: Cambridge University Press.

von den Driesch, A. 1976. *A guide to the measurements of animal bones from archaeological sites*. Peabody Museum Bulletin 1. Cambridge: Harvard University.

CHAPTER ELEVEN

The origins of the domesticated animals of southern Africa

Andrew B. Smith

1. Introduction

Herding societies of southern Africa were well-established by the time of the first European observers at the end of the fifteenth century. In fact, the cattle large herds of the Khoekhoen of the Cape were the main impetus for the Dutch settlement being set up in 1652, as these animals were wanted to refresh the Dutch East Indies ships coming from, and going to, the Far East.

Most of the pre-colonial societies of southern Africa were animal herders to a greater or lesser extent – those involved in "pure" pastoralism without any domesticated plants lived primarily on the drier western side of the subcontinent, while Iron Age farmers on the eastern side were agro-pastoralists whose diet consisted of a large proportion of sorghum or millet. The reason for this was partially owing to the favourable environment for keeping stock: good, tsetse-free grasslands and, for the most part, reasonably dependable rainfall. At the time of European contact cattle were the main source of wealth of the majority of these societies, but, as will be shown, archaeological evidence would suggest that small stock played a greater role in some earlier societies.

Several models of the spread of domestic stock to southern Africa have been put forward:

Model 1: the most commonly held idea is that African Iron Age people spread southwards from their original hearthland in the Cameroons bringing with them domestic plants and animals. They ultimately came into contact with San hunter–gatherers around the Caprivi strip and northern Botswana who acquired small stock from them through intermarriage with the incoming farmers. Groups of the former hunters retained their separate identity and split away from the donor group to become the Khoekhoen herders of the western and southern Cape.

Model 2: East African herders with pottery, stone bowls, sheep, and who practised cairn burial, moved far enough southwards to where they came into contact with San hunters of southern Africa. According to this model, domestic animals already existed in southern Africa before Iron Age farmers arrived. The movement of the East African herders may be difficult to follow since they were transhumant and less likely to leave much trace. Their archaeological signature could be masked by the dominance of much larger Iron Age sites in the landscape of countries like Tanzania, Zambia and Angola which are poorly known archaeologically.

Model 3: Khoesaan groups, defined both genetically and linguistically, lived in East Africa, obtained stock from Cushitic speakers and moved south.

2. Traditional breeds of domestic stock of southern Africa

There are several breeds of domestic animals that are considered unique to southern Africa, although early European breeding programmes, which began in the Cape as long ago as the end of the eighteenth century (Van Ryneveld 1942), may have tended to highlight differences, rather than similarities.

2.1. Sheep

Sheep breeds which are seen as being "traditional" include the so-called fat-tailed variety. In South Africa today, these are referred to as Namaqua Afrikaners (Epstein 1971ii:147) and Ronderib Afrikaners (ibid.:143). Early European settlers noted that none of the aboriginal sheep were wooled. Ronderib had two subsets: one with a coarse hairy coat, referred to as Steekhaar, and one with a shiny coat, called Blinkhaar (ibid.:146). Col. Robert Gordon, writing his journal in December 1778, refers to the latter as "blinkschapen" (Raper & Boucher 1988:200). In Namibia, the equivalent would be the Damara sheep (Epstein 1971:142). But not all the aboriginal breeds may have had fat tails. Gordon depicted a Hottentot sheep with a long thin tail (Cape Archives AG.7146.210/GA.5154.210) (Fig. 11.1), although, as Epstein (1971:149) suggests, this may have been a function of an early phase of fat accumulation in the tail (see depictions of Namaqua ewes, ibid.:plates 195–6). The fat accumulation in the tail was considerable, reaching as much as 7 kg. in some rare cases.

2.2. Goats

Goat remains have mostly been identified from Iron Age contexts, but are consistently outnumbered by sheep on those sites. They date to as early as the fourth to seventh centuries AD in the Transvaal (Voigt 1984). That dwarf goats may have existed in southern Africa is attested by the identification of one from QwaQwa in the Orange Free State (Brink & Holt 1992). The only goat remains identified in the western part of the subcontinent come from Bethelsklip in Namaqualand (Webley 1984) and are dated to 800±50 bp.

2.3. Cattle

Several breeds of cattle were to be found in pre-colonial southern Africa. The most important became the Africander cattle, which were bred by the Dutch settlers

Figure 11.1   Col. Gordon's eighteenth century depiction of a "Hottentot sheep" (Cape Archives AG.7146.210/GA.5154.210).

at the Cape for use as trek-oxen. These were derived from the Khoekhoe cattle at the Cape. Other breeds included Tswana, Damara, Ovambo, Nguni, Bapedi, Bolowana and Basuto. Of these, the Nguni has recently been selected by breeders as a strong disease-resistant animal, which is allowed to maintain its quality with minimal management by a form of Darwinian selection (allowing the best adapted to the environment to survive).

The small Basuto cattle may well have been part of a cattle population of southern Africa which was trypanotolerant, like the Ndama or Muturu of West Africa (see MacDonald & MacDonald Ch. 8 in this volume). Livingstone (Schapera 1960:16) in 1851 saw beautiful small animals belonging to the Makalolo on the edge of a tsetse area, although Chapman (1868:174) described the animals as "... very diminutive cows, standing very little higher than three feet ... but exceedingly good milkers ... carefully kept in the plains, or in parts known to be perfectly free of the fly". These are called Tongas by Falkner & Epstein (1957:68), and categorized as short-horned Sangas by Mason & Maule (1960:30).

2.4. Dogs

The skeletal remains of dogs are difficult to separate from jackals on African sites, as both canids are roughly the same size. Ina Plug (pers. comm.) says she is able to do it, and since she analyzed most of the fauna in Table 11.1, it is her identifications that are used (Plug & Voigt 1985, Plug 1996a). Dogs existed in early Iron Age contexts in Natal at Magogo and Ndondondwane between the sixth to

Table 11.1 Domestic animals from Iron Age sites in southern Africa.

Sites	Dates (century AD)	ovicaprines	cattle	pigs	dogs	chickens
Toteng (Botswana)	1st–14th	x	x			
Lotshitshi (Botswana)	3rd		x			
Mzonjani (Natal)	3rd		?x			
Ma 38 (Transvaal)	?3rd–4th	x				
Happy Rest (Transvaal)	4th	x	x			
Ma 4 (Transvaal)	?4th	x	x			
Nkope (Malawi)	4th–8th		x			
Magogo (Natal)	6th–8th	x	x		x	
Ficus, A, B, C (Transvaal)	6th–14th	x	x			
Lydenburg Head Site	6th–7th	x	x			
Msuluzi (Natal)	7th	x	x			
Divuyu (Botswana)	7th–9th	x	x			
Matlhapaneng (Botswana)	7th–10th	x	x			
Taukome (Botswana)	7th–10th	x	x		x	
Nqoma (Botswana)	7th–11th	x	x			
Matope Court (Malawi)	7th–14th		x			
Namichimba (Malawi)	7th–14th		x			
Le 6 (Transvaal)	8th	x	x			
Schroda (Transvaal)	8th–9th	x	x		x	
Ndondondwane (Natal)	8th	x	x		x	x
Nanga (Zambia)	8th–10th	x	x			
Salumano (Zambia)	8th–10th	x	x			
Bulila 1 (Zambia)	8th–10th	x	x			
Hippo Tooth (Botswana)	9th	x	x			
Pont Drift (Transvaal)	9th–12th	x	x		x	
K2 (Transvaal)	9th–11th	x	x		x	
Bosutswe (Botswana)	9th–15th	x	x		x	x
Mapungubwe (Transvaal)	10th–13th	x	x			
Commando Kop (Botswana)	10th–12th	x	x		x	
Toutswe	10th–16th	x	x		x	
Icon (Transvaal)	14th	x	x			
Rooikrans (Transvaal)	17th	x	x			
Rhenosterkloof (Transvaal)	17th	x	x			
Mgoduyanuka (Natal)	17th–19th	x	x			
Sh 16 (Transvaal)	18th–19th	x				
Sh 27 (Transvaal)	18th–19th	x				
Kekane (Transvaal)	19th–20th	x	x	x	x	?x

Table 11.2   Domestic animals from Namibia and the South African Cape.

Domestic animals in Namibia Sites	Dates	sheep	cattle
Falls rockshelter	190 BC–AD 383	x	
Snake rockshelter	AD 118–383	x	
Mirabib shelter	AD 430–652	x	
Geduld	AD 1213–1303	x	
≠Khîsa-//gubus	AD 1297–1436	x	x
Striped giraffe shelter	AD 1458–1657	x	
Domestic animals in the Northern Cape & Karoo			
Equus cave	751–255 BC	x	
Little Witkrans shelter	AD 123–391	x	
Limerock shelter 1	AD 395–607	x	
Blinkklipkop 1	AD 789–1017	x	
Doornfontein 1	AD 887–1023	x	
Domestic animals in the western Cape			
/Ai tomas	196 BC–AD 1954	x	
Spoegrivier cave	165 BC–AD 13	x	
Kasteelberg A	AD 75–674	x	
Tortoise cave	AD 263–545	x	
Diepkloof cave	AD 339–660	x	
Elands Bay Open	AD 547–1444	x	
Steenberg cove	AD 650–862	x	
Kasteelberg C	AD 674–957	x	
Elands Bay cave	AD 779–1162	x	
Kasteelberg B	AD 1036–1955	x	x
Dunefield midden	AD 1037–1479	x	
Bethelsklip	AD 1213–1303	x	
Heuningklip	AD 1225–1323	x	
Domestic animals in the Southern Cape			
Die Kelders cave	193 BC–AD 788	x	
Blombos cave	AD 3–119	x	
Nelson Bay cave	AD 4–315	x	
Hawston	AD 74–383	x	
Boomplaas cave	AD 252–682	x	
Smitswinkelbaai cave	AD 623–843	x	
Scott's cave	AD 667–1954	x	
Byneskranskop cave	AD 1329–1477	x	

eighth centuries AD, in Botswana at Taukome and Bosutswe in the seventh to tenth centuries AD, and in the Transvaal at Schroda in the eighth and ninth centuries AD.

While the general consensus is that the original domesticated dogs were derived from wolves in several parts of the northern hemisphere (Epstein 1971, Davis 1987, Morey 1994, etc.), the possibility of a later jackal introgression in Africa is not ruled out by Manwell & Baker (1983). All three canids: wolves, jackals and domestic dogs, have the same chromosome number (78), and can interbreed to produce fertile offspring, and such introgression "would explain the phenomenally high variation produced by selective breeding in modern dogs" (Clutton-Brock 1977:1340). The testing of the introgression hypothesis will depend on more work being done on the genetic distance between canids (Wayne 1993).

Many of the dogs of the southern part of the continent seen as being "African", such as Zulu dogs, are coursers of the whippet type, that could easily have been derived from greyhound types found further north (Epstein 1971:62–3).

2.5. Chickens

The appearance of the domestic chicken, *Gallus domesticus*, at Ndondondwane in Natal by the eighth century AD and in the Taukome phase of Bosutswe in Botswana in the ninth century AD (Table 11.1) is an indication of the importance of trade between the interior and the Indian Ocean, already established by Swahili traders before the ninth century (Horton 1987). A further clue to this trade is the intrusion of the house rat, *Rattus rattus*, in all levels at Bosutswe, with particularly large numbers in the later phases (Plug 1996a,b).

2.6. Pigs

Wild pigs, *Phacochoerus aethiopicus* are found at several early sites, but domestic suids only appear in the nineteenth century, suggesting that they were a European introduction.

3. The social conditions of early food producers in southern Africa

The archaeology of early food production in southern Africa tends to be confused by the assumption that not only were animals and plants introduced around 2,000 years ago, but that pottery was part of the "package", as it shows up around the same time. Thus, sites of this period with ceramics have automatically been assumed to be the habitation areas of food producers. While, in many cases, this may be true, it is by no means certain that it is always the case. Work in the southwestern Cape, for example, shows that there are sites with both small numbers of domestic animals and a few pottery fragments that were almost certain to have been occupied by hunters (Smith et al. 1991). Within the last 100 years in Africa and Arabia virtually all pastoralist societies had low status clients on the periphery. These were usually hunters, but occasionally they might be iron workers. These clients performed a number of tasks for herders, including acting as sentinels, as we see from this quotation from the van der Stel journal of September 1685: "... we find that these Sonquas are just the same as the poor in Europe, each tribe of Hottentots having some of them and employing them to bring news of the approach of a strange tribe" (Waterhouse 1932:122). Payment for services rendered to these people would have been in the form of animal products, such as milk, or sometimes non-breeding animals.

Equally, agro-pastoral people moved into southern Africa with an economy quite different from the aboriginal hunters of the region (Denbow & Wilmsen 1986). This did not mean there was no communication between the two. The very opposite almost certainly was the case. Hunters worked for the farmers, no doubt performing similar tasks as today, such as rain-making, as well as procuring wild meat products and honey. Domestic grains, milk, and an occasional slaughter beast would have been exchanged in return.

Thus, when dealing with the archaeological evidence we must be fully aware that there usually were different social and economic groups in the environment at any given moment. The archaeologist's task would be to separate these out. One way would be to keep the evidence for domestic stock and ceramics as independent variables.

4. Faunal evidence of early southern African livestock

There are a number of dates for the bones of domestic stock in southern Africa after 2000 BP, such as the complete cow mandible from the site of Mabveni, Zimbabwe dated to AD 180±120 (SR-43) (Thorp 1979). But this is not a single date since another date of AD 570±110 (SR-79) also comes from the same site. This is a good example of the problem of chronological controls inherent in the earliest appearance of domestic stock in the subcontinent, no doubt exacerbated by the fact that the population of the first food producers in the region was probably low, so sites of the earliest phases would be rare. From a non-Iron Age site in Zimbabwe, another date of 2140±60 bp was obtained from a hearth at Bambata Cave. In this same level, ovicaprid teeth were identified (Walker 1983).

Although sites with ceramics of the coastal Matola tradition of the early Iron Age have been dated to around 2,000 years ago in Mozambique, and in Natal by the third century AD, there is no evidence of herding. From Ma38, a Matola site in the Kruger National Park, Transvaal bones of ovicaprid size were found, but not securely identified, and at Silver Leaves impressions of millet were identified from the pottery. Henshilwood (1996) shows that sheep had reached the southernmost Cape at Blombos cave by 2000 bp. It is from other parts of the Transvaal in the fourth century AD that more secure dating of herding can be accepted. At Happy Rest and Ma4, sites of Huffman's "western stream", cattle, sheep and goat bones were recovered (Plug & Voigt 1985). Further west, in northern Botswana, cattle remains have been dated to the third century AD at Lotshishi (Wilson 1969) and at Toteng (Campbell 1992).

In non-Iron Age contexts in Namibia, ovicaprid bones have been dated to 1790±50 bp at Geduld (Smith & Jacobson 1995). Further south, in the northern Cape at Spoegrivier, a sheep first phalanx gave a date of 2105±65 bp (Sealy & Yates 1994), which supports the initial dating of the basal layer by Webley (1992) of 1920±40 bp. At Kasteelberg in the western Cape, a sheep thoracic vertebra was dated to 1630±60 bp (Sealy & Yates 1994), slightly later than the initial dates of 1860±60 bp and 1790±40 bp run on charcoal by Smith (1987).

5. Livestock in southern African rock art

Although southern Africa has some of the richest rock art in the world, domestic animals are only rarely represented in most places. The exception is the foothill region of the Drakensberg, where cattle-paintings are known in considerable numbers. This variable distribution reflects both the history of domestic animals in southern Africa, as well as the intensity of their importance in the sweetveld grassslands of Natal.

It is generally accepted that the artists were Bushmen, the aboriginal hunting people who occupied most of southern Africa before the expansion of Bantu-speaking people into the subcontinent some 2,000 years ago (although, see Van Rijssen 1984 for alternative views). It is further widely believed that most of the art was primarily a metaphor for the trance experiences of shaman-healers on out-of-body travel (Lewis-Williams 1981). Thus, there is a strong possibility that one should not view the art as a narrative form, but as a metaphor for more complex religious beliefs. Domestic animals may well have replaced wild animals as part of the symbolism once they became fixtures in the landscape, especially as they may have become used for ritual purposes as substitutes for wild animals, as is seen among the Kalahari Bushmen today (Yellen 1990).

The majority of the domestic animals depicted in the rock art of southern Africa are sheep and cattle, which come in two different genres: a) paintings, and b) engravings, with almost no overlap between them (Figs 11.2, 11.3). There are a few paintings of horses with European riders carrying firearms in conflict scenes from the Natal Drakensberg, as well as very late (unpatinated) engravings of mounted riders along the Orange River. Dogs may also be found in the rock art of the Drakensberg. Vinnicombe (1976:157) says that dogs constitute as much as 5 per cent of the domestic animals of the Drakensberg. She has illustrated several canid forms in association with people that are probably domestic dogs (ibid.: Figs 30, 83, 84, etc.). Ikeya (1994) has stated that, compared with traps and bows, dogs were previously only of secondary importance in hunting among the Kalahari Bushmen, but have recently increased in importance.

Paintings require reasonable "canvasses", so it is in the mountain areas, where one finds caves or rock overhangs in granite or sandstone walls, that one encounters the art. Engravings are most common in areas where large boulders show wind polish and patina that can be broken through to create images. While sheep and cattle depictions are found within both genres, their distribution is not always the same. For example, sheep paintings occur in the southwestern Cape, but no cattle-paintings have been found there. It is not obvious why such variability should have existed, unless the sheep-paintings predate the influx of significant numbers of cattle into the area. Certainly there were large numbers of cattle in the southwestern Cape by the time of the European observers in the sixteenth century (see Lancaster voyage of 1591 in Raven-Hart 1967:15). It has been noted (Wilson 1969) that the Khoekhoen used sheep in their rituals. Thus the painters may well have been reflecting the ritual value of these animals.

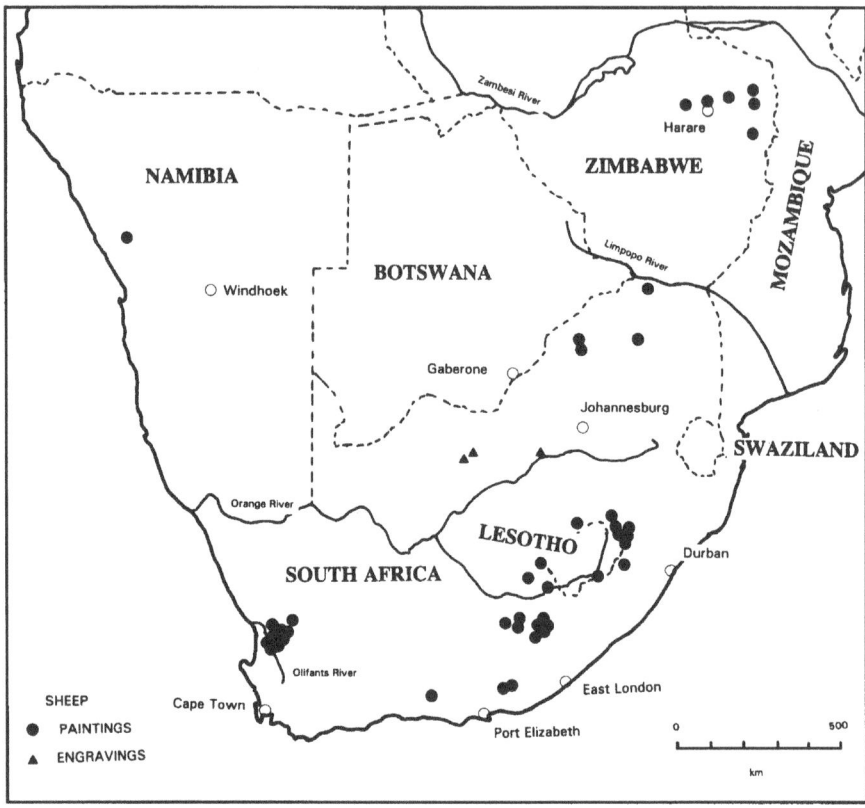

Figure 11.2   Distribution map of rock art depictions of sheep in southern Africa.

The paintings of sheep are in the mountains, away from the best pasture lands. The paintings are most plentiful in the area of the Oliphants and Koebee Rivers. In the latter area they constitute 10 per cent of the recorded paintings (Hollmann 1993). The breed of sheep are the long-legged, fat-tailed variety that were found in the subcontinent before European breeding programmes introduced wool-bearing animals from Spain.

Other areas where sheep-paintings are to be found include Zimbabwe (Goodall 1946) and the border area between Botswana and the northern Transvaal along the Limpopo River (Walker 1991, Eastwood & Fish 1995).

Cattle-paintings are more common than paintings of sheep in the eastern part of the subcontinent, although both exist. Of interest is that none of the cattle depicted show large humps, indicating that they were not zebu (*Bos indicus*) (see, for example, Fig. 11.4).

If, indeed, the painters were not themselves herders, this leads to speculation of the relationship between hunters and stock keepers. The discrete distribution of depiction of the two species in the area of the eastern Cape south of the Winterberg, is suggested by Hall (1986), as a function of successive stages in the assimilation,

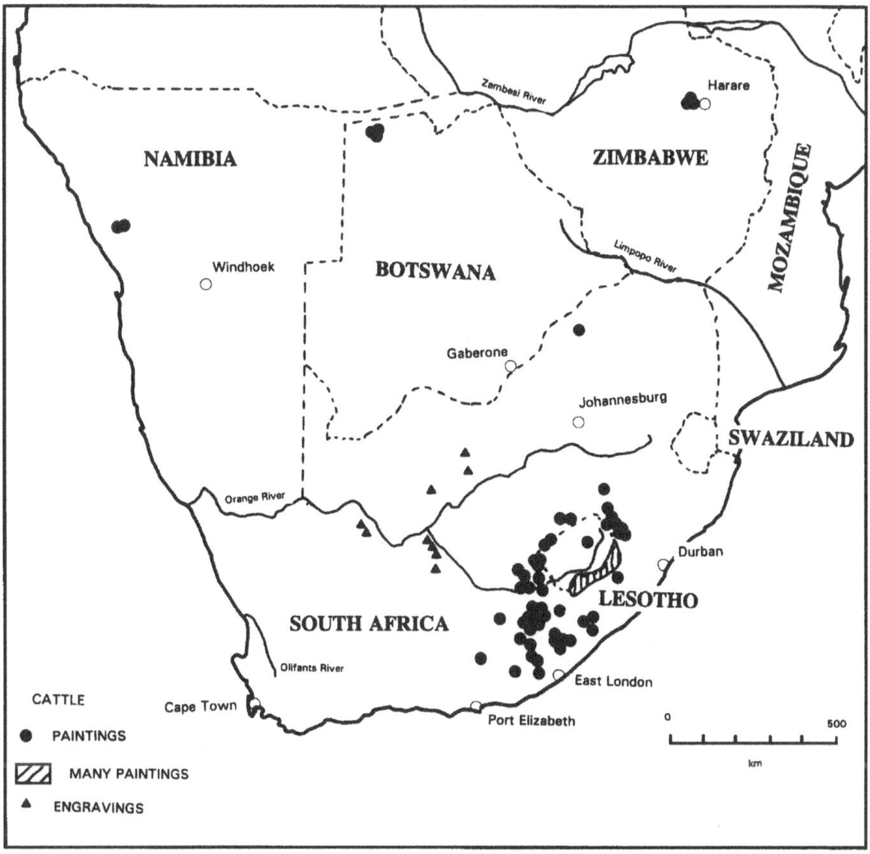

Figure 11.3   Distribution map of rock art depictions of cattle in southern Africa.

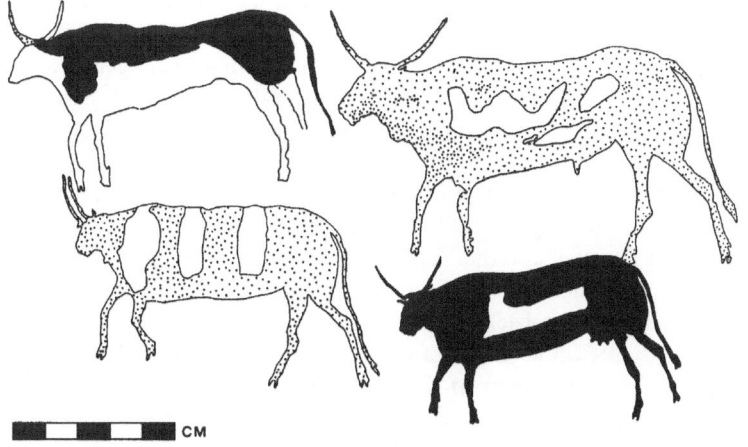

Figure 11.4   Cattle painting from Mpongweni, southern Natal Drakensberg (part on panel only). Cattle hide colours are red, white and mottled blue-white.

Figure 11.5   Illustration by Bell (1834) of "Griquas on Riding Oxen" (Africana Library, Johannesburg).

or otherwise displacement of hunters as they were pushed out of favourable pastoral habitats.

Overpainting of cattle on top of eland has also been found in the Winterberg, leading to the suggestion that this was a deliberate substitution of the domestic animal for an important wild animal (eland) used in rain-making rituals (Hall 1986:48).

Such information of a potential social nature can also be suggested in the depictions of coat colours of cattle in the Drakensberg (Fig. 11.4). That such distinctions were important is highlighted by the fact that the Zulus have over a hundred names to differentiate colour combinations and horn shapes (Krige 1950:187). Why shaman-painters would also go to the trouble of showing such detail, can only be surmised. It does imply a very close sharing of values that went along with the symbiotic role of Bushmen as ritual specialists for African agro-pastoralists (cf. Jolly 1995).

Not all relations were amicable. Campbell (1986) discusses trance symbolism in apparent narrative paintings of cattle raids by hunters on pastoralists. In one of the paintings, from Beersheba in the Natal Drakensberg, the raid has occurred among Boer herds, and the raiders with the stolen stock are being pursued by a commando carrying firearms on horseback.

Another aspect of cattle-paintings is that where individual animals have a stick or a bone through the septum of the nose. This was the technique used by the Khoekhoe to control animals used for riding or as draft (see Smith & Pheiffer 1993:plate 15, and Fig. 11.5). A number of instances of this has been recorded by Woodhouse (1987) from the foothills of the Drakensberg.

Changes in rock art style in the Brandberg of Namibia have been used to indicate social controls whereby shaman-painters become specialized at a time when an accumulation of dung appears in a painted shelter. This is interpreted as full-scale pastoralism which could only be a result of the painters controlling the pastoral production (Kinahan 1991). That the paintings are dated only by superposition, means there is a very tenuous connection between the dung accumulation and the paintings. One would also have thought that if the domestic animals became so ritually important they would have been more graphically represented in the art. Of the 10,000 images recorded in this particular ravine only 0.04 per cent can be suggested as sheep (Pager 1993).

6. Discussion

The foregoing compendium of early dated livestock remains are scattered over a large area, but do not reflect the amount of archaeological reasearch that has been done on the first food producers. Inevitably, not all sites produce bone that can be identified, so the associated cultural remains have been used to support the models of the spread of herding societies. In particular, pottery has tended to dominate the picture, and entire waves or "streams" of different pottery traditions have been suggested as carriers of food production (Phillipson 1977, Huffman 1979, 1982, etc) from further north. This is seen as the expansion of Bantu languages by agro-pastoral Iron Age people.

In essence, both Huffman and Phillipson have looked to East Africa as the source of domestic stock and ceramics. More recently, Denbow (1990), working in the Congo, has suggested that certain decorative traits on pottery found there can be matched with those from sites in Angola and northern Botswana, indicating yet another potential source of livestock, all within an assumed Iron Age context.

Non-Iron Age contexts of East Africa have produced herding economies with cultural traits that are somewhat different. From Kenya, close to the border with Tanzania, Robertshaw (1990) and Marshall (1990) have identified strong evidence for pastoralism at Ngamuriak, dated to 2135±140 bp and 1940±40 bp. Not only are there large numbers of cattle and ovicaprid bones, but the pottery is quite different from any found in Iron Age contexts further south. This thin ware has stylistic motifs that include lugs and spouts. The significance of the latter is that these are stylistic traits which are associated with the earliest non-Iron Age herding societies in the Cape (Rudner 1968), and, since the dates for the earliest domestic stock in the Cape seem to be earlier than that for the first Iron Age agro-pastoralists by as much as two centuries, this raises the possibility of an earlier connection with pastoral societies further north, possibly via northern Namibia.

One possible archaeological connection is with Bambata stylistic groups. At Bambata Cave, a single decorated spout was found (Schofield 1940) in a non-Iron Age context. While this is hardly conclusive evidence, the possibility of early herder movements into southern Africa by non-Iron Age pastoralists could be suggested as a working hypothesis. Under conditions of low nutrient status, or where herders have to move their stock within tsetse-free corridors during the dry season, such high mobility would work against the accumulation of archaeological material, rendering passage virtually invisible. An example of this can be seen in the interior of the southwestern Cape, away from the coast, where historically large herds of cattle were observed. No archaeological site with domestic animal bones has been found, in spite of systematic surveys over large areas (Hart 1987). Because of their agricultural diet and the need for fields, Iron Age people have tended to dominate the archaeological landscape. This may have masked a pastoralist signature.

Initially, Ehret (1982, 1984) suggested that the words for "ram", "milk ewe" and "cow" in Khoekhoe languages are derived from central Sudanic languages, and these words entered Bantu languages from the Khoekhoe. The inference from this would be that the Khoekhoe obtained their domestic stock from central Sudanic speakers before contact with Bantu-speaking people. Ehret suggests Kwadi-speakers of southern Angola as the contact point (an area virtually unknown archaeologically). While there is little value currently placed on the idea of contact with central Sudanic speakers, there are suggestions that Cushitic-speakers from East Africa may well have been the contact group. The Cushitic/Chadic word for "goat" or "general small ruminant" is kuru or guru, which often gets shortened to kuu. This is similar to the Khoe and Kwadi words *gu, meaning "sheep" (Ehret 1984:35). The distribution of Cushitic-speakers may well have been much further south than today. Ten Raa (1986:274) says that the Khoe-speaking Sandawe of East Africa have acquired their basic cattle vocabulary from the southern Cushitic-speaking Alagwa. Furthermore, ten Raa claims that this acquisition only occurred within the past ten generations, since this is when the Alagwa came into the area, and that it is highly probable this was not the first contact between the Cushitic-speaking Iraqw cluster and Sandawe. Thus, even though the Sandawe only obtained cattle for themselves as a result of Pax Germanica, this does not mean that they had no contact with cattle keepers in the area and further north for a long time prior to this.

In addition, the language of the Sandawe hunters of East Africa is a distant relative of Khoe in Botswana, which hints at Khoesaan-speakers being distributed over a much wider area to the north than is presently the case. The combination of a more southerly extent of Cushitic-speakers and Khoesaan being found much further north may well have been the mechanism for Khoe-speakers to obtain stock before the arrival of Iron Age farmers. This would explain why it seems that the Bantu-languages of southern Africa got their words for "sheep" from the Khoekhoen.

Support for this hypothesis has recently come from surveys being made in the Limopopo Valley in northern Transvaal. There, Warren Fish (Eastwood & Fish 1996) has found that fat-tailed sheep-paintings and thin-walled, undecorated pottery is quite different from any of the Iron Age ceramics previously identified from the area. This work adds to another idea of Ehret (1984:167) that there were Khoi-speakers in the area, but, unfortunately we still do not have any dates for the archaeological material as yet.

7. Conclusions

It must be obvious that, in spite of the amount of research that has gone into the origins of food production in southern Africa, there are major questions that still remain. Many of these are the result of lack of information from Angola and Tanzania, and even though much work has been done in Zambia (Fagan 1967, Phillipson 1977) there has been the tendency to go for the larger sites that have given us considerable information about Iron Age settlement. It is possible that these larger sites may have masked an earlier, less visible, pastoralist component.

Reasons for saying this, it must be admitted, are, for the most part, circumstantial. For example, sheep and ceramics have been identified from Geduld in northern Namibia (Smith & Jacobson 1995), but there is no obvious donor for either of these. The ceramics are thin-walled with a curious burnished corrugation design below the rim (known also from the Brandberg, cf. Kinahan 1991). However, such finely made ceramics appear without any known connection. Again, Angola would be a suggested source.

As mentioned above, the Limpopo area is another in which the Iron Age model of introduction of earliest domestic animals shows cracks, but what about possible timing differences between the appearance of sheep and cattle? As it stands, the Atlantic west coast route of early introduction just seems to have sheep at the bottom of any sequence, such as Geduld, Spoegrivier or Kasteelberg between 2000 BP and 1600 BP. If the linguistics are to be believed, Bantu-speaking people got the words for sheep from KhoeKoen. Does this mean that cattle were the main domestic animal introduced by Iron Age farmers into southern Africa? Cattle seem to increase in frequency in the southwestern Cape only after 1000 BP, with a ratio of 11 sheep to one cow (compared with more than 30:1 in the previous pastoral periods). Can we then further suggest that cattle were being introduced in larger numbers from Iron Age people along the eastern Cape coast at this time?

Another issue which needs addressing is the strong tendency to assume that domestic stock on an archaeological site inevitably meant the site was occupied by pastoralists. In the Kalahari today there are hunters who have small herds of goats around their camps. These animals are very hardy, and virtually take care of themselves. In addition, the goats are not used ritually, all ceremonial functions being served by the products of the hunt (Kent 1992). These animals are kept as a "cash" resource, being used to pay for commodities not available in the immediate environment (Ikeya 1993). There is some similarity to the Sandawe use of domestic stock, who equally see them as a cash resource, although now they are beginning to use them for ritual purposes, e.g. for rain-making ceremonies, replacing the wild animals that are not as readily available as before (Imogene Lim, pers. comm.).

One has to question whether hunters with small herds can be considered pastoralists. I would suggest that ritual incorporation of the stock is a crucial aspect of pastoralism, which further will go along with bride-wealth – rather than bride service as is associated with hunting societies in Africa. If, indeed, we can talk about the stock as performing different social and economic functions within groups that may be in contact with each other, we have the possibility of two quite distinct archaeological faunal assemblages: (a) with a characteristic hunting tool component, but with small numbers of domestic animals; (b) with large numbers of domestic animals. The possibility of one becoming the other is not impossible, but is not nearly as easy as has been implied by historians in the past (see Elphick 1977:30, and Smith 1990 for comment), or for indigenous development (Kinahan 1991).

References

Brink, J.S. & S. Holt 1992. A small goat, *Capra hircus*, from a late Iron Age site in the eastern Orange Free State. *Southern African Field Archaeology* **1**, 88–91.

Campbell, A.C. 1992. Southern Okavango Integrated Water Development Study: Archaeological Survey of Proposed Maun Reservoir. Gaberone: Govt. of Botswana, Ministry of Mineral Resources and Water Affairs.

Campbell, C. 1986. Images of war: a problem in San rock art research. *World Archaeology* **18**, 255–68.

Chapman, J. 1868. *Travels in the interior of South Africa* [2 volumes]. London: Bell & Daldy.

Clutton-Brock, J. 1977. Man-made dogs. *Science* **197**, 1340–42.

Davis, S. 1987. *The archaeology of animals.* London: Batsford.

Denbow, J. 1990. Congo to Kalahari: data and hypotheses about the political economy of the western stream of the early Iron Age. *The African Archaeological Review* **8**, 139–75.

Denbow, J.R. & E.N. Wilmsen 1986. Advent and course of pastoralism in the Kalahari. *Science* **234**, 1509–15.

Eastwood, E. & W. Fish 1995. *The rock art of the Venetia Limpopo Nature Reserve.* Soutpansberg Rock Art Conservation Group.

Ehret, C. 1982. The first spread of food production to southern Africa. In *The archaeological and linguistic reconstruction of African history*, C. Ehret & M. Posnansky (eds), 158–81. Berkeley: University of California Press.

Ehret, C. 1984. Historical/linguisitic evidence for early African food production. In *From hunters to farmers: the causes and consequences of food production in Africa*, J.D. Clark & S.A. Brandt (eds), 26–39, Berkeley: University of California Press.

Elphick, R. 1977. *Kraal and castle.* New Haven: Yale University Press.

Epstein, H. 1971. *The origin of the domestic animals of Africa.* New York [2 volumes]. New York: Africana Publishing.

Fagan, B.M. 1967. *Iron Age cultures in Zambia. Vol. 1: Kalomo and Kangila.* London: Chatto & Windus.

Faulkner, D.E. & H. Epstein 1957. *The indigenous cattle of the British Dependent Territories.* London: HMSO.

Goodall, E. 1946. Domestic animals in rock art. *Transactions of the Rhodesia Scientific Association* **41**, 57–62.

Hall, S.L. 1986. Pastoral adaptations and forager reactions in the eastern Cape. In *Prehistoric pastoralism in southern Africa*, M. Hall & A.B. Smith (eds). South African Archaeological Society, Goodwin Series **5**, 42–9.

Hart, T. 1987. Porterville survey. In *Papers in the prehistory of the western Cape, South Africa.* J.E. Parkington & M. Hall (eds), 403–23. Oxford: BAR International Series.

Henshilwood, C. 1996. A revised chronology for pastoralism in southernmost Africa: new evidence of sheep at *c.* 2000 b.p. from Blombos Cave, South Africa. *Antiquity* **70**, 945–9.

Hollmann, J. 1993. Preliminary report on the Koebee rock paintings, western Cape Province, South Africa. *South African Archaeological Bulletin* **48**, 16–25.

Horton, M. 1987. The Swahili Corridor. *Scientific American* **257**, 76–84.

Huffman, T.N. 1979. African origins. *South African Journal of Science* **75**, 233–7.

Huffman, T.N. 1982. Archaeology and ethnohistory of the African Iron Age. *Annual Review of Anthropology* **11**, 133–50.

Ikeya, K. 1993. Goat-raising among the San in the central Kalahari. *African Study Monographs* **14**, 39–52.

Ikeya, K. 1994. Hunting with dogs among the San in the central Kalahari. *African Study Monographs* **15**, 119–34.

Jolly, P. 1995. Melikane and upper Mangolong revisited: the possible effects on San art of symbiotic contact between southeastern San and southern Sotho and Nguni communities. *South African Archaeological Bulletin* **50**, 68–80.

Kent, S. 1992. The current forager controversy: real versus ideal views of hunter–gatherers. *Man* (n.s.) **27**, 45–70.

Kinahan, J. 1991. *Pastoral nomads of the central Namib Desert*. Windhoek: New Namibia Books.

Krige, E.J. 1950. *The social system of the Zulus*. London: Longmans, Green.

Lewis-Williams, J.D. 1981. *Believing and seeing: symbolic meanings in southern San rock paintings*. London: Academic Press.

Manwell, C. & C.M.A. Baker 1983. Origin of the dog: from wolf or wild *Canis familiaris*? *Speculations in Science and Technology* **6**(3), 213–24.

Marshall, F. 1990. Origins of specialized pastoral production in East Africa. *American Anthropologist* **92**(4), 873–94.

Mason, I.L. & J.P. Maule 1960. *The indigenous livestock of eastern and southern Africa*. Farnham Royal: Commonwealth Agricultural Bureaux.

Morey, D.F. 1994. The early evolution of the domestic dog. *American Scientist* **82**(4), 336–47.

Pager, H. 1993. *The rock paintings of the Upper Brandberg, Part 2. Hungorob Gorge*. Köln: Heinrich Barth Institut.

Phillipson, D.W. 1977. *The later prehistory of eastern and southern Africa*. London: Heinemann.

Plug, I. 1996a. Seven centuries of Iron Age traditions at Bosutswe, Botswana: a faunal perspective. *South African Journal of Science* **92**, 91–7.

Plug, I. 1996b. Domestic animals during the early Iron Age in southern Africa. In *Aspects of African archaeology, Papers from the 10th Congress of the PanAfrican Association of Prehistory and Related Studies*, G. Pwiti & R. Soper (eds), 515–22. Harare: University of Zimbabwe Publications.

Plug, I. & E.A. Voigt 1985. Archaeozoological studies of Iron Age communities in southern Africa. *Advances in World Archaeology* **4**, 189–238.

Raper, P.E. & M. Boucher 1988. *Robert Jacob Gordon: Cape Travels, 1777 to 1786* [2 volumes]. Johannesburg: Brenthurst Press.

Raven-Hart, R. 1967. *Before Van Riebeeck*. Cape Town: Struik.

Robertshaw, P.T. 1990. *Early pastoralists of south-western Kenya*. Nairobi: British Institute in eastern Africa.

Rudner, J. 1968. Strandloper pottery from South and South West Africa. *Annals of the South African Museum* **49**(2), 441–663.

Schapera, I. 1960. *Livingstone's Private Journals, 1851–53*. London: Chatto & Windus.

Schofield, J.F. 1940. A report on the pottery from Bambata Cave. *South African Journal of Science* **37**, 361–72.

Sealy, J. & R. Yates 1994. The chronology of the introduction of pastoralism to the Cape, South Africa. *Antiquity* **68**, 58–67.

Smith, A.B. 1987. Seasonal exploitation of resources on the Vredenburg Peninsula after 2000 BP. In *Papers in the prehistory of the western Cape, South Africa*, J.E. Parkington & M. Hall (eds), 393–402. Oxford: BAR International Series **332**.

Smith, A.B. 1990. On becoming herders: Khoikhoi and San ethnicity in southern Africa. *African Studies* **49**(2), 51–75.

Smith, A.B. & L. Jacobson 1995. Excavations at Geduld and the appearance of early domestic stock in Namibia. *South African Archaeological Bulletin* **50**, 3–14.

Smith, A.B. & R.H. Pheiffer 1993. *The Khoikhoi at the Cape of Good Hope: seventeenth-century drawings in the South African Library*. Cape Town: South African Library.

Smith, A.B., K. Sadr, J. Gribble, R. Yates 1991. Excavations in the south-western Cape, South Africa, and the archaeological identity of prehistoric hunter–gatherers within the last 2,000 years. *South African Archaeological Bulletin* **46**, 71–91.

Starkey, P.H. 1984. N'Dama cattle: a productive trypanotolerant breed. *World Animal Review* **50**, 2–15.

ten Raa, E. 1986. The Alagwa: a northern intrusion in a Tanzanian Khoi-San culture, as testified through Sandawe oral tradition. In *Contemporary Studies on Khoisan* [2 volumes], R. Vossen & K. Keuthmann (eds), 271–99. Hamburg: Helmut Buske.

Thorp, C. 1979. Cattle from the early Iron Age of Zimbabwe-Rhodesia. *South African Journal of Science* **75**, 461 only.

Van Rijssen, W.J.J. 1984. Southwestern Cape rock art – who painted what? *South African Archaeological Bulletin* **39**, 125–9.

Van Ryneveld, W.S. 1942. *Aanmerkingen over de verbetaring van het vee aan de Kaap de Goede Hoop, 1804*. Cape Town: Van Riebeeck Society, No. 23.

Vinnicombe, P. 1976. *People of the Eland*. Pietermaritzburg: University of Natal Press.

Voigt, E.A. 1984. Goats in the Transvaal. *South African Journal of Science* **80**, 484 only.

Walker, N. 1983. The significance of an early date for pottery and sheep in Zimbabwe. *South African Archaeological Bulletin* **38**, 88–92.

Walker, N. 1991. Rock paintings of sheep in Botswana. In *Rock art – the way ahead*, S.A. Pager (ed.), 183–95. Parkhurst: Proceedings of the Southern African Rock Art Research Association.

Waterhouse, G. 1932. *Simon van der Stel's journal of his expedition to Namaqualand, 1685–6*. Dublin: Hodges, Figgis.

Wayne, R.K. 1993. Molecular evolution of the dog family. *Trends in Genetics* **9**, 218–24.

Webley, L. 1984. *Archaeology and ethnoarchaeology in the Leliefontein Reserve and Surrounds, Namaqualand*. Unpublished MA thesis, University of Stellenbosch.

Webley, L. 1992. Early evidence for sheep from Spoeg River Cave, Namaqualand. *Southern African Field Archaeology* **1**, 3–13.

Wilson, M. 1969. The hunters and herders. In *The Oxford History of South Africa*, vol. 1, M. Wilson & L. Thompson (eds), 40–74. Oxford: Oxford University Press.

Woodhouse, H.C. 1987. Cattle used for transport in the rock art of southern Africa. *South African Archaeological Bulletin* **42**, 65–8.

Yellen, J.E. 1990. The transformation of the Kalahari !Kung. *Scientific American* **262**(4), 96–105.

PART THREE
Genetics and Breed Characterization

CHAPTER TWELVE

Genetics and the origins of African cattle

Patrick Cunningham

In recent millennia, the evolution of human societies in Africa has been heavily dependent on the continent's cattle resources. Today, and for the visible future, it is also a region that will be strongly reliant on its bovines.

The long symbiosis between cattle and people means that their genetic histories are also closely intertwined. The great diversity of African ecology presented both people and their livestock with parallel challenges. Of particular interest on the African continent is the great cline in rainfall, vegetation and disease challenge which spans the area from the Sahara south to the equatorial forest zone. Coupled with this is the fact that, as nowhere else in the world, running through this zone is a continuous interface between the two great cattle types of the world, *Bos indicus* and *B. taurus*.

A provisional inventory of African cattle breeds and types lists 120 known breeds, for 84 of which some systematic information is available, and of which a considerable number are regarded as being in danger of disappearance (FAO 1995).

Until quite recently, the array of methods that could be used to study genetic differences between these cattle populations was limited to observation of their differences in size, format, growth, behaviour, milk production and reaction to various diseases. This was augmented from about 1960 by the study of blood groups and protein differences (Baker & Manwell 1991). The scope and precision of genetic technology has been transformed in the past decade by the development of a repertoire of new methods, all focused on the study of the actual genetic material itself, DNA.

The group of projects reported on here had its origins in a desire to know more about the genetic background of *B. taurus* and *B. indicus* breeds. As economies begin to evolve beyond traditional subsistence agricultural structures, and as populations increase, one of the commodities for which demand rapidly expands is milk. As cattle are the principal agricultural resource in many of these societies, this demand translates into a desire to increase milk output. However, changing milk output levels by individual selection is a slow business. In the first place, it can be measured only in females, while most genetic change in a population comes about through selection on the male side, simply because the reproduction potential of individual males is so much higher. Since cows in tropical countries typically calve for the first time at four years or more, milk production can only be observed rather late in life. In addition, accurate measurement of milk production, and in particular of its content of fat and protein, requires relatively sophisticated, sustained and disciplined work. All of these factors mean that the improvement of milk output in most tropical breeds is likely to be a long slow process.

In contrast, several *B. taurus* breeds in developed countries have been subject to systematic selection for milk production for a very long time. This selection has been particularly effective in the last fifty years because of the introduction of artificial insemination. This makes it possible to progeny-test males with great accuracy and to greatly extend their reproductive potential. The result of these separate evolutionary histories is that while tropical breeds of cattle typically produce less than 1,000 litres of milk in a lactation, developed dairy breeds such as Holstein, Jersey or Brown Swiss can, if properly managed, produce up to ten times this amount.

In the early decades of the twentieth century, these and other high performing dairy breeds were introduced into many tropical countries. They generally succumbed to local diseases, climatic stress and frequently inadequate feed supply. The availability of artificial insemination, particularly of frozen semen, from the 1950s onwards led to renewed interest in using the developed dairy breeds for crossing on local stock. The results this time were generally much more encouraging. First cross animals exhibited some of the strengths of both *B. taurus* and *B. indicus*: higher production, combined with greater resilience in the face of climatic, feed and disease challenges. As a result, large-scale cross-breeding programmes, particularly using Holstein or Jersey, were undertaken in many countries. For example, in India the declared target was to produce 20 million crossbreds.

These developments produced new challenges. A particularly important one was the question of what to do after the first cross. Continued crossing to the exotic strain risked loss of all local adaptation, while crossing back to the local breed would undoubtedly lose the production potential. Strategies for continued balanced crossing systems, or the established of new synthetic breeds, can be expensive and difficult. The choice between these strategies depends on the nature of the genetic control of the various traits of importance. If this control is simply attributable to the additive effects of the different genetic sources, then balanced strategies are relatively straightforward. If, on the other hand, non-additive effects (i.e. effects specific to particular crosses) are important, then quite different, and more complicated, strategies may be called for.

In order to clarify these options, FAO commissioned a review of all cross-breeding experiments which have been undertaken, together with a study of the genetic basis for possible breeding strategies (Cunningham & Syrstad 1988). In this study, some 44 experimental datasets were reviewed. The broad conclusions were that the additive differences between *B. taurus* developed dairy breeds on the one hand and *B. indicus* breeds on the other was of the order of 100 per cent of mean production, while the non-additive effects were about 30 per cent. A range of strategies for the future was presented.

Arising out of these results, it was clear that it would be useful to know some-thing about the degree of genetic distance between the main B. *taurus* and B. *indicus* breeds. A project was therefore begun in 1989, with support from the EU STD programme, to use the latest DNA technologies for this purpose. Three breeds were chosen in India, two in East Africa, two in West Africa and six from Europe.

The evolution of the project says something about the rapid pace of technical development in this area. The initial intention was to use two types of genetic analysis: DNA–DNA hybridization, and Restriction Fragment Length Polymorphism (RFLP). The first was rapidly abandoned as being insufficiently precise. The second method is a way of pinpointing individual site differences in DNA sequences. It was rapidly superseded by the development of methods for comparing actual sequences, rather than simply sites within them. At the same time, the developing appreciation of the value of neutral repetitive DNA for measuring genetic variation presented additional opportunities. In particular, in recent years there has been an explosion in the number of dinucleotide repeat sequences (microsatellites) known. When the project began in 1989 some 20 microsatellite loci were available in cattle. Today, the figure is close to 2,000.

The methods used in these studies were

(a) the comparison of sequences, concentrating on mitochondrial DNA, which is inherited solely through the female
(b) comparison of sequences on the Y chromosome, which is inherited only through the male
(c) extensive analysis of differences within and between breeds at a range of microsatellite loci.

The parallel use of these three avenues permits an elegant dissection of male and female mediated gene flow, together with a tracking of the broad common genetic evolution of the population.

The project has produced some surprises. The first was that the Indian populations have genetic roots so far removed in time from those of the others that the most reasonable explanation is a separate domestication in India. The European popu-lations, despite their great physical differences, are little differentiated at the mito-chondrial DNA level. The African B. *indicus* populations all have B. *taurus* maternal inheritance, while the African B. *taurus* populations in many cases show evidence of B. *indicus* crossing. The papers presented here focus on the interpretation of these results.

The genetic results presented are a matter of fact and document the patterns of variation observed at the DNA level. In two broad respects they challenge conven-tional wisdom, by suggesting that all African cattle were originally B. *taurus* types and that most populations have considerably more hybrid origins than had been supposed. The results also suggest that a separate African domestication is not an unreasonable hypothesis, although the evidence is much less conclusive than for the Indian case.

The results need to be aligned with those coming from archaeology and from linguistic studies, and indeed from other fields too. For example, recent evidence on improved climatic conditions in the heart of the Sahara region in the critical millennia between 8500 and 6100 BP (Ritchie et al. 1985) could shed considerable light on possible paths of gene flow linking sub-Saharan Africa with Europe and Asia.

The development of the latest generation of genetic technology has in fact created a new, fascinating, and potentially very productive intersection of disciplines.

References

Baker, C.M.A. & C. Manwell 1991. Population genetics, molecular markers and gene conservation of bovine breeds. In *Cattle genetic resources*, C. Hickman (ed.), World Animal Science Series B7; 221–92. New York: Elsevier Science Publishers.

Cunningham, E.P. & O. Syrstad 1988. *Crossbreeding* Bos indicus *and* Bos taurus *for milk production in the Tropics*. FAO Animal Production and Health Paper, No. 68. FAO, Rome.

FAO 1995. *World watch list for domestic animal diversity*. Rome: FAO.

Ritchie, J.C., C.H. Eyles, C.V. Haynes 1985. Sediment and pollen evidence for an early to mid-Holocene humid period in the eastern Sahara. *Nature* **314**, 352–5.

CHAPTER THIRTEEN

Two Eves for *taurus*? Bovine mitochondrial DNA and African cattle domestication

Dan Bradley & Ronan Loftus

1. Genetic variation and archaeological inference

Patterns of genetic variation in animals and humans are influenced by both ancient and modern demographic processes, the signatures of which may often be recognized. The genetic investigation of extant populations may result in inferences about relationships and origins that have much to contribute to archaeological questions. In livestock species, the amount of potential DNA data is vast, both in sample size and in the amount of the genome which may now be assayed. A range of techniques examining different types of genetic change afford flexibility, allowing potential assessment of a range/scale of relationships almost as wide as the tree of life itself. Importantly, each type of change proceeds in time-related fashion and allows the establishing of a molecular clock through which time depths of past events may be estimated. Variation at the DNA level is the primary, basal level of genetic difference and is free from the environmental noise that blurs other data, such as that used in morphometric analyses.

2. Mitochondrial Eve

In 1987, with characteristic flair, a research group in Berkeley, California, published the headline-gaining results of a wide survey of human mitochondrial DNA (mtDNA) variation (Cann et al. 1987). They asserted that the ancestry of human maternal lineages converged to one African and she was estimated as having lived 100,000–250,000 years ago. Mitochondrial Eve was created and presented to the popular mind with a hard-edged scientific authority seen to be lacking in accounts of her biblical counterpart. The key anthropological question of the timing of the ancestral human exodus from Africa seemed to have been clearly answered. However, the last decade has seen some of that authority eroded as a series of methodological problems associated with the analysis of data such as that presented by the

Berkeley group have come to light. The tenure of mtDNA Eve in the garden of Eden has, at times, seemed tenuous.

A principal problem has been that methods of constructing, calibrating, and interpreting phylogenetic trees that perform well in other contexts prove to be inappropriate when applied to large volumes of relatively invariant mitochondrial data with its particular evolutionary quirks. More recently, however, a spate of phylogenetic debunking of mtDNA Eve has given way to more constructive approaches in the human genetics literature. It is clear that meaningful, if less assertive, archaeological inferences are indeed possible from mtDNA sequence comparisons within species.

The basis of the present study is mtDNA sequence from 90 cattle from three continents: 32 European, 28 African and 21 Indian animals (for sample sources, see Acknowledgements). DNA was extracted from blood and analyzed by PCR amplification and direct sequencing (Loftus et al. 1994, Bradley et al. 1996). The control region is the most variable portion of the mtDNA genome and 370 bp, or just over a third of its most informative sequence, was assayed in each sample. Figure 13.1 shows a schematic summary of the sequence data. The basic geographic pattern of sequence variation is clear: sequences from *Bos indicus* animals from the Indian subcontinent are markedly different from the rest of the sample and African and European samples exhibit some, somewhat consistent, differences from each other.

Notably the similarity of all sequences within Africa occurs despite the obvious presence of zebu ancestry in many of the individuals chosen. This is a result which is easily explained by the hybrid origins of these animals (Bradley et al. 1994, Loftus et al. 1994, Loftus & Cunningham (Ch. 14 in this volume)) and it is assumed here that the mtDNA chromosomes are representative of the earlier *Bos taurus* gene pool of that continent.

In all, no identical variants were found between continents but many were found within, including 20 identical European and nine identical African sequences (given as the first sequences in each group in Fig. 13.1).

3. Trees of knowledge

The estimation of a phylogenetic tree may provide a convenient graphic summary of DNA sequence data. Once constructed, using one of a number of alternative algorithms, a tree presents information through two inherent properties. First, the topology, or pattern of branching infers the order of phylogenetic relationships between samples, and secondly, the length of connecting branches provides some quantification of the genetic distances involved.

Figure 13.2 illustrates a phylogeny derived from the sequence data in Figure 13.1. This was constructed using average levels of sequence divergence between each continent, with the addition of bison data. The distinction of the Indian sequences is immediately obvious and the outstanding quantitative feature is the extent of difference between these and the other two groupings. This has previously been calculated to imply a time depth for *Bos taurus/Bos indicus* divergence that is

```
              0000000000 000000000000000111111111111111111111111111111111 2222222222223333300
              4444445555 555666677788899000111111222222222333333334445666899 2233445566990000103
              2457890135 678789456245492890236789124567890123578913744556 8912780504480124189

Europe  TAU  TCAGACCTTT*AGCATATAAGCTTTGTTCTTTGATGTTATCTATTATTTTCTATATTGAG*AACCCCATCGTAACGTTGC
        TAU  ----------*-----------------------C-----T----------------*------------------
        TAU  ----------*---------------------------------------------*-----------------A-
        TAU  ------T---*--T-------C----------------------------------*-----------------A---
        TAU  ----------*----------------------------------------------*-----------------A---
        TAU  ----------*---------------------------------------------*--T---------------
        TAU  ----------*------C---------------------------------------*----G------------
        TAU  ----------*-----------------------C---------------------*-------C----------
        TAU  ----------*-C--------------------------------A--*-------C-------C--
        TAU  C--A-----C*-----------------------------C--------------------T-------A-
        TAU  ----------*---------------------------------------------G-*-------------
        TAU  ----------*--------------C-------------------------------*--------------
        TAU  ----------*--T---------------------------C--------------*-G------------
        TAU  ----------*----------------------------T----------------*--------------
        TAU  ----------*----G----------------------------------------*--------------
        TAU  ----------*--T-----------C------------------------------*--------------
        TAU  ----------*----------------------T---C-----------------*--------------
        TAU  ------C--*----C---------------------------------C----*--------------
        TAU  ------C--*-------------------------------------------*--------------

Africa  TAU  -----T---*------------------C-------------------------*------C--------
        TAU  -----T---*------------------C---C---------------------*------C--------
        TAU  ------T---*-A---------------C-------------------------*------C--------
        TAU  -----T---*------------------C---AC-------------------*------C--------
        TAU  -----T---*------------------C-------------C-----------*------C--------
        IND  ------T---*---------C-------C-------------------------*------C--------
        IND  -----T---*------------------C-------------------------*------CT-------
        IND  -----TT-C-*-----------------C-------------------------*------C--------
        IND  ------T---*------------------C---A--------------------*------C--------
        IND  ----------*-----------------C-------------------------*----T-C--------
        IND  ----------*-----------------C---C--------------------*----T-C--------
        IND  ------TC-*-A--------C-------C-------------------------*---T-C---T-----
        IND  ------T---*--------T--------------------------------*------C-C-------
        IND  -T----T---*----------------------------------------*------C--------
        IND  ------T---*-----------------C-----------------------*------C---T-----
        IND  -----T---*-----------------C-------------T----------*------C--------
        IND  ------T---*-----------------C------------------G----*------C--------
        IND  -----T---*------G----------C------------------------*------C--------
        IND  ------T---*-----------------C-----------------------*-----T-C-------A-

India   IND  ----------*-AT---C--ATC--A-C--CCA-CAC------C----CC--*C-----AA-G--TT------GT---A-
        IND  -----T----*-AT---C--AT---A-C--CCA-CAC------C----CC--*C-----AA-G--TT-------T---A-
        IND  -----TT---*-AT------AT---A-C--CCA-CAC------C----CC--*C--C--AA-G--TT-----GT---A-
        IND  -----TT---*-AT---C--A-C--A-C--CCA-CAC------CC--*-----AA-G--TT-------T---A-
        IND  ------T---*-AT---C--ATC--A-C--CCA-CAC------C----CC--*C-----AA-G--TT------GT---A-
        IND  -----T----*-AT------AT---A-C---CA-CAC------C----CC--*C-----AA-G--TT------GT---A-
        IND  -----T----*-AT---C--A-C--A-C--A-CAC------C----CC-C*C-----AA-G-TTT-----GT---AT
        IND  -----T----*-AT---C--A-C--A-C--CCA-CAC------C----CC-C*C-----AA-G-TTT-----GT---AT
        IND  -----T----*-AT---C--AT---A-C--CCA-CAC------C----CC--*C-----AA-G--TT------GT---A-
        IND  -----T----*-AT---C--A-C--A-C--CCA-CAC------C----CC-C*C-----AAGG-TTT-----GT---AT
        IND  -----T----*-AT---C--AT--CA-C--CCA-CAC------C----CC--*C-----AA-G--TT------GT---A-
        IND  -----T----*-AT---C--AT--CA-C--CCA-CAC------C----CC--*C-----AA-G--TT------G----A-
        IND  ----T-----*-AT------AT---A-C--CCA-CACC-----C----CC--*C-----AA-G--TT------GT---A-
        IND  -----T----*-AT------A-C--A-C--CCA-CAC------C----CC-C*C-----AA-G-TTT-----GT---AT
        IND  -----T----*-AT---C--AT---A-C--CCA-CAC------C----CC--*C-----AA----TT-----GT---A-
        IND  -----T----*-AT---C--A-C--A-C--CCA-CAC------C----CC-C*C-----AA-G-TTT------T---AT

Bison   BIS  --TAT-----TGA-G--A-GAT-C---CTCC--GCAC-GC*******--C--*-------AG---TT-CTACG-TCC---
```

Figure 13.1 Sequence variations observed in 90 cattle and one bison control region. Only one example each of multiply represented sequences is shown. The first column indicates the morphology of the sampled animal with TAU indicating *Bos taurus* and IND indicating *Bos indicus* origin. One exception is the first African sequence listed (denoted TAU) which was observed in samples of both zebu and taurine type. Only variable sites, with sequence positions given above, are shown. Identity with the first sequence is denoted by a dash, substitution by a different base letter and deletions by asterisks.

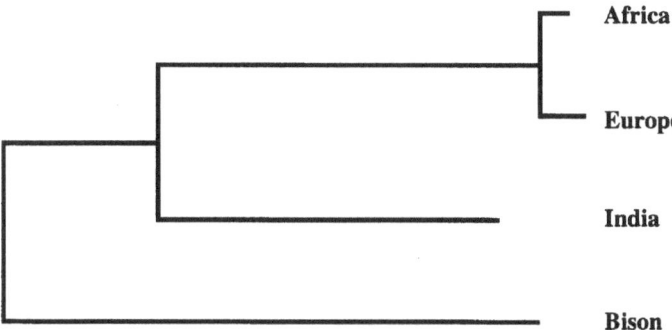

Figure 13.2 Unrooted phylogeny based on mean sequence divergence between continental populations, including a number of American bison sequences (Beech et al. 1995). The FITCH tree-drawing algorithm (Felsenstein 1993) and the Tamura-Nei distance (a=0.5) were used (Tamura & Nei 1993).

orders of magnitude greater than the domestic history of the species. This constitutes strong evidence for an independent domestication of zebu, possibly in Baluchistan (Meadow 1993, Loftus et al. 1994).

A second inferred phylogeny, which concentrates on the African/European relationship by excluding Indian sequences from the analysis, is given in Figure 13.3. The branching order is unlikely to concur in detail with the true phylogenetic relationships (this is virtually impossible to achieve in any phylogeny of a large number of closely related sequences). However, the major features, particularly the clustering of African and European variants into separate clades, are unlikely to be artefactual. In fact, the integrity of the two continental groupings are maintained when trees are drawn using a variety of different tree-drawing methods (Bradley et al. 1996).

Early assertions about human mtDNA Eve were based on interpretations of phylogenies that relied too heavily on the absolute accuracies of deduced tree structures. These studies foundered on the rocks of collegial scholarship as commentators pointed out that the particular result of any tree-drawing algorithm was often only one of a huge number of possible solutions, many of which could be superior to that initially presented (Penny et al. 1995). As with biblical accounts of genesis, wisdom is presently perceived as lying with allegorical rather than literal interpretation of such phylogenies.

In the present case, whereas the reconstruction is unlikely to be correct in detail, the integrity of the European/African division is attested to by a number of analyses and these may be treated as discrete genetic populations to which between-group comparisons may be applied and calibrated. An additional topological feature is that the tree has two foci from which star-shaped radiations emerge. This pattern is consistent with the occurrence of two past population expansions. In each continent there appears to be a central topological point, which represents in each the most common sequence and to which the majority of ancestral lineages appear to trace back. It is probable that domestication of a small number of initial livestock would

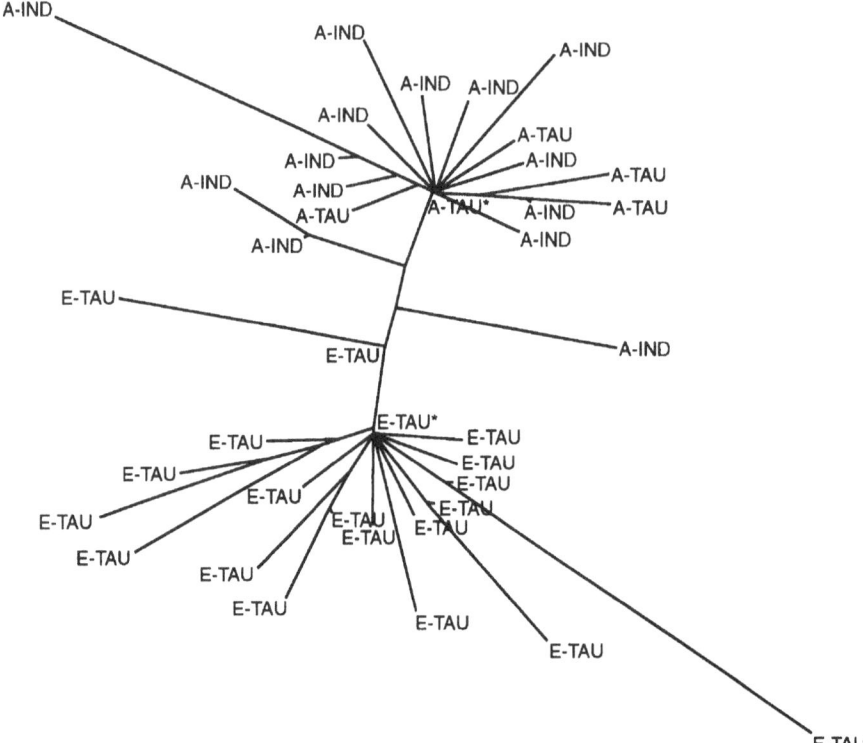

Figure 13.3   Unrooted phylogenetic tree constructed using the individual African and European sequences listed in Figure 13.1 as nodes and the Neighbour programme from Felsenstein (1993). Sequences obtained from African zebu are denoted as A-IND, those from African taurines as A-TAU and those from Europe as E-TAU. The two most commonly represented sequences are represented as A-TAU* and E-TAU* and, notably occupy central positions in the phylogenetic clusters of each continental group. A-TAU* was found in animals of both *Bos indicus* and *Bos taurus* morphology.

have been followed by a dramatic expansion of their offspring as the success of the novel technology spread. These genetic signatures of population expansions may be the direct result of the domestication process.

4. Time depth estimation

The timing of the demographic events detailed above may be estimated by examination of the extent of sequence divergence both within and between populations. However, the calibration of the molecular clock in mtDNA D-loop data is notoriously difficult. A number of complications arise from the biology of this molecule, most notably that different portions of the sequence are subject to different base-pair substitution rates in a manner that is only poorly understood. The D-loop molecular clock ticks erratically and hence resultant time depth estimates must be treated with appropriate caution.

From a comparison with bison sequence and an acceptance of 1 Myr BP as the divergence of its lineage from the *Bos* ancestor, the rate of change in the D-loop sequence fragment investigated here is estimated as 62.8 per cent per Myr (Bradley et al. 1996). This falls within the range of estimates (15 per cent–110 per cent per Myr) which a number of investigators have deduced for the more intensively studied human D-loop. Using this rate estimate, and the average mtDNA sequence divergence between the African and European continents, the time of the most recent common ancestor of taurine mtDNA chromosomes may be estimated at 22,000 years BP.

Thus there is quantitative evidence for a biological separation between ancestral African and European taurines that predates the domestication process, although the large error margin which must exist in such a calculation is noted. Additionally, there exist in the data, patterns of variation that are suggestive of dramatic population expansions of time depth which might be consistent with those that must have been associated with the domestication process. Notably, an estimation of both the time of ancestral lineage separation and that of population expansion suggests that the former predates the latter, a feature which is independent of specific molecular clock calibrations (Bradley et al. 1996).

Thus both qualitative and quantitative aspects of the sequence variation observed in African and European populations are consistent with two domestications, of moderately divergent *B. primigenius* strains, which separately expanded in numbers to give rise to the *B. taurus* herds of each continent. A likely positioning of the European ancestral domestication is in the known early centres of the Near East, and several authors have suggested independent African domestic locales (Gautier 1984, Wendorf & Schild 1994, Clutton-Brock 1989). However, it must be noted that the DNA evidence does not formally exclude other reconstructions which assume a single domestic origin.

First, it is possible that the calibration of the molecular clock could be sufficiently wide of the true rate such that the diversity described here may in fact be attributable to a single, post-domestic origin. Secondly, it is possible that two divergent mtDNA lineages may have persisted in a single or several ancestral Near eastern populations and that through stochastic sorting a different inheritance persisted in each of the two continental daughter groups. Thirdly, the seeming excess of modern mtDNA diversity could perhaps be attributed to the post-domestication adoption of local aurochs in either continent. If the early pastoralists adopted female wild oxen into existing herds it is conceivable that it may be their descendants which are represented in the modern mtDNA gene pools. Biologically separate domesticate ancestry need not necessarily imply culturally separate domestications.

The placing of the two *B. taurus* Eves in a single or separate gardens may yet be more firmly resolved, either by the addition of more extant mtDNA data, data from other sections of the genetic material or perhaps by using the emerging technology of ancient DNA analysis. MacHugh (1996) estimates a pre-domestic African/European separation from the results of a survey of microsatellite genetic variation in modern cattle. Also, the addition of a temporal dimension to the spatial analysis of bovine genetic variation is certainly feasible (Bailey et al. 1996), and direct analysis of both the wild and domestic contemporaries of ancestral *B. taurus* promises valuable information.

Acknowledgements

The authors would like to acknowledge the assistance of the following in the collection of blood samples: D.S. Balain and V.J. Shankar (NIAG, Karnal, Haryana, India), L.O. Ngere (University of Ibadan, Nigeria), A.M. Badi, J. Rizgalla and A. Hassan (APRA, Khartoum, Sudan). We are also grateful to D.E. MacHugh, E.P. Cunningham, P.M. Sharp, A.T. Lloyd, C.M. Meghen, C.S. Troy and the Irish National Centre for Bioinformatics for valuable assistance. This work was funded by the European Commission DGXII Science and Technology in Developing Countries Program.

References

Bailey, J.F. et al. 1996. Ancient DNA suggests a recent expansion of European cattle from a diverse wild progenitor species. *Proceedings of the Royal Society, London (Series B)* **263**, 1467–73.

Beech, R.N., J. Sheraton, R. Polziehn, C. Stronbeck 1995. GENBANK Release 44, Accession Nos BBU12936, BBU12946, BBU12948, BBU12955, and BBU12959.

Bradley D.G., D.E. MacHugh, E.P. Cunningham, R.T. Loftus 1996. Mitochondrial diversity and the origins of African and European cattle. *Proceedings of the National Academy of Sciences, USA* **93**, 5131–5.

Bradley, D.G., D.E. MacHugh, R.T. Loftus, R.S. Sow, C.H. Hoste, E.P. Cunningham 1994. Zebu-taurine variation in Y chromosomal DNA: a sensitive assay for genetic introgression in West African trypanotolerant cattle populations. *Animal Genetics* **25**, 7–12.

Cann, R.L., M. Stoneking, A.C. Wilson 1987. Mitochondrial DNA and human evolution. *Nature* **325**, 31–6.

Clutton-Brock, J. 1989. Cattle in ancient North Africa. In *The walking larder: patterns of domestication, pastoralism, and predation*, J. Clutton-Brock (ed.), 200–206. London: Allen & Unwin.

Felsenstein, J. 1993. *Phylip (Phylogeny Inference Package) Version 3.5c.* Washington, USA: University of Washington, Seattle.

Gautier, A. 1984. Archaeozoology of the Bir Kiseiba region, eastern Sahara. In *Cattle keepers of the eastern Sahara: the neolithic of Bir Kiseiba*, A.E. Close (ed.), 49–72. Dallas: SMU Press.

Loftus, R.T., D.E. MacHugh, D.G. Bradley, P.M. Sharp, E.P. Cunningham 1994. Evidence for two independent domestications of cattle. *Proceedings of the National Academy of Sciences, USA* **91**, 2757–61.

MacHugh, D.D. 1996. *Molecular biogeography and genetic structure of domesticated cattle.* PhD thesis, University of Dublin.

Meadow, R.H. 1993. Animal domestication in the Middle East: a revised view from the eastern Margin. In *Harappan Civilization*, G. Possehl (ed.), 2nd edn, 295–320. New Dehli: Oxford and IBH.

Penny, D., N. Steel, P.J. Waddell, D.D. Hendy 1995. Improved analyses of human mtDNA sequences support a recent African origin for *homo sapiens*. *Molecular Biology and Evolution* **12**, 863–82.

Tamura, K. & M. Nei 1993. Estimation of the number of nucleotide substitutions in the control region of mitochondrial DNA in humans and chimpanzees. *Molecular Biology and Evolution* **10**, 512–26.

Wendorf, F. & R. Schild 1994. Are the early Holocene cattle in the eastern Sahara domestic or wild? *Evolutionary Anthropology* **3**, 118–28.

CHAPTER FOURTEEN

Molecular genetic analysis of African zeboid populations

Ronan Loftus & Patrick Cunningham

1. Introduction

Despite receiving considerable attention in recent years, the origins of domesticated cattle in Africa remain unclear. The paucity of early archaeological remains
which distinguish wild and domesticated forms, in addition to the large amount of
genetic variability among current African cattle populations, have contributed much
confusion to determining the exact origin and chronology of events. Consequently,
a number of theories have been invoked offering alternative temporal frameworks
or even domestication centres to account for archaeological finds.

Cattle may be broadly subdivided into humped (*Bos indicus*) and humpless *(B.
taurus)* animals. The earliest representations of cattle in Africa would seem to be
humpless, and probably correspond to the earliest phase of pastoral economy in
the Sahara (*c.* 6500 bp) (see MacDonald Ch. 1 in this volume, and Muzzolini Ch. 6
in this volume). Their domestication in western Asia is documented from at least
2,000 years earlier. Whether the African animals were introduced from Asia (Mason
1984) or domesticated locally (Bradley et al. 1994) is still a matter of debate. Humped
or zebu animals are only found much later in the archaeological record. They appear
to have been introduced from Asia by migrating peoples, traders and Arab expansion in the region. While the study of humpless cattle has received considerable
attention from archaeologists, linguists and students of related disciplines, most in
search of a putative African domestication for the subspecies (see Clutton-Brock
1989), the origins and subsequent expansions of African zebu have generally been
the focus of less effort.

This paper examines molecular genetic data from studies conducted on contemporary African cattle. Before discussing the findings in detail, the current state of
thinking on the origins of African zebu populations is briefly reviewed (see Bradley
& Loftus Ch. 13 in this volume, for a more thorough discussion on the origins of
African humpless cattle).

Zebu have been divided into thoracic-humped, with the hump situated above the
1st to 9th thoracic vertebrae, and cervico-thoracic-humped, with the hump in the
posterior part of the neck between the 6th to 7th cervical and the 4th to 5th thoracic

vertebrae (Epstein 1971). This distinction can be a little ambiguous, especially in an African context where many animals exhibit either humps of intermediate position or very small humps, usually attributable to crossbreeding. Mason (1984) has used the term "zeboid" to describe cattle with varying degrees of zebu-like features. Although not entirely satisfactory, the use of this term circumvents the difficulty of assigning populations to particular morphotypes and may more accurately reflect the early history and crossbred nature of most humped cattle in the region.

It has been suggested that most early introductions of humped cattle to Africa were through the Horn of Africa from Arabia (Epstein 1971). Interactions between peoples of the Horn of Africa and peoples from southwestern Arabia during the first millennium BC to the first half of the first century AD would have provided an ideal opportunity for the introduction of such humped animals (Mason 1984). These early introductions may have had little influence on local animals, but as the number introduced and the frequency of introduction increased there is likely to have been an increased impact. Introduced zebu were crossbred with indigenous taurine cattle in Ethiopia where it is claimed they produced the Sanga – an intermediate between true zebu and taurine animals. Others spread across to West Africa through the Sudan, crossbreeding with local humpless animals to produce breeds such as the Fulani in West Africa. Zebu or Sanga are also known to have spread west and south where they eventually gave rise to breeds such as the Ankole of Uganda and the Nguni of South Africa.

The spread of zebu was facilitated by their greater adaptation to the arid climate of the region (Clutton-Brock 1989). Furthermore, zebu are considered to be better at regulating their body heat, have lower water requirements, hardened hooves and lighter bones enabling them to endure long migrations. This has led some authors to argue that their introduction enabled herders to practice long distance trans-humance, thereby facilitating pastoralist expansions and possibly indirectly assisting early Arab expansion in the Sudan and Somalia (Davis 1982, Marshall 1989). Whatever the truth of this, the current distribution of zebu, coupled with their extensive introgression into West African humpless cattle populations bears testament to their considerable success on the continent (Meghan et al. 1994).

The distribution of more recent introductions of zebu appears to have been directly influenced by Indian and Arab traders (Mason 1984). Breeds such as the Boran in Kenya and the Kenana in Sudan are seen as being the direct descendants of such relatively recent introductions.

This paper cites genetic evidence from two separate studies conducted on modern African cattle populations. Data are presented from both lines of investigation and subsequently interpreted in the light of current understanding on the origins and distribution of African zeboid populations.

2. Mitochondrial DNA analyses

In the first study, four breeds from diverse origins and geographic locations (Fig. 14.1) were selected: N'Dama – West African humpless longhorn directly descended from early humpless cattle; White Fulani – considered a cross between

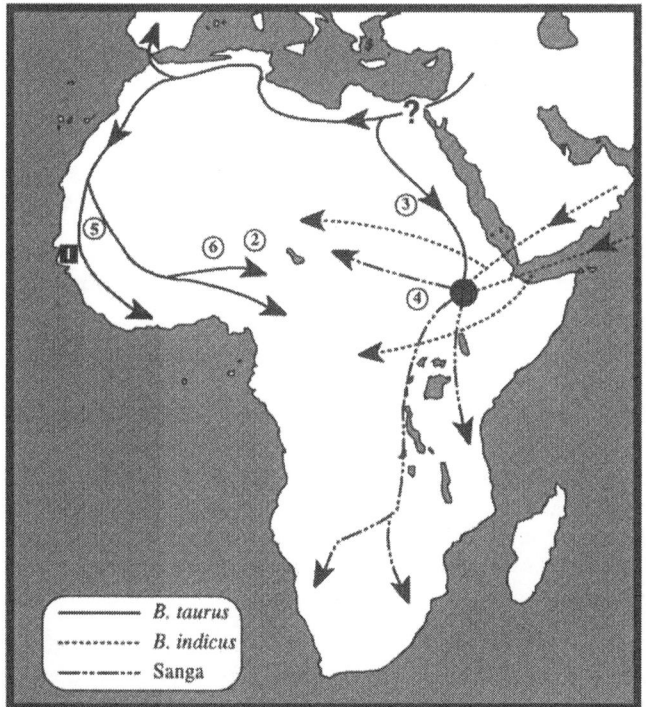

Figure 14.1 Geographical origins of breeds included in the study. Breeds of humpless morphology are indicated by boxes (1 N'Dama) whereas those of humped morphology are indicated by circles (2 White Fulani, 3 Butana, 4 Kenana, 5 Gobra and 6 Sokoto Gudali). The uncertain origin of African taurine populations is also indicated.

early zeboid introductions to West Africa and indigenous humpless longhorn cattle (Epstein 1971); Butana and Kenana – considered typical East African zebu breeds, very similar morphologically to zebu breeds found in Asia.

Mitochondrial DNA was isolated from a number of representative animals per breed and screened for inter- and intrabreed genetic variation. This locus is a particularly attractive tool for the population geneticist and was selected for a variety of reasons. First of all, it may be found in high copy number in all mammalian cells and is relatively easily extracted in pure form. It evolves on average 5–10 times faster than single copy nuclear DNA (even faster if particular regions such as the D-loop are considered) making it ideal for the study of closely related populations. It is maternally inherited (although there are a number of rare exceptions) and not subject to the processes of genetic recombination that may obscure its history. mtDNA therefore provides an unambiguous history of the matriarchal lineage. Furthermore, over the past decade whole mitochondrial DNA, or regions from it, have been sequenced in many species, both for intraspecific and interspecific comparisons. Consequently, the mode and tempo of mitochondrial evolution have been well characterized and are reasonably well understood (Hasegawa & Kishino 1989).

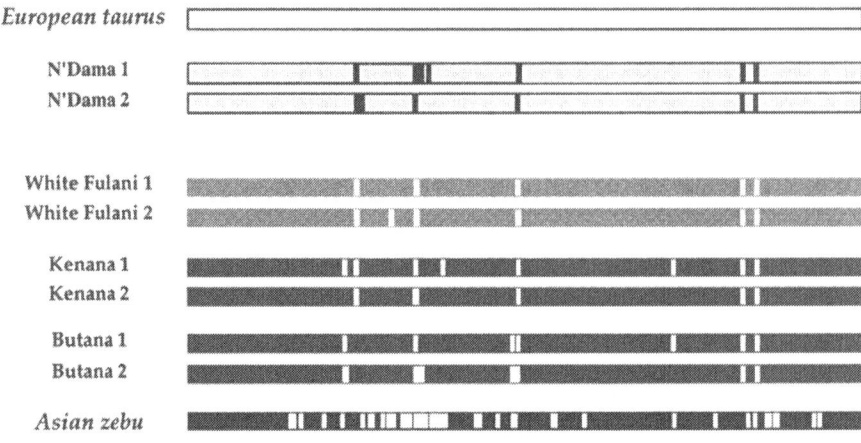

Figure 14.2   Diagrammatic representation of variable positions within the D-loop illustrating the non-random distribution of sites. The black area highlighted for Asian zebu (included for comparison) was selected for sequence analysis in additional individuals per breed. The differential shading in each of the boxes approximates the levels of zebu introgression found in the breeds sampled.

In the present study, a stretch of DNA within the mitochondrial genome known as the displacement loop, or D-loop, was selected for sequence analysis. This region, which regulates mtDNA replication and transcription has been shown to evolve at very high rates in humans (Vigilant et al. 1991) making it the most likely region to display genetic differences at the intraspecific level. Initially, total D-loops were sequenced from two animals per breed. This revealed a non-uniform pattern of mutation, most changes being concentrated around two principal hypervariable regions (Fig. 14.2). One of these, the most variable, was selected for study in a further five animals per breed.

The results, shown in a condensed form in Figure 14.2, demonstrate a remarkable similarity in all mtDNAs assayed, especially when compared with other mitochondrial investigations in natural populations (Avise 1994). Indeed, the lack of divergence between animals of humpless and humped morphology is quite surprising. D-loop sequences from Asian humped and European humpless breeds (Loftus et al. 1994) suggest a closer relationship between African and European than between Asian and African mitochondria. This is remarkable given the distinctly zebu character of at least two of the breeds sampled, Butana and Kenana, and will be elaborated on in the discussion. Phylogenetic trees were not constructed from the data presented given the apparent lack of diversity present among breeds.

In order to determine whether the discordance between morphological and mitochondrial data represented an artefact of mitochondrial dynamics, a further genetic system was investigated – the Y-chromosome. The system, which is transmitted through the male lineage, provides information on paternal ancestry in a manner analogous to mitochondrial DNA for the female lineage. Given that males can disseminate their genes more rapidly than females, especially in domesticated cattle, the study of the Y-chromosome should prove a better indicator of male introductions.

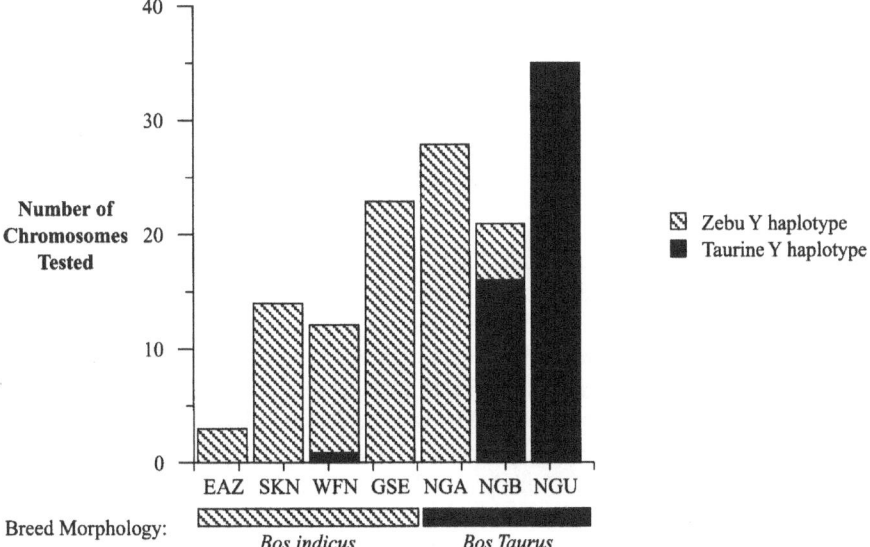

Figure 14.3 Number of zebu and taurine Y-chromosome haplotypes in the breeds sampled. The three letter code below the columns represent the following; EAZ, Butana and Kenana; SKN, Sokoto Gudali; WFN, White Fulani; GSE, Gobra; NGA, N'Dama-Gambia; NGB, N'Dama-Guinea Bissau; NGU, N'Dama-Guinea.

3. Y-chromosome analyses

In general, Y-chromosomes tend to be quite homogeneous intraspecifically and have not been used extensively for the assay of genetic variation at this level. However, in domesticated cattle populations, karyotypic studies have revealed differences in chromosome morphology and banding pattern between the subspecies; *B. indicus* display an acrocentric Y whereas taurine bulls have a submetacentric Y (Halnan 1989). Both variants have been described in African cattle. Furthermore, a Y-specific probe (btDYZ-1) was shown to detect polymorphism within a single European breed (Perret et al. 1990) and was therefore considered a likely candidate for the detection of variation here.

Genomic DNA from males of each breed was screened for polymorphism with the btDYZ-1 probe using the Southern blotting technique (Bradley et al. 1994). In addition to the breeds included in the mitochondrial study, a number of further populations and breeds were also surveyed, namely N'Dama from Guinea, Guinea Bissau and The Gambia, Gobra (Senegalese zebu) and Sokoto Gudali (typical West African shorthorn zebu). The resulting data, interpretable as a series of bands on an autoradiograph, demonstrate considerable differences between zebu and taurine cattle (Bradley et al. 1994). Generally, males of humped morphology show markedly different banding patterns from those of humpless cattle. In contrast, males within either the zebu or taurine sub-group displayed only minor differences from each other.

Figure 14.3 summarizes the data indicating the number of chromosomes analyzed. The relatively high levels of zebu Y-chromosome in certain N'Dama populations is thought to be attributable to introgression from zebu bulls (Bradley et al. 1994). Crossbreeding between N'Dama and zebu has been, and continues to be practised in some countries. Therefore it is not surprising that some background of zebu genes may be found in this breed.

4. Discussion

At first the results from both lines of investigation appear contradictory; mitochondrial DNA illustrates relatively little divergence between African zebu and taurine breeds whereas Y-chromosome haplotypes demonstrate considerable differences between animals of humped and humpless morphology. Indeed, morphometric and allozymic studies have already revealed considerable diversity between humped and humpless cattle (Epstein 1971, Manwell & Baker 1980). An interpretation consistent with both the observed variation between breed types and the present genetic data would suggest that all African breeds have *B. taurus* origins; that their taurine root can be seen in the maternal (mitochondrial) lineage; and that zeboid characters (both morphological and genetic) are the result of varying degrees of zebu introgression.

The current distribution of African zebu breeds may be seen as the result of a series of introductions from the Near East and Indian subcontinent (Fig. 14.1). Introductions, facilitated by the extensive trading links between the Arabian peninsula and Africa might have been prompted by the wish to introduce desirable gene combinations, but would always have been restricted by the need to cross either the Red Sea or the Isthmus of Suez. Consequently, the importation of large numbers of breeding animals would have presented considerable logistical problems, with early introductions likely to have consisted of only small numbers of cattle. Although the selective introduction of males would have presented early migrants or trading partners with a very effective way of disseminating zebu genes, it is unclear whether they were in fact so systematic. However, such an interpretation is consistent with the presence of zebu Y-chromosomes at relatively high levels in African humped cattle in parallel with the relative lack of zebu-type mitochondria.

Notwithstanding the likelihood of relatively small-scale zebu introductions, the continuation of such practices right up to recent times is difficult to reconcile with the total lack of zebu-type mtDNA in any of the zeboid animals surveyed. To more clearly understand this finding the data need to be interpreted in the light of the dynamics and peculiarities of mitochondrial inheritance.

While offering a relatively clear picture of the evolutionary history of a single genetic element, mitochondrial DNA represents only a component of the entire organismal genealogy. Owing to its uniparental mode of inheritance, the effective population size of mtDNA is one-quarter that of nuclear DNA and consequently it can be more sensitive to demographic processes than nuclear based loci. Relatively minor fluctuations in population size (e.g. bottlenecks) can result in the stochastic loss of rare mitochondrial lineages while not significantly altering nuclear variability. Furthermore, not all females in a particular population are expected to contribute

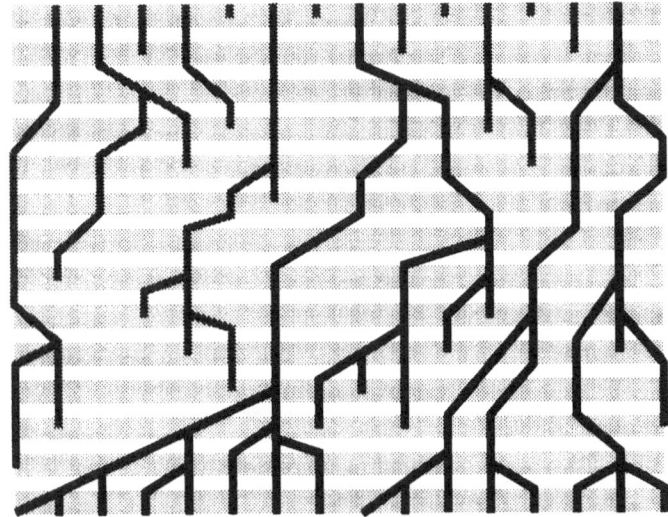

Figure 14.4   Mitochondrial lineage survivorship after 14 generations.

their mitotypes to subsequent generations. For example, in a population where females leave on average one surviving daughter per generation, any single mother has only a two per cent chance of contributing her mtDNA to a population one hundred generations later (Fig. 14.4). Sensitivity of this type to demographic processes has increased the popularity of mtDNA as a tool for the study of population evolutionary history, but it can result in a lack of concordance between mitochondrial and nuclear based phylogenies drawn from the same populations (Avise 1994).

In the current context, it is likely that only relatively small numbers of Asian zebu were introduced to Africa. This would have resulted in the presence of Asian zebu-type mitochondria at low frequencies in any emerging crossbred populations. Such relatively rare lineages could easily have been "pruned out" in subsequent generations, either through stochastic lineage loss, or as a result of population bottlenecks. Given the frequent outbreak of epidemics (e.g. the great rinderpest pandemic of the 1890s), and the constant threat of famines in the region, it would not be surprising that introduced mitotypes should become extinct.

In contrast to zebu mtDNA, the widespread distribution of the zebu Y-chromosome in Africa may be attributed to the effectiveness of males at spreading their genes. Early attempts at crossbreeding introduced cattle might have involved both sexes but males would always have had a far greater impact. Indeed, results from other studies point towards the presence of zebu-type alleles at intermediate frequencies in African zeboid populations and support the extensive introgression of Asian alleles (Manwell & Baker 1980, MacHugh 1996).

In summary, the data presented provide strong evidence for the crossbred nature of contemporary African zebu populations. They also serve to illustrate how molecular genetic analyses can assist in determining the early history and origins of domesticated animals. However, as a note of caution the results emphasize the need to look at a number of genetic systems and disciplines when addressing such questions. The sole use of one particular system or technique could lead to erroneous results, or an incomplete dataset, and the subsequent misinterpretation of events.

References

Avise, J.C. 1994. *Molecular Markers, Natural History and Evolution*, 1st edn. London: Chapman & Hall.

Bradley, D.G., D.E. MacHugh, R.T. Loftus, R.S. Sow, C.H. Hoste, E.P. Cunningham 1994. Zebu-taurine variation in Y-chromosomal DNA: a sensitive assay for genetic introgression in West African trypanotolerant cattle populations. *Animal Genetics* **25**, 7–12.

Clutton-Brock, J. 1989. *The walking larder: patterns of domestication, pastoralism and predation*. London: Unwin Hyman.

Davis, S. 1982. The taming of the few. *New Scientist*, 7 September 1982.

Epstein, H. 1971. *The origin of the domesticated animals of Africa, Vol. I*. New York: Africana Publishing.

Halnan, C.R.E. 1989. Karyotype and phenotype in cattle and hybrids of the genus. In *Cytogenetics of animals*, C.R.E. Halnan (ed.), 235–56. London: CAB International.

Hasegawa, M. & H. Kishino 1989. Heterogeneity of tempo and mode of mitochondrial DNA evolution among mammalian orders. *Japanese Journal of Genetics* **64**, 243–58.

Loftus, R.T., D.E. MacHugh, D.G. Bradley, P.M. Sharp, E.P. Cunningham 1994. Evidence for two independent domestications of cattle. *Proceedings of the National Academy of Sciences, USA* **91**, 2757–61.

MacHugh, D.E. 1996. *Molecular biogeography and genetic structure of domesticated cattle*. PhD thesis, University of Dublin.

Manwell, C. & C.M.A. Baker 1980. Chemical classification of cattle. 2. Phylogenetic tree and specific status of the zebu. *Animal Blood Groups Biochemical Genetics* **11**, 151–62.

Marshall, F. 1989. Rethinking the role of *Bos indicus* in sub-Saharan Africa. *Current Anthropology* **30**, 235–9.

Mason, I. (ed). 1984. *Evolution of domesticated animals*. London: Longman.

Meghan, C., D.E. MacHugh, D.G. Bradley 1994. Genetic characterization and West African cattle. *World Animal Review* **78**, 59–66.

Perret, J., Y.-C. Shia, R. Fries, G. Vassart, M. Georges 1990. A polymorphic satellite sequence maps to the pericentric region of the bovine Y-chromosome. *Genomics* **6**, 482–90.

Vigilant, L., M. Stoneking, H. Harpending, K. Hawkes, A.C. Wilson 1991. African populations and the evolution of human mitochondrial DNA. *Science* **253**, 1503–7.

CHAPTER FIFTEEN

Characterization of the Kuri cattle of Lake Chad using molecular genetic techniques

Ciaran Meghen, David MacHugh,
B. Sauveroche, G. Kana, Dan Bradley

1. Introduction

A DNA marker is a unit of measurable genetic variation that can be accessed reliably and repeatably. Genetic markers are carried by individual organisms through evolutionary time; it is these markers that geneticists use to address questions of population history and evolution. DNA markers are, in effect, molecular artefacts. The inferences that can be made with a particular marker or marker system depend in part on the rate at which variation accumulates. In general, functional genes are under strong selective constraint and vary slowly, whereas non-coding sequences and in particular repetitive sequences vary more rapidly. The different mutation rates of each system make it possible to examine population history at virtually any time depth.

Microsatellites are a relatively new class of genetic marker, although reported in 1984 (Tautz & Renz) they only became routinely available after the advent of the polymerase chain reaction, PCR (Saiki et al. 1988). They are found abundantly throughout the nuclear genetic material, with about 50,000 different microsatellite loci in eukaryotic genomes. Each one acts as an independently evolving character and it is believed that the majority of these loci are selectively neutral. Generated through errors in the DNA replication process, they consist of a short stretch of DNA that is characterized by the presence of repetitive motifs 1–6 base pairs in length, which are flanked by unique sequence. They show Mendelian inheritance and high mutation rates. Variation at these loci is characterized by differing numbers of repeat units (Litt & Luty 1989, Weber & May 1989).

Microsatellites were first used for linkage studies and parentage testing, more recently they have been added to the arsenal of molecular tools used in evolutionary and population work (Bruford & Wayne 1993). Their particular suitability for studies involving closely related populations has been established (Maggia &

Figure 15.1   Kuri cattle of Lake Chad.

Bousquet 1994) and MacHugh et al. (1994) have shown that microsatellites can be used to examine the recent molecular history of European cattle breeds. Work carried out on the Kuri cattle of Lake Chad, in conjunction with microsatellite data obtained from a broader survey, illustrates how these markers can be used to address questions of particular relevance to the livestock history of Africa.

The Kuri cattle of Lake Chad have been the focus of considerable scientific curiosity. The first published record was in 1851 and there have been numerous accounts since. The large body size and enormous inflated horns are distinctive features of the pure-bred Kuri (Fig. 15.1). It is clearly a taurine breed lacking the hump and other morphological characters typical of the neighbouring zebu. The Kuri are well adapted to the lacustrine environment in which they live; they spend appreciable time in the water grazing and they swim from island to island in search of new forage. Disease pressure is high yet these animals thrive with little veterinary intervention. They have good production qualities both as beef and dairying animals (Queval et al. 1971). Although the numbers of Kuri cattle are not well documented, with figures ranging from 10,000 (Doutressoulle 1947) to 400,000 head (LCBC/FAO n.d.), there are concerns that the breed is threatened by habitat loss and genetic erosion through uncontrolled cross-breeding with zebu (Bourzat et al. 1992).

Using molecular techniques we set out to address two particular questions concerning the Kuri: (a) What is the extent of cross-breeding between Kuri and other cattle breeds? (b) What are the phylogenetic affinities of this breed?

2. Materials and methods

Blood samples were collected from the Kuri and other breeds during a three month sampling mission to West Africa in 1992. Efforts to ensure that individuals sampled were not closely related involved selecting only a small number of animals from each herd and questioning the herders on the genealogy of each animal. Figure 15.2 shows sampling locations and the number of animals accessed.

Visualization of microsatellite loci is achieved by means of the PCR. PCR involves the exponential amplification of specific DNA sequences from within the total genetic material. These sequences are primed with oligonucleotides; pieces of artificial DNA about 20 base pairs long. The oligonucleotide primers are designed to match a portion of the unique sequence flanking both ends of the microsatellite. Using a temperature cycling procedure in the presence of the thermostable enzyme *Taq* Polymerase, the appropriate reaction buffer, dNTP's – the building blocks of DNA, and a radioactive label the primed microsatellite is preferentially amplified to the point where it becomes the dominant fraction of DNA in the reaction mixture. The microsatellite length variants or alleles are then size fractionated by polyacrylamide gel electrophoresis and visualized by autoradiography (Fig. 15.3). Incorporation of previously amplified samples of known size allows the unambiguous determination of allele lengths. Data is recorded in the form of di-allelic genotypes for each animal-microsatellite combination, where each allele is coded according to its actual length in base pairs. In this way 20 microsatellite loci were screened and data recorded for the 52 Kuri cattle sampled.

3. Results and discussion

3.1. Cross-breeding

Opinions differ considerably when discussing the extent and nature of cross-breeding between the Kuri and surrounding zebu. Mason (1976) suggests that the increasing number of crossbred animals is due to the spread of Kuri blood into zebu herds, others have expressed concern that the Kuri will become extinct as genetic entities as a result of extensive introgression of zebu genes (Bourzat et al. 1992, Hall 1993).

Zebu specific microsatellite alleles (MacHugh 1996), alleles not found in pure taurine populations, were detected in 10 of the 20 microsatellite loci tested (Fig. 15.4). Their enumeration allows a precise estimation of the proportion of zebu genes in the Kuri population. Admixture, which is a subtly different measure, indicates the level of genetic exchange required to account for the observed proportion of zebu

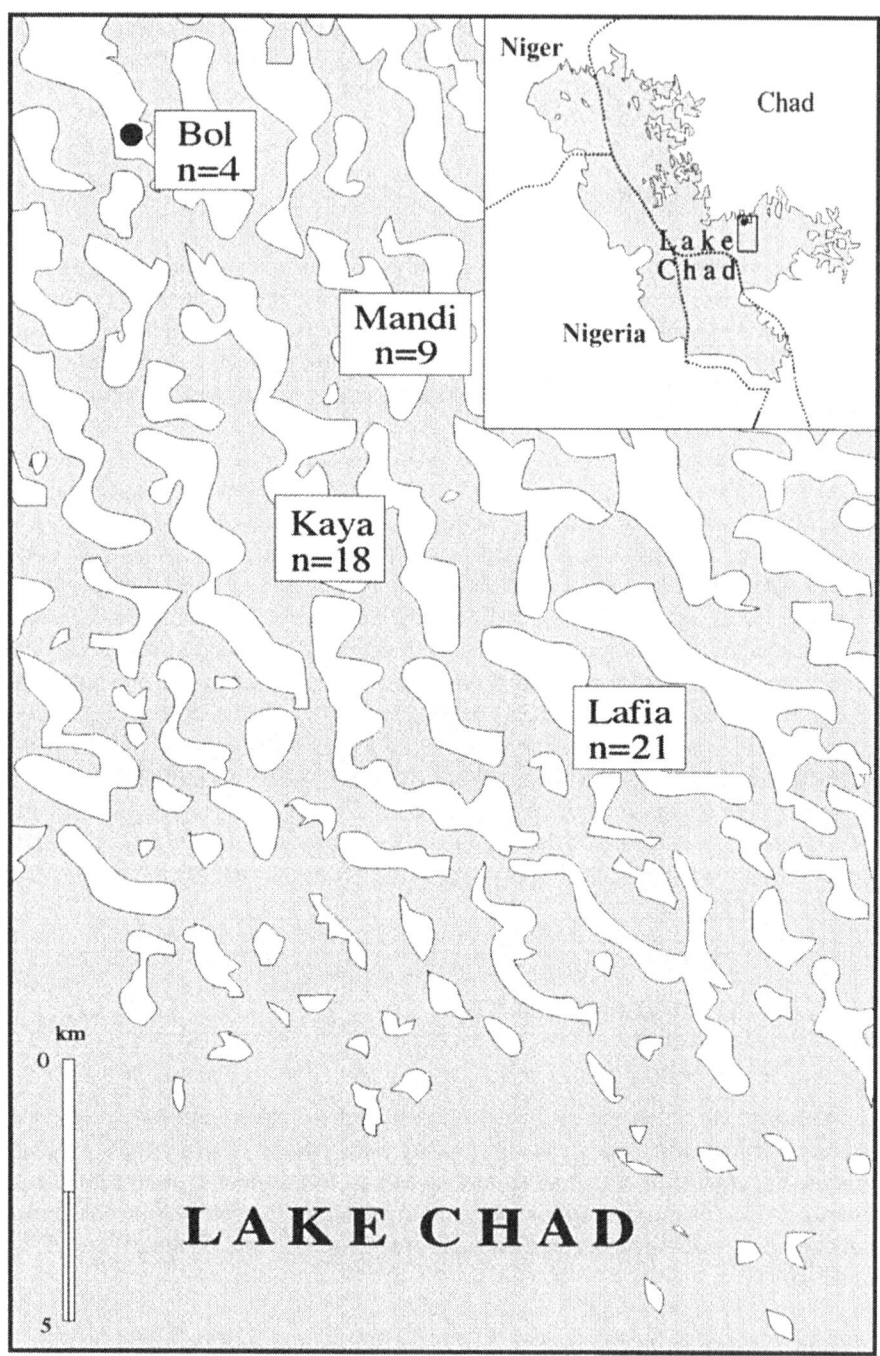

Figure 15.2 Lake Chad. Despite its apparent size there are few stretches of free water, the lake is largely composed of low islands and swamp. The map shows approximate sampling locations and the number of Kuri cattle accessed at each locality. Mandi, Kaya and Lafia are islands and Bol is a small town on the shore of Lake Chad.

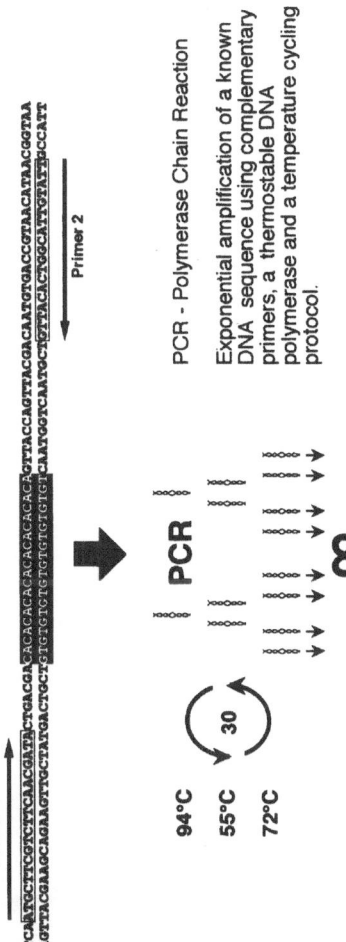

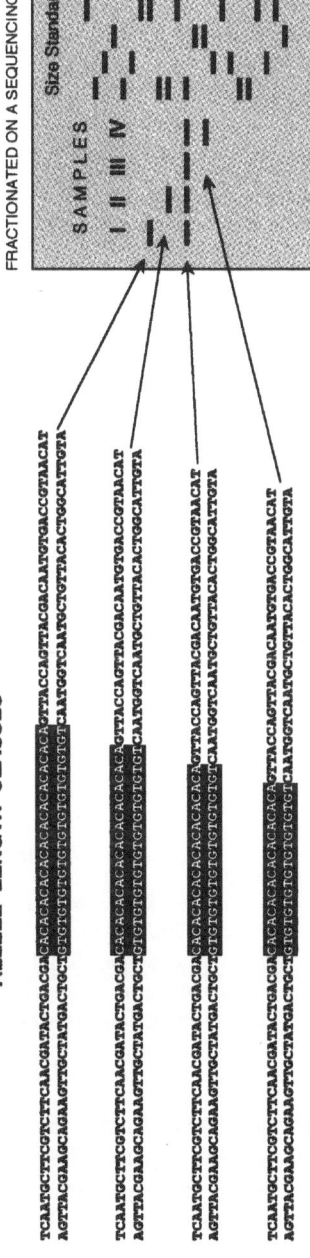

Figure 15.3 Schematic representation of the amplification and resolution of microsatellite alleles by Polymerase Chain Reaction (PCR) and Polyacrylamide Gel Electrophoresis (PAGE).

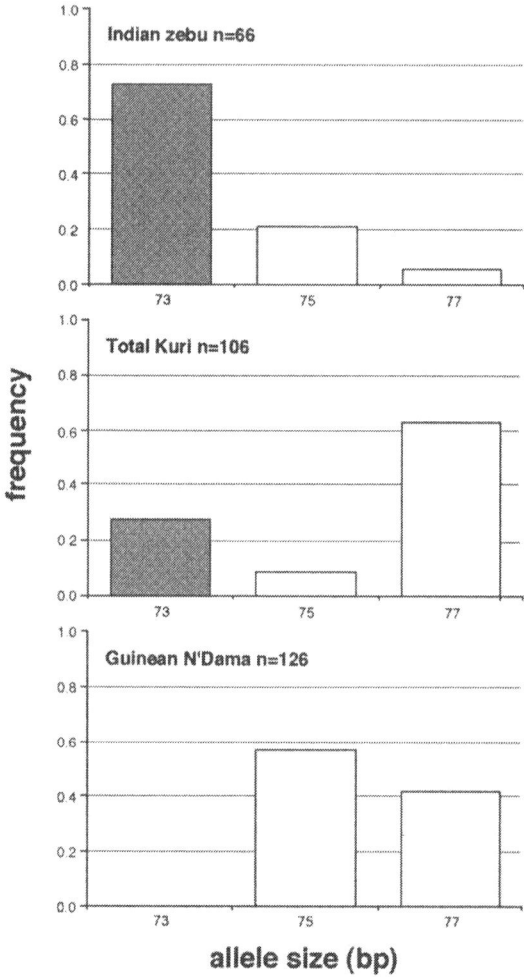

Figure 15.4   Allelic distributions for the microsatellite locus DL26S4 clearly show the 73 base pair allele to be a zebu specific allele. The sample size (n) equals the number of chromosomes tested, which is twice the number of animals sampled.

genes, this was calculated using the FORTRAN program ADMIX (Chakraborty et al. 1991). The program requires allele frequency data from three populations; the admixed population and two "ancestral" populations to represent the pure zebu and the pure taurine types. Comparable data for two such populations was available from the work of MacHugh (1996): 33 North Indian zebu and 63 Guinean N'Dama (taurine), for which the same loci had been typed, were included in the analysis. The results are shown in Table 15.1.

Table 15.1 Percentage of zebu genes in the Kuri and admixture proportions estimated using the FORTRAN 77 program ADMIX (Chakraborty et al. 1991).

Population	zebu genes %	admixture proportion ± s.e.
Total Kuri	22.51 %	47.00 ± 5.34 %
Bol	22.39 %	53.20 ± 5.63 %
Mandi	27.66 %	48.07 ± 5.90 %
Kaya	20.85 %	47.44 ± 4.81 %
Lafia	22.64 %	46.19 ± 5.90 %

From Table 15.1 the magnitude of zebu gene introgression can be seen, over 20 per cent of the Kuri nuclear DNA has been displaced by zebu genes and this is the result of almost 50 per cent genetic admixture with surrounding zebu. Similar work on five populations of N'Dama, another West African taurine breed, shows zebu gene introgression ranging from 0.2 per cent to 8.7 per cent (MacHugh 1996). Further work with six additional West African taurine breeds reveals that the Kuri are the most highly crossbred (C. Meghen, unpubl. data). These results give a numerical insight into the extent of cross-breeding in the Kuri cattle and exacerbate fears for the integrity of the breed. The particular state of the Kuri can be explained in the context of resistance to trypanosomiasis. Most taurine breeds of the region, Kuri being a notable exception, are trypanotolerant. This adaptive trait allows these breeds to thrive in an environment unsuitable for the trypanosusceptible zebu, the result is a greatly reduced contact pressure between the two cattle types. Lake Chad is a contact barrier between the Kuri and the zebu but evidently this barrier is not strong enough to prevent the substantial penetration of zebu genetic material into the Kuri gene pool.

3.2. Phylogeny

In the absence of data it is easy to generate hypotheses concerning the origins of the Kuri. For such an unusual breed, in form and ecology, it is little wonder that many have tried. Some of the first suggestions were reviewed by Malbrant et al. (1947) and are summarized here: Baron thought that the Kuri were a distinct race and referred to them as *Bos taurus bolensis*; Pecaud considered that their pale hides indicated close affinity to the White Fulani; Stewart believed the Kuri to be a remnant of the Egyptian Hamitic longhorns which had remained pure due to their isolation on the islands of Lake Chad; from photographs, Epstein initially categorized them as pseudo-zebu or sanga, and Curson and Thornton suggested that they might be the product of an African domestication some 5000–6000 bp, being derived from an Egyptian wild ox (*B. primigenius Hahni*). More recently Queval et al. (1971) pointed out that at the beginning of the Christian era the crossing of zebu with *B. primigenius ospithonomous* gave rise to the Sanga, and as such he could not account for the Kuri: the point being made was that the Kuri were unlikely to have hybrid origins. Epstein (1971), despite including the Kuri in the humpless longhorn clade, does not offer any suggestions as to their specific origin. The modern consensus is that the origins of the Kuri are unclear (RIM 1992, Hall 1993). However the Kuri's physical, behavioural and ecological character have continued to suggest that this is perhaps a very distinct genetic entity.

Microsatellite data can be used to produce genetic distances, the raw material of phylogenetic analyses. However there are a number of assumptions that must be met before such analysis can be considered appropriate. One of these is that the species or populations under investigation are not of hybrid origin or have not undergone secondary contact with a closely related form. From our investigations it is clear that these animals are markedly crossbred and as such the application of phylogenetic techniques to the Kuri would produce erroneous and possibly misleading results.

An alternative approach to the genetic origin of the Kuri is to examine the microsatellite variation by means of principal component analysis (PCA). Although this technique does not allow phylogenetic inferences to be made it does allow the investigator to examine the relationships between clades. The allele frequencies for the 20 microsatellite loci act as 20 correlated variables, PCA replaces these with a smaller number of variables or principal components (PC). These PCs are examined and used to provide a qualitative framework within which the position of the Kuri is explained. The PCA was carried out using seven European taurine breeds, five N'Dama populations from the far west of West Africa, five African zebu breeds, and three Indian zebu breeds in addition to the Kuri. From our analysis it emerged that the first PC, which accounted for 67 per cent of the variation within the microsatellite dataset, separates the Indian zebu breeds from all other breeds and populations. The European taurine and N'Dama populations cluster as discrete groups. The African zebu breeds are interspersed along the entire range of the European and African taurine distribution and the Kuri fall within the N'Dama cluster (Fig. 15.5a). The second PC, which accounts for a further 17 per cent of the variation, also separates taurine from Indian zebu but in this case the African zebu and the Kuri are more closely aligned with the Indian breeds (Fig. 15.5b). The third PC, accounting for 5 per cent of the variation, clearly distinguishes the European taurine from the N'Dama populations and in this dimension the Kuri are more closely grouped with N'Dama populations than are the African zebu (Fig. 15.5c). It is not, however, until one examines all three PCs simultaneously that a clear picture of the genetic patterns underlying the microsatellite data become apparent (Fig. 15.5d). In three dimensions it can be seen that the seven European taurine breeds, which cluster together, do not ally themselves to any of the other groups represented. This is interesting as it does not support the idea that the longhorn breeds of Europe share ancestry with the longhorn breeds of Africa. From the N'Dama cluster it can then be perceived that there is a continuum through the Kuri and the African zebu towards the Indian zebu. Furthermore, the positions of the different African zebu breeds correspond well with their geographical origin. Those breeds from Mauritania and Senegal appear closer to the N'Dama than do the White Fulani breed, sampled in Nigeria, and the breeds from Sudan and Kenya are most closely allied with the Indian zebu. In this analysis the

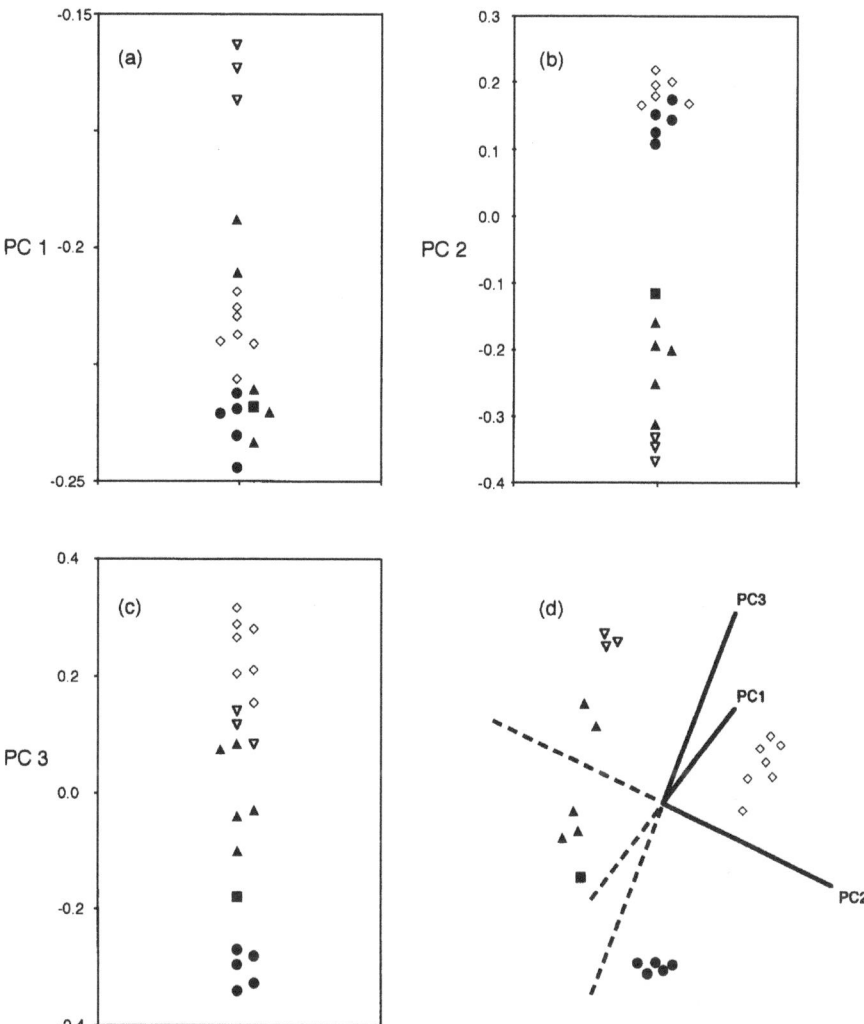

Figure 15.5 Dotplots a, b, and c show the separation and associations of the different cattle populations for each of the first, second and third principle components, accounting for a total of 89% of the detected variation. In part d, PC 1, 2 and 3 are combined to generate a 3-dimensional scatterplot.

Where: ▽ Indian zebu, ▲ African zebu, ◇ European taurine, ● N'Dama and ■ Kuri.

Kuri emerge as being akin to the African zebu. This observation is doubtless a result of the high level of crossbreeding already described. One could perhaps imagine a time when the Kuri were less crossbred and fell nearer to the N'Dama, but from this analysis there is no evidence to suggest that the Kuri were ever a highly divergent form.

4. Conclusions

Molecular data, and in particular microsatellites, offer new ways of examining African livestock history. Molecular techniques can generate an abundance of numerical data that can be used to challenge existing theories and to test new ones. This powerful tool can most effectively be employed when the zoological, archaeological and palaeontological groundwork has been laid down.

The Kuri cattle of Lake Chad make a good case study for the application of molecular analysis in issues of African livestock history; the origins of the Kuri are mysterious and fears of extensive crossbreeding have given rise to concern for the future of the breed. We show in quantitative terms that the breed is under threat of genetic assimilation, and also suggest that genetically the Kuri are not markedly divergent from another African taurine, the N'Dama.

References

Bourzat, D., A. Idriss, V. Zeuh 1992. La race Kouri. Une population bovine en danger d'absorption. *Animal Genetic Resource Information* **9**, 15–26.

Bruford, M.W. & R.K. Wayne 1993. Microsatellites and their application to population genetic studies. *Current Opinion in Genetics and Development* **3**, 939–43.

Chakraborty, R., M.I. Kamboh, M. Nwankwo, R.E. Ferrell 1991. Caucasian genes in American blacks: new data. *American Journal of Human Genetics* **50**, 145–55.

Doutressoulle, G. 1947. *L'elevage en Afrique Occidentale, Française*. Paris: Larousse.

Epstein, H. 1971. *The origin of domestic animals in Africa*. New York: Africana Publishing.

Hall, S.J.G. 1993. *Report of mission to Somba and Kuri cattle*. Rome: FAO.

LCBC/FAO. n.d. Regional project for conservation and development of an endangered bovine breed: the kuri cattle. Project proposal for discussion (1991), FAO Regional Office for Africa, Accra, Ghana.

Litt, M. & Luty J.A. 1989. A hypervariable microsatellite revealed by in vitro amplification of a dinucleotide repeat within the cardiac muscle actin gene. *American Journal of Human Genetics* **44**, 397–401.

MacHugh, D.E. 1996. Molecular biogeography and genetic structure of domesticated cattle. PhD thesis, Trinity College, Dublin.

MacHugh, D.E., R.T. Loftus, D.G. Bradley, P.M. Sharp, E.P. Cunningham 1994. Microsatellite DNA variation within and among European cattle breeds. *Proceedings of the Royal Society of London, Series B* **256**, 25–31.

Maggia, L. & Bousquet, J. 1994. Paternity exclusion in a community of wild chimpanzees using hypervariable simple sequence repeats. *Molecular Ecology* **3**, 469–78.

Malbrant, R., P. Receveur, R. Sabin 1947. Le boeuf du Lac Tchad. *Revue d'Elevage et de Médécine Vétérnaire des Pays Tropicales* **1**, 37–42, 109–29.

Mason, I.L. 1976. Report of the mission to the kuri cattle of Lake Chad. *Ark* **3**, 196–200.

Queval, R., J.P. Petit, G. Tacher, A. Provost, J. Pagot 1971. Le kouri: race bovine du Lac Tchad I. Introduction générale à son étude zootechnique et biochimique: origines etécologie de la race. *Revue d'Elevage et de Médécine Vétérnaire des Pays Tropicales* **24**, 667–87.

RIM 1992. Nigerian livestock resources, Vol. II National Synthesis. Resource Inventory Management/ Environmental Research Group Oxford.

Saiki, R.K., D.H. Gelfand, S. Stoffel, R. Higuchi, G.T. Horn, K.B. Mullis, H.A. Eerlich 1988. Primer-directed enzymatic amplification of DNA with a thermostable DNA polymerase. *Science* **239**, 487–91.

Tautz, D. & M. Renz 1984. Simple sequences are ubiquitous repetitive components of eukaryotic genomes. *Nucleic Acids Research* **12**, 4127–38.

Weber, J.L. & P.E. May 1989. Abundant class of human DNA polymorphisms which can be typed using the polymerase chain reaction. *American Journal of Human Genetics* **44**, 388–96.

CHAPTER SIXTEEN

Characterizations of African cattle, sheep and goats and their contributions to archaeological understanding

Stephen J.G. Hall

1. Introduction

The diversity of African livestock breeds has interested outside observers for a long time. Darwin (1868, 1:107) wrote:

> At the present day various travellers have noticed the differences in the [cattle] breeds in Southern Africa. Sir Andrew Smith several years ago remarked to me that the cattle possessed by the different tribes of Caffres, though living near each other under the same latitude and in the same kind of country, yet differed, and he expressed much surprise at the fact. Mr. Andersson has described the Damara, Bechuana and Namaqua cattle; and he informs me in a letter that the cattle north of Lake Ngami are likewise different, as Mr. Galton has heard is also the case with the cattle of Benguela. The Namaqua cattle in size and shape nearly resemble European cattle, and have short stout horns and large hoofs. The Damara cattle are very peculiar, being big-boned, with slender legs, and small hard feet; their tails are adorned with a tuft of long bushy hair nearly touching the ground, and their horns are extraordinarily large ... Mr. Andersson in his letter to me says that, though he will not venture to describe the differences between the breeds belonging to the many different sub-tribes, yet such certainly exist, as shown by the wonderful facility with which the natives discriminate them.

This evidence of biodiversity helped Darwin to defend his thesis that the form of living things can be permanently altered by selection. Can we understand the patterns and processes of livestock husbandry in the past by considering the livestock biodiversity of the present?

In the colonial and early postcolonial eras, large amounts of information were gathered in Africa on the geographical distribution and general appearance of breeds, particularly of cattle but also of the other livestock species and including characterizations of productive capacity (Doutressoulle 1947, Mason 1951, Joshi et al. 1957, Mason & Maule 1960, Epstein 1971). Much work has also been done on production systems (see for example ILCA 1989), especially the most spectacular forms of husbandry like pastoralism (Dahl & Hjort 1976). The ways in which these systems change in response to market and environmental conditions are becoming appreciated (Swift 1984, RIM 1992, Anderson 1993, Amanor 1995).

Naturally, the systems that have been studied have been the more economically important ones where investment can be easily justified. It is now realized (Rege & Lipner 1992, Scherf 1995) that many African livestock breeds are in danger of extinction, that this would mean an irretrievable loss of genetic resources, and that characterization of these breeds is necessary.

However, it is not entirely clear what is meant by characterization. One definition of breed characterization would be when the evolutionary relationships among a set of breeds are fully understood. This would enable the most distinctive breeds to be identified which may be, within the current paradigms of conservation biology, the most deserving of conservation (Hall & Bradley 1995). From another perspective, viewing a breed and its husbandry system as co-adapted, a description of the husbandry system is necessary as well as a catalogue of the breed's attributes. Finally, for those working within the paradigms of the science of animal breeding, a breed would be characterized when an understanding is achieved of the modes of inheritance of its attributes or qualities that are valuable under present or predicted circumstances.

These philosophies of characterization are very different. The first is a comparative approach, aimed at identifying the most distinctive breeds. The second leads towards an understanding of the relationship between the breed and its cultural, physical and biotic environment. The third focuses on a limited range of characters in a single breed, aiming to predict the response to selection so that a breeding programme can be properly planned (Hall 1996).

This chapter considers whether data on breed characterization can assist in the reconstruction of the prehistory of African livestock. Blench (1993) has argued from the distribution of present-day ruminant breeds in Africa that archaeology has yet to produce a convincing model to explain the synchronic situation. This is partly because it has so far proven to be extremely difficult to distinguish breeds in archaeological contexts. Questions archaeozoologists may want to ask that involve breeds might include, when, where and how did present-day breeds originate; what happened to breeds that have become extinct; and what were the characteristics of breeds and husbandry systems at particular times and places in the past? These questions are first considered in relation to recent developments in genetics and to characterizations by body dimensions, and by interviews with livestock producers. Two case studies are presented as a contribution to the development of a methodology for the rapid phenotypic characterization of African livestock biodiversity.

2. The origins of present-day breeds

Breeds originate by diverging from other breeds, but the challenge is to establish the period of divergence. Bradley et al. (1996) have shown how studies of mitochondrial DNA could be calibrated to estimate when numerical expansions of cattle populations occurred; these techniques have so far not been applied to livestock breeds. As a first stage, phylogenetic trees can at least indicate which breeds are closely related and as DNA data accumulate it may become possible to date the divergence of breeds.

Even with the most sensitive available tests (MacHugh et al. 1994) relationships have been revealed between breeds which may be of overall historical significance but which do not reflect the incidents of interest in the history of breeds. For instance, the Aberdeen-Angus and Jersey breeds of cattle are quite close in terms of microsatellites, but if the phylogeny of these breeds were defined solely by these variants the historical deductions possible would be woefully incomplete. The Jersey, along with the geographically close South Devon and the Guernsey, are distinct in that haemoglobin is polymorphic in those three breeds. The gene for haemoglobin type B is found as well as that for type A, which is the only one found in the Aberdeen-Angus and the other British breeds (Bangham 1957). Haemoglobin type B is associated with zebu cattle and this raises the possibility that the zebu contributed to the evolution of British cattle. Historical records support this; an "Indian bull" was used for breeding in Devon between 1795 and 1805 (Hall & Clutton-Brock 1988).

This is an illustration of the observation of Huelsenbeck et al. (1996) of "different genes providing significantly different estimates of phylogeny". Additionally, discordance, or lack of congruence, between phylogenies based on morphology and those based on molecular studies is very widespread (Patterson et al. 1993). It seems likely that it will be in livestock, or in humans, that studies of molecular–morphological congruence would be most likely to succeed because of the availability of genetic maps (Haley 1995). For livestock conservation, this is an important point as decisions whether or not to conserve particular breeds may be made on the basis of controversial phylogenetic inferences. For the present at least, in publications and discussion the precise basis of phylogenetic inferences must be carefully specified and alternative inferences properly discussed. For archaeology, discordances between phylogenies of livestock may be of themselves very informative. A phylogeny of African cattle breeds based on polymorphisms on the bovine Y chromosome suggests, as would be expected, a zebu connection. However, the discordance between this phylogeny and one based on mitochondrial DNA (mtDNA, which is maternally inherited) raises the possibility that the expansion of humped cattle into Africa was the result of the importation of bulls from further East and not of cows (Bradley et al. 1994, Bradley & Loftus Ch. 13 in this volume).

3. Extinctions of breeds

Generally, when one breed is absorbed into another this is because the superseding breed spreads into the area occupied by the declining breed and the latter loses its identity through its females being inseminated by males of the other breed (Hall & Clutton-Brock 1988). Presumably, when this happens, the number of mtDNA lineages in the superseding breed in that area will expand suddenly. It could thus be possible to date, at least relatively, when a superseding breed has absorbed different local breeds by studying geographical variation in number of mtDNA lineages. At present humpless shorthorns in West Africa are, following the humpless longhorns, being superseded by humped (zebu) breeds including the Keteku and the Borgou (Hall 1991, Hall et al. 1995 for references). mtDNA studies of the latter breeds in different parts of their range could indicate when local breeds such as the Pabli were lost (Blench 1993).

4. Husbandry systems of the past

The indispensability of an understanding of present-day livestock for interpreting archaeozoological findings is now clearly understood. For example, Pryor (1996) used his knowledge of the practicalities of sheep handling to infer functions for Bronze Age structures in lowland England. Halstead (1996) used present-day data on herding practices to achieve an understanding of Greek husbandry of the seventh to second millenia BC, differentiating large-scale pastoralism from small-scale mixed farming. Thorp (1995) deduced the age distribution of natural mortality in present-day Shona cattle, in order to interpret patterns of slaughter in Great Zimbabwe during the late Iron Age. This study revealed that cattle aged 18–30 months were deliberately slaughtered for the royal household, while older cattle were slaughtered at other sites (see also Reid 1996). The great value of modern understanding of animal environmental adaptation, breeding and husbandry, pathology and butchery for the interpretation of animal remains is illustrated by many recent studies including for example those recently reported relating to Lincoln, England (Dobney et al. 1996).

5. Biometric characterization

There are several ways in which simple measurements and descriptions of livestock can be used to characterize breeds. Lauvergne et al. (1993; Ch. 18 in this volume) present an "index of primarity", developed for the study of goats. The goat exhibits more easily coded, directly visible polymorphism than cattle or sheep. In this approach, populations with a high qualitative level of polymorphism in certain characteristics, including coat colour and horn and ear shape, are taken to be relatively unselected and panmictic (i.e. mating is not restricted). This index provides a rapid and objective way to classify populations. Despite this, the existence of polymorphism does not prove there is or has been no selection. Selection may favour the heterozygote, or different homozygotes may be favoured in different circumstances as recent work on the feral Soay sheep of a Scottish island (Moorcroft et al. 1996) illustrates. Following this thread, the idea of index of primarity could be applied to microsatellite studies.

In early 1990, 136 cows, 63 female goats and 71 female sheep were measured in Nigeria including examples of almost all recognized breeds (Hall 1991, RIM 1992). The original reason for doing this was to follow up British work on the relationship between ease of calving and external pelvic dimensions and to gather new descriptive data. Withers (shoulder) height, heart girth, and body length (from the withers to the pin bones, the most distal bony point of the body) were measured for all animals. External measurements of pelvic dimensions were made for cows, while sheep and goats were weighed. For the cows, comparable data were available from nine British breeds. British breeds tend to be long in the body, broad in the pelvis and short in the leg, while the opposite is true of Nigerian cattle. Selection has been for carcass conformation in the former (Hall & Clutton-Brock 1988), and it is possible that the long-legged, narrow and short-bodied form of the latter is an environmental adaptation.

In a quite unexpected way, these findings suggest a hypothesis for further investigation. Simple morphometric studies of the rock art depictions of livestock that are widespread in the Sahara might yield broad-brush information on the distribution of breeds. Smith (1968) said that "especially in the highly naturalistic paintings of cattle herds, information on non-skeletal materials, such as sexual dimorphism, body profile and coloration, are often available and should be invaluable to supplement the osteological remains from archaeological deposits". Published photographs of this rock art (see, for example, Lhote 1959, Muzzolini 1983) suggest that the cattle depicted, mostly humpless longhorns, were short-legged and long-bodied, more like present day British cattle than Nigerian. A test of the realism of the depictions could be based on a comparison of the body proportions of the wild animals also depicted, with those of the same species as they exist today.

6. Sheep and goats

In the case of small ruminants, comparison of the body dimensions of West African Dwarf (WAD) sheep and goats with their northern conspecifics suggested dwarfism might have evolved differently in the two species. The WAD goats appeared to be proportionate dwarfs, in that the body dimensions were in about the same proportions as in northern goats, while the WAD sheep had body proportions which were like those of a young northern sheep. Possibly, dwarfism arose by quite distinct mechanisms in the two species and in the WAD sheep it represents a kind of neoteny (retention of juvenile characteristics). Ancillary studies on the coat fibres of Nigerian sheep and goats (Hall et al. 1996) revealed the presence of substantial variation in the coat of sheep, and rather less in that of the goat. The WAD goat may have fewer sweat glands (associated with certain hairs) and its adaptation to the environment may involve a lessened dependence on sweating.

Morphometrics is a rapidly developing field (Rohlf & Marcus 1993) but it remains true that simple linear measurements are more effective than assessments of shape; measuring large numbers of unrelated individuals, recording only a few traits, is more productive than vice versa. Guidelines for sample sizes for livestock characterizations are given by Bruns (1992); for authoritative breed comparisons, measurement of 200 individuals per breed, the progeny of at least 20 sires, could suffice. Although awkward to obtain, body weights are very valuable and the correspondence with heart girths, though strong, is not complete (Dineur & Thys 1986). Cattle can be difficult to restrain for measurement and laborious to handle and it may be worthwhile developing techniques for the image analysis of video recordings, as described by Patterson (1990).

7. Biographic characterization

Many studies have been made of African ruminant husbandry systems. The performance of breeds has often been investigated by keeping records of a marked population of animals on particular farms that are enrolled in schemes run by governments or international agencies and accurate measures of production are possible (Peacock 1987). Conditions are uncontrolled in the sense that a mating plan is usually not imposed. However, the results from such studies are of little value in determining productivity in the traditional sector because pastoralists and village livestock producers rarely have the same priorities as university researchers. Indeed, there is usually a wide gap between research station results and traditionally managed herds (see RIM 1992 for a detailed study of this in the case of the White Fulani).

The alternative is therefore to work with traditionally managed herds using the Mature Breeding Female History (MBFH) method. This method allows researchers to estimate various productivity parameters through recording the parturition history of individual females (Swift 1984, RIM 1992). The kind of characterization that involves questioning owners about the history of individual animals is referred to here as biographic characterization. It is most effective with long-lived and valuable species such as camels and cattle; owners are often more insouciant about the breeding history of small ruminants and experience has shown that it is more reliable to record only the two most recent parturitions for sheep and goats.

Such questionnaire techniques need still to be reviewed and combined with a simulation study of how errors may arise and of their consequences for the accuracy of estimation of production parameters. For absolute estimates errors must of course be minimized, but in a comparative study, valuable results may still be obtained from an error-prone procedure as long as the incidence of errors does not differ between the subjects being compared. Given the urgency of the task of characterizing breeds, rapid methods should be discussed, standardized as far as possible and made available so results are comparable. If the discussion is interdisciplinary it may be possible to design questionnaires which provide data useful for many fields of study.

In this study (Hall et al. 1995), the productivities of two systems of husbandry of Somba cattle (a savanna West African shorthorn) were compared. Two questionnaires were applied, one to establish the general place of livestock husbandry in community life, and the other to establish the biographies of a sample of cows. An owner would be asked questions about a particular cow, selected at random from the herd. The questionnaires used are reproduced in the Appendix.

Table 16.1 Fecundity of cows.

	Calves per year	Mean calving interval (days)
Tata (traditional)	0.58	629
Ful6e (custodial)	0.61	598

Table 16.2 Production of calves.

	Total calves	Died young	Sold/given/ consumed	Still in herd	Other
Tata (traditional)	36	12 (33%)	5 (14%)	17 (47%)	2 (6%)
Ful6e (custodial)	52	9 (17%)	8 (15%)	31 (60%)	4 (8%)

In the traditional system, small groups of cows are kept at night in their owners' houses and are pastured during the day. In the modern system, Ful6e herdsmen look after herds of animals belonging to several owners. Juvenile survival appeared slightly better in the latter system, but statistical significance was not attained. In the latter system there was a tendency to retain calves in the herd rather than market them. Table 16.1 and Table 16.2 compare performances of Somba cattle kept under the *tata* (traditional) and Ful6e (custodial) systems;

Combining the two systems it is deduced that 100 fecund cows will give 59 calves per year of which 14 per cent (i.e. 8.12 calves) are not retained and are available for marketing. This could lead to an increase in the cattle population without an increase in the amount of meat marketed. Generally, productivity of Somba cattle is similar to other West African shorthorn populations. These findings, though obviously tentative because of the small number (11) of herds visited, have implications of wider interest. From the angle of environmental conservation the traditional system is probably more sustainable because it is unlikely to lead to overstocking, while from the archaeological viewpoint, this study raises the possibility that in certain systems the numbers of cattle kept and their physical size may be determined by the amount of space available in the houses.

8. Conclusions

A powerful methodological approach to understanding the evolution of livestock breed biodiversity is to use one discipline to generate hypotheses that can be tested by another. New DNA techniques and simple measurement and interview procedures can be applied to the characterization of breeds and husbandry systems. The resulting data and hypotheses can help in the interpretation of archaeological materials and lead to richer interpretation of prehistory.

Appendix

In view of the urgency of the task of characterizing African breeds and husbandry systems, and of the necessity of quick and effective methods, it is worth pointing out that relatively little fieldwork was required to carry out these studies. Both were conducted during the dry season. Periods of fieldwork were very short – six weeks for Nigerian livestock and one week for Benin cattle. Consequently the studies cannot in any sense be regarded as definitive, but as so much is still not known about African livestock breed biodiversity, and as in most countries local breeds are under threat, brief studies such as these may be all that it may be practicable to undertake.

Survey of Somba cattle in Benin: Questionnaire 1.

Details of village: Location of village:

Date: Time:

1. How many families in village?
2. How many people in village?
3. How many people own cattle?
4. Each owner has how many cattle?
5. How many cattle of each sex and how many suckling calves?
6. Are sheep and goats kept and are they owned by the same people who own the cattle?
7. Are the cattle grazed, tethered or stall fed, and at what time of year is each method practised?
8. Who does the herding?
9. Do the cattle receive veterinary care?
10. Do the cattle suffer disease?
11. Do the cattle receive supplements (minerals, salt)?
12. What is done with the dung?
13. Is there castration? How is it decided which bull calves to castrate?
14. Are the cattle used for work/draft?
15. Is theft a problem?
16. What ages are the breeding bulls in the herd/village?
17. What is the breed of the bulls that are used for service?
18. Are breeding bulls bought-in or are they bred in the village?
19. Is a stud fee charged?
20. Are animals sold to others for breeding or slaughter?
21. Are animals sold at the market?
22. Are the cows milked? Are milk or milk products sold at the market?
23. What price would they expect to pay if buying an animal (bull, heifer, cow, etc.).

Survey of Somba cattle in Benin: Questionnaire 2.

Cow case history

Date: Time: Village:

Name of cow:
Age of cow:
Number of incisor teeth:
Was this cow bought-in or bred within the village?
How many calves has the cow had:
Is she pregnant?
If pregnant, when will she calve:

Most recent calf
When did she last calve?
Sex of this calf:
Fate of this calf (still in herd; died; sold, etc.):
Was the calf sired by a bull of the village or from outside?
Why was that bull used for the mating?

Previous calf
When was this calf born?
Sex of this calf:
Fate of this calf (still in herd; died; sold, etc.):
Was the calf sired by a bull of the village or from outside?
Why was that bull used for the mating?

Previous calf
When was this calf born?
Sex of this calf:
Fate of this calf (still in herd; died; sold, etc.):
Was the calf sired by a bull of the village or from outside?
Why was that bull used for the mating?

Previous calf
When was this calf born?
Sex of this calf:
Fate of this calf (still in herd; died; sold, etc.):
Was the calf sired by a bull of the village or from outside?
Why was that bull used for the mating?

References

Amanor, K.S. 1995. Dynamics of herd structures and herding strategies in West Africa: a study of market integration and ecological adaptation. *Africa* **65**, 351–94.

Anderson, D.M. 1993. Cow power: livestock and the pastoralist in Africa. *African Affairs* **92**, 121–33.

Bangham, A.D. 1957. Distribution of electrophoretically different haemoglobins among cattle breeds of Great Britain. *Nature, London* **179**, 467–8.

Blench, R.M. 1993. Ethnographic and linguistic evidence for the prehistory of African ruminant livestock, horses and ponies. In *The archaeology of Africa. Food, metals and towns*, T. Shaw, P. Sinclair, B. Andah & A. Okpoko (eds), 71–103. London: Routledge.

Bradley, D.G., D.E. MacHugh, E.P. Cunningham, R.T. Loftus 1996. Mitochondrial diversity and the origins of African and European cattle. *Proceedings of the National Academy of Sciences of the USA* **93**, 5131–5.

Bradley, D.G., D.E. MacHugh, R.T. Loftus, R.S. Sow, C.H. Hoste, E.P. Cunningham 1994. Zebu-taurine variation in Y chromosomal DNA: a sensitive assay for genetic introgression in West African trypanotolerant cattle populations. *Animal Genetics* **25**, 7–12.

Bruns, E. 1992. Synthesis of research methodology. Appendix III. In *African animal genetic resources: their characterisation, conservation and utilisation. Proceedings of the Research Planning Workshop held at ILCA, Addis Ababa, Ethiopia 19–21 February 1992*, J.E.O. Rege and M.E. Lipner (eds), 125–34. Addis Ababa: International Livestock Centre for Africa.

Dahl, G. & A. Hjort 1976. *Having herds. Pastoral herd growth and household economy*. Stockholm studies in social anthropology 2. Stockholm: Department of Social Anthropology, University of Stockholm.

Darwin, C. 1868. *The variation of animals and plants under domestication*. [Popular Edition, 1905]. London: John Murray.

Dineur, B. & E. Thys 1986. Les Kapsiki: race taurine de l'extrême-Nord camerounais. 1. Introduction et barymétrie. *Revue d'Elevage et de Médecine vétérinaire des Pays tropicaux* **39**, 435–42.

Dobney, K.M., S.D. Jaques, B.G. Irving 1996. *Of butchers and breeds. Report on vertebrate remains from various sites in the City of Lincoln*. Lincoln Archaeological Studies No. 5. Lincoln: City of Lincoln Archaeology Unit.

Doutressoulle, G. 1947. *L'élevage en Afrique occidentale française*. Paris: Editions Larose.

Epstein, H. 1971. *The origin of the domestic animals of Africa*. New York: Africana Publishing Corporation.

Haley, C.S. 1995. Livestock QTLs – bringing home the bacon? *Trends in Genetics* **11**, 488–92.

Hall, S.J.G. 1991. Body dimensions of Nigerian cattle, sheep and goats. *Animal Production* **53**, 61–9.

Hall, S.J.G. 1996. Conservation and utilization of livestock breed biodiversity. *Outlook on Agriculture* **25**, 115–18.

Hall, S.J.G. & D.G. Bradley 1995. Conserving livestock breed biodiversity. *Trends in Ecology and Evolution* **10**, 267–70.

Hall, S.J.G. & J. Clutton-Brock 1988. *Two hundred years of British farm livestock*. London: British Museum (Natural History).

Hall, S.J.G., L.K. Gnaho, C. Meghen 1995. Une enquête sur la race bovine Somba au Bénin. *Revue d'Elevage et de Médecine vétérinaire des Pays tropicaux* **48**, 77–83.

Hall, S.J.G., A.J.F. Russel, H. Redden 1996. Coat fibres of Nigerian sheep and goats: a preliminary characterisation. *Small Ruminant Research* **84**, 1–7.

Halstead, P. 1996. Pastoralism or household herding? Problems of scale and specialization in early Greek animal husbandry. *World Archaeology* **28**, 20–42.

Huelsenbeck, J.P., J.J. Bull, C.W. Cunningham 1996. Combining data in phylogenetic analysis. *Trends in Ecology and Evolution* **11**, 152–8.

ILCA 1989. *ILCA in print. Publications du CIPEA 1975–1988*, A. Atlaw & M. Sahlu (compilers). Addis Ababa: International Livestock Centre for Africa.

Joshi, N.R., E.A. McLaughlin, R.W. Phillips 1957. *Types and breeds of African cattle*. FAO Agricultural Studies No. 37. Rome: FAO.

Lauvergne, J.J., D. Bourzat, P. Souvenir Zafindrajaona, V. Zeuh, A.-C. Ngo Tama (1993). Indices de primarité de chèvres au Nord Cameroun et au Tchad. *Revue d'Elevage et de Médecine vétérinaire des Pays tropicaux* **46**, 651–65.

Lhote, H. 1959. *The search for the Tassili frescoes* [English translation]. London: Hutchinson.

MacHugh, D.E., R.T. Loftus, D.G. Bradley, P.M. Sharp, E.P. Cunningham 1994. Microsatellite variation within and among European cattle breeds. *Proceedings of the Royal Society of London, Series B* **256**, 25–31.

Mason, I.L. 1951. *The classification of West African livestock. Technical Communication No. 7 of the Commonwealth Bureau of Animal Breeding and Genetics, Edinburgh*. Farnham Royal: Commonwealth Agricultural Bureaux.

Mason, I.L. & J.P. Maule 1960. *The indigenous livestock of eastern and southern Africa. Technical Communication No. 14 of the Commonwealth Bureau of Animal Breeding and Genetics, Edinburgh*. Farnham Royal: Commonwealth Agricultural Bureaux.

Moorcroft, P.R., S.D. Albon, J.M. Pemberton, I.R. Stevenson, T.H. Clutton-Brock 1996. Density-dependent selection in a fluctuating ungulate population. *Proceedings of the Royal Society of London, Series B* **263**, 31–8.

Muzzolini, A. 1983. *L'Art Rupestre du Sahara Central: Classification et chronologie. Le Boeuf dans la préhistoire africaine* [2 volumes]. Thèse de Troisième Cycle, Université de Provence.

Patterson, C., D.M. Williams, C.J. Humphries 1993. Congruence between molecular and morphological phylogenies. *Annual Review of Ecology and Systematics* **24**, 153–88.

Patterson, D.L. 1990. Obtaining objective measurements of animal conformation by video image analysis. *Proceedings of the 4th World Congress on Genetics applied to Livestock Production, Edinburgh 23–27 July 1990, XV*, 295–8.

Peacock, C.P. 1987. Measures for assessing the productivity of sheep and goats. *Agricultural Systems* **23**, 197–210.

Pryor, F. 1996. Sheep, stockyards and field systems: Bronze Age livestock populations in the Fenlands of eastern England. *Antiquity* **70**, 313–24.

Rege, J.E.O. & M.E. Lipner (eds) 1992. *African animal genetic resources: their characterisation, conservation and utilisation. Proceedings of the Research Planning Workshop held at ILCA, Addis Ababa, Ethiopia, 19–21 February 1992*. Addis Ababa: International Livestock Centre for Africa.

Reid, A. 1996. Cattle herds and the redistribution of cattle resources. *World Archaeology* **28**, 43–57.

RIM 1992. *Nigerian livestock resources, vols. I–IV*. Oxford: Resource Inventory and Management.

Rohlf, F.J. & L.F. Marcus 1993. A revolution in morphometrics. *Trends in Ecology and Evolution* **8**, 129–32.

Scherf, B. (ed.) 1995. *World watch list for domestic animal diversity*, 2nd edn. Rome: FAO.

Smith, P.E.L. 1968. Problems and possibilities of the prehistoric rock-art of northern Africa. *African Historical Studies* **1**, 1–39.

Swift, J. (ed.) 1984. *Pastoral development in central Niger: report of the Niger Range and Livestock Project*. Niamey: USAID and Republic of Niger Ministère de Développement Rural.

Thorp, C.R. 1995. *Kings, commoners and cattle at Zimbabwe tradition sites*. Museum Memoirs (New Series) No. 1. Harare: National Museums and Monuments of Zimbabwe.

CHAPTER SEVENTEEN

The characterization of indigenous goat types of Ethiopia and Eritrea

Workneh Ayalew, Christie Peacock, Nigatu Alemayehu, Alemayehu Reda, Bernard Rey

1. Introduction

The estimated goat population of Ethiopia and the newly-independent state of Eritrea is 18.1 million (FAO 1992). They are kept in a wide range of production systems reflecting the diversity of the Ethiopian and Eritrean environments. It was generally believed that the majority of the goat population is kept by pastoralists in the lowlands of the south, east and west, while the remainder is found in small flocks in agropastoral areas and on mixed highland farms. But recent survey work has revealed that the majority of goat farmers are found in the densely populated highlands where goats are of great economic importance.

To their owners goats are a valuable source of milk, meat, cash and security against the vagaries of the environment. Yet despite their economic importance to many of Ethiopia's and Eritrea's most vulnerable groups, little attention has been paid to their development through research and extension.

A comprehensive description of the diverse goat types found in Ethiopia and Eritrea has never been made. Descriptions of localized goat types were attempted during the colonial era in Eritrea and Somaliland. Marchi (1929) and Gadola (1947) described the variety of small goats kept by Eritrean peoples such as the Beni-Amer, Assaorta and Bilena. Salerno (1939) described a short-haired goat of the Danakil desert, and the white and variegated coloured goats of the Hararghe highlands and lowlands (Roetti 1938). Girardon (1939) mentions the Bati goat of Wollo valued for its skin, the Arusi goat and the goats of the western lowlands of Ethiopia. Drake-Brockman (1912) described the Somali goats he encountered in Somaliland. In all 17 goat types were described during the period 1920–50 (Mason & Maule 1960).

A systematic description of the goat types and management systems should be considered a prerequisite for planning the rational use of the national goat genetic resources. In addition, breed characterization is the first step in the urgent task of genetic resource conservation (FAO 1992, ILCA 1992).

To make a first attempt at characterizing the goat types of the region the British non-governmental organization, FARM-Africa, sponsored a National Goat Breed Survey of Ethiopia and Eritrea in 1990.[1] The objectives of the survey were: (a) to identify and characterize the indigenous goat types in Ethiopia and Eritrea, (b) to describe the traditional goat husbandry practices in different production systems, and (c) to develop and test a method for the rapid survey of indigenous livestock.

2. Materials and methods

The National Indigenous Goat Breed Survey was carried out in three consecutive studies covering different regions during January 1990 to June 1993. A standard survey method was followed in each case to ensure comparable data.

Each region surveyed was stratified by altitude and ethnic group. Sample sites were selected at 500 m intervals along altitude transects spanning the known ethnic range. The survey included only one visit to a sampling site at which whole flocks of goats were sampled until about 500 goats had been physically measured. A total of 21 qualitative and quantitative variables were selected from the FAO (1986) goat breed descriptor list:

Morphological variables:
Qualitative: sex, dentition, beard, ruff, wattles, head profile, coat colour pattern, coat type, ear form, horn shape, horn orientation.
Quantitative: body length, chest girth, height at withers, pelvic width, ear length, horn length, body weight.
Other variables:
Parturition history of does as recalled by owner, i.e. total number of parturitions, single births, multiple births.
Source of animal whether home-bred, purchased, given, etc.

Although it is customary to describe breeds in terms of mature females, whole flocks were sampled to obtain information on the size of flock owned and its flock structure.

Goats were measured using a plastic tape to the nearest unit centimetre and weighed using a 100 kg × 500 g spring balance suspended from a tripod. Weights were recorded to the nearest unit kilogram. Flocks were sampled early in the morning before grazing. To facilitate the survey, local Ministry of Agriculture veterinary personnel accompanied the survey team to vaccinate or treat goats brought to be measured.

A semi-structured questionnaire was used to collect information on housing, feeding, breeding management, product use and the traditional use of goats in each site.

The analytical goal was to determine the morphological resemblance or dissimilarities within an assemblage of animals, the classification being done at the population level. Multivariate analysis techniques were used to explore the factors of dissimilarity within a population, and eventually reorganize a heterogeneous set of observations into relatively more homogenous groups from the total sample population. The unit of analysis was the population of mature female goats at each site characterized by the mean of the quantitative variables and the average frequency of qualitative variables.

Body weight, chest girth, ear length and horn length were converted into ratios with height to reduce the effect of environment. Variables were screened for discriminating power. A total of 35 variables were finally selected. 103 sites were included in the analysis with 13 additional sites considered only for supplementary observations, owing to the small number of animals recorded, implying high variability for means and frequencies.

The 35 variables were first transformed into as many independent principal components using the procedure ANCOMP of ADADSAS software (ADDAD 1990). The first 15 principal components, accounting for 94 per cent of the total variance, were considered to develop the classification by cluster analysis. The set of 103 observations (populations means for sample sites) against the 15 principal components was clustered by a single-linkage, agglomerative, hierarchical and non-overlapping (SAHN) technique which is the most widely applied method of clustering in biological systematics (Sneath & Sokal 1973, Aldenderfer & Blashfield 1984). The Mahalanobis' distance was the similarity coefficient used to develop the classification tree from which the desired number of clusters was obtained.

3. Results and discussion

The classification tree or dendogram (Fig. 17.1) allows the exploration of the general pattern of relationships between the observation units. Heuristic decisions were taken to determine the number of clusters. The choice was largely subjective but efforts were made to find one with the most meaningful biological interpretation. The initial results of the cluster analysis were mapped for consistency. Population means and frequencies were computed for the newly assembled populations. A few (9) misclassified sites, forming one cluster, were reclassified into the 14 other clusters. The geographical representation of that reclassification was pertinent.

The dendrogram shows four major groups of goat populations:

Nubian family: this family includes the lowland goats of western Eritrea with distinctly Roman-nose facial profile. It appears closely related to the goats found in Egypt, Syria, Iraq and Iran today. These goats are known to be good milk producers.

Coastal goats: some of these goats are found in northeastern Ethiopia and along the Red Sea coastline; these are thought to have originated from Yemen and Saudi Arabia and entered Ethiopia from across the Red Sea. Their distribution extends along the Rift Valley in a wide range of agro-ecological zones in central and southern Ethiopia.

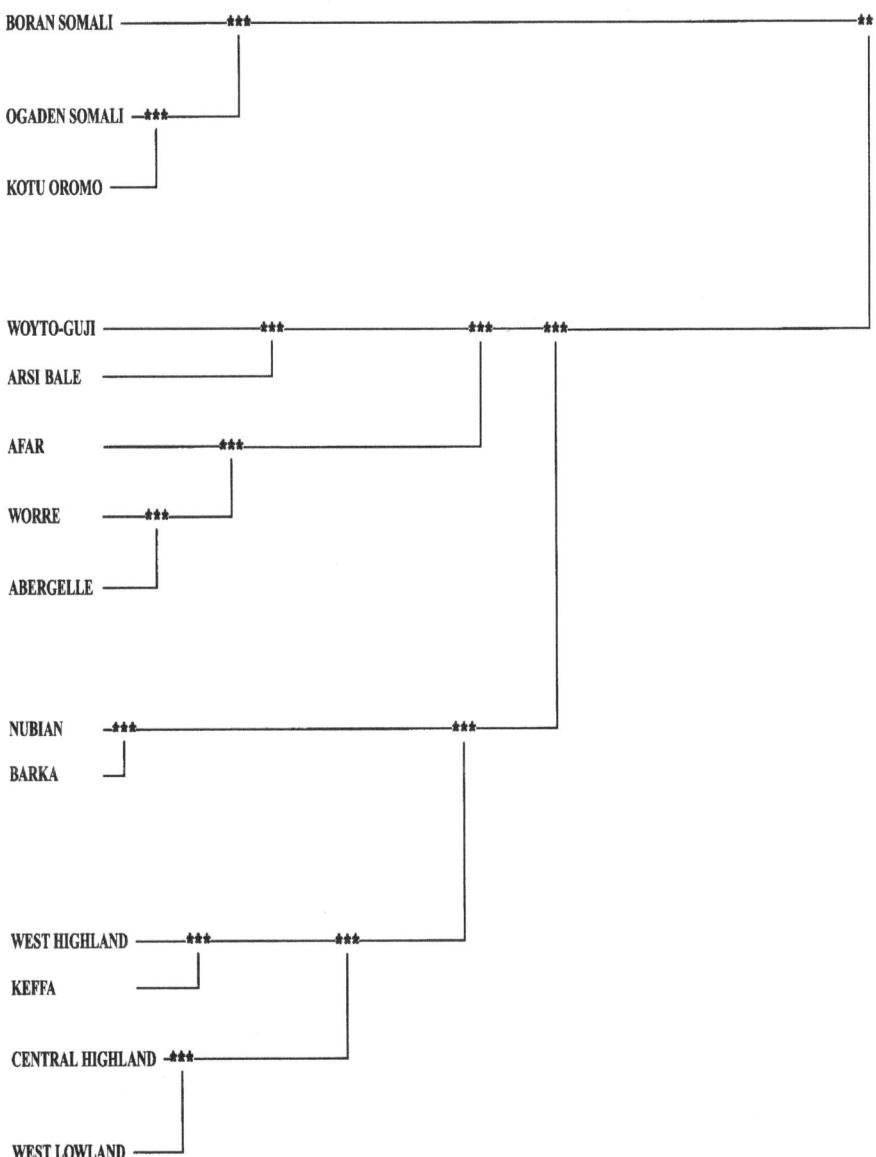

Figure 17.1 Hierarchical classification tree.

Somali family: this includes the lowland and mid-altitude goats of southern and southeastern Ethiopia. They are probably related to goats from Arabia into Somalia. They are extensively used for milk production and have good adaptive characteristics to arid environments.

Small East African family: this family includes a large heterogeneous population of goats from central highlands and the adjacent areas to the west.

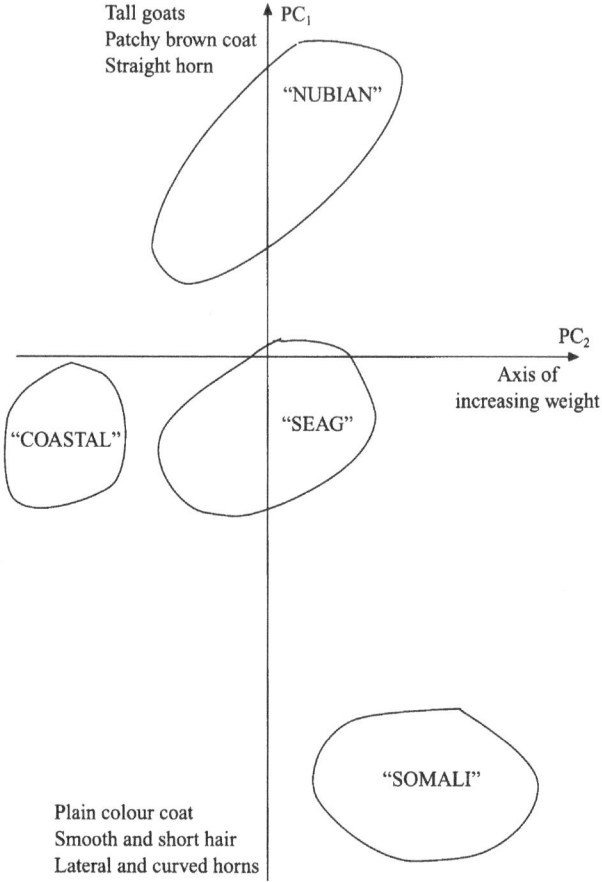

Figure 17.2   INERTY of clouds of four major clusters in the co-ordinate system of PC1 and PC2.

They have diverse production functions. Mixing and interbreeding of goats in the densely populated highlands perhaps created the heterogeneity of highland goat types we observe today.

Figure 17.2 represents the INERTY clouds of those four major families of goats in the co-ordinate system defined by axes 1 and 2 of the Principal Components Analysis. A 14-cluster solution was found to be most meaningful. The names given to the clusters identified (Table 17.1) have tried to reflect geographical location as much as possible rather than owning ethnic groups. Population estimates and the key identifying features of the identified goat types are presented in Table 17.1. Table 17.2 shows the average flock composition of sample populations. The higher relative proportion of breeding females seems to indicate more extensive use of goat milk. Table 17.3 presents average measurements of body size for mature males and females.

Information, reported by owners, on the reproductive performance of each doe sampled enabled the estimation of average litter size for the identified goat types.

The results of this exploratory survey pose more questions than they answer. Can these results be explained by evidence from other disciplines? How can the results of this survey be interpreted in the context of applied animal breeding and conservation of animal genetic resources?

Field observations suggest that goat populations in lowland pastoralist areas are more homogenous in phenotypic characteristics than populations found on the highland smallholder farms. The Somali and Nubian family of goats are quite distinct in morphology and milk production ability. On the other hand, the small flocks in the highlands are more heterogeneous. Small flocks are prone to rapid drifts in genetic constitution of populations because of migration and marketing. It is more difficult to maintain genetic attributes in smaller flocks, especially in low income societies where a high proportion of the flock will be regularly sold.

The identification, description, classification and naming of livestock types is the first step in the assessment of a country's livestock resources. The physical description of types can only dimly reveal the genetic relationships between individuals. It would be ideal if the relationships of the goat types identified phenotypically could be confirmed through analysis of their DNA.

Performance assessment must be considered a prerequisite for the sound planning of livestock development in any country. In addition the classification and characterization of livestock types is necessary before indigenous genetic material can be preserved for future generations.

The limited quantitative and qualitative measurements made during this survey have allowed the preliminary description of the phenotypic characteristics of the goat types of Ethiopia. What are the next steps?

Bruns (1992) suggests that once breeds are known there are three levels of research studies required for genetic characterization of known breeds:

Macro level studies: on-farm surveys to describe the population structure and make a detailed physical description of the breeds.

Meta level studies: for phenotypic characterization of breeds a more detailed on-farm performance monitoring is needed. This should be designed to allow performance (productive and adaptive traits) comparisons between breeds.

Micro level studies: comprehensive on-station characterization studies for the estimation of genetic parameters.

From this it will be possible to design sound breed improvement strategies, although limited resources may allow for only the most promising goat types to be selected and studied in detail. The Nubian, Barka and Somali goats have promising milk type attributes. Abergelle goats are fast growing. The western Lowland, Keffa and Woyto-Guji goats are prolific.

Table 17.1 Goat types: name, key identifying features, distribution and estimated population.

Goat type	Key identifying features	Distribution	Population
Nubian	tall, markedly convex facial profile, long ears, hairy	lowlands of western Eritrea (Gash & Setit)	200,000
Barka	tall, hairy thighs, predominantly white coat colour, long ears	lowlands of western and southwestern Eritrea (Gash & Setit and Akordat)	600,000
Worre	short, concave facial profile, short horns	highlands of northern Eritrea (Keren and Sahil)	500,000
Afar	concave narrow face, prick ears, leggy, long, upward pointing horns, patchy coat colour	coastal areas of Eritrea and northeastern Ethiopia (Afar region)	1,000,000
Abergelle	stocky, mostly reddish brown, males with magnificent spiral horns	mid-altitude of southern Tigray and northern Wollo, along Tekeze valley	300,000
Central Highland	medium size, broad face, thick horns, reddish-brown colour	central highlands west of the Rift Valley in Tigray, Wollo, Gondar and Shoa	6,000,000
Western Highland	tall, coarse hair, white and/fawn colour	highlands of western Ethiopia (south Gondar, Gojam, eastern Wellega and Illubabor)	3,000,000
Western Lowland	short, straight facial profile, mainly white and fawn, ruff on males	lowlands of western Ethiopia (Metekel, Asossa and Gambela)	400,000
Keffa	small, red or black colour, short neck, prick ears	Keffa, parts of Illubabor and south Shewa	1,000,000
Arsi-Bale	large, often hairy, patchy brown, white or black coat colour	Highlands of Arsi, Bale and south Shewa	600,000
Woyto-Guji	medium body size, small heads, straight or concave face, short ears, brown, red and black colour, shiny smooth coat	South and North Omo, northern Borena and southern Sidama (Guji)	900,000
Kotu Oromo	small, straight to concave face, frequent polledness (37%), brown, black or white coat	highlands of eastern and western Hararghe	1,000,000
Issa Somali	medium size, mainly white, short hair	northern and eastern parts of Ogaden and Dire Dawa	1,500,000
Boran Somali	large, brilliant white colour, short smooth hair	Ogaden and most of Borena	1,500,000

Table 17.2 Flock composition of the identified goat types.

Goat type	Total sample size	Flock composition, per cent			
		total females	breeding females	intact males	castrates
Nubian	569	94.5	76.1	5.0	0.4
Barka	855	95.0	77.6	4.9	0.1
Worre	651	96.5	87.8	3.5	0.0
Afar	2,653	93.6	79.2	6.1	0.2
Abergelle	457	83.8	60.8	13.6	2.3
Central highland	6,036	71.3	48.4	22.2	6.5
Western highland	3,278	73.3	43.0	23.0	3.7
Western lowland	1,005	79.0	47.7	21.4	0.7
Arsi-Bale	4,652	74.4	50.0	22.1	3.5
Woyto-Guji	8,880	74.0	52.8	22.9	3.1
Keffa	3,370	70.1	45.2	25.3	4.4
Kotu Oromo	1,649	71.1	48.9	24.6	4.3
Issa Somali	3,432	78.7	64.6	17.2	4.1
Boran Somali	4,039	85.0	67.5	13.6	1.4
Total	41,526				

A comprehensive description of the diverse goat types found in Ethiopia and Eritrea has never been made. A systematic description of the goat types and management systems should be considered a prerequisite for planning the rational use of the national goat genetic resources. In combination with historical and archaeological data it could also provide new insights into the evolution of present-day patterns of livestock production.

Table 17.3 Average body size of mature males and females by goat type.

Goat type	Sex	Height at withers (cm)	Body weight (kg)	Chest girth (cm)	Ear length (cm)	Horn length (cm)
Nubian	M	74.1+4.6	30.0+4.2	75.0+4.8	19.7+3.9	16.8+3.1
	F	70.1+3.4	34.3+5.4	74.3+3.8	20.1+3.6	14.6+2.9
Barka	M	74.3+7.2	45.3+14.1	83.0+9.5	18.1+5.3	20.0+9.9
	F	67.9+4.3	33.8+5.3	73.9+4.8	18.2+3.1	13.9+3.1
Worre	M	61.7+4.4	26.9+6.0	68.9+4.7	15.1+2.3	16.1+3.6
	F	61.0+3.1	24.9+4.3	67.8+3.8	13.9+2.8	13.2+2.9
Afar	M	64.5+2.9	31.3+3.7	74.6+3.8	12.4+0.7	29.8+6.8
	F	60.9+3.3	23.7+3.4	67.4+3.4	12.3+1.8	17.4+3.9
Abergelle	M	71.4+3.5	33.6+5.9	79.5+2.9	13.0+0.8	37.0+9.1
	F	65.0+2.8	28.4+3.5	71.2+3.8	12.7+0.8	19.6+5.7
Central highland	M	76.3+5.0	43.0+7.7	84.6+5.8	13.5+0.9	23.4+5.1
	F	67.9+3.2	30.1+5.4	74.1+4.4	13.1+1.1	13.7+3.5
Western highland	M	80.7+6.5	48.4+9.9	87.2+7.9	14.6+1.6	20.8+4.8
	F	70.8+4.7	33.0+6.0	75.8+4.5	14.7+1.6	12.8+3.6
Western lowland	M	67.2+5.0	35.5+10.2	77.0+9.2	14.1+1.6	18.5+7.2
	F	63.5+3.8	33.9+6.9	75.9+5.2	13.8+1.5	12.8+3.6
Arsi-Bale	M	73.2+6.9	42.4+9.6	85.2+7.0	14.1+1.3	23.7+7.2
	F	66.1+3.5	30.4+4.5	74.9+4.0	14.0+1.3	12.5+3.3
Woyto-Guji	M	72.9+5.0	39.0+6.3	80.8+6.6	12.5+1.3	17.6+7.2
	F	66.4+3.5	28.8+5.0	72.5+4.2	12.5+1.0	10.8+3.7
Keffa	M	75.6+6.8	40.5+8.4	82.7+5.9	13.3+1.1	20.1+5.5
	F	66.7+4.0	28.2+5.2	72.2+4.5	13.0+1.0	11.6+3.6
Kotu Oromo	M	71.5+7.2	41.9+7.2	80.6+7.9	14.4+1.4	21.4+6.7
	F	62.5+3.5	29.1+4.5	72.8+4.5	13.0+1.1	13.1+3.4
Issa Somali	M	64.9+5.5	32.8+6.5	72.8+4.7	12.1+2.2	19.6+6.9
	F	61.8+4.1	27.8+6.0	70.4+4.7	12.8+1.8	12.2+4.2
Boran Somali	M	75.8+4.2	42.3+7.4	82.3+4.9	14.8+1.7	13.5+6.2
	F	69.4+3.3	31.8+5.4	74.4+4.0	14.6+1.7	9.0+3.8

Note

1. FARM Africa, a British non-governmental organization sponsored this study as part of its Dairy Goat Development Programme undertaken in collaboration with the Ministry of Agriculture, Alemaya University of Agriculture, Awassa College of Agriculture and several international and indigenous

NGOs. The International Livestock Centre for Africa (ILCA) provided assistance in study design and data analysis. The Alemaya University of Agriculture has assigned three graduate students to take up the survey for their thesis research.

References

ADDAD 1990. *Manuel de reference ADDADSAS.* Paris: Association pour le development et la Diffusion de l'Analyse des Données, 22 rue Charcot.

Aldenderfer, M.S. & R.K. Blashfield 1984. *Cluster analysis.* Sage University Paper Series on Quantitative Applications in the Social Sciences, 07-044. London: Sage Publications.

Alemayehu, R. 1993. *Characterisation (phenotypic) of indigenous goats and goat husbandry practices in east and southeastern Ethiopia.* MSc thesis, Alemaya University of Agriculture, Ethiopia.

Bruns, E. 1992. Synthesis of research methodology. In *Animal genetic resources: their characterisation, conservation and utilization (Appendix II).* J.O.E Rege & M.E. Lipner (eds), 125–34. Research planning workshop, ILCA, Addis Ababa, Ethiopia, 19–21 February 1992. Addis Ababa: ILCA.

Drake-Brockman, R.E. 1912. *British Somaliland.* London: Hurst & Blachlett.

FAO 1986. Animal Genetic Resources Data Banks. 2. *Descriptor lists for cattle, buffalo, pigs, sheep and goats.* Animal Production and Health Paper No. **50**(2), 96–129. Rome: FAO.

FAO 1992. *Production Yearbook,* vol. 45. Rome: FAO.

Gadola, A. 1947. *Zootecnia profilassi e igiene zootechnica in Africa Orientale, Roma.* Roma: Istituto Superiore di Sanità, Fondazione Emanuele Paternò.

Girardon, C.A. 1939. Le risorse zootechniche nei Governo dello Scioa. *Collezione Studii Coloniale Istituto Africano Italiano Sez, Milano* **7**, 5–89.

ILCA 1992. *Animal genetic resources: their characterisation, conservation and utilisation.* J.O.E. Rege & M.E. Lipner (eds), Research planning workshop, ILCA, Addis Ababa, Ethiopia, 19–21 February 1992. Addis Ababa: ILCA.

Marchi, E. 1929. *Studi sulla pastorizia della colonia Eritrea.* Firenze. Istituto Agricolo Coloniale Italiano.

Mason, I.L. & J.P. Maule 1960. *The indigenous livestock of eastern and southern Africa.* Technical Communication No. 14. Commonwealth Bureau of Animal Breeding and Genetics, Edinburgh. Farnham Royal: Commonwealth Agricultural Bureaux.

Nigatu, A. 1994. *Characterization of indigenous goat types of Eritrea, northern and western Ethiopia.* MSc thesis, Alemaya University of Agriculture, Ethiopia.

Roetti, C. 1938. Considerazione Zootechniche dell'Ethiopia. *Rivista Militoria Médicine Vétérinaire* **2**, 476 only.

Salerno, A. 1939. Indagini preliminari sul patrimonio zootechnico dell'Harar e problemi che lo riguardano. *Agricoltura coloniale* **33**, 82–100.

Sneath, P.H.A. & R.R. Sokal 1973. *Numerical taxonomy.* San Francisco: W.H. Freeman.

Workneh, A. 1992. *Preliminary survey of indigenous goat types and goat husbandry practices in southern Ethiopia.* MSc thesis, Alemaya University of Agriculture, Ethiopia.

CHAPTER EIGHTEEN

Using morphobiometric indices to map goat resources in Africa

J.J. Lauvergne, D. Bourzat, F. Minvielle

1. Introduction

To use morphobiometric indices to map the genetic goat resources in Africa, the genetic structure of the populations in question must be known. Work by Muzzolini (1993) and Bouchel & Lauvergne (1995), analyzing how Africa has been populated by domestic goats, has revealed that the genetic structure of the goat populations is most likely to be of a primary type. A primary-type genetic structure is the state of a species' population during the first stage following its domestication, prior to its fragmentation by man into genetic isolates such as standardized breeds and selected lines.

In such a population, characterized by a great variety of visual attributes such as coat colour, wool texture, etc., previous authors have easily distinguished biometrical differences in size, limb to trunk ratio, convexity of the nose, auricle development, etc., which enabled them to attempt a morphobiometric classification for goats.

After summarizing the concept of primarity and its application to the African goat population, this chapter reviews the various attempts at morphobiometric classifications by previous authors and examines developments concerning the utilization of a number of measurements for the calculation of morphobiometric indices utilized for the mapping of goat resources.

These indices have been established through a collaboration in 1992, between INRA (Laboratoire de Génétique factorielle), CIRAD EMVT (LRVZ of Farcha, Chad) and the research organizations of three French-speaking countries in Africa (Chad, Cameroon and Nigeria) who were participating in the Regional Research Project on Small Ruminants (PRRPR).

2. Concept of primarity and assessment

2.1. Definition of the state of primarity

The characteristics of the so-called state of genetic primarity for a domesticated population have already been described for the horse by Buffon (1753), for the pigeon by Darwin (1859) in the first chapter (Variation under domestication) of *The Origin of Species by means of natural selection*, and also for a number of species, by Geoffroy-Saint-Hilaire (1861). Primarity is characterized by a great variability in visible effects contrasting with the uniformity observed in wild species.

Mason (1951), an ethnographer of animal breeding, felt that a category for non-uniform populations or geographical terms meaning the "cattle of such and such a place" or even the "breed of such or such a place" should be created to designate these "ethnic" entities. The term "traditional population" was later suggested by Lauvergne (1982) following the example of the term "traditional cultivar", used by Fraenkel (1971) in the classification of plant genetic resources. The expression was subsequently adopted in 1986, for the title of the 47th INRA symposium: Traditional populations and first standardized breeds of Ovicaprinae in the Mediterranean (Lauvergne 1988a), after it had been previously proposed and approved by the European Cooperative Network on Sheep and Goat Production, at an FAO meeting in Thessaloniki (Greece), Lauvergne (1985). This term, however, was later replaced with the expression "primary population" by Lauvergne (1993).

2.2. Primarity assessment

An initial assessment of the visible genetic variability in a traditional population was made by Lauvergne et al. (1987) on Common Provence goats in the Forcalquier arrondissement of the Alpes-de-Haute-Provence (France). Following Lauvergne's suggestion, the same grid was later utilized on 13 sites along the northern shore of the Mediterranean, by researchers who were taking part in the 47th INRA symposium (1988), to describe the variability in the primary populations of Macedonia (Boyazoglu et al. 1988), Sardinia (Branca & Casu 1988), Bulgaria (Djorbineva et al. 1988), northern Spain (Dunner & Caçon 1988), Corsica (Franceschi & Santucci 1988), Malta (Gruppetta et al. 1988), the Upper Roya region (Matrès & Benadjaoud 1988) and in a Rove population (Sadorge & Benadjaoud 1988).

These studies only consisted of a comparison of locus profiles with present alleles. No global quantification of primarity was provided. This global quantification was, however, completed later by Machado et al. (1992) using traditional indices based on the principle of counting the loci present as evidenced by the presence of at least one mutant allele. These indices were adopted under the name of indices of primarity by Lauvergne et al. (1993). Their formula, symbol and name are given in Table 18.1.

2.3. State of primarity in African populations

Studies carried out by Lauvergne et al. (1993) revealed that the primarity indices in the goat populations of Chad and Cameroon, considered in a pilot study, were not significantly different from the value of 1, given the panmixy generally observed in Africa. This tended to indicate that the African populations are mostly of the primary

Table 18.1 Formula, symbol and *index of primarity* name (formerly *indices of traditionality*).

Symbol and formula	Name	
	in Machado et al. (1992)	*in* Lauvergne et al. (1993)
IPS = $n_s^{(1)}/N_s^{(2)}$	index of traditionality loci with visible effect in segregation (in per cent)	index of primarity loci with visible effect in segregation (0 to 1)
IPpa$^{(3)}$ = number of identified phenotypes at *Agouti* locus /number of possible phenotypes at this locus	index of traditionality phenotypes at the *Agouti* locus	
IPa = $(n_a^{(4)}-1)/(N_a^{(5)}-1)$		index of primarity alleles at the *Agouti* locus (0†to†1)

$^{(1)}$ n_s number of loci with visible effect in segregation
$^{(2)}$ N_s number of loci with visible effect identified in the species
$^{(3)}$ symbol proposed in the current article
$^{(4)}$ n_a number of alleles at A locus present in the populations
$^{(5)}$ N_a number of alleles at the A locus already identified in the species

Table 18.2 IPS (indices of primarity loci with visible attribute in segregation) for a primary type goat population in Brazil (Ceara state), the Mediterranean and Africa (Cameroon/Chad).

Identified loci with visible effect in the goats	state of segregation of the loci*		
	(1)	(2)	(3)
Agouti (A^b, A^l, A^a) $^{(4)}$	1	1	1
Brown (B^b, Br$^+$) $^{(4)}$	1	1	1
Frosting (Fr^D, Fr$^+$) $^{(4)}$	1	1	1
Roan (Rn^R, Rn$^+$) $^{(4)}$	1	1	1
Beard (Bd^b, Bd$^+$) $^{(5)}$	1	1	1
Hair (HL^l, HL$^+$) $^{(5)}$	1	1	1
Horns (Ho^P, Ho$^+$) $^{(5)}$	1	1	0
Wattles (Wa^w, Wa$^+$) $^{(5)}$	1	1	1
IPS	8/8 = 1	8/8 = 1	7/8 = 0, 875

* 0 = non segregating, 1 = segregating
$^{(1)}$ state of Ceara (Brazil): Machado et al. (1992)
$^{(2)}$ 13 samples from the northern shore of the Mediterranean Sea, from Bulgaria to Spain (nine articles presented at the Colloque INRA No. 47)
$^{(3)}$ a zone in northern Cameroon and another in Chad, Lauvergne et al. (1993)
$^{(4)}$ genetic nomenclature of Lauvergne et al. (1989) (1990)
$^{(5)}$ genetic nomenclature of Lauvergne (1989)

Table 18.3 IPpa (index of primarity *identified phenotypes at Agouti locus*) for goat populations of Brazil (Ceara state), the Mediterranean and Africa (Cameroon/Chad).

identified phenotypes in primary populations	frequency and absence/presence					
	(1)		(2)		(3)	
	%	(4)	%	(4)	%	(4)
1. eumelanic	23.0	1	26.9	1	20.8	1
2. red cheek	0.7	1	5.0	1	–	0
3. eumelanic and tan	1.1	1	11.7	1	0.4	1
4. eumelanic and tan light belly	9.4	1	8.4	1	0.9	1
5. anterior mantle	0.2	1	2.4	1	5.8	1
6. posterior mantle	2.5	1	5.0	1	5.0	1
7. badger face	12.8	1	2.5	1	8.8	1
8. wild (with or without lists)	37.4	1	3.9	1	24.8	1
9. phaeomelanic	8.0	1	22.1	1	13.7	1
no identified	4.8		12.2		20.4	
IPpa		1.00		1.00		0.89

(1) state of Ceara (Brazil), Machado et al. (1992)
(2) 13 samples from the northern Mediterranenan shore, from Bulgaria to Spain (nine articles presented at the Colloque INRA No. 47)
(3) two zones, in northern Cameroon and in Chad, Lauvergne et al. (1993)
(4) absence/presence of the phenotype (0/1)

or subprimary type. Tables 18.2 and 18.3 provide some estimates of two indices of primarity (IPS and IPpa) among various populations considered *a priori* as being primary in Brazil, the northern shore of the Mediterranean, Chad and Cameroon.

3. Goat classifications based on morphobiometric criteria

In the analysis undertaken by Bouchel (1996), eight criteria were taken into account on the grids intended for the classification of the African goat. Among these, five were anatomic criteria: withers height, ear development, phaneroptics, craniology and slenderness (longipes and brevipes characters, according to Geoffroy-Saint-Hilaire (1861) later incorporated into the trigram of Baron (1893)). Such criteria have been used by 35 authors to establish "ethnic" nomenclatures. As a whole, the descriptive criteria were those most commonly used for classification purposes, especially the withers height and the relative proportions (descriptive trigram or slenderness). These were followed by the criterion of auricle development.

It should be noted, however, that a measurement such as the total withers height has the drawback of including the development of various skeleton constituents and could therefore possibly be confusing: for example, for two goats having the same withers height, one may exhibit a substantial trunk with short legs while the other exhibits a slender trunk with long legs. Similarly, the auricle length must be related to the body size to have any significance.

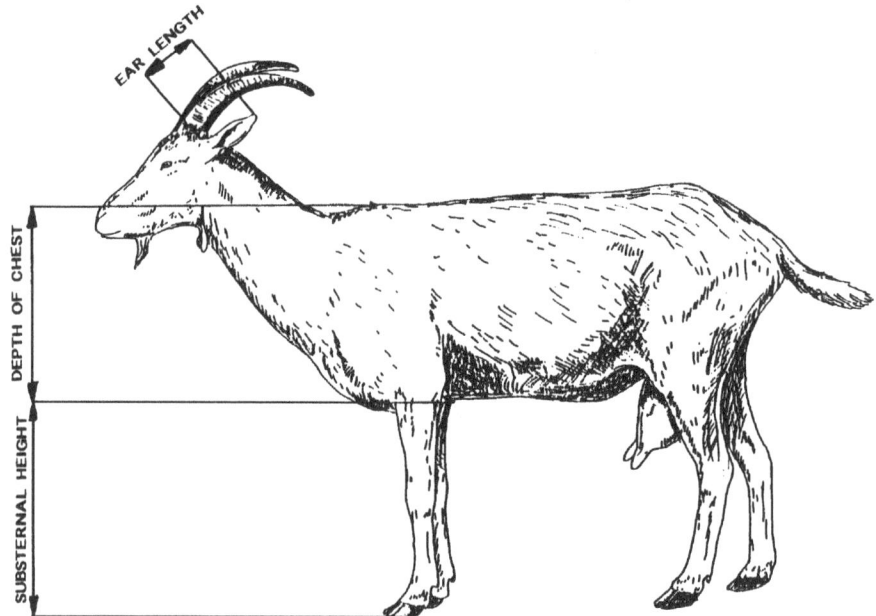

Figure 18.1   Measurements used for plotting morphobiometric indices on goats in Africa.

4. Current morphobiometric research

4.1. Definition of morphobiometric indices

As previously said, the concept of slenderness conveyed by the words longipes, for long-legged animals, and brevipes, for short-legged animals (Geoffroy-Saint-Hilaire, 1861) was used again in Baron's descriptive trigram (1893). It takes into consideration the relative lengths of the various bone segments. To be of value, however, this idea needs to be quantified as by Bourzat et al. (1993), who introduced indexes that are the ratios of measurements such as the volume of the substernum cavity, the thorax depth and the auricle length (Fig. 18.1). The index definitions are given in Table 18.4.

4.2. Study of indices of variation

The possible evolution of the indices with age needs to be studied to classify certain data with given age groups. A corresponding variation study was undertaken with Rove goats, originating from the northern Mediterranean shore, in Provence. Practical ages were established from this study for the index values at maturity for both ISS (Bouchel et al. 1997) and ATI (Lauvergne et al. 1997).

4.3. Applications to goat mapping in Africa

Even before the notions of relative slenderness and auricle length were quantified, authors had used a number of biometrical criteria for the mapping of the

Table 18.4 Formula, symbol and morphobiometric index names used for the goat in Africa (Lauvergne et al. 1993, Zeuh et al. 1995).

Symbol and formula		Name	
English	French	English	French
$ISS = SH^1/DC^2$	$IGs = VSS^4/PT^5$	index of substernal slenderness	indice de gracilité sous-sternale
$ATI = EL^3/DC$	$IAt = LO^6/PT$	auricle thorax index	indice auriculaire thoracique

[1] SH: substernal height
[2] DC: depth of chest
[3] EL: ear length
[4] VSS: vide sous-sternal
[5] PT: profondeur de thorax
[6] LO: longueur d'oreille
[7] translation from French to English based on Straszewska & Ollivier's Dictionary (1993)

African goat, for example, Wilson (1991) who produced a small-scale map of the continent showing the demarcation border between zones populated by large goats, small goats and dwarf goats.

4.3.1. First methodological study   A first mapping of these indices was produced by Bourzat et al. (1993) who studied two distinct zones, separated by 600 km, in which they expected to find unmixed groups of both types of goats – longipes and brevipes – based on previous authors' reports. This proved to be indeed the case. A possible contact zone for study would therefore be located between these two zones.

4.3.2. Pilot study in Chad – 1992–4   Still operating under the authority of the PRRPR and its satellite laboratories (including the LRZV of Farcha and the LG of INRA), the next step was to undertake a pilot mapping study, located on this previously determined contact zone. Sixty sites across the southwestern quarter of Chad were thus selected, including a total of 2,785 adult goats. These were examined in the course of a field study lasting for more than two years (1992–4). The analyses of the results have already been submitted for publication (Zeuh et al. 1995) or are in preparation (Lauvergne et al. 1996). These analyses have demonstrated the existence of two subpopulations, defined by both ISS and ATI, each characterized by two clearly distinct histograms that are reproduced in Figure 18.2.

After having recorded the points representing both subpopulations on the computer generated map, it appeared that they were located in clearly distinct zones, with a mean ISS of 1.04 and a mean ATI of 0.5 for the southern sub-population and with a mean ISS of 1.49 and a mean ATI of 0.75 for the northern sub-population (Fig. 18.3).

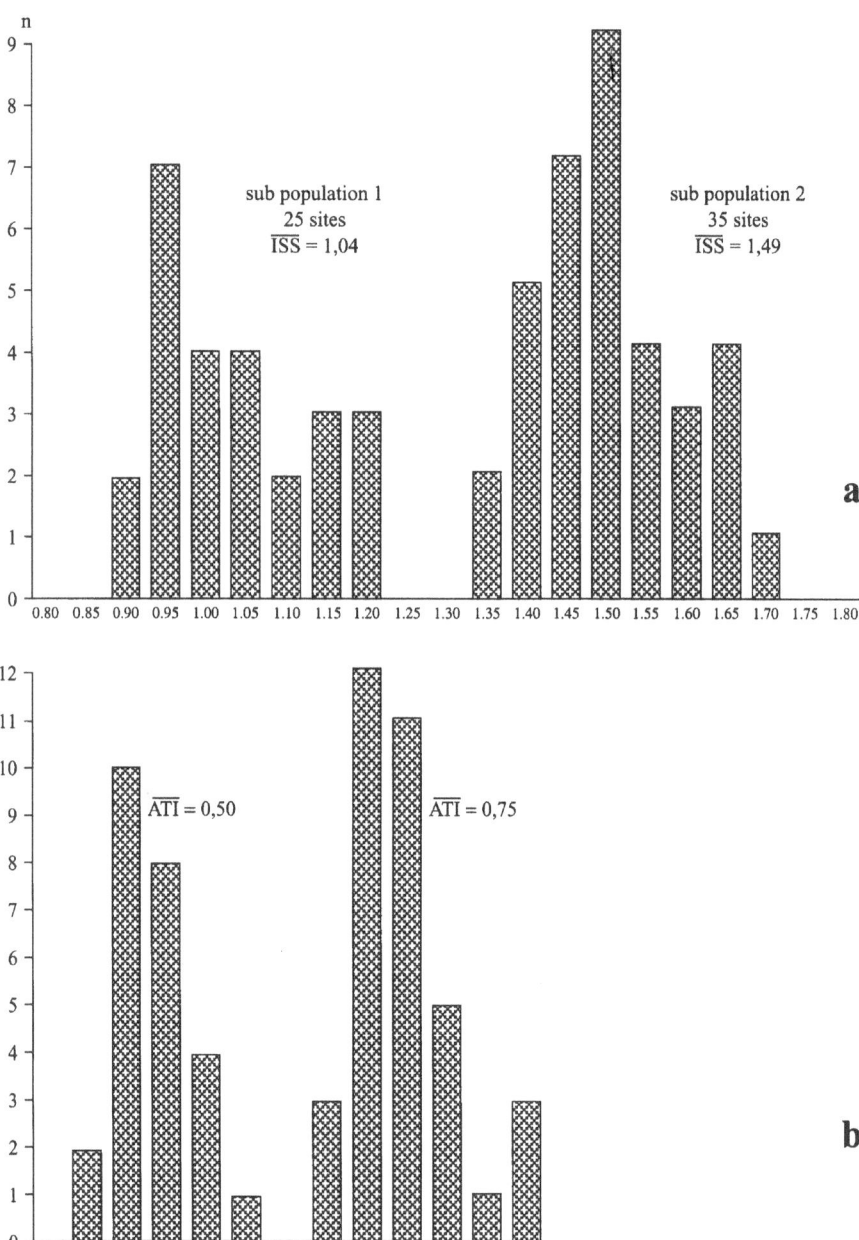

Figure 18.2   Distribution of the mean values of ISS and ATI on 60 sites of southwestern Chad according to Zeuh et al. 1995.
(a) ISS: index of substernal slenderness;
(b) ATI: auricle thorax index.

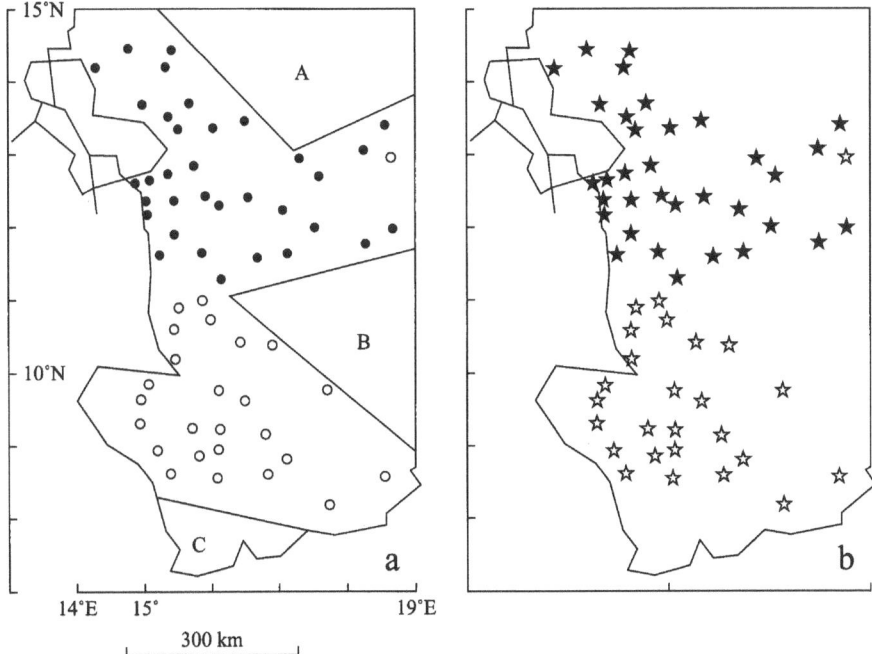

Figure 18.3 Location of subpopulations of ISS and ATI.
(a) ISS: subpopulation 1 (m = 1,04) clear circles, subpopulation 2 (m = 1,49) black points;
(b) ATI: subpopulation 1 (m = 0,50) clear stars, subpopulation 2 (m = 0,75) black stars.

4.3.3. Discussion The division reported by previous authors who were more or less using the slenderness criterion (longipes character/brevipes character) as well as by those authors who were taking into account the auricle development seems to be confirmed by the data from Chad (Lauvergne et al. in prep.). A number of questions remain:

(a) Why do both the longipes character and the relative length of the auricle increase when moving north?

(b) How is it that the division between both subpopulations is so clear-cut in the region of the 11th parallel?

(c) What happens further south, where dwarf populations are reported?

(d) Does the same phenomenon occur when going east?

To answer the first question, it may be postulated that both the increase of the longipes character and of the auricle length, in animals whose trunk and head developments remain unchanged, result from intense natural selection pressure attributable to climate, especially mean temperature. Longer legs may better protect the animals from ground heat reverberation, while longer ears offer a wider surface for the export of metabolic calories. The south/north gradient may be clearly explained by the fact that the thermic equator in this region is located further north, by at least 17 degrees of latitude (2,000 km) and slightly to the north of Lake Chad (Godard & Tableaud 1933).

Table 18.5 Mean and standard deviation of histograms for subpopulations showing the mean values for the *index of substernal slenderness* (ISS) and *auricular-thorax index* (ATI) on 60 sites of southwestern Chad (Zeuh et al. (1995), Lauvergne et al. (1996)).

	Subpopulation		ISS		ATI	
No.	location	n	m	σ	m	σ
1	south	25	1.04	0.10	0.50	0.03
2	north	35	1.49	0.09	0.75	0.06

Table 18.6 Index of substernal slenderness (ISS) regression per degree of latitude north in each of the zones occupied by subpopulations 1 (south) and 2 (north), from Zeuh et al. (1995).

subpopulation	zone	increase of ISS for one degree of latitude northward	t
1	1 (south)	0.088	5.404***
2	2 (north)	0.043	2.693*

* P ≤ 0,05
*** P ≤ 0,001

As far as the second question is concerned, it is conceivable that the clear-cut distinction between both subpopulations results from a sampling deficiency: gap B in Figure 18.3a, which is situated east of the longitude 16° East. Both sides of the 11th parallel, in southwestern Chad, might be located on the frontier if the data from this area were as clearly delimited as the other data.

To clarify what happens further south, one has to examine the regression of ISS in relation to the latitude, which was studied by Zeuh et al. (1995).

A slight and hardly significant regression can be observed for subpopulation 2 (in the north), between latitudes 11°5′ and 15° north, whereas for subpopulation 1 (in the south), the slenderness index decreases significantly, by about 0.09 on average, with each new latitude degree south. Should the phenomenon proceed at the same pace going south, one is bound to encounter dwarf goats, such as those described by Herre (1939), Epstein (1953, 1971), Pagot (1965) or Bourzat (1985). Unfortunately, no comparable index measurements were used by these authors. It remains to be determined where such a regression will stop and what values of ISS and ATI are likely to be attained.

The extension of the observed partition to the east in the 11th parallel region in Chad is likely to be a reality. Whereas no data concerning Arabia is available, descriptions concerning the Indian subcontinent, whose southern end exists between 10° and 27° of latitude north, by authors such as Acharya (1982) and Hasnain (1985), mention a great variation in the IGs and IAt of the Indian and Pakistani populations. Unfortunately, no index values have been established so far.

5. Conclusion

This research has demonstrated that the use of appropriate and fairly simple morphobiometric indices would make it possible to map the genetic goat resources of Africa, in a manner that completes and develops the vision of previous authors and introduces quantitative methods. This type of more sophisticated resource mapping will make it possible to analyze more clearly the historical, geographical and climatic factors responsible for the present-day situation.

A number of questions, however, remain unanswered, such as:

(a) Where should the limit be drawn between subpopulations?
(b) Does such a clear-cut transition such as that in the southwestern part of Chad exist everywhere or is it merely a sampling artefact?
(c) How should the situation be represented on the map?
(d) What is the relation between ISS and AIT when subpopulations do *not* overlap?
(e) What is the exact influence of external pressures, thought to be mainly climatic, on such morphobiometric index variations?

Moreover, the hypothesis concerning the effect of temperature on the formation of the subpopulations remains to be verified. The phenomenon should be studied along the length of the thermic equator.

Abbreviations used in this chapter

ADCR	Association de défense des caprins du Rove
ATI	auricle thorax index
CIRAD/EMVT	Centre de coopération Internationale en Recherche Agronomique pour le Développement/élevage et Médecine Vétérinaire des pays Tropicaux
EDE	Etablissement de l'élevage
GLM	General Linear Model
IAt	Indice auriculaire thorax
IGS	Indice de gracilité sous-sternale
INRA	Institut National de la Recherche Agronomique (France)
ISS	index of substernal slenderness
LO	Longueur d'oreille
NLIN	Non Linear
OVC	Ovicaprinae de Provence
PT	Profondeur du thorax

References

Acharya, R.M. 1982. *Sheep and goat breeds of India.* FAO Animal Production and Health Paper No. 30.

Baron, R. 1893. *La morphologie universelle. Leçon d'ouverture du cours de M. Robur Lanâo.* Luxembourg: Imprimerie Justin Schroell.

Bouchel, D. & J.J. Lauvergne 1996. Le peuplement de l'Afrique par la chèvre domestique. *Revue d'Elevage et de Médécine vétérinaire des Pays Tropicaux* 49, 80–90.

Bouchel, D., J.J. Lauvergne, E. Guibert, F. Minvielle 1997. Étude morpho-biométrique de la chêvre du *Rove*. I. Hauteur au garrot (HG), profondeur de thorax (PT), vide sous-sternal (VSS) et *indice de gracilité sous-sternale* (IGS). *Revue d'Elevage et de Médécine vétérinaire des Pays Tropicaux* **148**, 37–46.

Bourzat, D. 1985. La chèvre naine d'Afrique Occidentale: Monographie. *Groupe de Recherche sur les Petits Ruminants et les Camélidés, Document du Groupe No. SRC 4*. Addis Ababa: ILCA.

Bourzat, D., P. Souvenir Zafindrajaona, J.J. Lauvergne, V. Zeuh 1993. Comparaison morpho-biométrique de chèvres au Nord Cameroun et au Tchad. *Revue d'Elevage et de Médécine vétérinaire des Pays Tropicaux* **46**, 667–74.

Boyazoglu, J.G., J. Hatziminaoglou, J.J. Lauvergne 1988. Visible genetic profiles of the *Macedonian* goat. See Lauvergne (1988b), 105–12.

Branca, A. & S. Casu 1988. Visible genetic profiles of the Sardinian goat. See Lauvergne (1988b), 135–43.

Buffon J.-M. Leclerc, Comte de 1753. *Histoire générale et particulière avec la description du Cabinet du Roi. Tome IV. Discours sur les animaux*. Paris: Imprimerie Royale.

Darwin, Charles 1859. On the origins of species by means of natural selection. London: John Murray. In J.W. Burrow, 1968: *Darwin: the origin of species*, 71–100. Harmonsworth, Middlesex: Penguin.

Djorbineva, M.K. Alexieva Snejana, T. Hinkovski, J.J. Lauvergne 1988. Visible genetic profile of the Sakkhar goat in Bulgaria. See Lauvergne (1988b), 97–104.

Dunner, S. & J. Cañon 1988. Visible genetic profiles of the goat in northern Spain. See Lauvergne (1988b), 159–67.

Epstein, H. 1953. The dwarf goats of Africa. *East African Agricultural Journal* **18**, 123–32.

Epstein, H. 1971. II. The dwarf goats of equatorial Africa. In *The origin of the domestic animals in Africa* [2 volumes], 210–235. New York: Africana Publishing Corporation.

Fraenkel, O.H. 1971. *The significance, utilization and conservation of crop genetic resources*. Rome: FAO.

Franceschi, P. & P. Santucci 1988. Visible genetic profiles of the Corsican goat. See Lauvergne (1988), 141–51.

Geoffroy-Saint-Hilaire, I. 1861. *Acclimatation et domestication des animaux utiles*. Paris: Librairie agricole de la Maison Rustique.

Godard, A. & M. Tableaud 1993. *Les climats. Mécanismes et répartition*. Paris: Armand Colin.

Gruppetta, A., C. Renieri, M. Silvestrelli, F. Valfré 1988. Visible genetic profiles of the goat in Malta. See Lauvergne (1988b), 113–121.

Hasnain, H.U. 1985. *Sheep and goats in Pakistan*. FAO Animal Production and Health Paper No. 56. Rome: FAO.

Herre, W. 1939. Beitrag zur Kenntnis der Zwergzeigen. *Der Zoologische Garten* (NF) **15**(1/2).

Lauvergne, J.J. 1982. Genética en poblaciones animales despuès de la domesticacion. concecuencias por la concervacion de las razas [Genetics in Animal Population since Domestication. Outcomes for Breeds Conservation]. *2do Congreso Mundial Genet aplic. prod. ganaderia*, Madrid, **6**, 77–87.

Lauvergne, J.J. 1985. *The use of visible genetic profiles for the identification of domestic goat populations*. FAO Goat Subnetwork, Thessaloniki, Sept 26–27. 1–10. Dpt of Animal Genetics, INRA, Jouy-en-Josas.

Lauvergne, J.J. 1988a. Méthodologie proposée pour l'étude des *Ovicaprinae* méditerranéens en 1986. In Lauvergne (1988b), 77–94.

Lauvergne, J.J. (ed.) 1988b. *Populations traditionnelles et premières races standardisées d'Ovicaprinae dans le Bassin Méditérranéen (Traditional populations and first standardized breeds of Ovicaprinae in the Mediterranean)*. Colloq. INRA No. 47. Paris: INRA.

Lauvergne, J.J. (ed.) 1989. *Standardized nomenclature for sheep and goats, 1987*. Proceedings of the Cognosag Workshop July 1987. Paris: BRG.

Lauvergne, J.J. 1993. Breed development and breed differentiation. In *Data collection, conservation and use of farm animal genetic resources*, D. Simon, Doris Buchenauer (eds), 53–64. CEC Workshop and training course, Hanover, Dec 7–9, 1992.

Lauvergne, J.J., D. Bourzat, P. Souvenir Zafindrajaona, V. Zeuh, A.-C. Ngo Tama 1993. Indices de primarité de chèvres au Nord Cameroun et au Tchad. *Revue d'Elevage et de Médérine vétérinaire des Pays Tropicaux* **46**, 651–65.

Lauvergne, J.J., C. Renieri, A. Audiot 1987. Estimating erosion of phenotypic variation in a French traditional goat population. *Journal of Heredity* **78**, 307–14.

Lauvergne, J.J., D. Bouchel, F. Minvielle, E. Guibert 1997. Étude morpho-biométrique de la chèvre du *Rove*. II. Longueur d'oreille (LO) et *indice auriculaire thorax* (IAT). *Revue d'Elevage et de Médécine vétérinaire des Pays Tropicaux* **148**, 501–10.

Lauvergne, J.J., V. Zeuh, D. Bourzat, P. Souvenir Zafindrajoana in preparation. *Cartographie des ressources génétiques caprines du Tchad du Sud-Ouest. II. L'indice auriculaire* (IAt).

Machado, T.M., J.J. Lauvergne, P. Souvenir Zafindrajaona 1992. Le scénario du peuplement brésilien depuis la découverte. *Archivas Zootecnia* (Cordoba) **41**, 55–466.

Martrès, J.-P. & A. Benadjaoud 1988. Visible genetic profiles of the Haute Roya goat. See Lauvergne (1988b), 153–8.

Mason, I.L. 1951. *A world dictionary of breeds, types and varieties of livestock*. Farnham Royal: CAB.

Muzzolini, A. 1993. L'origine des chèvres et des moutons domestiques en Afrique. Reconsidération de la thèse diffusionniste traditionnelle. *Empuries* **2**, 160–71.

Pagot, J.R. 1965. *Chèvres naines de la Côte d'Ivoire*. S.L., S.N., Lib, code. AVM-L10-PO164.

Renieri, C., R. Rubino, F. La Tessa, F. Muscillo, G. Sarrica, G. Zarriello 1988. Visible genetic profiles of the goat in southern Italy. See Lauvergne (1988b), 123–34.

Sadorge, A. & A. Benadjaoud 1988. Visible genetic profiles of the *Rove* goat. See Lauvergne (1988b), 169–74.

Straszewska, S. & L. Ollivier 1993. *Dictionary of animal production terminology*. Amsterdam: Elsevier.

Wilson. R.T. 1991. *Small ruminant production and small ruminants genetic resources in tropical Africa*. FAO Animal Production Health Paper No. 88. Rome: FAO.

Zeuh, V., J.J. Lauvergne, D. Bourzat, P. Souvenir Zafindrajoana 1995. Cartographie des ressources génétiques caprines du Tchad du Sud-Ouest. I. *L'indice de gracilité sous-sternale* (IGS). *Revue d'Elevage et de Médécine vétérinaire des Pays Tropicaux* **48**.

CHAPTER NINETEEN

Indigenous domesticated dogs of southern Africa: an introduction

Sian Hall

1. The archaeological background

Dogs were most probably introduced into southern Africa by Bantu-speaking agriculturalists and/or Khoi pastoralists when they migrated into southern Africa around 2,000 years ago. Dog skeletal remains, and more indirect evidence such as acid-etched bone fragments and gnawed bones from scavenged carcasses, indicate the presence of dogs on many Iron Age, and some Stone Age, sites.

The earliest conclusive evidence for dogs comes from the early Iron Age site known as Diamant in the Northern Province dated to AD 570 (Plug 1996). This site is grouped among those belonging to the Kutama Tradition (Huffman 1989). The Kutama Tradition spread from the tropical forest belt into southern Africa and reached its southernmost distribution in the region of the Kei River at about AD 700 where further movement was hindered by ecological factors (Huffman & Herbert 1996, Prins 1996). It is therefore not surprising that dogs also appear on these more southern Iron Age settlements of KwaZulu–Natal and the eastern Province at about the same time (Prins 1993, Voigt & von den Driesch 1984, Plug & Voigt 1985).

During the later Iron Age, at the beginning of the present millennium, the ancestors of the present Sotho and Nguni-speaking people arrived on the subcontinent through East Africa (Huffman 1989). Although the earliest sites associated with Nguni and Sotho people have not yet yielded any dog skeletal remains it is almost certain that dogs would have accompanied these immigrants via East Africa into the southern subcontinent. A working hypothesis would be that the earlier western stream immigrants introduced a small spitz-type dog from the equatorial forest region – very similar to the Basenji and other extant spitz-types from this region – as well as a more slender pariah/hound, while later eastern stream Bantu-speakers introduced a slender leggy pariah/grazoid hound – typical of the arid and semi-arid region of northeastern and North Africa – into southern Africa. It is too early to test this hypothesis but both slender greyhound-type dogs and what appear to be smaller spitz-types have been identified on Iron Age sites spanning both the first and second millennia AD (Voigt 1983, Plug & Voigt 1985). San rock art paintings of domesticated dog clearly depict both types of dog. Greyhound types are most often depicted in hunting scenes with men, while the smaller spitz-type are often shown accompanying women (for instance, see Vinnicombe 1976, Woodhouse 1990).

Positively identified dog skeletal remains are virtually absent from Khoisan hunter-gatherer and pastoralist sites. Notable exceptions to this are the Cape St Francis skull dated to around AD 800 (Voigt 1983) and skeletal remains found on hunter–gatherer deposits in the eastern Orange Free State (Wadley pers. comm.). But inconclusive evidence for dogs has been found at the later Stone Age pastoralist sites of Die Kelders (*c.* AD 360), Nelson Bay Cave (*c.* AD 450) (Deacon 1986), and Kasteelberg (*c.* AD 700) (Klein & Cruz-Uribe 1989) in the southwestern Cape Province. If these remains are those of *Canis familiaris* their presence would provide a strong argument that Khoi pastoralists introduced dogs into South Africa independently of Iron Age farmers.

1.1. Dogs of later Stone Age peoples

The simple notion that Khoisan populations, including pastoralists, originally obtained all their dogs from Iron Age farmers in South Africa (Gallant 1996) needs more critical assessment.

It is true that Khoisan populations obtained dogs through trade networks with Iron Age farmers. Archaeological evidence certainly suggests that active trade and interaction between Bantu-speaking farmers and Khoisan populations took place during most of the last 2,000 years (Denbow 1984). There is also good historical evidence that San hunter–gatherers actively traded and received dogs from their Bantu-speaking neighbours in the not-so-distant past (Jolly 1994).

The many San paintings of dogs throughout the subcontinent (Vinnicombe 1976, Woodhouse 1990) demonstrate that dogs rapidly became significant in the economy and cosmology of hunter–gatherers. The style and archaeological associations of some of these paintings – such as those at Zimri shelter near Clanwilliam – suggest a minimum age of at least AD 1000 (see Yates et al. 1994), although the majority of these paintings, such as those in the southeastern mountains, are most probably more recent.

1.2. Dogs of the Khoi pastoralists

Definite statements that Khoi pastoralists originally had no dogs until they later obtained dogs from the "Nguni" (Gallant 1996) are clearly incorrect. Nguni-speaking farmers only arrived in southern Africa at the beginning of the second millennium AD (Maggs 1989) at a time when dogs were already clearly in the possession of Khoi pastoralists. In fact, Khoe-speaking hunter–gatherers most probably obtained dogs from early Iron Age farmers in northern Botswana around 2000 BP, or even earlier, along their migratory route. It is also equally possible that herders later traded dogs from early Iron Age communities in other areas of contact, such as in the eastern Cape. As early Iron Age populations had different origins and social organization from later Nguni-speaking farmers (Maggs 1989) it would be chronologically incorrect to refer to dogs kept by early Khoi – even if obtained from early Iron Age farmers – as "Nguni" dogs.

It is also conceivable that some dogs entered southern Africa via the east coast of Africa. Between AD 900 and AD 1400 an Islamic trade network dealing in rare and valuable goods such as gold and ivory from the east coast of Africa extended into Zimbabwe, the northeastern parts of the South African interior (Phillipson 1985), and further south along the coast, possibly as far south as modern-day Durban (Whitelaw 1996). It is known that Islamic traders took dogs along with them on their sailing vessels (Southall 1975). These dogs most probably aided in keeping down the vermin population, in obtaining meat, and may also have functioned as guard dogs. After AD 1500 the Islamic trade network was taken over, and dominated by, the Portuguese. The evidence for both Islamic and Portuguese presence is evidenced by characteristics now to be found in the indigenous dog population that almost certainly originates from Portuguese and Middle eastern gazehounds, which, incidentally, also share a common ancient ancestry with indigenous Iron Age and Stone Age dogs (Epstein 1971). These foreign hounds may have been traded, stolen, or even incorporated into the indigenous population as shipwreck survivors, along the east coast.

2. Historical background

Perhaps the earliest historical reference to indigenous dogs in southern Africa is that recorded by the Portuguese explorer Vasco Da Gama in 1497. Referring to a group of San he encountered at St Helena Bay on the east coast of southern Africa he writes "They have many dogs like those of Portugal, which bark as do these" (Boonzaier et al. 1996:54). Later, mostly between 1700 and 1800, various inland explorers and travellers such as Gordon, Le Vaillant and Lichtenstein also observed dogs among indigenous groups (Forbes 1965). Many such sightings were not well documented leaving us with little historical information on the exact appearance, and other characteristics of these dogs.

One of the exceptions is the traveller Burchell who, in 1811, gave a good description of dogs encountered among Bushmen near the Orange River, describing them as: "these dogs were mostly a small species, entirely white, with erect pointed ears ... they are of a race perhaps peculiar to these tribes" (Burchell 1811).

Anderson is another who gave detailed descriptions of dogs from Namibia and Botswana, and wrote of them in this way:

> The chasse of the leopard, by both colonists and natives, is commonly conducted on foot, the hunters being accompanied by dogs, of which the more there are the better, as they are the greatest safeguards from this fierce and agile beast; and, though the native fox-like breed are awful-looking creatures I have never found any others equal to them for daring or pertinacity. (Andersson 1969)

And although, like the above, some descriptions were far from complimentary referring to these dogs as ugly animals and looking like jackals or foxes, most reports praise their virtues, their exceptional courage and loyalty and their hunting prowess (Herrman 1937, Stuart & Malcolm 1986). This gives us a good idea of their character, while paintings by traveller artists such as Bains and Daniells give some idea of the dogs' morphology which conforms to the archaeological material and rock paintings of two basic morphological types. However, some early ethnographers provided the most detailed information regarding these dogs.

3. Ethnography

Soga (1905), an early ethnographer in the eastern Province, gives some of the best descriptions of indigenous dogs and their specific functions. Soga described four types of dog to be found among Xhosa-speakers, namely the iTwina, iBaku, Inqeqe, and the iNgesi (the English greyhound). The only comparable description of dogs among indigenous groups by an ethnographer is that given by Bryant (1967) concerning the iSiqha kept by Zulu-speakers. Both ethnographers believed that indigenous dogs, even then, were threatened with extinction. In fact, Soga wrote that the iTwina had largely disappeared. Given the availability of ethnographic material the initial objectives of my research was to establish whether these dog types still occur and to determine their present status.

Initial investigation indicated that dogs morphologically similar to those identified on archaeological sites, described by observers, and depicted in historical paintings

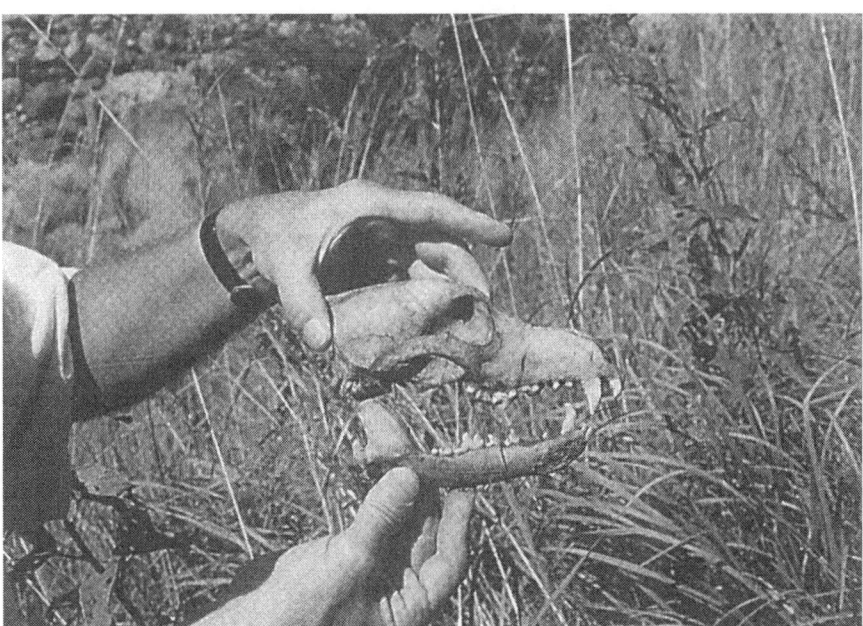

Figure 19.1   Skull of a typical iSiqha Zulu dog shot in Umgeni Valley, KwaZulu–Natal Midlands.

Figure 19.2 Xhosa-speaker's iBaku puppy scavenging on a low-protein maize meal diet close to the homestead.

Figure 19.3 Typical iBaku with sharp, elegant muzzle and long coat.

Figure 19.4 Xhosa-speaker's iTwina.

are still kept by small-scale subsistence farmers throughout southern Africa – especially in areas formally designated as homelands. Surprisingly, a significant population of the supposedly extinct iTwina exists in the eastern Cape, and large numbers of iSiqha (or iSica) are still kept by rural Zulu-speakers.

In fact, indigenous dogs continue to be important to indigenous groups and feature in traditional life cycles. They are particularly associated with hunting and young boys' puberty and initiation ceremonies. In some isolated areas, such as in part of the eastern Cape Province, dogs are implicated in the veneration of ancestors. Dogs may also be implicated in witchcraft accusations (Berglund 1976) while it is well known that traditional healers often include their body parts in traditional medicines or *umuthi*.

4. Characteristics

Preliminary investigation has shown these dogs to be well adapted to the African environment. Not only do they have a high resistance to African tropical diseases and pests, such as tick-borne biliary (canine babesiosis), and have much lower protein requirements than western imports, but they are also uniquely constructed to negotiate and survive the African landscape.

There are a number of different sub-types of dogs to be found among small-scale subsistence farmers in various parts of southern Africa. However, it appears that the spitz-type has been lost, leaving us with only the slender greyhound-pariah. These vary slightly depending upon factors such as environment, social and cultural preferences, history of a group or area, and the specific function for which the dog is intended. For instance, iSiqha is the most common dog to be found among Zulu-speakers, but within the iSiqha there are two slightly different forms. In the higher altitude regions of the KwaZulu–Natal Drakensberg and Midlands the iSiqha tends to be a more robustly-built hound with a rather thick, woolly coat and semi-erect ears. These characteristics enable this dog to cope with, and act as efficient hunters, on the broken, rocky and precipitous mountainous terrain, while also protecting it better from the bitterly cold conditions during the winter months when snowfalls are not unusual. In the lower altitude and warmer river valleys and coastal areas – especially the area to the north of the Tugela – the iSiqha is more slenderly-built with less bone, often taller, with a very short coat and upright ears. All of these features maximize heat loss and facilitate sprinting after game in the sandy thorn savanna.

This observation is in line with pariah dog populations elsewhere in the world (Clutton-Brock 1987), and is indicative of the genetic diversity within a type where selective breeding is not rigorously practiced. In fact, indigenous groups may sometimes classify their dogs according to function, although a certain morphological type of dog is better suited to particular functions.

The iSiqha and related types of indigenous dog are almost certainly the descendents of the Iron Age dog, and appear to be morphologically similar to the remains from archaeological sites. However, among the Xhosa-speakers in the eastern Cape, and Venda and Sotho-speakers of the Northern Province a type of dog exists that demonstrates ancestry from both the typical Iron Age dog and the eastern gazehounds, and northern desert hounds such as the Sloughi, which were most probably introduced during the Islamic and Portuguese trade.

The iTwina, for example, is a medium-sized, very gracile and elegant hound, with a short and sleek coat, elegant long muzzle and upright ears. A similar dog is to be found in Venda. They strongly resemble eastern or Mediterranean gazehounds such as the Pharaoh Hound, the Cirneco dell'etna and Podenco Canario. So strong is the influence of these exotic imported genes that some indigenous iTwina and Venda hounds even exhibit the recessive liver coloration of these Mediterranean hounds. And of such quality are they that they catch the eye of experienced western dog enthusiasts as possible candidates for the show ring.

5. Function

The co-occurence of domestic dog remains with the earliest evidence in southern Africa for house rat (*Rattus rattus*) on the eighth century site of Ndondonwane (Voigt & von den Driesch 1984) graphically illustrates their probable function as vermin controllers soon after being introduced onto the subcontinent. In fact, this function is still adhered to in rural and peri-urban areas and is perhaps their single most important functional attribute. Although the contribution of African dogs in controlling the rat and mice population in the close vicinity of grain storage bins are often acknowledged it is only those dogs which have proven themselves in the hunt of larger game that are praised.

African dogs are exceptional hunters, especially as sprinters over short to medium distances, although they are also able travel long distances with their owners. Dogs in the rural setting have also been trained to herd and guard domestic stock, the so-called "goat dogs" of Botswana are perhaps the best known in this regard although I have also observed herding along the foothills of the Drakensberg.

In fact, domestic dogs were almost certainly one of the factors that contributed towards altering the South African environment during the past 2,000 years. With dogs a new hunting strategy was introduced to the subcontinent. Game previously stalked with a bow and arrow was now more effectively hunted in collaboration with dogs – so much so that successful hunting could also be conducted by utilizing just clubs and spears. This type of hunting was perfected by Zulu-speakers who developed their well known cattle-horn formation of warfare out of these hunting methods. Dogs together with various economic activities must therefore have had a definite impact on the faunal diversity and richness in the catchments of settlements.

Today, uncontrolled hunting with indigenous dogs on protected land has led to the great loss of rare and endangered animals, even when used only with traditional weapons. But the most damage occurs when dogs are used together with firearms – then the results can be devastating. So effective are indigenous dogs as hunters that they have been described as one of the greatest threats to biodiversity in southern Africa (Ledger pers. comm.).

But in spite of these negative associations the indigenous dog should not be blackballed as a pest. They are versatile workers and companions in both the indigenous and western context. They contribute in providing protein to the rural diet that would otherwise be lacking in protein and monotonous. They protect the herds from natural predators such as leopard, wild dog, and jackal, and they act as the alarms and guards of marauders in the homestead. In the western context they have proved themselves in obedience and working competitions, being highly intelligent, quick and alert. They also mature much earlier than many western breeds, and being small but tough they can adapt to both the farm or the town house.

6. Conclusion

We are fortunate enough to still have the descendants of the Iron Age dog. What a pity it would be if, like the spitz-type, they were allowed to join the packs of indigenous dogs that have followed the ever-widening track to extinction. Urbanization and loss of tradition in a rapidly changing Africa does not bode well for the indigenous dog. Many African people now prefer western breeds, regarding them as appropriate status symbols and indicative of their personal success. English greyhounds especially are replacing the indigenous dog and are contaminating indigenous gene pools.

The African dog is a perfectly crafted work. There is nothing exaggerated about it. Exaggerated forms only come with interference from humans. These dogs are sensibly built specifically for the African environment by the African environment, a model that was constructed over thousands of years. In their physical form they embody a people's history, and can be regarded as a cultural heritage. Before these irreplaceable assets are lost I hope to incorporate outstanding specimens into a controlled breeding programme to ensure their survival, and the Iron Age dog may, someday, take its place in the international show-ring.

References

Andersson, C.J. 1969. *Notes of travel in South Africa.* Cape Town: Struik.

Berglund, I. 1976. *Zulu thought-patterns and symbolism.* Cape Town: David-Philip.

Boonzaier, E., C. Malherbe, P. Berens, A. Smith 1996. *The Cape herders: a history of the Khoikhoi of southern Africa.* Cape Town: David Philip.

Bryant, A.T. 1967. *The Zulu people as they were before the white man came.* Pietermaritzburg: Shuter & Shooter.

Burchell, W.J. 1811. *Travels in the interior of Southern Africa.* London: Oxford University Press.

Clutton-Brock, J. 1987. *A natural history of domesticated animals.* Cambridge: Cambridge University Press.

Deacon, J. 1986. Human settlement in South Africa and archaeological evidence for alien plants and animals. In *The ecology and management of biological invasions in southern Africa. Proceedings of the national synthesis on the ecology of biological invasions,* I.A.W. Macdonald, F.J. Kruger, A.A. Ferrar (eds). Cape Town: Oxford University Press.

Denbow, J.R. 1984. Prehistoric herders and foragers of the Kalahari: the evidence for 1,500 years of interaction. In *Past and present in hunter-gatherer studies,* C. Schrire (ed.), 175–93. Orlando: Academic Press.

Epstein, H. 1971. *The origin of the domestic animals of Africa* [two volumes]. New York: Africana Publishing.

Forbes, V.S. 1965. *Pioneer travellers of South Africa: a geographical commentary upon routes, records, observations, and opinions of travellers at the Cape.* Cape Town: A.A. Balkema.

Gallant, J. (ed.) 1996. Exploring the pre-history of the Rhodesian Ridgeback. *Dogs in Africa* [special issue], **November,** 14–16.

Herrman, L. 1937. *Travels and adventures in eastern Africa by Nathaniel Isaacs.* Cape Town: The Van Riebeeck Society.

Huffman, T.N. 1989. Ceramics, settlements and late Iron Age migrations. *African Archaeological Review* **7,** 155–82.

Huffman, T.N. & R.K. Herbert 1996. New perspectives on eastern Bantu. *Azania* [special volume] **XXIX–XXX,** 27–36.

Jolly, P. 1994. *Strangers to brothers: interaction between southeastern San and southern Nguni/Sotho communities.* Unpublished MA thesis, University of Cape Town.

Klein, R.G. & K. Cruz-Uribe 1989. Faunal evidence for prehistoric herder-forager activities at Kasteelberg, western Cape Province, South Africa. *The South African Archaeological Bulletin* **XLIV**(150), 82–97.

Maggs, T. 1989. The Iron Age farming communities. In *Natal and Zululand: from earliest times to 1910 a new history,* A. Duning & B. Guest (eds). Pietermaritzburg: University of Natal Press.

Phillipson, D.W. 1985. *African archaeology.* Cambridge: Cambridge University Press.

Plug, I. 1996. Domestic animals during the early Iron Age in South Africa. In Exploring the pre-history of the Rhodesian Ridgeback. J. Gallant (ed.). *Dogs in Africa* [special issue], **November,** 14–16.

Plug & Voigt, E.A. 1985. Archaeozoological studies of Iron Age communities in southern Africa. *Advances in World Archaeology* **4,** 189–238.

Prins, F.E. 1993. *Aspects of Iron Age ecology in Transkei.* Unpublished MA thesis, University of Stellenbosch.

Prins, F.E. 1996. Climate, vegetation and early agriculturist communities in Transkei and KwaZulu–Natal. *Azania* [special volume] **XXIX–XXX**, 179–86.

Soga, J.H. 1905. *The Ama-Xosa: life and customs.* Lovedale Press: London.

Southall, A. 1975. The problem of Malagasy origins. In *East Africa and the Orient: cultural syntheses in pre-colonial times*, H.N. Chittick & R.I. Rotberg (eds), 192–251. New York: Africana Publishing.

Stuart, J. & D.M. Malcolm 1986. *The diary of Henry Francis Fynn.* Pietermaritzburg: Shuter & Shooter.

Vinnicombe, P. 1976. *People of the Eland.* Pietermaritzburg: University of Natal Press.

Voigt, E.A. 1983. *Mapungubwe: an archaeozoological interpretation of an Iron Age community.* Pretoria: Transvaal Museum.

Voigt, E.A. & A. von den Dreisch 1984. A preliminary report on the faunal assemblage from Ndondondwane, Natal. *Annals of the Natal Museum* **26**(1), 95–104.

Whitelaw, G. 1996. Towards an early Iron Age worldview: some ideas from KwaZulu–Natal. *Azania* [special volume] **XXIX–XXX**, 37–50.

Woodhouse, H.C. 1990. Dogs in the rock art of southern Africa. *South African Journal of Ethnology* **13**(3), 117–24.

Yates, R., A. Manhire, J. Parkington 1994. Rock painting and history in the Southwestern Cape. In *Contested images: diversity in Southern African rock art research*, T. Dowson & D. Lewis-Williams (eds), 29–60. Johannesburg: Witwatersrand University Press.

PART FOUR
Linguistics and Ethnography

CHAPTER TWENTY

African minor livestock species

Roger M. Blench

1. Introduction

1.1. The study of "minor" domesticates

Historical studies of the domestication and diffusion of livestock, such as Boettger (1958), Zeuner (1963) or the contributors to Mason (1984a) often give Africa somewhat short shrift. This is especially the case for so-called "minor" species; i.e. any species other than cattle, sheep and goats. The absence of iconographic or literary records and the patchy coverage of archaeology has often led researchers to conclude that little can be said. However, methods do exist of filling these historical lacunae, in particular the use of linguistics and comparative ethnography. The recent development-oriented literature on ruminant breeds summarized in Blench (1993) provides synchronic distributions of major species and races.

Africa represents an elaborate mosaic of production systems and livestock species other than the principal ruminants (Blench 1997). Although cited as "minor" species, animals such as the donkey or camel can play a major role in the economic life of ordinary rural householders. They are, however, of no significant interest to major donor agencies and research is often confined to enthusiastic individuals. As a result, there are often startling lacunae in our knowledge of, say, the history of the domestic pig in Africa. The only author to consider some of these species in detail was Lagercrantz (1950) who reviews the literature on cats, pigeons, ducks, geese and turkeys. Ruminants, donkeys, pigs, chickens and bees are described in other chapters in this volume (Chs 21, 22, 27). The history of the horse in Africa has been discussed elsewhere extensively (Law 1980, Seignobos 1987, Blench 1993) and will not be further treated here. This chapter[1] synthesizes current knowledge of the history of the residual species – "minor" domesticates of Africa.

The use of productivity data from livestock kept under village conditions is beginning to be used to interpret archaeozoological material (see Thorp 1995, Hall, Ch. 19 in this volume). For this reason, the data that exists has been summarized for the species described in this chapter. Minor species have often been given short shrift in this respect and the information summarized here should therefore be treated as tentative.

1.2. The domestication process

Sheep, goats, chickens and pigs arrived in Africa fully domesticated and although local races have developed there can be no further interaction with their wild relatives. However, for indigenous African fauna, domestication remains a dynamic process, both in terms of interaction with wild populations and continuing experimentation with new species (Blench 1997). The donkey was almost certainly domesticated in Africa and there is evidence for some introgression of genetic material from wild ass populations in historic times (see Blench, this volume, Ch. 21). With the probable elimination of the last Somali wild asses this process has come to an end. On the other hand, the guinea-fowl is part of the indigenous avifauna of Africa which has been only partly domesticated. In some regions, guinea-fowl are kept in the compound, grow fat and have little tendency to fly away, but their wild ancestry is reflected in their habit of laying eggs scattered in the bush, rather than in a single place.

The process of taming wild animals, especially birds, is already well documented for Ancient Egypt and there is substantial evidence for it in the ethnographic literature. Iconographic records make plain that numerous species were either wild-caught and tamed or actually domesticated. Some of these, such as the crane, are no longer known as domestic animals. In some cases, cachet attaches to the taming of wild animals so that taming does not act as a prologue to the domestication process. The Romans in North Africa are shown as using domesticated cheetahs for hunting while hyena-taming is found across Sahelian Muslim Africa, usually as a type of circus act. However, experimentation continues in sub-Saharan Africa, and there are modern records of unusual domesticates, wild-caught animals "finished" in captivity. Two species of wild fauna that are in transition between husbandry and domestication are the giant rat (*Cricetomys*) and the African Land Snail.

An important aspect of the relationship between man and livestock relevant to the minor species is the keeping of pets. Species regarded as edible in some parts of Africa are kept as pets or working animals elsewhere. In many places these two categories would be regarded as distinct, but even affectionate man–animal relations do not stop pets being regarded as protein. Dogs are commonly used for hunting or to guard property, and are sometimes eaten. Even cats, which are usually semi-feral and whose existence on the margin of villages is tolerated because of their vermin destroying habits, are eaten in some communities.

2. Mammals

The history of individual species can be tracked broadly through archaeological and historical sources although more detailed regional information comes from local traditions and lexicographic data.

2.1. Camel

The one-humped dromedary is originally an Asian domesticate (Epstein 1971, Wilson 1984) although wild camels were known in North Africa in the Pleistocene.

Camels are present during the Quaternary in the Maghreb but are usually thought to have subsequently died out and been re-introduced in the Graeco–Roman period. This view has been supported by a number of authors, notably Bulliet (1990). Shaw (1979) has provided a history of this debate and vigorously canvassed a contrary view, that the camel was present continuously in the Maghreb, but that its presence was at a low-level and therefore less archaeologically visible. He argues that the camel survived through from the Pleistocene and is therefore indigenous to North Africa. In the absence of archaeological evidence from more recent periods, most scholars do not accept this.

Occasional representations suggest that the camel was brought to Egypt as an exotic at an early period (Brewer et al. 1994:104). The most striking evidence of this is the camel vertebrae discovered in a 1st Dynasty cemetery at Helwan (ibid.). Finds of camel-hair and ceramic models of camels confirm that at least some camels were kept in Egypt and it is now thought that the introduction of the camel in large numbers may be associated with the Assyrians (c. 500 BC). It may be that only with the opening up of long-distance trade routes through desertic regions that the camel came into its own under the Ptolemies.

Whatever its antiquity in North Africa, the camel appears to be represented significantly in the Maghreb only in the first few centuries BC. In North Africa it appears, above all, as a plough-animal and to carry loads (Morales Muniz et al. 1995). In the case of sub-Saharan West Africa, the camel is almost certainly more recent. Bones dating to between AD 250 and AD 400 have been found in the Middle Senegal Valley and bones and camel dung have been identified at Qasr Ibrim, in Egypt in the early first millennium BC (MacDonald & MacDonald, Ch. 8 in this volume). Muzzolini (Ch. 6 in this volume) refers to the extensive rock art evidence for camels and some striking images from Chad have recently been published (Boccazzi et al. 1995).

Unusually, the archaeological materials cited above predate the historical record. The first reference to camels in West Africa is by Al-Yaʿqubi, writing in AD 889–890 who mentions camel nomads, the Anbiya, living south of Sijilmāsa (Levtzion & Hopkins 1981:22). After this date, there are numerous references to camel in Arab writers, mentioning both its use for packing and for irrigation (in Zawīla). Camels were also sacrificed in rituals to establish the location of gemstones according to Al-Bakrī (Levtzion & Hopkins 1981:86). Further east, Al-Idrīsī mentions the Zaghawa people as eating sun-dried camel meat (ibid. 114). The Kano chronicle, referring to present-day northern Nigeria, mentions that the first ruler to own camels was Abdullahi Burja, in about AD 1440.

Linguistic evidence for the camel in West Africa is reviewed in Blench (1995). In west–central Africa, there are two sources of words for camel, loans from Berber and from Fulfulde. Versions of Berber *lɣm are common through from northern Nigeria to Chad, whereas in Adamawa, Fulfulde ngelooba is usually borrowed. Skinner (1977:179ff.) discusses the history of the *lɣm root. He notes that it is probably a borrowing from the Arabic *gml root (also borrowed into English) and that the Fulfulde term is probably another version of the same root, perhaps borrowed directly from Arabic al-gml.

More problematic is the antiquity of the camel in the Horn of Africa. Archaeological finds of camel materials from this area are summarized in Esser & Esser (1982) and Banti (1993). These authors have argued for its early domestication in the Horn of Africa, from wild camels in the Arabian peninsula. There are several studies of the linguistic evidence or terminology in the Horn of Africa. Most detailed is Bechhaus-Gerst (1991/2) who has explored the vocabulary associated with the camel in Beja. She notes that Arabic sources for Beja camel terminology are few and probably late, concerning only details of saddling leatherwork. Heine (1981) points to the regular reconstruction of terms connected with camel production for example the word for "camel-bell" in proto-Sam, i.e. Somali-Boni-Rendille. It could therefore have spread across from Arabia in "pre-Arabic" times and thence up the Red Sea coast to Egypt and North Africa. The "riding camel" *r-k-b*, shows up as a loan *into* the Sam languages. Banti (1993) has reviewed the considerable linguistic evidence suggesting a deep embedding of camel terminology in the cultures of this region.

2.2. Dog

The ancestry of the domestic dog remains uncertain and a number of canids may be implicated in present-day types (Clutton-Brock 1984). European and New World dog remains go back to 10,000 BP. The dog is not native to Africa and was introduced at an unknown period in the past. According to Brewer et al. (1994:114ff.) dogs were known in pre-Dynastic Egypt (Merimde Beni Salame at 6800 BP). They are represented in the rock art of the eastern and western Deserts and so could have been brought across the Sahara in prehistoric times. Earlier this century it was argued that the jackal had played a part in the ancestry of the African pariah dog, a theory that is generally discounted today (Epstein, 1971i).

Three basic types of dog are recorded in Ancient Egypt, the pariah dog, the greyhound and the mastiff. The greyhound was divided into two types, the *tesem* and the *saluki*, the *tesem* being the lean, tall, prick-eared dog represented in many wall-paintings. The *tesem* seems originally to have come from further south, from Nubia and Punt, although where they evolved remains uncertain. Mastiffs were brought into Egypt from Mesopotamia during the Middle Kingdom period but seem not to have persisted (Hauck 1941, Epstein 1971i:3–184 and Brewer et al. 1994:117).

The pariah and the greyhound appear to have spread out from North Africa over much of the continent. The social and cultural importance of dogs in African culture, as well as the antiquity of their domestication in the Near East suggests that they should be at least as old as other domestic stock in Africa. Archaeological evidence for the antiquity of dogs in sub-Saharan Africa is somewhat sparse (see MacDonald & MacDonald Ch. 8, and Smith Ch. 11 in this volume). Despite this, all other types of circumstantial evidence suggest they are of considerable antiquity. The pariah is the common dog found all over Africa, whose distribution is shown in Figure 20.1.

The greyhound seems to have spread widely in Africa, although in many places it crossed with the pariah, thereby diluting its distinctive body shape. The approximate distribution of greyhounds and mastiffs is shown in Figure 20.2.

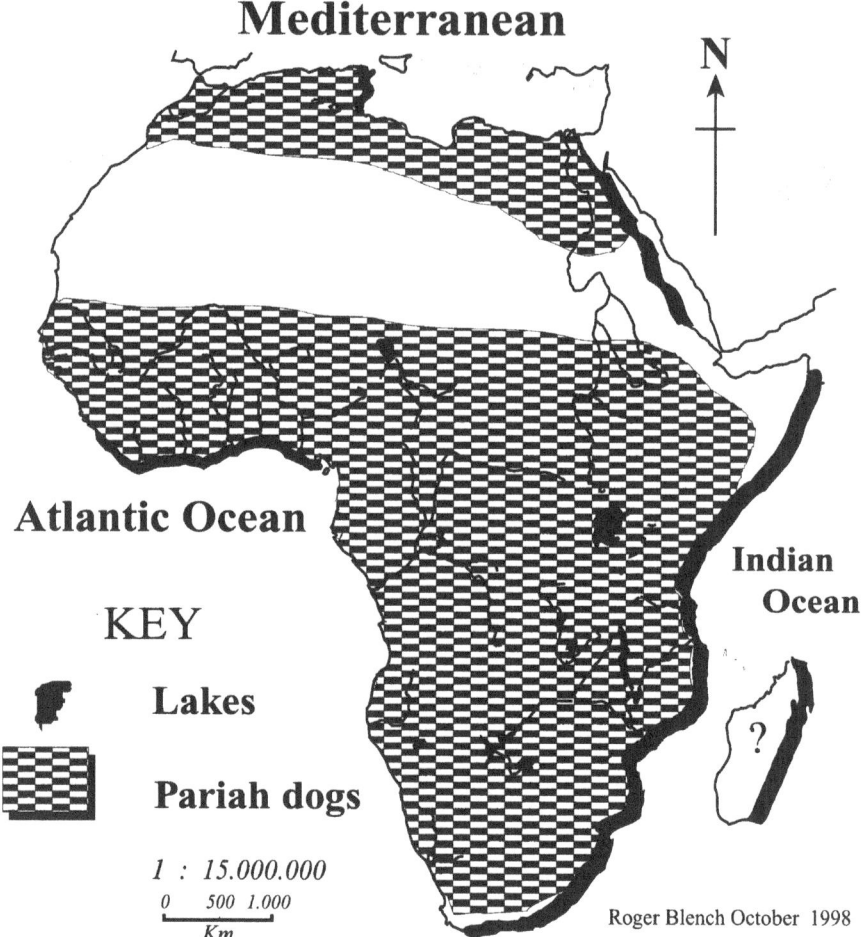

Figure 20.1   Distribution of pariah dogs in Africa.

Exotic breeds were brought in during the colonial period and have in many places crossed with local dog races to produce distinctive breeds such as the Rhodesian ridgeback. The Basenji, a non-barking dog in west–central Africa is not usually held to constitute a distinct race (Dollman 1937).

2.2.1. Social role of dogs   Dogs are kept for a variety of functions, notably hunting, guarding, to fulfil social obligations and as food. Frank (1965) has exhaustively reviewed the ethnographic literature on domestic dogs in Africa. The use of dogs to pay brideprice is widely attested throughout Africa. Young men who are getting married often buy dogs from other communities as part of the payment to their prospective father-in-law. Dogs are used to herd other domestic animals, especially sheep, in North Africa and as far south as the rangelands of eastern

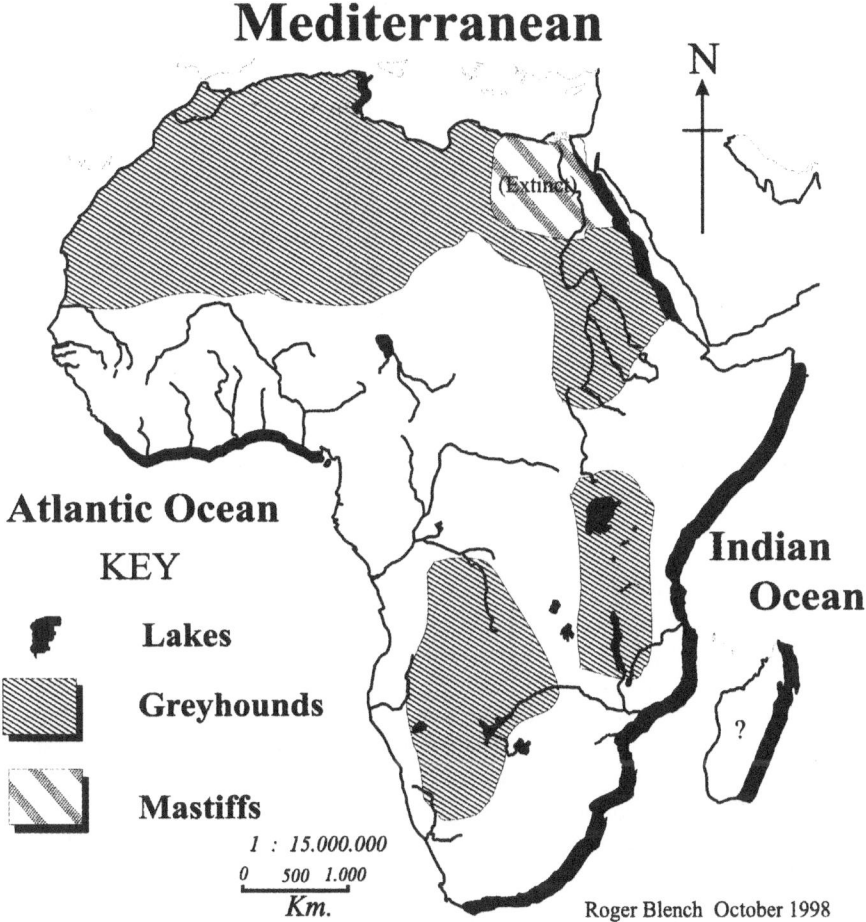

Figure 20.2 Distribution of greyhounds and mastiffs in Africa.

Sudan. The culture of "sheep-dogs" seems never to have spread further into sub-Saharan Africa.

2.2.2. Cynophagy In some parts of Africa, eating dog meat is regarded with horror, whereas elsewhere the flesh is regarded as a delicacy. Because this is a distasteful subject for many European authors and because both Livestock Departments in African countries and international agencies such as FAO do not usually regard dogs as livestock, the prevalence and nutritional significance of eating dogs has probably been under-rated. A study of the trade in dogmeat in Nigeria found a flourishing commerce from north to south as well as specialized butcheries passing outside the usual market system (RIM 1992ii). A similar situation is documented for northern Cameroon by Thys & Nyssens (1983). Such descriptions are rare and the trade is almost certainly more widespread than this fragmentary documentation suggests.

Discussions of dog eating are uncommon in the literature, but a lively debate about cynophagy in the Sahara was conducted in the 1950s (Canard 1953, Bureau 1954, Chalumeau 1954). Other reviews of material on dog eating in Africa can be found in Mauny (1954) and You (1955). Where dogs are eaten, they are regarded as a prestigious food, often with unspecified magical properties. Among the minority peoples of Nigeria, for example, where the community is called for collective farmwork by an individual it is common to reward those who come with roast dog. By contrast, in Muslim areas eating dog meat is regarded as wholly unacceptable and where dogs are slaughtered in northern towns, it is usually in secrecy.

2.2.3. Linguistic evidence   Using linguistic evidence to uncover the diffusion of the domestic dog has a specific problem; the tendency for names for dog to be phonaesthetic. Barth (1862ii) observed long ago that the widespread similarities in names for dog in Africa argued for a single broad introduction into Africa. Sasse (1993) has shown that terms for "dog" show remarkable similarities all over the world probably reflecting the early and rapid diffusion of the domestic dog.

2.3. Cat

Domestic cats are kept in all parts of Africa, and are used to hunt vermin and for medicinal and magical purposes. In some places, like dogs, they have become semi-feral. Domestic cats are usually considered to have developed from *Felis sylvestris libyca*, still found wild through much of arid North Africa (Robinson 1984). Cat remains are found in Jericho as early as 7000 BC and in Egypt at 4000 BC but there seems to be no way to establish whether these are domestic or simply tamed wild cats (Brewer et al. 1994:108). The Egyptians are likely to have brought the cat into domestication gradually, with full domestication by 1000 BC. There is no evidence on the date or means whereby it spread south of the Sahara, although today it is found throughout the continent. Cats can survive in a feral state and there are reports of such cats being tamed again, especially by children (see references in Lagercrantz 1950:59–60).

There are virtually no archaeological records of the domestic cat in sub-Saharan Africa, apart from a find at Jenné-Jeno and even in this case it is unclear that it could be adequately distinguished from wild or feral types (MacDonald 1995b). Cats are well embedded in the culture of Arab North Africa and it is assumed that they spread as commensals both across the Sahara and down the Nile into sub-Saharan Africa after 1000 BC. Although cats are usually considered as forbidden for food, there are widespread reports of their consumption for magical purposes.

The ethnographic literature on the domestic cat has been reviewed by Lagercrantz (1950:54–66). Cats are nowhere common, yet they are found throughout the continent as Lagercrantz (ibid. Map. 10) shows. Cats seem to thrive and there have almost certainly been multiple importations from different sources. There appears to be at least one very ancient stratum of cat populations, since the cat, like the pig, is common among the Omotic and Nilo–Saharan populations of the Ethiopia–Sudan borderland who have until recently been rarely exposed to trade. Strikingly, among the Dogon in Mali, the cat is considered as belonging to the inhabitants of the aboriginal people of their country. At the same time, European traders introduced cats all around the coast and Muslim traders brought cats across the Sahara, while the Indian Ocean trade brought Persian cats to all the ports of East Africa.

There have been some limited linguistic studies, notably Skinner (1977) and Blench (1995). Skinner (1977:181) argued that the Cushitic and Chadic lexemes were cognate and thus cats were of great antiquity among Afroasiatic speakers. This seems difficult to believe as there is no evidence for such early transmission of the domestic cat through this region. Blench (1995) shows that words for cats are highly diverse and heavily influenced by phonaesthetic factors (in other words the names are influenced by imitations of cat sounds): this makes the usual methods of historical linguistics difficult to apply.

2.4. Other mammals

2.4.1. Rabbit   The rearing of rabbits was introduced into Africa in the late nineteenth century through Christian missions and the colonial agricultural service. Rabbit production is now well established in many parts of Africa, and breeding stock can be bought in most major urban markets. Little has been written about traditional rabbit production in Africa, with the exception of Matthewman (1977).

If rabbits give birth while still pubescent, at 3–5 months, the progeny are less likely to survive owing to an inadequate milk supply from the dam. The kindling interval can be reduced to 4–5 weeks if the feeding is maintained at a high level. Some producers say that a younger doe will give birth to more kittens and have more litters if she is mated before full sexual maturity. A principle of rabbit production is separating the buck from the pregnant doe. The buck should be kept apart until the kittens are weaned, otherwise he is likely to eat them. If mating is too frequent, the breeding life and progeny output of the doe will decline.

Most rabbits kept in compounds are fed maize or cereal-bran with beer residues and greenery. The greenery is either grasses, in the wet season, or tree browse, during the dry season. These feeds are then supplemented with an irregular supply of kitchen scraps and tuber peelings. Local rabbits in Africa seem to have adapted rapidly to a low quality diet and minimal attention. Despite disease problems and faults in husbandry techniques, rabbit-rearing is an enterprise that is spreading rapidly.

2.4.2. Giant rat   The giant rat (*Cricetomys gambianus* Waterhouse) is part of the native wild fauna of West Africa. It has been in domestication since the early 1960s, and perhaps previous to that. Several writers have discussed the biology feeding and carcass composition of the giant rat (Ajayi 1974, Tewe & Ajayi 1979, Ajayi & Tewe 1980) and there is also a short discussion of the extension of giant rats in village-level production in Matthewman (1977). The step to domestication was originally based on the capture of wild litters which were then fattened and slaughtered. However, more recently, many producers have taken to breeding rats in captivity. The productivity of the giant rats has been studied by Tewe et al. (1984) and they give the following values for various reproductive parameters (Table 20.1).

Table 20.1   Reproductive parameters of the giant rat.

Adult weight (g)	692–1220
Female breeding age (wks)	20–24
Oestrous cycle (days)	4–5
Gestation period (days)	28
Weaning age (days)	26
Litter size	1–5
Birth weight (g)	16–28
Weaning weight (g)	66–186
Age at maturity (wks)	24
Killing-out percentage	51.5

Source: Tewe et al. (1984).

These figures suggest that the giant rat has many advantages as a domesticate, in particular a high turnover of breeding stock and a killing-out percentage comparable with the rabbit.

2.4.3. Guinea-pig   The guinea-pig, *Cavia porcellus*, is rarely included in any discussion of African livestock and almost nothing has been published about their introduction, spread and traditional management. Matthewman (1977), who studied backyard stock in the Ibadan area of Nigeria, found that guinea-pigs were relatively common, and in one village some 10 per cent of households kept them. Guinea-pigs are known in most parts of West Africa but are only sporadically kept, with no consistent pattern of distribution.

Guinea-pigs were originally domesticated in the Andean region of South America and they were probably introduced to Africa by missionaries and colonial agricultural officers, as they seem to have been known in a few areas for some time. A Nupe name for guinea-pig, *etsu nasara*, "the white man's rat" was recorded in Nigeria in 1914 (Banfield 1914). The Hausa name is *beran Masar*, "rat of Egypt" although it is unlikely that guinea-pigs were traded across the desert.

Guinea-pigs first give birth at 6–12 wk of age. After the first birth, the female will give birth every month thereafter. There is no controlled breeding and the male and female are permanently together even during birth and suckling. The size of litters is between two and four, and mortalities are rare among the new-born. Variations in productivity do not seem to be the result of breed differences but to differing husbandry and hygiene conditions. Animals reared in a confined space seem to thrive least well.

2.4.4. Water-Buffalo   The domestic water-buffalo is kept in the Nile Valley for dairying and draught power. It is especially adapted to working in flooded fields, but is also used for threshing and water-raising. It is divided into two local races, the Beheri of the Delta and the Saidi of upper Egypt (FAO 1977:236). Its introduction to Egypt seems to have been from the Near East, since it is represented in Syria and Palestine in the late pre-Christian era. It is usually assumed to have been brought by Islam, although there is no direct evidence of this. Dates for its introduction to Africa vary between the sixth and ninth centuries AD. Although occasional attempts to export it to other North and East African countries have been made it has nowhere persisted (Epstein 1971i). There is a history of attempts to introduce buffaloes into the countries of eastern and southern Africa (FAO 1977:239ff.) but all of these have failed. In Mozambique, however, buffalo were introduced in 1966 and some 1,000 animals are still kept on private farms (Goe 1996).

2.4.5. Elephant Marshall (Ch. 10 in this volume) refers to evidence for domestication of the African elephant, *Loxodonta africana*, at Meroe some 2,000 years ago. At the site of Musarwarat es Sofra (first century AD), northeast of Khartoum, pens have been identified for domesticated elephant (Shinnie 1967:95). The African elephant is generally thought to be much more difficult to tame than the Asian elephant and this represents a considerable achievement but one which did not persist.

2.4.6. Experimental domestications and commensals
Grasscutter – The grass-cutter or cane rat, *Thryonomys swinderianus*, has been the object of various domestication initiatives, particularly in Nigeria and Benin Republic, some of which are described in Matthewman (1977) and Ajayi & Tewe (1980). Experimental work at the University of Ibadan in Nigeria suggests that domestic cane-rat colonies should be economically viable, but the practice has yet to spread to farmers in the same way as domesticated giant rats.

Antelope – Attempts to farm individual antelope species have a long history in eastern and southern Africa, going back to the colonial period. Although individual farm owners have successfully kept species such as eland, it has not so far proven possible to breed them economically (Field 1984).

Hedgehog – In eastern Bauchi State, northeastern Nigeria, wild hedgehogs, *Atelerix albiventris*, were recorded being caught and fattened to eat (RIM 1992ii).

Rock hyrax – In Kaduna State northern Nigeria, the practice of catching wild rock hyraxes and fattening them for eating was recorded (RIM 1992ii).

European rats and mice – The house rat is not strictly a domesticate but a commensal whose spread appears to follow that of humans. The spread of such a rat has been well documented in Oceania, where the Polynesian rat appears to have travelled from island to island in the canoes of the Austronesians as they expanded across the Pacific. Armitage (1994) has summarized recent findings concerning the early spread of the black rat in Eurasia. Surprisingly, the European Black rat, *Rattus rattus*, is reported by Plug & Dippenaar (1979) and Plug (1996) from the ninth century site at Ndondowane, Natal in South Africa. Armitage (1994:235) attributes this to Arab traders, but this is chronologically problematic as Natal is remote from their routes and it does not allow sufficient time for the rat to spread overland from the East coast ports. More likely, therefore, is the possibility that the rat came with the Graeco–Roman traders whose pottery has been recently uncovered in Tanzania (Juma 1996).

In West Africa, the Black rat, *Rattus rattus*, the Norway Rat, *Rattus norvegicus* and the House Mouse, *Mus musculus* are all present today (Rosevear 1969, Happold

1987:127–8). The last two are fairly rare, being only recorded as commensals in large coastal towns. However, the Black rat has been noted as gradually spreading northward in this century and is now well-adapted to inselberg habitats even in arid areas. Black rats have been studied in Ghana (Ewer 1971) and Khartoum, Sudan (Happold 1967).

The black rat is a major pest on stored grain and indeed can only survive in conjunction with human settlement. In non-Muslim areas they are frequently eaten and towns such as Benin have whole market sections devoted to different types of raw and prepared rat. In Kano city, northern Nigeria, small amounts of food are left out for domestic rats apparently to deter them from infiltrating the family food stores.

3. Birds

3.1. General

The domestic poultry found in Africa are chickens, pigeons, Muscovy ducks, guinea-fowl and turkeys. Attempts to introduce Rouen ducks and geese have generally been unsuccessful. Chickens are by the far the most important poultry species, numerically and in terms of social and economic significance. Williamson (Ch. 23 in this volume) and MacDonald (1992, 1995a) examine the history of the chicken in Africa in greater detail and it is not further discussed here.

There is relatively little information on the productivity of local breeds and virtually no studies of poultry kept under village conditions. The most useful comparative material was compiled in central Mali by Kuit et al. (1986) and Wilson et al. (1987) and some comparative data can also be drawn from the studies of Wilson (1979) in Sudan.

3.2. Guinea-fowl

The crested or helmet guinea-fowl, *Numida meleagris galeata*, Pallas, is part of the native fauna of West Africa. It is distributed from Senegambia to Cameroon and is also found in a part of western Zaïre. It was presumably domesticated long ago, although the larger domestic races closely resemble their wild counterparts. There are several wild species and genera of guinea-fowl in West and East Africa, notably *N. meleagris meleagris* in Sudan and Ethiopia, but apparently only *N. meleagris galeata* has been domesticated (see Donkin 1991, Map 1.). Wild guinea-fowl are still regularly trapped as a source of food and their eggs are raided in the bush. Mongin & Plouzeau (1984) present an overview of recent scholarship on the guinea-fowl worldwide, while Ayeni (1983) summarizes existing information for West Africa. Donkin (1991) is an "ethnogeographical" study of the guinea-fowl that synthesizes a great deal of scattered material, especially on the iconography of the guinea-fowl in the Mediterranean.

One of the more puzzling aspects of the history of the guinea-fowl is a residual population found in the foothills of the Middle Atlas in Morocco (Hartert 1919). This is claimed to be a new subspecies, *N. meleagris sabyi*, which is said not to be feral nor simply an isolated population of *galeata*. However, it is known as *djaj el hend* ("hen of the Indies") or *habeshr* ("of Ethiopia" standing for Africa in general) to the local Berber populations. This makes it seem more than probable that it does represent escaped domestic birds, although presumably this question could be resolved through DNA analysis.

There is no evidence that *N. meleagris* is found wild in Egypt and it may be that what few ornithological records there are represent feral escapes (Houlihan & Goodman 1986:82–3). Representations of guinea-fowl occur from the pre-Dynastic period onwards, but with a somewhat shaky command of detail that suggests the artists were not very familiar with this bird. During the Ptolemaic period there is a record of them being carried in cages in a pageant, suggesting their continued rarity. There are no certain finds of *domestic* guinea-fowl in sub-Saharan sites, although remains attributable to either wild or domestic guinea-fowl seem common enough in West Africa (cf. MacDonald & Macdonald Ch. 8 in this volume). The problem would appear to be one of osteological differentiation. The fact that guinea-fowl seem to have been imported into Europe from the fifth century BC onwards suggests a fairly early domestication in Africa (Mongin & Plouzeau 1984). Poultry are poorly represented in early African historical sources, but Ibn Sa'īd mentions guinea-fowl in Jaja, i.e. medieval Borno (Lewicki 1974:91). The more abundant recent historical and ethnographic references are collated in Donkin (1991).

Guinea-fowl rearing is mostly associated with semi-arid West Africa, although various subspecies of guinea-fowl are found in all ecozones. Five colour-types are distinguished, based on plumage, although these interbreed freely (Okaeme 1982:36). White guinea-fowl are common in parts of West Africa, an indicator of long-established domestication, since this would decrease their fitness to blend in with vegetation in the wild. In the subhumid zone, domesticated guinea-fowl are rarer, and it is more common to buy "wild" eggs from hunters or Fulɓe women and then put them under chickens to hatch. The keets remain in the village with the hen until they are mature. They are then marketed or slaughtered before they escape back to the wild. Guinea-fowl are both territorial and monogamous and a high number of males are necessary to prevent the females laying infertile eggs. Domestic guinea-fowl are not usually provided with housing and roost in nearby trees at night. Donkin (1991:66) cites ethnographic sources suggesting that guinea-fowl were tamed but not domesticated in other parts of Africa as prestige possessions.

Measuring the productivity of village guinea-fowl is not easy because they tend to scatter their eggs in the bush and raise the keets outside the compound. Kuit et al. (1986) in Mali give some reproductive parameters for traditional management in a village environment in West Africa (Table 20.2). Ayeni & Ajayi (1983) explore the significance of guinea-fowl as an animal protein supplement and Ayorinde & Ayeni (1986) compare the seasonal performance of local and exotic guinea-fowl under station conditions in northwestern Nigeria.

In the wild, females lay some 15–20 eggs per breeding season, but in captivity they may lay as many as 50–100. Ayeni & Ajayi (1983:164) recorded guinea-fowl in captivity begin to lay eggs at between 28 wks and 32 wks and the eggs take 27 days to hatch, with low egg fertilities of around 30 per cent.

Table 20.2   Comparative productivity parameters for guinea-fowl.

Parameter	Kuit et al. (1986)
Mean egg weight (g)	37.3
Mean eggs per clutch	9.6
Hatchability %	44
Mortality to one month (%)	49

Domestic guinea-fowl are rarely fed but generally allowed to find their own food. Their diet is a mixture of seeds and other vegetable matter and insects. Ayeni (1983:143) gives a breakdown of the typical diet of wild guinea-fowl in the Lake Kainji region of northwestern Nigeria. Guinea-fowl respond to supplementation with grain in the dry season. Grain can also be scattered simply to build up the attachment of the birds to a particular compound.

3.3. Pigeon

Charles Darwin first identified the rock-pigeon, *Columba livia*, as the ancestor of the domestic pigeon, and his insight has been affirmed by recent research. Its domestication is discussed by Zeuner (1963) and Hawes (1984) who argue that pigeon-keeping may have begun in Persia and spread to Egypt. Domestic pigeons have been known for some 3,000 years, and the practice of attracting semi-feral pigeons to stay near the household is probably equally ancient. The ethnographic literature on pigeons in Africa has been reviewed by Lagercrantz (1950:66–74).

Representations of pigeons in Ancient Egypt are often not clearly distinguishable from the turtle-dove, *Streptopelia turtur*, which is known to have been tamed and sometimes force-fed (Houlihan & Goodman 1986:99ff.). The antiquity of pigeon-keeping in west–central Africa is unclear, as the grey pigeon is part of the indigenous fauna of the region. Al-ʿUmarī reports the peoples of the "Sudan" kept pigeons (or doves?) in the fourteenth century (Levtzion & Hopkins 1981:267).

Lagercrantz (1950:Map 11) represents the distribution of managed domestic pigeons in Africa and Figure 20.3 is an adapted and updated version of this.

The patchiness of distribution is hard to interpret. Although there is some link with the trans-Saharan trade, pigeon-keeping in west–central Africa forms a very uneven pattern. There is some evidence that pigeons were also introduced in coastal areas by Europeans, which may explain their presence in Ghana and at the mouth of the River Congo. The distribution of managed pigeons across the Congo basin may in part be connected with the Arab trade routes, but this makes it difficult to understand why the practice of pigeon-keeping is virtually unknown in the Great Lakes region which experienced extensive Arab contact. Peoples such as the Ila and Mbala people of Zambia, who seem not to have been in contact with the coast trade, keep pigeons in cotes atop tall platforms (Smith & Dale 1920i:134).

This usual system for keeping pigeons in Africa is semi-feral, and only occasionally are more elaborate production systems used. Even the concept of ownership is somewhat fluid, since pigeons can be "lured" from one cote to another by

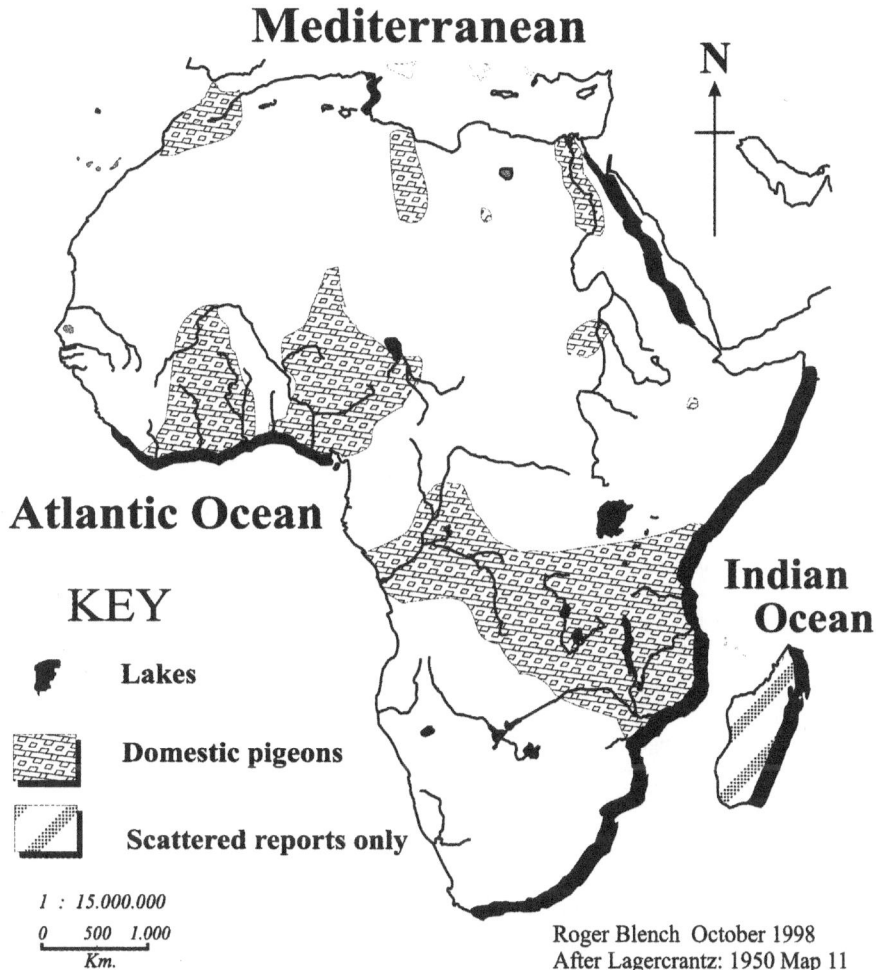

Figure 20.3   Distribution of domestic pigeons in Africa.

putting out sugared water or grains. Pigeons are housed in purpose-built pigeon-cotes made of pottery or mud but not otherwise confined. In Egypt, depictions of pigeon-cotes first appear in the Graeco–Roman period (Husselman 1953) although it has been suggested that pigeon domestication took place earlier (Keimer 1956).

Large-scale Egyptian columbaria are made from mud and thatch, and these may contain as many as 1,000 pigeons (Hornell 1947). The semi-feral system of production and the mud columbaria in west–central Africa resemble strongly those still used in Egypt. Shelters for pigeons vary quite widely in different areas. In the towns, purpose-built cages are made from scrap wire and plywood. These are placed in the centre of the compound, and the faeces drop to the ground through the wire floor and are swept up every morning. In the old city of Kano pigeons are housed in clay pots suspended from the rafters of buildings.

Table 20.3 Comparative productivity parameters for African domestic pigeons.

Parameter	Wilson (1979)	Kuit et al. (1986)
Mean egg weight (g)	16.6	14.4
Mean eggs per clutch	2	2
Age at first egg (days)	132	
Clutches per year	8.4	8
Mean annual egg output	17	15
Hatchability %	81	68
Mortality to leaving nest (%)		25
Weight of mature males (g)		338
Weight of mature females (g)		290

Table 20.4 Flock structure for African domestic pigeons.

n=1020

Age-Class	Sex	%
Adults	male	38.6
	female	39.3
Immature	male	4.1
	female	3.7
Nestlings		14.5

Source: Kuit et al. (1986: Table V).

The training of pigeons to carry messages, a practice widespread in Europe, is also known throughout the Middle East and in Egypt (Lagercrantz 1950:67). This is rare in sub-Saharan Africa, but carrier-pigeons seem to have been used in the inter-city warfare between Muslims in the Lake Chad area in the nineteenth century. Rabeh, the slaver who controlled much of this area in the later part of the century, built carrier pigeon towers into some of his forts.

Pigeons, once paired up at sexual maturity, remain with their mate for life. A female starts to lay clutches of eggs at approximately 6 weeks of age. The brooding period is seven days during the dry season and ten days in the wet season. Chicks normally leave the nest after 14 days and mature body size is attained by 6–8 wk. Pigeons generally raise two or more broods a year and each clutch always consists of one male and one female chick.

Pigeons are not kept for their eggs which are anyway laid in small clutches, but are raised solely for meat. The productivity of pigeons is difficult to measure as pigeons are kept in such low-input systems and the owners rarely note either the fertility of individual animals or the hatchability of eggs. Table 20.3 shows estimates for some productivity parameters in traditional production systems in Sudan and Mali and Table 20.4 shows a flock structure in Mali.

The flock structure is strikingly different from chickens as proportions of males to females are approximately equal and the numbers of nestlings and immatures much lower. This is because pigeons lay smaller clutches, but a greater percentage reach adulthood as losses to predation and disease are much lower. In addition, adult offtake is divided equally between the sexes as opposed to chickens where cocks are slaughtered for preference.

Pigeons have traits that make them poorly adapted to intensive production:

- monogamous matings;
- altricial young;
- considerable space requirements for flight.

This suggests that semi-feral production systems in use at present are very efficient in terms of the low level of inputs required.

Linguistic evidence from the names of pigeon in the vernaculars of West Africa is discussed in detail by Blench (1995). The widespread Hausa name *tàantabàřaa*, is borrowed from Twareg, supporting the hypothesis of a trans-Saharan introduction. Barth (1862ii:201) says, somewhat mysteriously, "This domestic pigeon has, beyond a doubt, been introduced into Negroland by the Sonɣai" although he gives no reason for this beyond the obvious resemblance between the Songhay and Hausa names for pigeon.

Otherwise, there is a great variety of names, presumably reflecting the fact that the pigeon is part of the indigenous fauna. A few names actually attribute to the pigeon an Egyptian origin, such as Mandara "cock of Egypt" or the Margi "bird of Egypt". Although this is not necessarily a reliable guide, in this case, it seems likely that the culture of pigeon-keeping travelled across the desert with the caravan trade.

3.4. Geese

The common domestic goose of Europe and North Africa, *Anser anser*, is a domestic form of the greylag goose (Zeuner 1963, Crawford 1984). The earliest evidence for domestic geese is in Ancient Egypt, where there are Old Kingdom representations of geese confined in poultry yards or being herded. The pictures are sufficiently imprecise so that it is unsure whether the greylag goose or the white-fronted goose, *Anser albifrons*, is being depicted. Both were trapped and eaten, and on occasion force-fed to increase their plumpness for the table (Houlihan & Goodman 1986:54ff.). However, if the white-fronted goose was domesticated, this practice did not persist, whereas the rearing of the greylag goose spread westwards along the North African littoral, and into the Near East and Europe. Despite its belligerent nature, the Egyptian goose, *Alopochen aegyptiaca*, was apparently also domesticated for the table in the Old Kingdom (Houlihan & Goodman 1986:64). A domestic goose bone has been excavated at Carthage, although from a relatively late period, the sixth to seventh centuries AD (Levine 1994).

References to geese in Africa have been reviewed by Lagercrantz (1950:82–7). Lagercrantz concludes that almost all reports of greylag geese are connected with direct European or Arab contact. It is generally assumed that the domestic goose did not cross the Sahara, although the Songhay in Mali do have geese (Rouch 1954:21) and this may result from contact with Morocco. Al-ˤUmarī mentions that under Mansa Musa, the ruler of Mali in the fourteenth century, the peoples of the "Sudan" kept geese (Levtzion & Hopkins 1981:267). It may be that geese were kept in a few communities at the end of the trans-Saharan trade routes. Equally likely, however, is the possibility that these are not *Anser* but another species of goose domesticated or tamed. The spur-winged goose, *Plectopterus gambensis*, has been recorded in Mali at San, Bamako and Segou, kept as a backyard species (MacDonald, pers. comm.) as well as in northeast Nigeria (RIM 1992ii).

3.5. Duck

The common domestic duck in Africa, the Muscovy duck, *Cairina moschata*, was first brought from South America by the Portuguese in the sixteenth century. Clayton (1984) and Donkin (1989) describe the domestication and spread of the Muscovy duck. A related wild species, *C. hartlaubi*, is indigenous to the forest zones of West Africa.

The ethnographic literature on ducks in Africa has been reviewed by Lagercrantz (1950:74–82) but since many sources conflate European ducks, *Anas*, with *Cairina*, these are not as useful as for other species. Lagercrantz's Map 12 shows the distribution of references to ducks in his sources. Jeffreys (1956) describes early sources relating the spread of the Muscovy duck in West Africa, although he drew the mistaken conclusion that it was a pre-Columbian introduction. Muscovy ducks spread inland relatively early, although many Muslims regard ducks as unclean. The Eurasian domestic duck, or Rouen duck, derived from the green-headed mallard (*Anas platyrhyncos*), has been brought into Africa on an experimental basis but does not usually enter into village production.

Ducks are invariably kept in free-range systems and scavenge for their food. Although they prefer to be near water, they seem to tolerate the dry season in the semi-arid zone successfully. Ducks have not been reared intensively and have rarely attained major importance in the household production system, because they are too susceptible to disease and predators. Ducks are rarely taken to market and are most often eaten as a protein supplement or served to unexpected visitors. Duck eggs are not usually collected or eaten as it is generally said that the females should raise as many young as possible in view of high mortalities.

Data available for the reproductive parameters of the Muscovy duck under traditional management are limited to Kuit et al. (1986) summarized in Table 20.5. Egg output is low compared with chickens, and ducklings suffer from heavy predation losses.

3.6. Turkey

The turkey is of North American origin, and was first taken to Europe in the sixteenth century. It is usually considered to have been introduced to Africa only in colonial times. Lagercrantz (1950:87–91) has reviewed references to its presence and shows that these are almost entirely associated with coastal settlement. Turkeys are scattered in rural areas of semi-arid Africa and are usually produced to sell to wealthy Christian families at Christmas time.

Table 20.5 Productivity parameters for traditionally managed Muscovy ducks.

Parameter	Wilson (1979)	Kuit et al. (1986)
Mean egg weight (g)	66.8	69.3
Age at first egg (days)	213	–
Mean eggs per clutch	10.8	13.3
Clutches per year	4.7	2–3
Mean annual egg output	50	30–40
Hatchability %	84	51
Weight of mature drakes (kg)	–	3.07
Weight of mature ducks (kg)	–	2.04

In West Africa, the turkey is widely known by the name *tolotolo* or some variant thereof. This has usually been assumed to be a recent ideophonic construct but somewhat surprisingly a recent study of domesticates in Mixe–Zoque languages of central America has recorded very similar terms, for example Jicaque *tolo* (Wichmann 1997). This may suggest that some turkeys were brought by the Portuguese along with their Amerindian name.

Turkeys kept free-range are allowed to scavenge for their food, but even under village conditions they must be supplemented with grain if they are to stay healthy. Turkeys are thus relatively expensive to produce (RIM 1992ii). Moreover, many producers were discouraged by outbreaks of unknown diseases and there is no reservoir of traditional expertise to draw on to prevent this.

3.7. Other birds

Many bird species can be captured wild and then tamed and experimentation has a long history in the continent. The Ancient Egyptians seem to have been experts in this process, as a wide variety of species are represented in wall-paintings and engravings, appearing as offerings or being fed for the table (Houlihan & Goodman 1986).

3.7.1. Ostriches The ostrich, *Struthio camelus*, is distributed throughout arid and semi-arid Africa and still survives even in heavily hunted environments such as the Egyptian Desert (Goodman & Meininger 1989:113). The four extant wild races are interfertile and have become less genetically distinct following crosses with feral ostriches moved around for domestication purposes. Rock engravings representing the hunting of wild ostrich are known from the Badarian period in Egypt. The process of domestication may have begun with the corralling of wild ostriches in hunting preserves (Houlihan & Goodman 1986:3ff.). Whether the ostrich was truly domesticated is disputed but by the Ptolemaic period they were used to pull carts in ceremonial processions. During the Byzantine era ostriches were bred on farms for their feathers and this practice seems to have continued up until the nineteenth century.

This process of domesticating ostriches for their feathers may be associated with Islam, for early twentieth century livestock censuses in Borno, northeastern Nigeria, record ostriches raised for their feathers and skin. Early colonial livestock census forms invariably included a column for ostriches, but the practice of keeping them seems to have disappeared in the 1940s. Ostrich farming has a long history in southern Africa and the high price of feathers in the nineteenth century led to a substantial production in some communities. The trade subsequently went into decline as the fashion for feathers fell away, but it has recently revived as a meat production operation (Siegfried 1984).

Ostrich eggs are widely used as ornaments in Islamic buildings in West Africa and along the Blue and White Niles. Lagercrantz (1950:380–86) has reviewed the distribution of ostrich eggshell ornaments in Africa. In Mali, for example, many mosques have ostrich eggs on the spires of the roof. This conception seems also to have been present on the North Africa littoral and to have penetrated Europe in the Middle Ages as the ostrich egg occasionally appears in paintings from the fifteenth century onwards.

3.7.2. Turtle-dove   The turtle-dove, *Streptopelia turtur*, was domesticated in Egypt by the 5th Dynasty (*c.* 2400 BC) and appears to have been kept both for the table (when it was force-fed) and as a pet (Houlihan & Goodman 1986:103–6). Unlike other Egyptian domesticates this one seems to have taken hold and domestic doves spread both around the Mediterranean and across the Sahara. Bynon (1984:253) quotes the Ghat (Berber) name as *taturturt* and further connects this with the Latin *turtur*. Doves are presently kept in some semi-arid areas of west–central Africa as pets.

3.7.3. Cranes   The common crane, *Grus grus*, was well known in Ancient Egypt and by the 4th Dynasty (*c.* 2500 BC) and appears to have been already domesticated (Houlihan & Goodman 1986:84–6). It is commonly represented as force-fed and this may be because the flesh would be otherwise unpalatable. The demoiselle crane, *Anthropoides virgo*, seems also to have been domesticated and is represented in mixed flocks with the common crane from the 5th Dynasty onwards. This practice seems not have survived into the Islamic period.

3.7.4. Peacocks   In Islamic areas of sub-Saharan Africa, peacocks are a well known household pet of wealthy families, and are often kept in the courts of rulers. This practice appears to be first recorded in the Kingdom of Jaja (present-day Kanem) by Ibn Saʕīd in the thirteenth century (Levtzion & Hopkins 1981:187) and is very much the case in present-day Nigeria (RIM 1992ii).

3.7.5. Crows   In the course of the survey of Kano in northern Nigeria, a single instance of an individual keeping pied crows (*Corvus albus*) in an enclosure was recorded. These crows were sold for medicinal purposes (RIM 1992ii).

4. Other species

4.1. Reptiles and amphibians

4.1.1. Crocodiles The practice of keeping crocodiles has been recorded from most parts of Africa. In some traditional religions in West Africa, for example among the Nupe of Nigeria and the Dogon of Mali, sacred crocodiles were kept in ponds and given offerings (Nadel 1954:27–8). More commonly, however, young crocodiles are kept in small hand-dug ponds and are eaten for medicinal purposes. Crocodile farming has recently become established in the hinterland of Mombasa, Kenya.

4.1.2. Tortoises The ordinary savanna tortoise, *Geochelone sulcata*, is well known in semi-arid West Africa and very large tortoises were kept in some of the courts of the Muslim Emirs. In Mali, the Dogon people raise giant tortoises for meat and Wilson et al. (1987) report on the productivity of tortoises raised in urban settings in Mali. Three species of the much smaller *Kinixys*, the hinged tortoise are also found in Africa (Villiers 1958:131ff.). Tortoise meat is reported to have medicinal virtues and tortoises are sometimes caught in the bush and raised for food.

4.1.3. Soft-shelled turtles and terrapins There are four species of soft-shelled turtle, and six species of terrapin recorded in the fresh waters of Africa (Villiers 1958).

English	Family	Genera
Soft-shelled turtles	Trionychidae	Trionyx, Cycloderma, Cyclanorbis
Terrapins	Pleurodira	Pelomedusa, Pelusios

In Sahelian west–central Africa, turtles and terrapins are captured for food and are also kept as pets in water sources, where their function is to keep the water clean. Turtles placed in water-pots to eat mosquito larvae and clear other possible worm infestations are common in the region of Lake Chad. Food scraps are thrown down the wells to feed the turtles. Apart from these hygienic functions, turtles are also reared for meat. At Okotiama, in the Niger Delta in southern Nigeria, turtles are reared for sale: they are kept in large bowls of water and are fed on raw palm kernel and fresh fish (RIM 1992ii).

The marine green turtle, *Chelonia mydas*, is intensively reared for its meat and shell on Réunion island, as well as in the Caribbean and on the Torres Strait Islands (Reme 1980).

4.2. Molluscs – snails

The African land snail, *Achatina* sp., has long been exploited in the more humid regions of West Africa. The snails are usually collected from the bush in the rainy season, along with other edible molluscs. Martinson (1929) describes local collection systems in Ghana and Prunier (1945) has made a short but valuable compilation of information and references. Stahl (1993) describing the Punpun phase of the Kintampo complex in Ghana (c. 3500 BP) notes heavy exploitation of *Achatina* for food.

Prunier (1945) recommended that snail farming (heliculture) be tried in West Africa and this actually began in the 1970s, as a response to the declining supplies of snails from the forest. Snail farms are usually found in humid West Africa close to large towns and there is a luxury trade in *Achatina* to northern cities.

The land snail grows slowly: it takes three years to reach sexual maturity, five to attain a normal mature weight and seven to reach 400 g. The average snail found in markets has a shell weighing 85 g and meat weighing 250 g. Snail eggs are eaten in some places, rather like the roe of fish. None the less, its growth characteristics have made it attractive to producers in Europe and some snail-farms in Wales have recently experimented with *Achatina*.

Small-scale snail farms are usually made up of rectangular cement enclosures about a metre high, covered in wire netting to prevent the snails escaping. The whole enclosure is surrounded by a ditch to prevent predatory ants from entering and eating the snails. Fresh green leaves are placed in the enclosures occasionally, but no other husbandry measures are taken.

5. Conclusion

This preliminary study of "minor" African domestic animals has highlighted important lacunae in the prehistory of the region. To recover their history, a synthesis of distributional, linguistic and ethnographic data must be employed. Archaeology has so far contributed little, principally because it tends to concentrate on larger species, especially the ruminants. In part, this is because smaller fauna are less often preserved or recorded and because there are too few specialists skilled in identifying non-mammals or mammalian microfauna. In the case of recently introduced species, these rarely occur in the sites with a time depth of greatest interest to archaeologists. However, in understanding the subsistence strategies of African populations outside the strictly pastoral areas, small but populous species may have an importance similar to cattle in the arid zones.

Evidence from both Ancient Egypt and from recent surveys suggest that the process of domestication is not fixed, and that individuals or cultures continue to experiment with new species and techniques. The failure of certain Egyptian domesticates to persist or spread to the rest of Africa (e.g. the white-fronted goose, *Anser albifrons*) suggests that the high labour inputs needed to maintain certain species made them viable only under rather specific economic conditions, perhaps in the context of royal or aristocratic households. Synchronic cases of small, local and recent domestication abound in Africa and it is clear that domestic stock can exist in a dynamic relationship with wild fauna. At the same time, the rapid spread of recent exotics such as the rabbit argues that new domesticates may meet the evolving needs of urban dwellers and that the pressure of modernization can even accelerate this process.

Note

1. I am grateful to Kevin MacDonald, Stephen Hall and Paul Starkey for reading and commenting on this paper. Much of the literature search and fieldwork in the section on experimental domesticates was originally conducted under the auspices of the Nigerian National Livestock Resource Survey (RIM 1992). Other unreferenced ethnographic and historical materials come from the author's own field notes, 1974–97.

References

Ajayi, S.S. 1974. Preliminary observations on the biology and domestication of the African giant rat (*Cricetomys gambianus* Waterhouse) in Nigeria. *Mammalia* **39**, 343–64.

Ajayi, S.S. & O.O. Tewe 1980. Food preference and carcass composition of the grass cutter (*Thryonomys swinderianus*) in captivity. *African Journal of Ecology* **18**, 133–40.

Armitage, P.L. 1994. Unwelcome companions: ancient rats reviewed. *Antiquity* **68**, 231–40.

Ayeni, J.S.O. 1983. Studies of the grey-breasted helmeted guineafowl (*Numida meleagris galeata*, Pallas) in Nigeria. *World's Poultry Science Journal* **39**(2), 143–51.

Ayeni, J.S.O. & S.S. Ajayi 1983. Wildlife protein: guineafowl (*Numida meleagris galeata*, Pallas) as animal protein supplement. *Nigerian Field* **47**(4), 156–66.

Ayorinde, K.L. & J.S.O. Ayeni 1986. The reproductive performance of indigenous and exotic varieties of the guinea-fowl (*Numida meleagris*) during different seasons in Nigeria. *Journal of Animal Production Research* **6**(2), 127–40.

Banfield, A.W. 1914. *Dictionary of the Nupe language.* Shonga: The Niger Press.

Banti, G. 1993. Ancora sull'origine del cammello nel Corno d'Africa: osservazioni di un linguista. In *Ethno, Lingua e Cultura*, A. Belardi (ed.), 183–223. Roma: Calamo.

Barth, H. 1862. *Collection of vocabularies of Central African languages.* Gotha: Justus Perthes.

Bechhaus-Gerst, M. 1991/2. The Beja and the Camel. Camel-related lexicon in *tu-beḏauwie. Sprache und Geschichte in Afrika* **12/13**, 41–62.

Blench, R.M. 1993. Ethnographic and linguistic evidence for the prehistory of African ruminant livestock, horses and ponies. In *The archaeology of Africa. Food, metals and towns*, T. Shaw, P. Sinclair, B. Andah, A. Okpoko (eds), 71–103. London: Routledge.

Blench, R.M. 1995. A history of domestic animals in northeastern Nigeria. *Cahiers de Science Humaine*, ORSTOM, **31**(1), 181–238.

Blench, R.M. 1997. *Neglected species, livelihoods and biodiversity in difficult areas: how should the public sector respond?* Natural Resource Perspective Paper 23. London: Overseas Development Institute.

Brewer, D.J., D.B. Redford, S. Redford. 1994. *Domestic plants and animals: the Egyptian origins.* Warminster: Aris & Phillips.

Boccazzi, A., A. Scarpa Falce, S. Scarpa Falce 1995. Segnalazione di uno sito dell'enneri Korossom (Tibesti nord-orientale). *Sahara* **7**, 85–8.

Boettger, C. 1958. *Die Haustiere Afrikas.* Jena.

Bulliet, R.W. 1990. *The camel and the wheel*, 2nd edn. New York: Columbia University Press.

Bureau, R. 1954. Manger du chien! ... et du chat? *Bulletin Liaison Saharienne* **V**(16), 15–16.

Bynon, J. 1984. Berber and Chadic: the lexical evidence. In *Current progress in Afro-Asiatic linguistics: Papers of the Third International Hamito-Semitic Congress*, J. Bynon (ed.), 241–90. Amsterdam: John Benjamins.

Canard, M. 1953. La cynophagie au Sahara. *Bulletin Liaison Saharienne* **IV**(15), 2–8.

Chalumeau, P. 1954. Cynophagie. *Bulletin Liaison Saharienne* **V**(17), 77–84.

Clayton, G.A. 1984. Muscovy duck. In *Evolution of domesticated animals*, I.L. Mason (ed.), 340–44.

Clutton-Brock, J. 1984. Dog. In *Evolution of domesticated animals*, I.L. Mason (ed.), 198–211.

Crawford, R.D. 1984. Geese. In *Evolution of domesticated animals*, I.L. Mason (ed.), 345–9. London: Longman.

Dollman, G. 1937. The Bansenji dog. *Journal of the Royal African Society* **36**(CXLIII), 148–9.

Donkin, R.A. 1989. *The Muscovy duck, Cairina moschata domestica. Origins, dispersal and associated aspects of the geography of domestication.* Rotterdam: Balkema.

Donkin, R.A. 1991. *Meleagrides: an historical and ethnogeographical study of the guinea fowl.* London: Ethnographica.

Epstein, H. 1971. *The origin of the domestic animals of Africa* [2 volumes]. New York: Africana Publishing.

Esser, M. & O. Esser 1982. Bemerkungen zum Vorkommen des Kamels im östlichen Afrika im 14. Jahrhundert. *Sprache und Geschichte in Afrika* **4**, 225–38.

Ewer, R.F. 1971. The biology and behaviour of a free-living population of black rats (*Rattus rattus*). *Animal Behaviour Monographs* **4**, 127–74.

Field, C.R. 1984. Potential domesticants: Bovidae. See Mason (1984), 102–6.

FAO 1977. *The water buffalo.* Rome: FAO.

Frank, B. 1965. *Die rolle des hundes in afrikanischen kulturen.* Wiesbaden: Franz Steiner.

Goe, M. 1996. Working Paper 3: Livestock Development. In *Family Sector Development Project, Republic of Mozambique.* [Unpublished Report]. Rome: International Fund for Agricultural Development.

Goodman, S.M. & P.L. Meininger (eds) 1989. *The birds of Egypt.* Oxford: Oxford University Press.

Happold, D.C.D. 1967. Guide to the natural history of Khartoum Province. Part III – Mammals. *Sudan Notes and Records* **48**, 111–32.

Happold, D.C.D. 1987. *The mammals of Nigeria.* Oxford: Clarendon Press.

Hartert, E. 1919. On the guinea-fowl of Morocco. *Bulletin of the British Ornithologists' Club* **39**, 68–9, 85–7.

Hauck, E. 1941. *Die Hunderassen in Alten Ägypten.* Zeitschrift für Hundeforschung, 16. Leipzig.

Hawes, R.O. 1984. Pigeons. In *Evolution of domesticated animals*, I.L. Mason (ed.), 351–6.

Heine, B. 1981. Some cultural evidence on the early Sam-speaking people in East Africa. *Sprache und Geschichte in Afrika* **3**, 169–200.

Hornell, J. 1947. Egyptian and medieval pigeon-houses. *Antiquity* **21**, 182–5.

Houlihan, P.F. & S.M. Goodman 1986. *The birds of Ancient Egypt.* Warminster: Aris & Phillips.

Husselman, E.M. 1953. The dovecotes of Karanis. *Transactions and Proceedings of the American Philological Association* **84**, 81–91.

Jeffreys, M.D.W. 1956. The muscovy duck. *The Nigerian Field* **21**(3), 108–111.

Juma, A.M. 1996. The Swahili and Mediterranean worlds: pottery of the late Roman period from Zanzibar. *Antiquity* **70**, 148–54.

Keimer, L. 1956. The pigeons and pigeon cotes of Egypt. *Egyptian Travel Magazine* **20**, 2–29.

Kuit, H.G., A. Traore, R.T. Wilson 1986. Livestock production in central Mali: ownership, management and productivity of poultry in the traditional sector. *Tropical Animal Health and Production* **18**, 222–31.

Lagercrantz, S. 1950. *Contributions to the ethnology of Africa.* Studia Ethnographica Upsaliensia, I. Lund: Håkan Ohlssons.

Law, R. 1980. *The horse in West African history.* Oxford: Oxford University Press.

Levine, M.A. 1994. The analysis of mammal and bird remains. In *Excavations at Carthage, the British Mission. Vol. II, I.* H.R. Hurst (ed.), 314–17. Oxford: Oxford University Press for the British Academy.

Levtzion, N. & J.F.P. Hopkins 1981. *Corpus of early Arabic sources for West African history.* Cambridge: Cambridge University Press.

Lewicki, T. 1974. *West African food in the Middle Ages.* Cambridge: Cambridge University Press.

MacDonald, K.C. 1992. The domestic chicken (*Gallus gallus*) in sub-Saharan Africa: a background to its introduction and its osteological differentiation from indigenous fowls (Numidinae and Francolinus sp.). *Journal of Archaeological Science* **19**, 303–18.

MacDonald, K.C. 1995a. The faunal remains (mammals, birds, and reptiles). In *Excavations at Jenné-Jeno, Hambarketolo, and Kaniana (Inland Niger Delta, Mali). The 1981 Season.* S.K. McIntosh (ed.), 291–318. Berkeley: University of California Press.

MacDonald, K.C. 1995b. Why chickens?: the centrality of the domestic fowl in West African ritual and magic. In *Animal symbolism and archaeology*, K. Ryan & P.J. Crabtree (eds), 50–56. Philadelphia: MASCA/University of Pennsylvania Press.

Martinson, R. 1929. The local edible snail industry. *Gold Coast Department of Agriculture Year Book 1928*, 231–5.

Mason, I.L. (ed.) 1984a. *Evolution of domesticated animals.* London: Longman.

Mason, I.L. 1984b. Goat. In *Evolution of domesticated animals*, I.L. Mason (ed.), 85–99.

Matthewman, R.W. 1977. *A survey of small livestock production at the village level in the derived savanna and lowland forest zones of south west Nigeria.* Reading: Study No. 24, Department of Agriculture, University of Reading.

Mauny, Raymond 1954. La cynophagie en Afrique occidentale. *Notes Africaines* **64**, 114 only.

Mongin, P. & M. Plouzeau 1984. Guinea-fowl. In *Evolution of domesticated animals*, I.L. Mason (ed.), 322–5.

Morales Muniz, A., J.A. Riquelme, C. Liesau von Lettow-Vorbeck 1995. Dromedaries in antiquity; Iberia and beyond. *Antiquity* **69**, 368–75.

Nadel, S.F. 1954. *Nupe religion.* London: Kegan Paul.

Okaeme, A.N. 1982. Guineafowl production in Nigeria. *World's Poultry Science Journal* **38**(1), 36–9.

Plug, I. 1996. Domestic animals during the early Iron Age in southern Africa. *Aspects of African archaeology. Papers from the 10th Congress of the Pan-African Association for prehistory and related studies*, G. Pwiti & R. Soper (eds), 515–22. Harare: University of Zimbabwe Publications.

Plug, I. & N.J. Dippenaar 1979. Evidence of *Rattus rattus* (house rat) from Pont Drift, an Iron Age site in northern Transvaal. *South African Journal of Science* **75**, 82 only.

Prunier, R. 1945. L'escargot de Guinée. *Farm and Forest* **VI**(1), 34–5.

Reme, A. 1980. Quelques problèmes sanitaires et pathologiques dans l'élevage intensif de la tortue marine (*Chelonia mydas*, L.) à la réunion. *Revue d'Elevage et de Médécine Vétérinaire des Pays Tropicaux* **33**(2), 177–92.

RIM 1992. *National Livestock Resource Survey* [6 volumes]. Abuja: Final Report to Federal Department of Livestock and Pest Control Services, Federal Government of Nigeria.

Robinson, R. 1984. Cat. In *Evolution of domesticated animals*, I.L. Mason (ed.), 217–25.

Rosevear, D.R. 1969. *The rodents of West Africa.* London: British Museum.

Rouch, J. 1954. *Les Songhay.* Paris: Presses Universitaires de France.

Sasse, H.-J. 1993. Ein weltweites Hundewort. In *Sprachen und Schriften des antiken Mittelmeerraums. Festschrift für Juergen Untermann zum 65. Geburtstag*, F. Heidermanns, H. Rix, E. Seebold (eds), 349–66. Innsbrucker Beiträge zur Sprachwissenschaft, 78. Innsbruck: Institut für Sprachwissenschaft.

Seignobos, C. (ed.) 1987. *Le Poney du Logone.* Paris: Institut d'Élevage et de Médécine Vétérinaire des Pays Tropicaux.

Shaw, B.D. 1979. The camel in ancient North Africa and the Sahara: history, biology and human economy. *Bulletin de L'Institut Fondamental D'Afrique Noire* **41**(4), 663–721.

Shinnie, P.L. 1967. *Meroe: a civilization of the Sudan.* New York: Praeger.

Siegfried, W.R. 1984. Ostrich. In *Evolution of domesticated animals*, I.L. Mason (ed.), 364–6.

Skinner, N. 1977. Domestic animals in Chadic. In *Papers in Chadic Linguistics*, P. Newman & R.M. Newman (eds), 175–98. Leiden: Afrika-Studiecentrum.

Smith, E.W. & A.M. Dale 1920. *The Ila-speaking peoples of Northern Rhodesia.* London: Macmillan.

Stahl, A.B. 1993. Intensification in the West African late Stone Age: a view from central Ghana. In *The archaeology of Africa. Food, metals and towns*, T. Shaw, P. Sinclair, B. Andah, A. Okpoko (eds), 261–73. London: Routledge.

Tewe, O.O. & S.S. Ajayi 1979. Utilisation of some common tropical food stuffs by the African giant rat (*Cricetomys gambianus* Waterhouse). *African Journal of Ecology* **1**, 165–73.

Tewe, O.O., S.S. Ajayi, E.O. Faturoti 1984. Giant rat and cane rat. See Mason (1984), 291–3.

Thorp, C.R. 1995. *Kings, commoners and cattle at Zimbabwe tradition sites.* Museum Memoirs (New Series) No. 1. Harare: National Museums and Monuments of Zimbabwe.

Thys, E. & O. Nyssens 1983. Preparation et commercialisation de la viande canine chez les Vame-Mbreme, population animiste des monts Mandara (Nord-Cameroun). In *Proceedings of the International Colloquium on Tropical Animal Production for the Benefit of Man* 511–17. Antwerp.

Villiers, A. 1958. *Tortues et crocodiles de L'Afrique Noire Française*, Initiations Africaines 15. Dakar: IFAN.

Wichmann, S. 1998. A conservative look at diffusion involving Mixe-Zoquean languages. In *Archaeology and Language, II*, R.M. Blench & M. Spriggs (eds), 297–323. London: Routledge.

Wilson, R.T. 1979. Studies of the livestock in South Darfur, Sudan. (VII). Production of poultry under simulated traditional conditions. *Tropical Animal Health and Production* **11**, 143–50.

Wilson, R.T. 1984. *The camel*. London: Longmans.

Wilson, R.T., A. Traore, H.G. Kuit, M. Slingerland 1987. Livestock production in central Mali: reproduction, growth and mortality of domestic fowl under traditional management. *Tropical Animal Health and Production* **19**, 229–36.

You, R. 1955. Cynophagie. *Notes Africaines* **66**, 40 only.

Zeuner, F.E. 1963. *A history of domesticated animals*. London: Hutchinson.

CHAPTER TWENTY-ONE

A history of donkeys, wild asses and mules in Africa

Roger M. Blench

1. Introduction

While it is probably poor practice to award regions of the world marks for originating domesticates it is worth noting that Africa is responsible for four species of domestic animal in common use today, the donkey, the cat, the guinea-fowl and (probably) cattle. Of these, only cattle have attracted substantial attention from archaeozoologists, although the Near East lens through which much of their work is viewed has probably acted to obscure as much as to illuminate. To fill at least one of these lacunae, this paper focuses on reconstructing the history of the domestication of the donkey.[1]

Although donkeys are both widespread and economically important to their owners, they are rarely studied and are not usually subject to any improvement, development or loan schemes (Svendsen 1986). Donkeys are not conventional sources of meat, and their uses for packing and traction do not fit within the stereotyped perspectives of livestock agencies. None the less, they are essential to the subsistence strategies of many communities in semi-arid regions, relieving families of repetitive and energy-consuming tasks (Fielding & Pearson 1991). Moreover, they stay healthy on a varied and often poor-quality diet and require little management.

The early history of the donkey in Africa is also notable for a near-absence of substantive archaeological data. The use of the domestic donkey is well documented in Egyptian wall-paintings and other iconography. Elsewhere in the continent, although there are representations of wild asses in rock art, the domestic donkey is remarkable chiefly for its absence. Historical references to the donkey in the West African Sahel are collected together by Lewicki (1974:88–9) and Levtzion & Hopkins (1981).

One strategy to fill this lacuna is the use of linguistics. Terms for donkeys and asses have been recorded in numerous African and Near eastern languages. Compiling these terms and tracing the links between them makes it possible to extend some hypotheses both about the process of domestication and the routes along which the donkey spread. Combined judiciously with modern ethnographic data this can be used to partially reconstruct the prehistory of the donkey in Africa.

2. Biogeography

The wild ass, *Equus asinus africanus*, is indigenous to the African continent and is usually divided into a chain of races of subspecies spreading from the Atlas mountains eastwards to Nubia, down the Red Sea and probably as far as the border of present-day northern Kenya (Groves 1966, 1986, Haltenorth & Diller 1980:109). The extent to which the wild ass penetrated the interior of Africa is controversial, but it is generally considered unlikely that it ever occurred in sub-Saharan regions. Groves (1986) argues that the wild ass extended into the Near East in ancient times and co-existed with the onager, *E. hemionus*.

Figure 21.1 shows the actual range of the wild ass in the 1990s (Kingdon 1997) superimposed on the hypothetical former distribution prior to Roman depredations in North Africa. Four notional races, *atlanticus*, *africanus*, *taeniopus* and *somaliensis* are located on the map approximately where they are shown in earlier studies (e.g. Haltenorth & Diller 1980). However, two of these, *atlanticus* and *taeniopus* have been rejected more recently and indeed the proposed *atlanticus* race turns out to have been based on misidentified zebra bones (Kingdon 1997:311).

The main features differentiating races of wild ass are the amount and type of stripes and the shoulder crosses. However, their characterization may be somewhat blurred, since populations that survived into historical times have almost certainly crossed with feral donkeys, leading to a merger of characteristics. Groves (1986) presents arguments for making further differentiation within these groups.

Of these races, the wild ass of the Atlas mountains became extinct by AD 300 and is known only through depictions (Haltenorth & Diller 1980:109). Civil war in both Somalia and Eritrea may mean that the fragile populations marked have disappeared or are severely threatened. There are two doubtful populations of wild ass near Siwa oasis in Egypt and further south towards the Sahara proper. There are breeding animals conserved today in Basle zoo and the Hai Bar Reserve in the Negev desert, and these may well be the last remaining genetically pure populations.

Ethnographic reports cited by Groves (1986:34) appear to suggest the presence of wild asses in the Tibesti and Ahaggar. These may well be feral donkeys or populations substantially interbred with the domestic donkey. The populations on the island of Soqotra are certainly feral donkeys (Haltenorth & Diller 1980). The wild ass is limited to the semi-arid regions through its susceptibility to humidity, but the southern range of the domesticated donkey can be extended by careful management.

The original motive for domesticating the donkey is unknown, and it is not certain that it would necessarily reflect its common usage today, as transport for people and goods. It may have been domesticated for its meat or milk, with its use for portage a later development. Certainly the fact that the wild ass and the donkey have remained interfertile suggests that there was little breeding and selection. This may reflect a management system based on the seasonal corralling of wild animals, rather like reindeer management among the Saami today. Such management systems were practised through much of Sahelian West Africa into the present century and were probably once considerably more common.

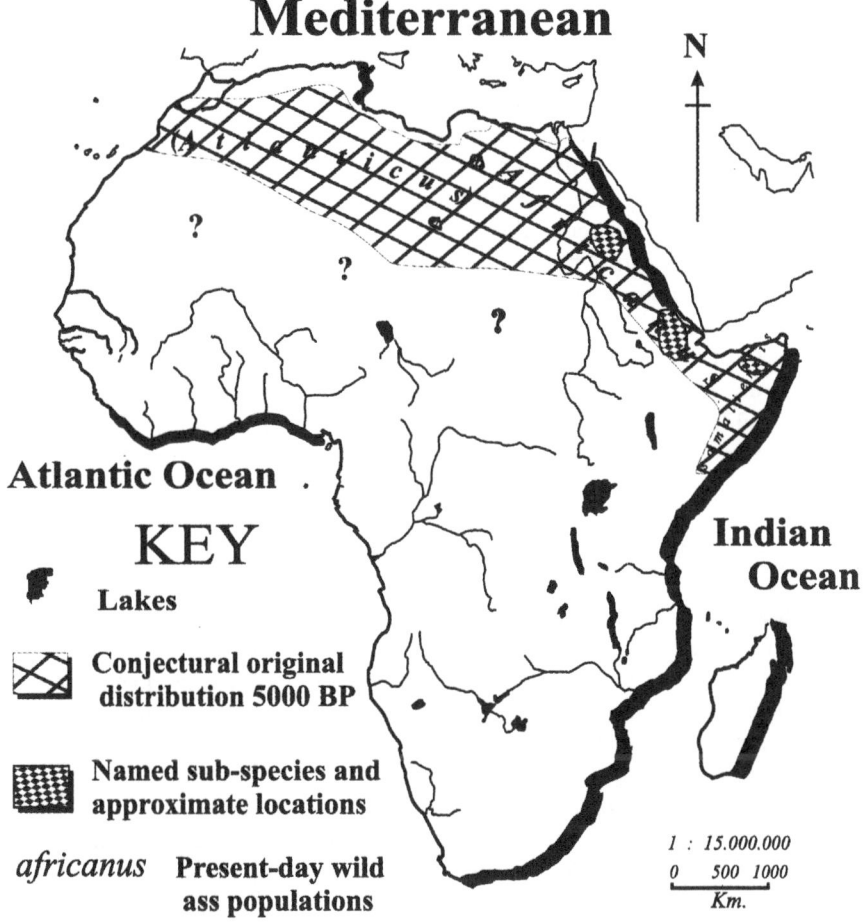

Figure 21.1   Wild asses: current and historical distribution.

The precise wild progenitor of the domestic donkey is disputed. Eisenmann (1995) has recently reviewed the competing arguments. Groves (1966, 1986) set out the possible ancestors of the donkey and pointed out that present-day wild populations are no longer sufficiently homogeneous to resolve the question unambiguously. Groves argues that we cannot exclude the Middle East as the original locus of domestication, although there is no positive evidence in favour of this. The Nubian wild ass is cited in many textbooks and other races may also have contributed to the gene-pool (Epstein 1984:176). Bökönyi (1991) argued that domestication took place in Egypt and Clutton-Brock (1992) notes that the skeletons of three domestic donkeys have been found in an Egyptian tomb dated to 4500–4000 BC. There are comparably early skeletons in the Near East but whether these are domestic remains uncertain (Eisenmann 1995:11).

Haltenorth & Diller (1980) describe the characteristics of the African races of wild ass and it does seem that there is some correspondence between local forms of the donkey and the phenotypes of the wild ass race.[2] For example, *Asinus somaliensis* is notable for the leg rings, on both fore and hind legs. The Somali donkey is described as having "zebra markings" on the legs (Mason & Maule 1960:14). In contrast, West African donkeys usually have distinct shoulder crosses (shown in representations of the Atlas wild ass) but rarely any leg markings.

Earlier writers considered that the Asiatic onager may have played some part in the descent of the African donkey, and one remarkable wall-painting from Thebes dated to the 18th Dynasty does show some onagers apparently pulling a chariot (Epstein 1971ii:397). However, it is generally considered unlikely that this was more than an exotic curiosity, especially as onager x ass crosses are sterile.

3. Donkeys in use

Donkeys are kept in Africa for four reasons: (a) as work animals; (b) for breeding; (c) for milking; (d) to eat.

Of these, work is by far the most important. Donkeys are used mainly as pack animals, either for carrying loads or for riding. Less commonly, they are used in traction, for example, pulling carts or ploughs, although in sub-Saharan Africa both of these are post-European introductions (Fielding 1987). In Ancient Egypt, asses were used both for treading seed into the furrow and for threshing, but there seem to be no modern reports of these practices. A review of some of the existing literature is given in Clutton-Brock (1992) although this focuses principally on horses.

Breeding donkeys can be a profitable business in certain regions of the Sahel. Below a certain isohyet, the reproduction of donkeys becomes increasingly problematic, owing to humidity-related infections. It is therefore more practical for donkey-users to buy animals from further north and replace them at the end of their working life. Sahelian countries such as Niger and Mali have a considerable trade selling donkeys, usually males, to communities further south.

Although asses' milk has an important symbolic value due to its prominence in certain Near-Eastern texts, the milking of donkeys in Africa is rare and of little economic importance. The western Maasai are reported to milk donkeys (Epstein 1971ii:386) and donkeys' milk is used in magical remedies in parts of West Africa. The main reason for this is probably the low-management systems that obtained until recently; donkeys were not milked because of the labour of catching them regularly.

The extent to which donkeys are eaten is probably greatly underestimated, since this is something of a taboo area for many observers. None the less, the wild ass has been hunted to near-extinction for its meat and eating equids is common in many Eurasian pastoral systems. In West Africa, the trade in donkeys for meat is essentially of old, sick or exhausted animals that have been used as work animals in the villages of the semi-arid zone. Because of its ambiguous status, the trade in donkeys remains poorly documented.

Islam prohibits the consumption of donkey meat and many Christian and traditionalist groups also refuse to eat it. Ibn Baṭṭūta, travelling in the Empire of Mali in 1352 noted with distaste the consumption of donkeys (Ibn Batoutah 1893–1922iv:423–4 also Levtzion & Hopkins 1981:297). Fernandes (1938:76) describes the Berber nomads of Mauretania as eating donkey in the early sixteenth century. Donkey meat was still eaten in the Malian Gourma at the turn of the century (Desplagnes 1907:228).

Further south, in the more humid regions of west–central Africa, the donkey is an exotic to which no culinary taboos attach. In Nigeria there is a thriving trade in donkeys reaching southern markets and this is probably replicated along the West African coast (RIM 1992). Formerly much of the trade had been in smoked meat, as donkeys bought in intermediate markets were slaughtered and skinned and the meat then prepared by drying and smoking. This practice seems to have largely disappeared, and the trade is confined to live donkeys. The meat is sold as donkey meat locally, but is sometimes passed off as the more expensive beef outside the area. In East Africa there are also reports of eating donkeys. The Kamba people in Kenya are recorded as actually fattening donkeys for consumption and some of the other cultivators close to the Maasai may also eat donkeys (Epstein 1971ii:387).

4. Productivity of donkeys under traditional management

Fielding (1988) has reviewed existing productivity data for female donkeys worldwide. Studies on the productivity of donkeys under traditional management in sub-Saharan Africa are sparse, consisting principally of Wilson (1980) for two different systems in Mali, Wilson et al. (1984) for the Twareg pastoral herds of Niger and RIM (1992) for northern Nigeria. This latter study has the most comprehensive data and the largest sample size; its findings are therefore quoted here as indicative (Table 21.1).

The mean age at first foaling, 56.9 months, is substantially higher than in temperate countries where about three years is considered usual (Fielding 1988:163).

Table 21.1 General reproductive parameters of donkeys in Nigeria.

Category	Value	SD	n
Mean age of breeding female	96.3 mths	29.0	77
Mean age at first foaling	56.9 mths	16.6	76
Foaling interval	25.5 mths	–	12
Mean number of previous parities	2.1	1.3	77

Source: RIM (1992).

Donkeys in Nigeria are allowed to mate freely when herded, but restrictions on access to males when jennies are used for work can mean that oestrus is overlooked. Estimates from the literature suggest that the length of the oestrous cycle is about 24 days and the length of the oestrus itself 6–7 days. Donkeys are usually seasonal breeders in temperate regions but in the tropics they come into oestrus throughout the year. Variations in the annual pattern of foaling are most likely to reflect nutritional differences. Donkeys have a gestation period of almost precisely a year (374 days in the estimates quoted in Fielding 1988). The body condition of breeding females never deteriorates so far as to inhibit fertility, and an even conception pattern reflects their ability to thrive on the poorest of diets (Borwick 1970).

These figures provide numerical confirmation of many generalizations about donkeys, both in terms of their hardiness and productivity. However, they should be used with caution as they represent the system in one specific region of Sahelian Africa. The degree of variation within Africa as a whole may be considerable.

5. Archaeology, history and ethnography

Osteological records of domestic donkeys begin in Egypt in the fourth millennium BC from the site of Maadi (Midant-Reynes 1992). There are clear representations of working donkeys by the middle of the next millennium (Epstein 1971:392, Brewer et al. 1994:99). At about the same period there are textual records of large herds of donkeys, many of which were used for portage. Under the Pharaoh Pepi II (c. 2270 BC) trading expeditions to Punt (Ethiopia) consisted of caravans with pack donkeys (Kitchen 1993). The extent to which the donkey departs from its wild relative can be tracked through Egyptian wall-paintings, where the dark shoulder-stripe of the ass gradually disappears from the donkeys as the Old Kingdom gives way to the Middle Kingdom (Brewer et al. 1994:100). Donkeys from the second millennium BC occur at Shaqadud in the Butana grasslands of Sudan (Peters 1991). The historical and archaeological evidence for domestic donkeys in the Maghreb is reviewed by Camps (1988). Donkeys were found in the faunal assemblages at Carthage in the Roman period (first to fourth centuries AD) (Levine 1994). Kaache (1996) reviews the evidence for donkeys in Morocco; there are possible finds of ass bones at the "neolithic" sites of Dar-es-Soltane and Tangier but no certain representations in rock art.

5.1. Wild asses and donkeys in rock art

Donkeys can only be distinguished from wild asses if they are shown in use; representations are not therefore evidence of domestication but only of their presence. Representations of asses or donkeys are sparse outside of a few scattered petroglyphs in the Saharan Atlas and the Mathendous (southern Libya). A recent review of west–central Saharan rock art suggests that there are virtually no representations of wild asses or domestic donkeys (Muzzolini 1995). Similarly, and perhaps more surprisingly, there appear to be no representations of asses or donkeys in the Horn of Africa (Phillipson 1993:350). Only further research will show if the intention was to depict donkeys. As usual, dating rock art is highly problematic; the following occurrences are given as a basis for further biogeographical and archaeozoological studies.

5.1.1. Algeria   Capderou (1995, Fig. 4.) depicts a very clear head of an ass in the Ksour mountains of the Saharan Atlas. Muzzolini (1995, Fig. 426) illustrates a female ass with her young in a rock engraving at El Richa, Saharan Atlas, assigned to the Bubaline school (*c.* 5000–2000 BC).

5.1.2. Libya   At Messak in southern Libya, a rock engraving (post 1500 BC ?) shows a donkey with pointed legs (Lutz & Lutz 1995, Fig. 6). The ritual importance of the wild ass is well-illustrated in a Bubaline period engraving from Mathendous, Tassili in Ajjer (southern Libya) given by Muzzolini (1995, Fig. 436) which shows two men wearing asses' head masks apparently committing sodomy.

5.1.3. Egypt   Winkler (1938–39) identified wild asses in the rock art of the Eastern Desert.

5.2. Archaeology

Archaeologically, there are few certain records of domestic donkeys in sub-Saharan Africa. The earliest record of a donkey in West Africa is at Siouré in Senegambia (MacDonald & MacDonald Ch. 8 in this volume). The stratigraphy of this site appears to be reliable and the donkey bone is dated to AD 0–250. After this, the next donkey bones occur at Akumbu in Mali with a date of AD 600–1000. However, these are extremely rare, even in sites such as Tegdaoust, where there have been extensive finds of other domestic species. Bearing this in mind, it is curious that bones identified as *Equus asinus* at MK40 in Mali are dismissed by Gaultier (1991) as "intrusive".

The picture for eastern Africa is much richer. Marshall (Ch. 10 in this volume) summarizes the evidence which suggests that there were domestic donkeys near the Nile confluence as early as the fourth millennium bp. A site in northern Ethiopia without radiocarbon dates has been assigned a comparable antiquity.

The scarcity in West Africa may relate to a problem of identification. There is considerable evidence for the widespread use of ponies in west–central Africa, a cultural pattern that evolved from the adaptation of North African horses to the ecology of the sub-Saharan region (Blench 1993). West African ponies are extremely small and it remains to be demonstrated that they have been reliably distinguished from donkeys and mules. Eisenmann (1986) has published extensively on the distinction between horses, asses, mules and donkeys but not all archaeozoologists working on Africa have made use of the criteria she has established.

Equid teeth have been recovered from excavations in central Nigeria from rock shelters at Kariya Wuro (Allsworth-Jones 1982) and Rop. The Rop teeth, in particular, which are dated to the first millennium BC, have been identified as those of a wild ass or donkey (Sutton 1985). This seems unlikely, unless either the stratigraphy at Rop is misleading or these are in fact pony teeth.

5.3. Historical sources

Historical sources on the spread of the donkey are exiguous. The Arabic sources for west–central Africa mention donkeys several times (all references from Levtzion & Hopkins 1981). Al-Bakrī (p. 81) noted the use of donkeys to carry salt in the Kingdom of Ghana and Al-ʿUmarī (p. 263) commented on their small size in the Empire of Mali. However, donkeys pass unnoticed in Ethiopian historical chronicles (Pankhurst 1968). When European trading voyages begin there are a few scarce references. Donkeys and mules from Persia were apparently first landed at the Cape by the Dutch East India Company in 1689[3] (Boettger 1958). Little is known of their subsequent history, but it seems likely that the Boer farmers were the initial agents of their spread into the interior.

The few rock art depictions and the sparsity of references to donkeys in textual records presumably relates to their low status. Rock art in the Sahara focused on the high prestige horse and later the camel. However, it may also be that the spread of donkeys was at first slow and scattered and its importance developed with the evolution of long-distance trade.

6. Linguistic evidence

Another way of approaching the history of the donkey is through vernacular names in the languages of sub-Saharan Africa. Two authors, Skinner (1977) and Bender (1988) have looked at the potential for reconstruction in specific language groups, respectively Chadic and Omotic. Tourneux (1987) discusses names for equids in "Afrique Centrale" as part of an investigation of the antiquity of the pony in this region. Blench (1995) is an exploration of the terminology for donkeys in the Lake Chad area. This section attempts to identify some of the principal roots for "ass/donkey" in African languages and advances some hypotheses about the implications to be drawn from this data. Donkeys may be represented by a ramified terminology; there can be separate terms for wild ass, jenny, young donkey, etc. These are often quite obscure words and lexicographers not specialized in livestock do not always record them. Further research may thus reveal connections and extensions of root forms not at present apparent.

The principal base forms identified are:

#*kuur-*	Widespread in Africa
#*harre*	Ethiopian languages
#*d-q-r.*	Cushitic languages
#*aɣyul*	Berber
#*aʒəd*	Berber

#*kuur-*

Bender (1988:152) reconstructs proto-Omotic **kur* for ass, although to judge by some Omotic citations this probably had a long vowel. Words of this general

Table 21.2 #k-r root.

Phylum	Family	Branch	Language	Form
Afroasiatic	Omotic	Gimira	Benc Non	kur$^{2-3}$
		Mao	Hozo	kuuri
		Southern	Karo	uk'ulí
	Cushitic	Eastern	Borana	bukura°
			Saho	okáalo
	Chadic	West	Karekare	kóoróo
		Central	Vulum	kùré
		Masa	Peve	koro
		East	Nancere	kurá
Nilo–Saharan	C. Sudanic	Sara	Mbay	kòro
	Saharan		Kanuri	kóro

°young donkey.

formula run through Cushitic and Chadic as well as Omotic and it seems reasonable to assume that the Omotic form gave rise to the others. However, many Omotic languages also have the common Cushitic *harre*. Traces of the *#kuur-* root are found through much of Afroasiatic, notably Chadic languages. Its presence in Nilo–Saharan languages such as Kanuri, suggest that it was carried across Central Africa as part of the westward expansion of Cushitic (Table 21.2).

There is no trace of the *ħarre* root in Chadic, which suggest that when speakers of proto-Chadic split off from Cushitic, asses were still being managed on a semi-wild basis. *#kuur-* has remained the dominant lexeme in most of Chadic.

#ħarre

This is an extremely widespread root through the Horn of Africa, and appears virtually unchanged in numerous East Cushitic and Omotic languages. This suggests that it is probably a widespread loanword and should *not* be reconstructed to proto-Cushitic. The Ethio–Semitic languages have a different word, cognate with the Near eastern Semitic root *ḥ-m-r*, arguing that the ancestral speakers of these languages already had a domestic donkey when they crossed the Bab el Mandeb.

The most probable source for *ħarre* are the Oromoid words for "zebra". Zebras are not part of the fauna of the highlands but they are widespread in the lowlands south of the Ethiopian Plateau and are very familiar to pastoral groups such as the Borana. Borana has *harre dida* for zebra, with *dida* meaning "outdoors" or "open air". The term *harre* was probably originally a word for zebra in lowland Oromoid and was transferred to donkey once it was fully domesticated. The zebra would then become the "donkey of the plains". Formations such as Konso *harr-etita* for "zebra" would be calques of the Borana expression, already using the borrowed word for donkey. The development of the donkey as pack animal is probably reflected in the Beja *harri* "anything ridden, from a camel to a train".

In the Horn of Africa, an old root for the wild ass *#kuur-* was largely displaced by *#harre* when the domesticated donkey developed economic significance. The term *#harre* was probably borrowed from terms in lowland Oromoid originally applied to "zebra".

#d-q-r.

Surprisingly, the Agaw terms and those in West Rift (southern Cushitic) seem to be related despite their considerable geographical separation. The dV- initial syllable is not a prefix in either group and the words look too similar for this to be merely coincidence.

Table 21.3 #d-q-r root.

Family	Branch	Language	Form
Cushitic	Agaw	Bilin	dəxʷara
	West Rift	Iraqw	daqwaay

It has been suggested that this form is derived from southern Cushitic "zebra", for example, Iraqw *dakeeti*, but this is not very convincing.

6.1. Ancient Egyptian

The principal form recorded for Ancient Egyptian, ḥ' is too reduced to be certain of its affiliations. It may be related to either of the Semitic roots set out below.

6.2. Semitic

There are two widespread base forms in Semitic, *#ḥ-y-r* and *#ḥ-m-r*. These may ultimately be related, but both are attested synchronically in many languages. Table 21.4 and Table 21.5 show a short series of witnesses for these base forms;

These widespread roots suggest that wild ass was familiar to proto-Semitic speakers and that it was transferred early to the donkey.

Table 21.4 *#ḥ-y-r* base form.

Branch	Language	Form
	Ugaritic	pḥl
Canaanite	Classical Hebrew	ḥayr
Arabic	Classical Arabic	ḥayr
South Arabian	Mehri	ḥayr/ḥəyeer
Ethio–Semitic	Amharic	ahɨyya

Table 21.5 *#ḥ-m-r* root.

Branch	Language	Form	Gloss
	Ugaritic	ḥmr	
Arabic	Classical Arabic	ḥimaar	
	Shuwa Arabic	ḥumaar	
South Arabian	Epigraphic	ḥmr	wild ass
	Soqotri	ʃmálhen	
Ethio–Semitic	Gurage Caha	əmar	

6.3. Berber

There are two principal Berber roots, *#ayyul* and *#aʒəɗ*. Neither of these have any proven connection with any other Afroasiatic terms and probably represent ancient names for the North African wild ass transferred to the donkey at an unknown period.

6.4. Summary

The linguistic evidence suggests that individual branches of the Afroasiatic language phylum seem all to have quite distinctive lexical items for wild ass/donkey. In most cases, the speakers would have been familiar with the wild ass, and so would have named this creature in the pre-domestication era. Only the *#k-r* root is widespread in Central Africa and seems to have been carried from the Cushitic-speaking regions in the Horn of Africa to the Lake Chad region (hence the loans into Nilo–Saharan languages). This is consonant with the hypothesis that the donkey was taken into domestication several times around the fringes of the Sahara.

7. Patterns of spread of the domestic donkey

The spread of the domestic donkey can be divided into two key phases: the diffusion of domestic donkeys prior to European contact and the subsequent era. These two eras are not, as is common, distinguished by documentation; indeed, there are many lacunae in the historical record. The main differences are shown in Table 21.6.

Each of these call for some comment.

7.1. Documentation

By and large there are no records describing the spread of the donkey in the period prior to European contact. Arabic chronicle material describing this region refers to the donkey as already domesticated. Later texts in European languages usually refer to the presence of the donkey, not to its introduction.

7.2. Land or sea routes

The diffusion of the donkey in pre-European contact times, seems to have been strictly via land; most notably across the Sahara, but usually simply spreading

Table 21.6   Patterns of diffusion of the domestic donkey.

Spread prior to European contact	Spread post European contact
Sparse documentation though some graphic representation	Some historical documents
Donkeys spread only by land	Donkeys also spread by carriage in ships
Donkeys spread from farmer to farmer	Donkeys also spread through projects, state institutions, etc.
Slow	Rapid

gradually from area to area. However, once the donkey became seen as a product-ive animal for all of semi-arid Africa, it seems to have been brought to southern Africa in ships, hence its disjunct distribution.

There is a reference to so-called "Muscat" donkeys in Tanzania in the 1950s (Mason & Maule 1960:16). These were light-coloured donkeys associated with the Arabs and may thus have been brought from the Gulf region or from Egypt where they have a long tradition of use.

7.3. Informal versus formal diffusion

In the past, donkeys diffused principally from farmer to farmer or were sold by occupationally specialized pastoralists, as in West Africa. However, they have been spread in the present century as part of broad agricultural strategies associated either with the nation–state or with aid agencies. Most importantly, they have been recommended for traction in regions with light, sandy soils and the industrial manu-facture of axles for donkey-carts has also given their diffusion among farmers considerable stimulus. In the light of this, it is ironic in many ways that in southern Africa today they are seen by the authorities principally as a pest (Starkey 1995).

The informal diffusion of donkeys continues even today; the clearing of savanna forest south of the Sahel and the consequent decline in tsetse challenge has permit-ted donkeys to spread southwards. Donkeys can survive on unspecialized diets and can find food in the peri-urban wastelands surrounding many African towns. Sim-ilarly, deforestation and land degradation leads to decreased biodiversity; donkeys can feed on the shrubs that persist under these conditions.

The use of donkeys is closely related to road infrastructure and the price of rural transport. In Nigeria, for example, the oil boom era led to massive importation of small pickups and these came to be the preferred means of transporting farm pro-duce to market. Indeed prices of both vehicles and fuel were so low that many farmers sold their donkeys and breeders in the semi-arid region turned to other enterprises. However, once the recession set in at the end of the 1980s, the eco-nomics of motorized rural transport became more doubtful and farmers became anxious to acquire donkeys again. Having receded in Nigeria, the donkey is once again spreading (RIM 1992ii).

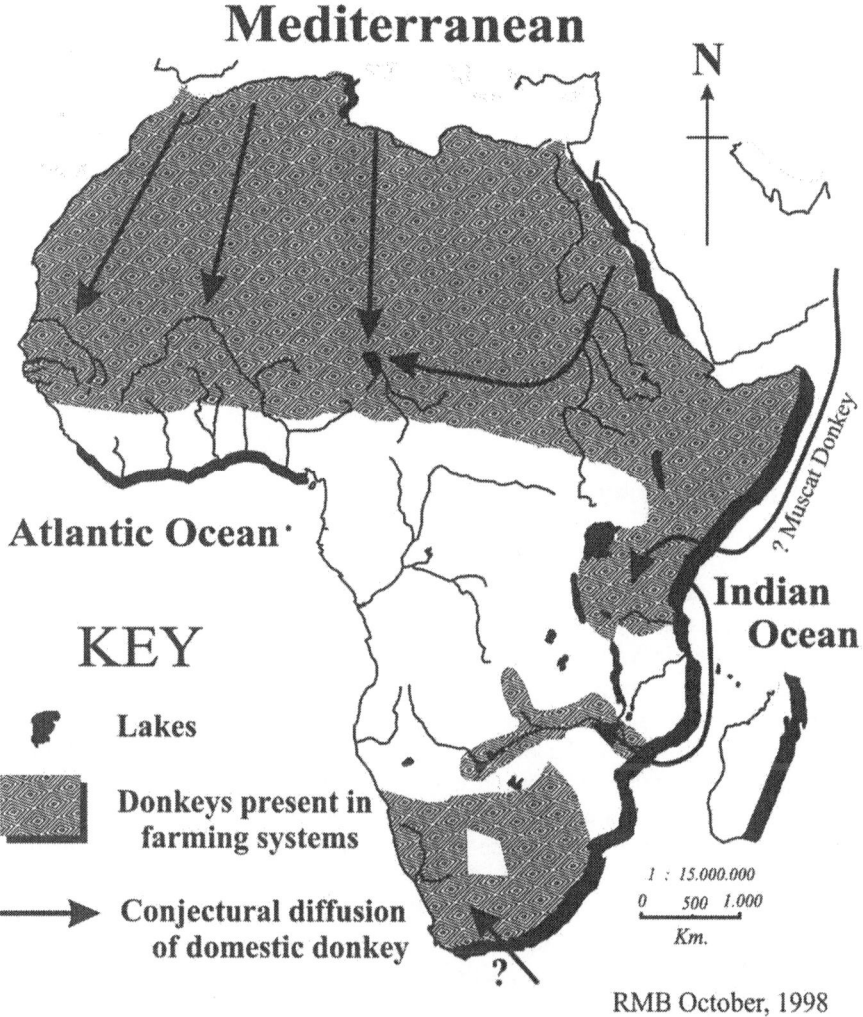

Figure 21.2   Distribution of donkeys in Africa in 1997.

8. Mules and hinnies

If the history of the donkey is known only very partially, the history of mules is almost completely invisible. Mules are the F_1 cross between a horse and a donkey and are valued for their hybrid vigour, but are generally infertile. Mules are usually produced from a male donkey with a female horse, and hinnies from the reverse pairing. Mules are presently used throughout North Africa and Ethiopia. They are very much associated with Arab culture and the reconstruction of "mule" in West Semitic languages suggests that they represent an ancient practice in the Near East.

Mules are difficult to detect archaeologically (i.e. their bones can often not be reliably distinguished from donkeys and horses). Although techniques *are* available (see Eisenmann 1986), it is safe to say that these have rarely been applied in Africa. In the light of this, only linguistics and ethnography have some potential for recovering their history.

A unique Egyptian wall-painting from the New Kingdom (*c.* 1570 BC) appears to represent a pair of hinnies pulling a chariot (Brewer et al. 1994: Fig. 8.3). However, they are almost unknown in the rest of the continent. Doutressoulle (1947:264) notes that there are mule races in Senegambia and Guinea, apparently brought from Algeria. Further east, in Niger and Nigeria, mules are not bred, apparently because it is thought to be unnatural to intentionally produce a sterile animal (RIM 1992). Where the donkey still represents a valuable possession, this is a rational strategy.

9. Conclusions and further research

The donkey certainly originated with the African wild ass, although it may have been domesticated several times in regions of its former range no longer represented by its present-day distribution. This appears to be confirmed by studies of terms for donkey in various African language families. Egypt remains the most likely centre for its early development for agricultural work, although without further archaeological data from outside the Nile Valley this is uncertain.

Although at least one archaeological site appears to confirm the donkey crossed the Sahara some 2,000 years ago, it may have been as a rare exotic, since both bones and rock-paintings are otherwise scarce. It is probable that donkey use only took off in West Africa with the rise of the long-distance caravan trade. However, there appears to be strong evidence for an east–west link suggesting that donkeys could have reached Lake Chad across the Sahel. Given the early dates for donkeys in the Ethiopia–Sudan region this would be quite reasonable.

To understand the broader parameters of donkey use and its role in the economic system of its owners, studies of productivity under traditional management such as those reported in §4 need to be replicated in other parts of the continent and stratified both according to ecological zone and production strategy. To understand the past we need to know considerably more about the donkey in the present.

Notes

1. I would like to thank Paul Starkey for general discussions as well as assistance with updating the map of current donkey distribution. Kevin MacDonald kindly helped me with the archaeological and rock art references as well as commenting on the whole paper and Stephen Hall pointed me in the direction of the special issues of *Ethnozotechnie*. Catherine Baroin kindly sent me an advance copy of her paper "*L'âne, ce mal aimé*" presented at the Méga-Tchad Colloquium, Orléans, October 1997.
2. I am grateful to Juliet Clutton-Brock for the stimulus to pursue this point.
3. A date of 1656 is given by Joubert (1995) but without supporting evidence.

References

Allsworth-Jones, P. 1982. Kariya Wuro faunal report. *Zaria Archaeology Papers* **4**, 6–9.

Baroin, C. in press. L'âne, ce mal aimé. In *Comptes-Rendues de la Colloque Méga-Tchad*, Orléans, 15–17 October 1997. Paris: ORSTOM.

Bender, M.L. 1988. Proto-Omotic: phonology and lexicon. In *Cushitic-Omotic: Papers from the International Symposium on Cushitic and Omotic Languages*, M. Bechhaus-Gerst & F. Serzisko (eds), 121–62. Hamburg: Buoke.

Blench, R.M. 1993. Ethnographic and linguistic evidence for the prehistory of African ruminant livestock, horses and ponies. In *The Archaeology of Africa. Food, metals and towns*. T. Shaw, P. Sinclair, B. Andah, A. Okpoko (eds), 71–103. London: Routledge.

Blench, R.M. 1995. A History of domestic animals in northeastern Nigeria. *Cahiers de Science Humaine* **31**(1), 181–238. Paris: ORSTOM.

Boettger, C. 1958. *Die haustiere Afrikas*. Jena.

Bökönyi, S. 1991. The earliest occurrence of domestic asses in Italy. *Equids in the Ancient World*, Volume II. R.H. Meadow & H.P. Uerpmann (eds), pp. 178–216. Wiesbaden: Ludwig Reichart.

Borwick, R. 1970. *Donkeys*. London: Cassell.

Brewer, D.J., D.B. Redford, S. Redford 1994. *Domestic plants and animals: the Egyptian origins*. Warminster: Aris & Phillips.

Camps, G. 1988. Âne. *Encyclopedie berbère, Cahier* **14**, 647–54.

Capderou, M. 1995. Raha nta' Sidi Brahim et autres regions rupestres de la région de Moghrar Tahtani (Monts des Ksour, Atlas Saharien, Algérie). *Sahara* **7**, 93–6.

Clutton-Brock, J. 1992. *Horse power: a history of the horse and donkey in human societies*. Cambridge, Mass.: Harvard University Press.

Desplagnes, L. 1907. *Le plateau central nigérien, une mission archaeologique et ethnographique au Soudan français*. Paris.

Doutressoulle, G. 1947. *L'élevage en Afrique occidentale française*. Paris: Editions Larose.

Eisenmann, V. 1986. Comparative osteology of modern and fossil horses, asses and half asses. In Meadow & Uerpmann (1986), 67–116.

Eisenmann, V. 1995. L'Origine des ânes: questions et réponses paléontologiques. *Ethnozootechnie* **56**, 5–26.

Epstein, H. 1971. *The origin of the domestic animals of Africa* [2 volumes]. New York: Africana Publishing.

Epstein, H. 1984. Ass, mule and onager. In *Evolution of domesticated animals*, I.L. Mason (ed.), 174–84. London: Longman.

Fernandes, V. 1938. *Description de la côte d'Afrique de Ceuta au Sénégal*. P. de Cenival & Th. Monod (eds). Paris.

Fielding, D. 1987. Donkey power in African rural transport. *World Animal Review* **63**, 23–30.

Fielding, D. 1988. Reproductive characteristics of the jenny donkey – *Equus asinus*: a review. *Tropical Animal Health and Production* **20**, 160–66.

Fielding, D. & R.A. Pearson 1991. *Donkeys, mules and horses in tropical agricultural development*. Edinburgh: University of Edinburgh.

Gautier, A. 1991. Mammifères fossiles Holocenes du Sahara Malien. In *Paléoenvironnements du Sahara: Lacs holocénes à Taoudenni (Mali)*, N. Petit-Maire (ed.), 173–6. Paris: Editions du CNRS.

Groves, C.P. 1966. Taxonomy. In *Sull'Asino Selvatico Africano*, C.P. Groves, F. Ziccardi, A. Toschi (eds), 2–11. Supplement to Richerche di Zoologia Applicata alla Caccia, Vol. 5, 1.

Groves, C.P. 1986. The taxonomy, distribution and adaptations of recent equids. See Meadow & Uerpmann (1986), 11–51.

Haltenorth, T. & H. Diller 1980. *A field guide to the mammals of Africa, including Madagascar*. London: Collins.

Ibn Batoutah 1893–1922. *Voyages d'Ibn Batoutah*. C. Defremery & B.R. Sanguinetti (eds and trans.). Paris.

Kaache, B. 1996. L'Origine des animaux domestiques au Maroc: état des connaissances. *Préhistoire et Anthropologie Mediterranéennes* **5**, 85–92.

Kingdon, J.A. 1997. *The Kingdon field guide to African mammals.* San Diego: Academic Press.

Kitchen, K.A. 1993. The land of Punt. In *The archaeology of Africa. Food, metals and towns*, T. Shaw, P. Sinclair, B. Andah, A. Okpoko (eds), 587–608. London: Routledge.

Levine, M.A. 1994. The analysis of mammal and bird remains. In *Excavations at Carthage, the British Mission, Vol II, Pt. I.* H.R. Hurst (ed.), 314–17. Oxford: Oxford University Press *for* the British Academy.

Levtzion, N. & J.F.P. Hopkins 1981. *Corpus of early Arabic sources for West African history.* Cambridge: Cambridge University Press.

Lewicki, T. 1974. *West African food in the Middle Ages.* Cambridge: Cambridge University Press.

Lutz, R. & G. Lutz. 1995. Spears and Ovoids in the rock art of Messak Sattafet and Mellet. *Sahara* 7, 89–96.

Mason, I.L. & J.P. Maule 1960. *The indigenous livestock of eastern and southern Africa.* Technical Communication No. 14 of the Commonwealth Bureau of Animal Breeding and Genetics, Edinburgh. Farnham Royal: Commonwealth Agricultural Bureaux.

Meadow, R.H. & H.P. Uerpmann (eds) 1986. *Equids in the Ancient world.* Wiesbaden: Ludwig Reichart.

Midant-Reynes, B. 1992. *Préhistoire de l'Egypte. Des premiers hommes aux premiers pharaons.* Paris: A. Colin.

Muzzolini, A. 1995. *Les images rupestres du Sahara.* Toulouse: Préhistoire du Sahara, 1.

Pankhurst, R. 1968. *Economic history of Ethiopia, 1800–1935.* Addis Ababa: Artistic Printing Press.

Peters, J. 1991. The faunal remains from Shaqadud. In *The late prehistory of the eastern Sahel*, A.E. Marks & A. Mohammed-Ali (eds), 197–235. Dallas: Southern Methodist University Press.

Phillipson, D.W. 1993. The antiquity of cultivation and herding in Ethiopia. In *The archaeology of Africa. Food, metals and towns*, T. Shaw, P. Sinclair, B. Andah, A. Okpoko (eds), 344–57. London: Routledge.

RIM 1992. *Nigerian National Livestock Resource Survey* [6 volumes]. Abuja, Nigeria: Report by Resource Inventory and Management Limited (RIM) to FDL&PCS.

Skinner, N.A. 1977. Domestic animals in Chadic. In *Papers in Chadic linguistics*, P. Newman & R.M. Newman (eds), 175–98. Leiden: Afrika-Studiecentrum.

Starkey, P. (ed.) 1995. *Animal traction in South Africa: empowering rural communities.* Gauteng: Development Bank of southern Africa.

Sutton, J.E.G. 1985. The antiquity of horses and asses in West Africa: a correction. *Oxford Journal of Archaeology* 4(1), 117–18.

Svendsen, E.D. (ed.) 1986. *The professional handbook of the donkey.* Sidmouth: The Donkey Sanctuary.

Tourneux, H. 1987. Les noms des equides en Afrique Centrale. In *Le Poney du Logone*, C. Seignobos (ed.), 169–205. Paris: Institut d'Élevage et de Médécine Vétérinaire des Pays Tropicaux.

Wilson, R.T. (1980). *Livestock production in central Mali: structure of the herds and flocks and some related demographic parameters.* ILCA Programme Document, Bamako.

Wilson, R.T., K. Wagenaar, S. Louis 1984. Animal production. In *Pastoral development in central Niger*, J.J. Swift (ed.), 69–144. Final report of the Niger Range and Livestock Project. Niamey: Government of the Republic of Niger.

Winkler, H.A. 1938–39. *Rock-drawings of Southern Upper Egypt* [2 volumes]. London: Egypt Exploration Society.

CHAPTER TWENTY-TWO

A history of pigs in Africa

Roger M. Blench

1. Introduction

The history of the domestic pig in Africa is highly controversial. Its ancestor, the wild pig, *Sus scrofa*, is native to North Africa, and its range extends along the Atlantic coast at least as far as the Rio de Oro. The Maghreb race is sometimes known as *Sus scrofa barbarus* and there was in addition a Saharan race known as *sahariensis* (Epstein 1971i:314). A more recent classification conjoins these into a single race *Sus scrofa algira* (Groves 1981:29). There is no positive evidence for the domestication of the pig in Africa although this was argued by some writers in the early part of this century (Epstein 1971ii). *Sus scrofa* gave rise to all domesticated pigs and it continues to thrive in the wild.

The wild pigs of Africa are the warthog, *Phacochoerus aethiopicus*, the giant forest hog, *Hylochoerus meinertzhageni* and the bush-pig, *Potamochoerus porcus*, none of which mate with domestic pigs and have thus made no contribution to its characteristics (Haltenorth & Diller 1980). Archaeologically, however, these species may be difficult to distinguish from bones of domestic pigs.

The history of pigs in sub-Saharan Africa is blurred by the circumstance that very large numbers of European pig breeds were brought to all parts of the continent with European contact, both as part of undocumented subsistence strategies (as with Portuguese introductions) and in conjunction with missionary and colonial agricultural development projects. In contrast to ruminants, these introductions thrived in what was an unfamiliar disease and climatic regime and crossed freely with resident populations. The genetic heritage of today's African pig populations is thus extremely mixed.

Another aspect of the history of pigs in Africa that is crucial when contrasting its distribution today with evidence for its former extension; the presence of Islam. Islam forbids Muslims to eat pork and this is usually interpreted as a prohibition on any sort of contact with pigs. Muslims living in multi-religion communities have tolerated pigs kept by others and indeed often make use of pigs or pork for magical purposes (see examples in Epstein 1971ii:330). Where Islam becomes dominant all pig production is forbidden and this has been responsible for the disappearance of pigs from a wide swathe of Africa in historic times. The Ethiopian Christian church, whose dietary prohibitions are usually based on the Old Testament, also bans the eating of pork and for this reason the pigs in Ethiopia are confined to the non-Christian regions in the west of the country. Some scholars believe that there was also a pork taboo in Ancient Egypt, accounting for the relative scarcity of pig remains in the Old and Middle Kingdoms compared with the pre-dynastic period. Brewer et al. (1994:96) summarize the arguments for and against this hypothesis.

For reasons unconnected with their economic importance, pigs have been relatively little researched in Africa. Doutressoulle (1947) and Mason & Maule (1960) both devote only a couple of pages to pigs, recounting frankly anecdotal material. During its years of operation, the International Livestock Centre for Africa in Addis Ababa excluded all research on pigs. This may be connected with prejudice against pigs from potential donor agencies but it also reflects a questionable belief that pigs compete with human beings for food. As a consequence, information about the distribution, productivity and genetic affiliations of African pigs is patchy and unreliable.

As in so many areas of African economic prehistory, Murdock (1959) made a number of innovative suggestions in relation to the antiquity of pig-keeping. He argued that the degree of ritual embedding in a culture can be taken as evidence of relative antiquity and gives several examples of the importance of pigs in ritual life. Epstein (1971ii:346) seems also to have leant to this view although he does not state it unequivocally. This chapter seeks to re-evaluate some of these arguments in the light of more recent research and to argue that the pig *is* a major species of African domestic stock. Cultural and linguistic evidence suggests that pigs were present in much of west–central and Equatorial Africa and that this fact has been obscured by researchers' stereotypes of its recent introduction.

2. Breed types

The pigs of sub-Saharan Africa are conventionally divided into two major types; the so-called "indigenous" pig and the introduced exotic breeds. The indigenous types are usually black or pied with medium, semi-erect, swept-back ears, a straight tail and a long snout. These are now found only in remote areas, especially in hill regions where there has been less opportunity to mate with incoming exotics. Wall-paintings and pottery representations suggest that the pigs of Ancient Egypt were black and that the piglets had striped underparts, suggesting a closeness to the wild boar. Although authors such as Mason (1988) group the West African "indigenous" pig such as the Ashanti Dwarf (Ghana) and the Bakosi (Cameroon) into the "West African", an Iberian type, it has nowhere been demonstrated that all the pigs of this type are indeed of Portuguese origin.

The exotic pigs in Africa that arrived in the colonial period came originally from Europe, America and the Far East. Almost all modern piggeries use exclusively exotics, especially Large White, Landrace, Duroc and Hampshire (RIM 1992ii, for Nigerian evidence). Most of the pigs found among smallholders in South Africa proper are of two European breeds, the Windsnyer and the Kolbroek. There are, however, residual populations of Chinese-style lard pigs in northeast Zimbabwe and adjacent Mozambique, apparently brought from Macau via the Portuguese Indian Ocean trade (Ellert 1993).

Under traditional management the exotics were allowed to breed uncontrolled with local races. The crossbred progeny of exotic and indigenous pigs take on the characteristics of the former, as the exotic breeds are highly prepotent. When cross-bred stock mate they produce a high proportion of progeny resembling the indigenous founder stock. After a few generations only traces of the exotic influence will be left.

3. Archaeological evidence

Pigs are usually thought to have been domesticated in Anatolia and the earliest archaeological finds of pigs date back to 7000 BC (Epstein & Bichard 1984). Pigs presumed to have been domesticated were kept in the Ancient Near East and Egypt from the end of the fifth millennium BC. So many bones have been recovered from the pre-Dynastic site of Merimde that it is assumed the population was raising pigs on a large scale (Brewer et al. 1994:97). None of this, however, constitutes direct evidence for the original locale of pig domestication; indeed the domestication of these early swine is often inferred from the contexts of their bones rather than their osteology. It would be appropriate, therefore, to keep open the possibility that pigs were either domesticated several times in this region or that they were indeed domesticated in Africa, as earlier authors thought.

Archaeological evidence for the distribution of the domestic pig in Africa might tactfully be described as slight. Iconographic evidence for pig production in the Egyptian Old and Middle Kingdoms is scarce, although by the period of the New Kingdom era there is not only an increase in depictions but also evidence of intensive pig production at Amarna (Kemp 1989:256).

In sub-Saharan Africa the situation is still less encouraging. Only two sites reviewed in this book contain bones of domestic pig, and one of these, Nkile in Zaïre, is nineteenth century (see Van Neer, Ch. 9 in this volume). Voigt & von den Driesch (1984) report a tentatively identified domestic pig from the ninth century site at Ndondowane in Natal. As Plug (1996) points out, this has yet to be confirmed in other southern African sites. There are a number of reasons for this lack of evidence:

(1) The ethnographic evidence suggests that pig-keeping is only sporadic even in areas where it is well-established.
(2) Detailed archaeozoological studies on large bone assemblages in the areas suggested are very rare.
(3) Only certain skeletal elements are diagnostic of *Sus*, for example, the teeth. Other bones may easily be confused with various wild suids.

It may well be that if archaeozoologists are more sensitized to the possibility of domestic pig bones then more will be identified.

4. The geography of past and present African pig populations

Pigs are generally not kept by nomadic pastoralists since they cannot survive by grazing for more than part of the year. They depend on grown food and are thus usually kept by settled farmers. As a result, they did not spread across the Sahara from North Africa with other domestic ruminants (Blench 1993, 1995). Their diffusion over long distances is often through being transported in boats, as in Oceania. They are likely to have reached sub-Saharan Africa by being taken down the Nile and could then have spread to west–central Africa overland along a corridor from Darfur to Lake Chad.

4.1. Pigs in North Africa

Pigs were once very widespread from Egypt along the North African littoral and along the Nile, as is amply attested by iconographic and archaeological evidence. Gilman (1975) shows convincingly that the domestic pig was the major source of food for the neolithic populations of Tangier, Morocco (i.e. 4000–1000 bc). At Carthage, for example, *Sus scrofa* was present in the Punic era, but becomes very common during Roman times (Levine 1994). The pig was a significant domestic animal among the Berbers, but the spread of Islam from the seventh century onwards confined pig-keeping to increasingly marginalized communities (Mouliéras 1905). Epstein (1971ii:330) quotes evidence that some Berber groups kept pigs in the Maghreb into recent times. Moreover, in Egypt, the avowedly Christian Copts still keep some pigs although these are tending to disappear in the present political climate. There were substantial importations of Mediterranean swine races along the North African littoral from the early periods of European trade and these importations expanded in the colonial era, accelerating the disappearance of indigenous races.

4.2. Pigs and the Guanche

Prior to their invasion by Spanish mariners in the fifteenth century, the Canaries were inhabited by the Guanche people, whose closest ethnolinguistic affiliation was to the Berbers of North Africa. Under what circumstances they arrived there is unknown since at European contact they had no seagoing vessels. The four major islands of the Canaries developed quite distinct cultures, suggesting that they had been in isolation from one another for some time. Archaeological work in the Canaries has given earliest settlement dates of *c.* 2000 BP although some material culture finds suggest rather earlier links with the mainland (Gonzalez & Tejera Gaspar 1990).

The Guanche and their culture were eliminated or absorbed by the Spaniards and there are only the rather scattered records of their customs and economy that can be gleaned from early travellers' accounts. The Guanche were pig producers and the pig also played a role in their ritual life on some islands. Pig bones have been recorded on all islands except Lanzarote and Fuerteventura (Mercer 1980:117).

"Wild" pigs, almost certainly feral, are recorded as part of the fauna of most of the western islands (Mercer 1980:13). Pigs must therefore have been brought to the Canaries with the first Berber incursions, well before the spread of Islam along the North African littoral.

4.3. Pigs on the Nile

Pigs were widespread and an apparently popular domestic species in Ancient Egypt. The first records of pigs in Upper Egypt are at Toukh in the second half of the fourth millennium (Epstein 1971ii:340, Epstein & Bichard 1984). Strikingly, pigs seem to have been used for work in Ancient Egypt, both treading and threshing seed in the eighteenth Dynasty, a practice also confirmed by Herodotus (Zeuner 1963:262). Pigs seem to have spread down the Nile at least as far as Sennar, where they are still kept (Spaulding & Spaulding 1988). Such pigs can be dated at least to the medieval period although they are probably older. Ethnographic records suggest that domestic pigs are very common in the Omotic-speaking regions ("Prenilotes") of the Ethiopian–Sudan borderland (Murdock 1959:173). The statement by Phillipson (1993:352) that pig "is not now kept in Ethiopia" is somewhat misleading, since it applies only to the highland areas. Fleming (1965) argued on linguistic grounds that pig-keeping was ancient among Omotic speakers in Ethiopia. Bechhaus-Gerst (Ch. 24 in this volume) extends his argument for the central Sudan.

4.4. Pigs in west–central Africa

Pigs are kept in a wide belt of the forest–savanna region of West Africa today although they are barely discussed in standard livestock texts. Some of the pigs in West Africa were introduced at an early period by the Portuguese, "unimproved Iberian swine", as Epstein has it. Jollans (1959) is one of the first authors to draw attention to the "native" pigs of the Ashanti in Ghana. He notes that they are both numerous and popular with farmers as well as being hardy and able to survive on a varied diet. These characteristics are also found among pig populations in isolated areas of south–central Nigeria (Adebambo 1982, RIM 1992ii).

These populations may well be pre-Portuguese to judge by their degree of establishment. The linguistic evidence seems to distinguish sharply between those regions where pigs have Portuguese-derived names and those where they do not (see §6, and Appendix). Pig-keeping may therefore have spread across the savannas of Central Africa from the Ethiopian borderlands to central Nigeria in medieval times. An ethnographic link between these two regions is probably represented by the pigs of the Nuba people of Kordofan (Epstein 1971ii:332). Barth (1857–8) mentions that feral pigs were common in Chad in the nineteenth century, which would provide the appropriate geographical link. The black, hairy pigs found today in remote areas of the Nigerian Middle Belt may also be a relic of this practice. Connah (1981:185) recovered a fired clay figurine at Daima in northeast Nigeria that might be a pig, though no bones have been definitely identified.

Although there is little or no documentation relating to the introduction and spread of pigs in the early period of European contact it is fairly certain that the

Portuguese played an important role in the coastal areas. Williamson (pers. comm.) observes that along the sea-coast of the Bight of Benin, most languages have borrowed their term for pig from Portuguese.

4.5. Pigs in Equatorial Africa

Pigs cannot thrive in dense tropical forest, but with small cleared areas they appear to adapt to conditions of intense moisture. The first European incursions into West Equatorial Africa encountered pig-keeping in a strip stretching from southern Cameroon, through the Congo and down into Angola. This is shown as the "Angola extension" on Figure 22.1. If the report of domestic pig bones in Natal (§3) is confirmed by other finds, then to explain their presence a convincing route must be established. The path by which they could have reached South Africa is at present unknown but it is likely to have been an extension of the equatorial pig-keeping zone. This is shown on Figure 22.1 as passing along the valley of the Zambezi, simply because this is a well established pig rearing area today.

As elsewhere, the system of production is essentially semi-feral. Descriptions of pig-keeping in Equatorial Africa are given in Adamantidis (1951) and Merckx (1956). Pigs are largely left to find their own food and kept attached to the residential unit with occasional specially prepared food. They can be confined in the growing season if there is danger of damage to the crops. When they must be slaughtered they are hunted down. Pigs played an important role in the ritual of some societies. Among the Fang of Gabon and adjacent regions, Tessmann & Wasmuth (1913) records that the domestic pig was important in the Sso cult.

4.6. Pigs in the Senegambian region

Murdock (1959:266) pointed out that there is a nuclear area of apparently ancient pig production among the speakers of Atlantic languages near the Gambia river. Bernatznik (1933) seems to have been the first to document the importance of the pig in this region although it is clearly possible this nucleus is of Portuguese origin. It is striking, however, that words for "pig" in the languages of this region do not resemble Portuguese terms, suggesting at least the possibility that this "island" of pig production has a distinct origin, perhaps connected with the Guanche in the Canaries.

4.7. Pigs in eastern and southern Africa

It is usually accepted that there were no pig populations in eastern and southern Africa prior to European contact despite a single report of domestic pig remains from a ninth century site in Natal (Voigt & von den Driesch 1984). The pig populations of eastern and southern Africa today are usually assumed to result from European introductions (Holness 1974). Mason & Maule (1960) imply that this was solely a nineteenth century phenomenon, but an earlier date is likely, as the Indian Ocean trade brought southeast Asian pigs as well as Mediterranean breeds. The Portuguese brought pigs to the East African coast via Goa, and these diffused inland along now-defunct trade-routes from ports such as Sofala on the Mozambique coast

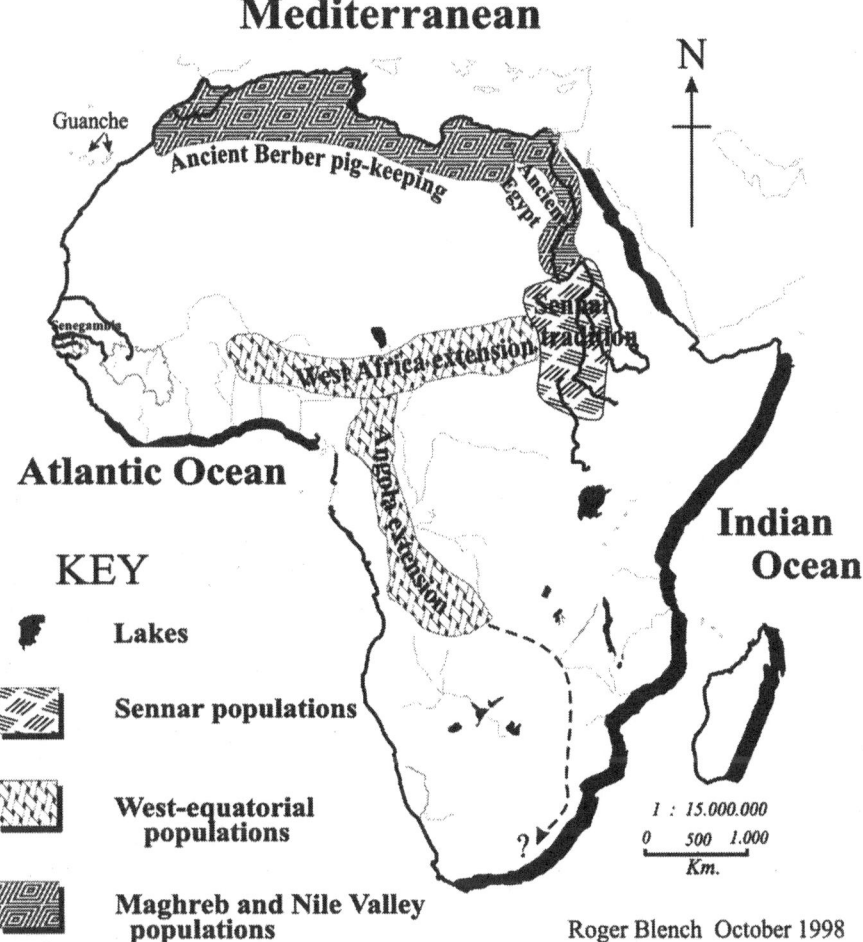

Figure 22.1   Historical distribution of pig-keeping in Africa.

(Ellert 1993). They presumably then spread slowly northwards, since mid-nineteenth century travellers recorded the presence of pigs far inland. This contact with Macau seems to have been responsible for the isolated populations of Chinese lard pigs found today in northeast Zimbabwe (Ellert 1993).

4.8. Summary

The precise historical distribution of pigs in Africa must remain guesswork until substantially more archaeological and genetic data are available. Figure 22.1 presents a synthesis of the broad areas where pigs seem to have been present prior to European contact. It should be compared with Figure 22.2 which shows the modern-day distribution of pigs in Africa.

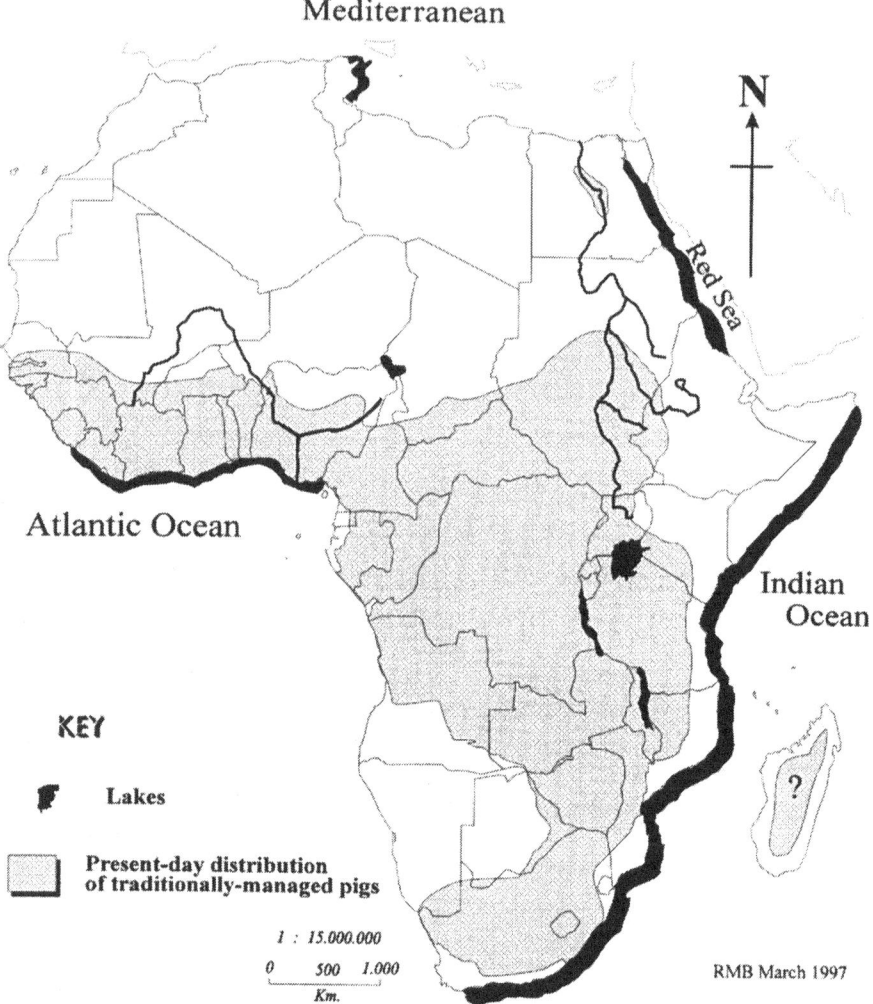

Figure 22.2   Modern-day pig-keeping in Africa.

5. Management and husbandry

Of all the major species of domestic livestock in Africa, the pig is the least well known. Compared with the extensive bibliography for cattle, sheep and goats, there are relatively few descriptions of smallholder pig production systems. Pigs are exceptional in the broad context of African livestock production, since significant numbers are kept today in semi-intensive production systems, usually near to towns. None the less, the majority of pigs are still kept in villages under traditional management.

Pig production systems can be divided into two main categories; herding/scavenging systems where the pig finds most of its own food, and intensive systems where the majority of the food consists of household scraps or specially grown food. True herding systems were common in Europe until recently. In England, pigs were moved into acorn woods for some months and in Spain, cork forests, *dehesas*, are still used to feed pigs for part of the year. Similar systems are frequently mentioned in the Bible and other Near eastern texts, and to judge by references in Egyptian materials were common both in the Nile Valley and North Africa.

Swineherding seems not to have spread to sub-Saharan Africa. Pigs are either more intensively managed or else allowed to scavenge freely with no management at all. In areas where there are no farm crops to damage, such as in the Niger Delta mangroves, pigs are allowed to forage freely and are shot by hunters when they are needed for a sacrifice (RIM 1992ii). The absence of swineherding may be related to relatively low population densities in Africa. Where population densities are high, pigs must be either penned or watched if they are not to damage crops or property. Economies of scale then suggest that a large number of pigs are herded together to economize on labour.

5.1. Smallholder production

The most common system of pig production is smallholders who keep pigs on a "backyard" basis. Pigs differ from ruminant livestock in two important ways: they eat the same sort of food as human beings, and they are extremely destructive in their search for food, uprooting tuber crops and tearing down fences. The consequences are that they must be fed specially prepared food for some part of the year and they are often confined even when goats, for example, are free range. Pigs can only be allowed to scavenge for food when there are no crops in the ground or the fields are protected by strong fences.

Pigs are usually housed, rather than tethered like goats and sheep, as they are more selective feeders. Pigs have not traditionally been favoured by livestock schemes because they are deemed to compete with humans for food. This may be true where pigs are fed cassava grown purposely for them, but the great majority of pigs depend on residues from beer-brewing or tree-crop processing that would otherwise be wasted. Table 22.1 summarizes the principal systems of pig production in Africa:

Table 22.1   Systems of pig production.

	Characteristics			
	Housing	Ownership	Feeding	Breeding
Scavenging	None	Often communal	None	Uncontrolled
Herded	None	Individual	Seasonal diet supplements	Uncontrolled
Semi-intensive	Semi-permanent construction from local materials	Individual smallholders	Household waste and sometimes specially grown cassava	Uncontrolled or use of local stud boars
Intensive (Modern)	Modern pens made of concrete with zinc roofing	Urban-based entrepreneurs and businessmen	Agro-industrial by-products	Only selected boars used for stud

6. Linguistic evidence

One of the most compelling pieces of evidence for the early spread of the domestic pig from the Nile to other regions of Africa both east and west is provided by linguistics. Despite the hypothesis of an Iberian introduction, borrowed vernacular terms for pig have a very restricted distribution. This is in contrast to terms for crops such as cassava or citrus, whose Portuguese provenance is clear (Williamson 1970).

Spaulding & Spaulding (1988) and Bechhaus-Gerst (Ch. 24 in this volume) have made preliminary compilations of the evidence. However, it is clear that the pig terms spread considerably further into West Africa than they record. The Appendix (#-kutu. pig) presents an expanded version of vernacular names for pig with West African data included. There is a chain of terms stretching from eastern Burkina Faso to the Sudan–Ethiopian borderlands that appears to be unrelated to European introductions. Manessy (1972), in an investigation of names for domestic animals in Gur languages, maps a distinct frontier between terms derived from Portuguese *porco* and terms of no known origin. He also observes that although today's term in Hausa, *aleedi*, is borrowed from Yoruba, the nineteenth century term was *gursunu*, resembling those found across the Sahelian belt.

7. Conclusion

This chapter proposes some very radical revisions to the conventional picture of African pig-keeping. It suggests that the domestic pig spread to sub-Saharan Africa relatively early, certainly following the Nile Corridor, and thence as far as South Africa. The evidence for this is cultural and linguistic rather than archaeological; however, the temptation to ascribe suid bones to wild pig species may be partly responsible for this lacuna.

The linguistic evidence is compelling; it suggests strongly that the small black pigs of the interior of Africa were indeed part of an ancient pig-keeping culture that spread from the Nile across Central Africa, southwards with the Bantu expansion and westwards into Ghana and Burkina Faso. The rise of Islam drove pig production into isolated regions throughout northern Africa and the Sudan. The semi-feral nature of the production system and the relative success of imported breeds has led to the virtual disappearance of "indigenous" breeds in many areas and the illusion that all village pigs are descendants of Portuguese imports. The pig, the "democratic philosopher of the Medieval Sudan" (Spaulding & Spaulding 1988) can now be highlighted as a significant and ancient element in African subsistence strategies.

Appendix 22.1. Names for pig in west–central Africa

#-kutu. pig

Phylum	Family	Branch	Language	Gloss
NS	Koman	Anej	kuturu	
NS	ES	Nyimang	kudur	
		Old Nobiin	kutun	
		Dair	kid'aŋ	
		Gule	kuturu	
		Lumun	kutura	
		Temein	kudur	
		Uduk	ḳuthar	
NS	Maba	Aiki	gìrwà	wart-hog (?)
NS	Saharan	Kanuri	godú	warthog
NS	Kadu	Kamdang	b-oḍuruk pl. k-aḍuruk	
		Kadugli	kuḍuru	
NC	Kordofanian	Orig	kàdìrú	
	Kordofanian	Talodi	b/k-uduru	
	Kordofanian	Otoro	kudur	
NC	Benue–Congo	Nupe	kutsũ	
NC	Kwa	Fon	agurusa	
NC	Gur	Dagbane	kurutʃu	
NC	Bantu	#CB	#-gùdú	wild pig
AA	Omotic	Kefa	gudinoo	
AA	Semitic	Sudan Arabic	kadruuk	
AA	Chadic	Hausa	gursunu	

NS=Nilo–Saharan   NC=Niger–Congo   AA=Afroasiatic

This root appears in Nilo–Saharan, Niger–Congo and Afroasiatic language phyla and is sometimes applied both to the warthog and the bush-pig (*Potamochoerus porcus*). Manessy (1972:314) points out that the chain of lexemes connecting to the Gur languages can be traced through dialect and obsolete terms cited in Koelle. Cited by Gregersen (1972:86) who used this as evidence for a proposed "Kongo–Saharan" grouping (wrongly, given that it is clearly a widespread cultural loan). Gregersen (op cit.) also mentions Greenberg's suggestion that the Saharan form was loaned into *PB. Schadeberg & Elias (1979:84) mention that this root has been loaned into Sudanese Arabic to give *kadruuk*.

References

Adamantidis, D. 1951. Organisation et exploitation des élevages porcines à la colonie. *Bulletin agricole du Congo Belge* **42**, 1007–32.

Adebambo, O.A. 1982. Evaluation of the genetic potential of Nigerian indigenous pigs. *Proceedings of the Second World Conference on Genetics applied to Livestock Production*, 543–53. Madrid, Spain.

Barth H. 1857–8. *Travels and discoveries in North and Central Africa: being a Journal of an expedition undertaken under the Auspices of HM's Government in the Years 1849–1955* [5 volumes]. London: Longmans.

Bernatznik, H.A. 1933. *Aethiopien des Westens* [2 volumes]. Wien.

Blench, R.M. 1993. Ethnographic and linguistic evidence for the prehistory of African ruminant livestock, horses and ponies. In *The archaeology of Africa. Food, metals and towns*, T. Shaw, P. Sinclair, B. Andah, A. Okpoko (eds), 71–103. London: Routledge.

Blench, R.M. 1995a. A history of domestic animals in northeastern Nigeria. *Cahiers de Sciences Humaine*, ORSTOM **31**(1), 181–238.

Brewer, D.J., D.B. Redford, S. Redford 1994. *Domestic plants and animals: the Egyptian origins*. Warminster: Aris & Phillips.

Connah, G. 1981. *Three thousand years in Africa*. Cambridge: Cambridge University Press.

Doutressoulle, G. 1947. *L'élevage en Afrique occidentale française*. Paris: Editions Larose.

Ellert, H. 1993. *Rivers of gold*. Gweru: Mambo Press.

Epstein, H. 1971. *The origin of the domestic animals of Africa*. [2 volumes]. New York: Africana Publishing.

Epstein, H. & M. Bichard 1984. Pig. In *Evolution of domesticated animals*, I.L. Mason (ed.), 145–62. London: Longman.

Fleming, H.C. 1965. *The age-grading cultures of East Africa: an historical inquiry*. PhD Anthropology, Michigan: Ann Arbor.

Gilman, A. 1975. *The later prehistory of Tangier, Morocco*. Cambridge, Mass.: Peabody Museum of Archaeology and Ethnology.

Gonzalez, A.R. & A. Tejera Gaspar 1990. *Los aborigenes canarios*. Oviedo: Ediciones ISTMO.

Gregersen, E.A. 1972. Kongo–Saharan. *Journal of African Languages* **11**(1), 69–89.

Groves, C. 1981. *Ancestors for the pigs: taxonomy and phylogeny of the genus Sus*. Technical Bulletin No. 3. RSPS, Canberra: Department of Prehistory.

Haltenorth, T. & H. Diller 1980. *A field guide to the mammals of Africa, including Madagascar*. London: Collins.

Holness, D.H. 1974. The role of the indigenous pig as a potential source of protein in Africa – a review. *Rhodesian Journal of Agricultural Research* **73**(3), 59–62.

Jollans, J.L. 1959. A preliminary report on the indigenous pig of Ashanti, Ghana. *Journal of the West African Science Association* **5**, 133–45.

Kemp, B. 1989. *Ancient Egypt: anatomy of a civilization*. London: Routledge.

Levine, M.A. 1994. The analysis of mammal and bird remains. In *Excavations at Carthage, the British Mission. Vol. II*, Pt. 1, H.R. Hurst (ed.), 314–17. Oxford: Oxford University Press *for* the British Academy.

Manessy, G. 1972. Les Noms d'Animaux domestiques dans les langues Voltaiques. In *Langues et Techniques, Nature et Societé. Approche Linguistique*, vol. 1, 301–20. Paris: Klincksieck.

Mason, I.L. 1988. *A world dictionary of livestock breeds, types and varieties*, 3rd edn. Wallingford, UK: CAB International.

Mason, I.L. & J.P. Maule 1960. *The indigenous livestock of eastern and southern Africa*. Technical Communication No. 14 of the Commonwealth Bureau of Animal Breeding and Genetics, Edinburgh. Farnham Royal: Commonwealth Agricultural Bureaux.

Mercer, J. 1980. *The Canary islanders: their prehistory, conquest and survival*. London: Rex Collings.

Merckx, C. 1956. Élevage du gros et petit bétail dans la Province de Léopoldville. *Bulletin agricole du Congo Belge* **47**, 31–73.

Mouliéras, A. 1905. *Une tribu zénète anti-musulmane au Maroc*. Paris.

Murdock, G.P. 1959. *Africa: its peoples and their culture history*. New York: McGraw Hill.

Phillipson, D.W. 1993. The antiquity of cultivation and herding in Ethiopia. In *The archaeology of Africa. Food, metals and towns*, T. Shaw, P. Sinclair, B. Andah, A. Okpoko (eds), 344–57. London: Routledge.

Plug, I. 1996. Domestic animals during the early Iron Age in southern Africa. *Aspects of African Archaeology. Papers from the 10th Congress of the PanAfrican Association of Prehistory and Related Studies*, G. Pwiti & R. Soper (eds), 515–22. Harare: University of Zimbabwe Publications.

RIM 1992. *Nigerian National Livestock Resource Survey* [6 volumes]. Report by Resource Inventory and Management Limited (RIM) to FDL&PCS, Abuja, Nigeria.

Schadeberg, T.C. & P. Elias 1979. *A Description of the Orig language.* Tervuren: MRAC.

Spaulding, J.L. & J. Spaulding 1988. The democratic philosophers of the Medieval Sudan: the pig. *Sprache und Geschichte in Afrika* **9**, 247–68.

Tessmann, G. & E. Wasmuth 1913. *Die Pangwe: völkerkundliche Monographie eines westafrikanischen Negerstamme* [2 volumes]. Berlin: Hansa-Verlag für Modern Literatur.

Voigt, E.A. & A. von den Driesch 1984. Preliminary report on the faunal assemblages from Ndondondwane, Natal. *Annals of the Natal Museum* **26**(1), 95–104.

Williamson, K. 1970. Some food plant names in the Niger Delta. *International Journal of American Linguistics* **36**, 156–67.

Zeuner, F.E. 1963. *A history of domesticated animals.* London: Hutchinson.

CHAPTER TWENTY-THREE

Did chickens go west?

Kay Williamson

Cock Ugbon-alele am I.
I obtain this nickname through
   The crown on my head
And my generosity ...
   And the great help
I give men.
   I offer myself for their sacrifices
And I measure the day
   And night into equal parts
And tell them the time
   For labour and for rest.

I meet my wives
   In every public place,
I am Cock Ugbon-alele
   The great.
Richard Ayeberẹmọ Freemann (1972)

1. The origin of the domestic fowl or chicken

The domestic fowl originates from the junglefowl, the four species of which are distributed in southern Asia from Pakistan to Indonesia. Stevens (1991) summarizes the situation as follows. The Red Junglefowl (*Gallus gallus*) has the widest distribution and is thought to be the main progenitor of the domestic fowl, though some genes have probably been introduced from other species by hybridization, particularly in Asiatic breeds. Those who favour a single or monophyletic origin refer to the domestic fowl as *G. gallus*, while those who believe in a multiple or polyphyletic origin call it *Gallus domesticus* or *G. gallus domesticus*, depending on whether it is regarded as a species or a subspecies.

2. The spread of the domestic fowl in Eurasia and America

There is no doubt that the domestic fowl originated in Asia, so its distribution throughout the world has to be traced from there. Until recently it was believed that it was first domesticated around the Indus Valley about 2000 BC, because chicken bones were found in the upper levels of Mohenjo-Daro, and it has been assumed that it diffused from there to both east and west. There is a difficulty here in that the Indus Valley is apparently within the range of the Grey Junglefowl (*G. sonnerati*) and not *G. gallus*. West & Zhou (1988) presented extensive evidence that northern China has far older sites than India, dated as far back as 6000 BC, yielding remains of domestic fowl, and they argue that it was first domesticated in southeast Asia, in the range of *G. gallus*, and carried northeast into China; it could not have been domesticated near the Chinese sites, because they are climatically unsuitable for the wild junglefowl and there is no evidence that the climate of northern China has changed dramatically in the last 10,000 years. West & Zhou (1988: Fig. 1) show that there are 13 other sites in Asia and as far west in Europe as Greece which are older than Mohenjo-Daro; others, from China to Romania, are roughly contemporary, while those further west in Europe are later. West & Zhou's argument for a southeast Asian origin has been supported by Fumihito et al. (1994), who have shown convincingly that all modern chicken genes can be derived from the subspecies of *Gallus* found in northeast Thailand.

Carter (1971) provides a wide-ranging survey of the spread of the domestic fowl in Eurasia and the Americas, noting features that distinguish Asian from European breeds. He observes that Asian features appear in Polynesia and also in large parts of South America. He argues that while the Spanish undoubtedly introduced their European chickens to parts of the Caribbean and Central and South America, where they were often known by their Spanish name *gallina*, these coexisted with Asian types that were older and bore non-Spanish names such as *hualpa* and *takara/karaka*; Spanish sources distinguish between the Spanish-introduced *gallina de Castille* and the older *gallina de Mexico*. He concludes that there were a number of pre-Columbian introductions from different parts of Asia, evidenced by similarity of names as well as appearance, and suggests that archaeological collections of unidentified bird bones might be re-examined. His final plea is as valid today as in 1971:

> We desperately need ethnological reporting of details of chickens in the hands of native people as well as a wealth of information on names, uses, attitudes, customs, and mythology related to chickens.

Langdon (1990) more specifically argues that the Araucanian or blue-egg chicken was the result of mutations from specifically Japanese types of fowl, that they reached South America by sea from Japan, and thence were carried to Easter Island. He also notes that two distinct types of fowl were reported in Polynesia by early European visitors; a small cock-fighting breed and a large culinary breed.

3. Archaeological evidence for the domestic fowl in Africa

The archaeological evidence for the domestic fowl in Africa is summarized by Clutton-Brock (1993:62): the earliest evidence is a sketch of a cockerel on an ostracon from the tomb of Rameses IX (1156–1148 BC). Chickens were not common in Egypt until the Ptolemaic period (332–330 BC). In West Africa they have been excavated from the Iron Age site of Jenne-jalo in Mali, dated to about AD 500–800. In East Africa they are recorded from two Iron Age sites in Mozambique, and in southern Africa from eighth century Iron Age sites; Plug (1996:517) confirms that they are found in early Iron Age sites, but were not apparently common.

MacDonald (1992:304) has drawn attention to the neglect of questions concerning the archaeological study of the introduction and spread of the domestic fowl in Africa, and has provided detailed criteria for distinguishing between the bones of chickens, guinea-fowl and francolins. He also notes the small size of the chickens in Malian Iron Age sites, "comparable to modern bantam breeds or wild junglefowl", and noted (p. 306) that a type of small-sized chickens has been reported from Rwanda. He observes that in Europe smaller size fowl have also been reported from Iron Age sites, and that the size of European chickens increased suddenly in early Roman times both inside and outside the Roman Empire. He suggests that smaller size is probably linked with chickens being left to forage for their own food, as is still the case in traditional villages in Africa.

It is likely that, as believed by MacDonald (1992:307), chickens may have been present in Africa well before the earliest date yet attested by archaeology, and that their presence may have been overlooked because of the difficulty of distinguishing their bones from those of similar species.

MacDonald & Edwards (1993) report the remains of a chicken in an apparently ritual context at Qasr Ibrim in Egypt, dated to the late fifth or early sixth century AD. MacDonald (1995) states that the chicken reached West Africa comparatively late, during the middle of the first millennium, and raises the interesting question as to why it became so crucial in mythic, ritual and iconographic contexts after this late arrival. He suggests first that it "was introduced at a time of social complexity when its symbolic power could initially be manipulated and established by elites", and secondly that its "exotic appearance and behaviour easily created a new and important niche for them in West Africa's animal mythology" (1995:55). While this appears a reasonable conclusion from the present archaeological evidence, the linguistic evidence discussed below still suggests an earlier introduction of the chicken into West Africa.

4. Linguistic evidence for domestic fowls in West Africa

4.1. Method

Linguistic evidence for the introduction of domestic animals is useful if it is carefully used. If a word for a particular creature can be reconstructed to a proto-language, it is likely that the animal was known to the speakers of that proto-language, unless there has been a shift of meaning from that of a similar creature. If there was indeed a name in the proto-language for the creature, which continues to be familiar to the speakers of the daughter languages, the word should appear in the daughter languages showing the regular sound shifts that appear in other words in those daughter languages. Thus, to decide if a language retains an original word from its proto-language, it is necessary that the regular sound shifts have been carefully worked out.

If, on the other hand, the creature was introduced later, when the proto-language had already begun to split into its daughter languages, and its original name was introduced with it, a similar word is likely to appear in the various daughter languages without showing the regular sound shifts. This word will, furthermore, be similar to that of the source language, and will also appear in neighbouring languages that do not share the same proto-language. To distinguish between the two cases, retention from the proto-language and borrowing at a later stage, is by no means always easy, especially when the proto-language has not been reconstructed.

I shall now look at the names for "chicken/domestic fowl" and "cock/rooster" in the three language phyla which are represented in West Africa: Niger–Congo, Afroasiatic, and Nilo–Saharan.

4.2. Linguistic evidence in Niger–Congo

The discussion follows the branches of the Niger–Congo family tree, basically as presented in Bendor-Samuel (1989). The detailed classification into groups within each table generally follows that of the *Ethnologue language family index* (Grimes & Grimes 1993), except in the case of Nigerian languages, which generally follow Crozier & Blench (1992).

Forms in particular languages are given in bold. Reconstructions by authors using regular sound correspondences are preceded by *. Quasi-reconstructions based on inspection of similar forms are preceded by #. Formulae for widespread roots, whether the result of inheritance from a proto-language or of borrowing, are represented in capital (upper-case) letters.

4.2.1. Kordofanian (Table 23.1)   The most widely spread form consists of a root #akaro preceded by a singular prefix k-. If Kordofanian has lost root-initial consonants, it is conceivable that this represents the root which will be given below, the formula KKR. The only Rashad form available shows a root something like -durik. (See Appendix for Tables 23.1 to 23.20.)

4.2.2. Mande (Table 23.2)   The most widespread root for "fowl" is reconstructed by Dwyer (1987/88) as proto-Mande *togo. Where appropriate, the various linguistic forms of this root will be summarized in a formula TK. Table 23.2 shows Susu-Yalunka tɔyɛ, suggesting original ɔ-ɛ vocalism; I would therefore suggest revising Dwyer's reconstruction to #tɔgɛ. It is easy to assume #tɔgɛ > #tɛyɛ > tɛɛ, as consistently in southwestern Mande; or #tɔgɛ > tɔyɔ > tɔɔ, as in the Tura–Dan–Mano group. This is the only root in western Mande, except for sīsé in Mandekan. The different phonological forms correlate rather neatly with the group boundaries and look as if they could well have developed regularly from a proto-form in western Mande.

In eastern Mande, #tɔgɛ is found only in the Tura–Dan–Mano group as tɔɔ. We must therefore consider the possibility that it is borrowed from adjacent western languages. In Dwyer's (1989) map, Tura is isolated from Dan–Mano; the common root must therefore date back to a time prior to the separation of Tura. The adjacent western languages are southwestern, with téé, and Mandekan, with sìsé; neither of these is likely to be borrowed as tɔɔ. If #tɔgɛ was borrowed into proto-Tura–Dan–Mano, therefore, it must have been at an early stage, before the development of the téé and sìsé forms. It is thus, linguistically, quite possible that the root #tɔgɛ occurred in proto-Mande, at least before proto-Tura–Dan–Mano had lost touch with western Mande. If the referent was always the fowl, this implies a considerable time depth for the fowl in this part of West Africa. Proto-Mande is externally well separated from the rest of Niger–Congo and is currently classified as one of the earliest branchings from proto-Niger–Congo, but internally is not as differentiated as some other branches of Niger–Congo. Lexicostatistic percentages go as low as 17 per cent (Dwyer 1989), which suggests a time depth of some thousands of years. Such time depths are not supported by the existing archaeological evidence for the fowl in West Africa, though MacDonald (1992:307) believes that "the origins of chicken in Africa may date to appreciably before the *c.* AD 500–800 dates for the Jenne-Jeno chicken".

The next most widely spread root is kɔɔrɔ or kɔtɔ, which will be referred to by the formula KKR. In the meaning "fowl" it is restricted to the three eastern groups (Samo, Bisa, and Busa) of southwestern Mande in eastern Mande, all spoken in areas surrounded by non-Mande languages. As we shall see, this root recurs in non-Mande groups and there is no reason to consider it to be as ancient in Mande as #tɔgɛ. The root also occurs as "cock" in scattered languages; not only Bobo kokor- in eastern Mande, but Sembla kukee, Susu koŋkoːre, and Kono koːɲɛ in northwestern Mande. Its scattered distribution in northwestern Mande suggests it is a loanword in these groups.

Other restricted roots for "fowl" are (nã)-nõn in Bobo, which may be a semantic extension of an ancient Niger–Congo root for "bird" (formula NN) and #maị in Mwan and Yaouré, apparently a local innovation. "Cock" is most commonly "fowl male", but in northwestern Mande Yalunka and Manding (Mandinka, Bambara) show #dondon-.

The proto-Mande for "bird" has been reconstructed as *kuani (Westermann 1927:185). This will be given the formula KN.

4.2.3. Atlantic (Table 23.3)   Atlantic (Table 23.3) presents a much more varied picture. The internal diversity of Atlantic is much greater than that of Mande, as shown by lexicostatistics (Sapir 1971), and no reconstruction of the whole family has yet been attempted. In southern Atlantic the Mel languages have a consistent root #tɔkɔ, weakening to tsɔgɔ, sɔk, and sɔɔ, which I relate to the TK root; without knowledge of the sound correspondences within Mel (t = ts = s is not among the sample correspondences presented by Dalby (1965)) it is not possible to say whether these are regular developments from a proto-Mel #tɔkɔ or developments from a more recent loanword. Given the fact that the Mel languages are in contact with the Susu-Yalunka group of Mande, for which we have already noted a root #tɔgɛ > tɔyɔ, it appears likely that this root was borrowed from that group at an early stage, since tɔkɔ, with a voiceless intervocalic consonant, is likely to reflect an earlier stage than tɔgɛ or tɔyɔ. Possibly, therefore, the earliest Mande form, no longer reconstructible from within Mande itself, was #tɔkɛ, becoming #tɔkɔ by the time it was borrowed into Mel. Lexicostatistic percentages within Mel go as low as 14 per cent (Sapir 1971), and are thus very roughly comparable with those in proto-Mande; if the root was indeed borrowed at a proto-Mel stage, a greater time depth than can be attested archaeologically for the fowl in West Africa is again suggested.

Limba, within southern Atlantic but excluded from Mel, shows tɛ, which looks like a loan from neighbouring southwestern Mande languages.

Dalby (1965:14–15) presents a word for "chicken-coop" in Bullom and Temne (to which can be added Sherbro from Pichl (1963:6)) among his examples of regular sound correspondences between different subgroups of Mel, and therefore implies that the word can be reconstructed to proto-Mel:

	sg.			pl.	
Temne	kə	–	bənthi	tə	–
South Bullom		–	binti	thi	–
Sherbro		–	bịnti	ti	–

This in turn would imply that the keeping of chickens in coops goes back to ancient times. It should be noted, however, that Bullom and Temne are neighbours, so that in the absence of more widespread regularly corresponding forms the possibility of more recent borrowing cannot be ruled out.

In northern Atlantic, the Cangin languages show an otherwise unknown #pan- or #pan-bet. Several Bak languages show a #gok root which recalls the KKR root in Mande, and may equally be a loan. Basari and Bedik share a root #tyarɛ. Other roots are quite miscellaneous.

The first element of Fulfulde gertogal/gertooɗe (with many dialectal variants), resembles gerlal/gerle "bush-fowl, francolin" (Sow 1971:69), and it is tempting to speculate that both words are actually old compounds, the second element of gertogal being the #tɔgɛ root noted in Mande. But Anneke Breedveld (pers.comm.) explains that in Fulfulde:

words with a grade C suffix, i.e. with the suffix -gal instead of -al, -wal, or -ngal, and words of which the derivation is transparent, are always deverbal. Words with a grade C suffix like -gal in particular carry instrumental meaning vis-à-vis the verb stem from which they are derived. Now the derivation of the word gertoogal is not transparent, so ... my first guess would be that its morphological derivation can be paraphrased as "bird which is used for X", where X would be the meaning of the verb stem. It is morphologically plaus-ible that the verb stem yert- "bury" is the stem in gertoogal: it is less certain whether this is also the semantic history of the word. Are fowls used in burial rituals? If so, how? The morpheme -oo- of the middle voice (often indicating reflexive meaning) is also not easily explained. If there is a relation with the word gerlal, it is very old and not transparent. It might be derived from the same verb stem with the exhaustive extension -l-. The paraphrase of the morphological constituents would then be "bird which is object of burial".

For "cock", the dontoŋ root noted in Mande recurs in Fulfulde as a loanword; de Wolf (1995) gives forms like dontoŋiri (Fuuta Jallon) and ndontoori (dialects from Mali to Sokoto), where the word is integrated into the NDI class (which contains male domestic animals) with a Grade B suffix -ri (instead of -di or -ndi), which is typical for loanwords (A. Breedveld, pers.comm.). Other Fulfulde roots for "cock" include agugumri/asgumri (eastern dialects), (N)gori (Pulaar, Fuuta Tooro, Sokoto), (n)jakaraari (Nigeria, Cameroon), and nuunuuri (castrated, Fuuta Jallon). A #gihan root surfaces in Bak and #ŋane in Tenda.

While arguing from negative evidence is not conclusive, the absence of a consistent root for "fowl" in northern Atlantic, and therefore also in proto-Atlantic, makes it unlikely that the fowl was known to the speakers of either proto-Atlantic or proto-northern Atlantic.

4.2.4. Ijoid (Table 23.4)   No root can be reconstructed for Ijoid as a whole. Defaka, which is much less closely related to the rest of the branch, has a root òkùnà (borrowed into its intimate neighbour Nkọrọ), a form of the KN root, which is distinct from yèì "bird". In Ịjọ, the most widespread root is *òfóní "bird", which can be qualified as a "ground" or "town" bird to distinguish it from a "bush" or wild bird. Eastern Ịjọ has a different root #ɔɓíɔ́ꜜkɔ́; the only similar form known is Banen (Bantu A.44) "Biokko" (Clarke [1848] 1971: 11). Inland Ịjọ has àwɹ̤yɛ̀ and àwʊ̀mɛ̰̂, also meaning "bird", whose source is unknown. The cock is always the "male fowl". Overall, it is clear that proto-Ijoid had no specific word for "fowl", and it is therefore unlikely that speakers were acquainted with it when they settled in the Niger Delta. If #ɔɓíɔ́ꜜkɔ́ is really connected with the Banen form, this suggests an introduction from the east.

4.2.5. Dogon (Table 23.5)   Dogon shows an unusual root #eɲe for "fowl"; its only conceivable connection is with the Y- root for "bird" found in Defaka and Yoruboid; the modern Dogon word for "bird" is sasáː.

4.2.6. Kru (Table 23.6)   Western Kru has two distinct roots; #hapɛ, which seems to be unique, and #sọ, which is also unique unless it is connected with the #tɔkɛ root of Mande; this is unlikely because of the nasalization, which has not been found in TK. Eastern Kru has #kɔkɔ forms of the KKR root, also found in Grebo for "cock", which has already been noted in Mande as a likely loanword, occurring sometimes as "fowl" and sometimes as "cock". In Kru as in Mande this word occurs more to the east. We may hypothesize that the KKR root has moved across West Africa from east to west, first with the meaning "cock" but in some groups generalizing to "fowl".

4.2.7. Gur (Table 23.7) Manessy reconstructs a root *ko "poule" for proto-Oti-Volta (1975:308) and *ko, *kol "poule" for proto-central Gur (1979:104). In the absence of any particular discussion of the word, it seems he has reconstructed only the first syllable which is common to #kotata/kotera of eastern Oti-Volta and #kolo forms found in the Gurma and northern Grusi groups; and he perhaps considers forms like #kokolo to involve a secondary reduplication. Alternatively, the alignment of forms in Table 23.7 suggests that the longer form #kokolo is the original, and that the #kolo forms are the result of shortening. In Eastern Mande we have already noted kɔɔrɔ and kɔtɔ, which could be shortened from kɔkɔrɔ or kɔkɔtɔ, as well as Bobo kokoro "cock". As noted above, these KKR forms are scattered in eastern Mande, whereas the Gur root is among those appearing to Manessy (1979:25) as "authentiquement voltaïques". I therefore conclude that this root is old in Gur and has been borrowed into Mande, chiefly eastern Mande, and Kru, mainly eastern Kru; it has hardly been noted in Atlantic further to the west. This further suggests that the root has been moving from east to west.

Returning to Gur, the same root is found in the Kirma–Tyurama group and even in Senufo, outside central Gur. Possibly, therefore, a case could be made for reconstructing it even to proto-Gur. This would make it a very ancient root in West Africa. Swadesh's et al. (1966) glottochronology of Gur give time depths of 3,000 or more years, once again far older than any archaeological record of fowls in West Africa. Interestingly, however, this is a *different* ancient root from the #tɔkɛ root of Mande borrowed into southern Atlantic, which does not appear in Gur. We thus have two ancient roots in West Africa, TK in the northwest and KKR more to the east.

4.2.8. Adamawa-Ubangi (Tables 23.8–9) Current thinking on classification (cf. Blench 1993a, and in press) is that, as first pointed out by Bennett & Sterk (1977), Gur originally formed a continuum with Adamawa–Ubangi, stretching across the savanna in a great arc. The two families, sometimes with Kru included, have therefore been linked as North Volta–Congo. Their original continuity has been broken by the intrusion of Chadic, presumably from the northeast, and Benue–Congo, which appears to have expanded from around the Niger–Benue confluence. Ubangi is much more closely internally related than Adamawa, and it is probably to be considered as a single branch of Adamawa which has expanded to the east, reaching as far as Sudan.

Table 23.9 suggests that it is possible to reconstruct a root to proto-Ubangi. At first sight it would appear to have the form #kɔndɔ, as seen in the Ngbandi and Zande branches. The forms with prenasalized initial consonant seen in the Sere-Ngbaka–Mba and Banda branches can easily be explained as the remains of a nasal prefix that is no longer functioning (the Ubangi languages which still have noun classes mark them by suffixes, not prefixes) and which has voiced the following consonant *k to [g]. Bennett (n.d.:24) reconstructs *ngu̠ "chicken" for Mundu–Gbanzili (apparently equivalent to the Ngbaka group), and comments that, like all the prenasalized stops, *ng is rare. If we leave out the forms of Gbanzili and 'Baka, which could be loans from Bantu, and the apparent non-cognates of 'Dongo-ko and

Ama-lo, the second consonants of the Gbaya, Sere, Banda, Ngbandi and Zande groups all suggest an alveolar, occurring in different groups as [r], [l], [t] or [nd]. There are three possible explanations for the [nd]; that it results from a nasalized vowel preceding [t] which voices to [d] (nasal epenthesis), in which case the reconstruction would be #kV̰tV, or that it results from a nasalized vowel preceding [n] followed by an oral vowel (stop epenthesis), in which case the reconstruction would be #kV̰nV, or that a final nasal caused the nasality of both vowels before it was lost, in which case the reconstruction would be #kVtVn̰ > #kV̰tV̰; nasality was then lost on one or both vowels in different groups. I tentatively adopt the third explanation because it helps to explain forms in other language groups.

The vowels are more difficult to reconstruct. If the first was *ʊ, as suggested by Bennett (n.d.) for Mundu–Gbanzili, that could easily give rise to the [u o ɔ] reflexes. The [a] reflexes are less easy to understand, unless the [wa] of Bangando is close to the original and was simplified to yield either [a] or a back vowel. The second vowel is more varied, and may have been affected by suffixes that can no longer be separated from the stem. The final reconstruction could be #kuạtVn, formula KTN. This is surprisingly close to Westermann's (1927:185) reconstruction for "bird", #**kuani**. Either the resemblance is accidental, or both Mande and Ubangi reflect an ancient Niger–Congo root KTN which originally meant "bird"; when the fowl reached the speakers of proto-Ubangi in Central Africa their old word for "bird" was specialized to mean "fowl", and with this meaning it then spread to other groups.

The Adamawa words are more varied and cannot so easily be reconstructed to a single major root. This is consistent with the more ancient time depth of this branch as compared with Ubangi. The forms in the Duru, Leko, and Mumuye groups appear to be cognate with the KTN root found in Ubangi. But it is notable that their medial consonant is only [n], or vowel nasalization after its loss in the Leko group, instead of the variety of alveolar consonants found in Ubangi. If a form like #kon- had been inherited by Ubangi, it is hard to see how voiceless [t] could develop from [n], as in the Banda group. The alternative is to consider that the root is older in Ubangi and that one form of it was borrowed into some Adamawa groups. This implies an east-to-west movement for the root, at least in the meaning "fowl".

A second root of the form #yibe (formula YB) is found in Tula, Jen, and Munga, and perhaps in Waja. Another form of YB, #yaab-, occurs in the Yungur group, and also in a Fali language, Yɛk gopri, spoken in Cameroon (Kropp Dakubu 1977:227). Unfortunately, I do not have data for other languages near to Adamawa Fali, which would show if the YB root is widespread in that area. As we shall see in Table 23.19 and Map 23.2, this root also occurs in areas of Chadic which are adjacent to Adamawa.

Finally, the Mbum–Day languages show a consistent root #kạka, which appears to be a reduced form of KKR.

The presence of three major roots and a number of minor ones suggests that no proto-Adamawa root for "fowl" can be reconstructed, and that it is therefore likely that the bird reached the speakers of Adamawa after their languages had begun to diverge. One root (KN) is similar to the KTN of the Ubangi languages, but it appears that the Ubangi forms could not all be derived from Adamawa, whereas the Adamawa forms could all be derived from Ubangi; this suggests that at least one introduction of the fowl was from Ubangi to neighbouring Adamawa languages, and therefore represents an east-to-west movement.

4.2.9. Kwa (Table 23.10)   The Gbe language cluster shows forms of the root #kokolo (KKR) which we have met in Gur; this suggests a southwest movement of the word, and probably the bird, from Gur-speaking to Kwa-speaking areas. The alignment in Table 23.10 suggests that languages which show a word of roughly the shape #a-kɔkɔ have lost the last syllable, while those of the shape #-kɔla/karo/wolo have lost the first syllable. A clear example of first syllable loss (after loss of the final syllable) is seen in the correspondence of Anyin-Baule ákɔ́ to Akan àkúkɔ̀; the high tone of the lost first syllable of the root has shifted to the prefix. Heine (1968:220) reconstructs proto-Togo remnant *o-əulo, citing the Animere and Akposo-Igo forms, but it is here suggested that these are part of the #-kɔla/karo/wolo set and thus derived from #kokolo. The Ga-Dangme forms probably belong to the #a-kɔkɔ set, with changes.

A second set is formed by the #kusɛ/kɔsɔ/koʃi/tʃasì forms. Possibly these are related to the #a-kɔkɔ forms, but if so they have undergone some distinctive changes. They are symbolized KS.

4.2.10. Benue-Congo (Tables 23.11–18)   As noted above, Benue–Congo is largely intrusive between Gur and Adamawa–Ubangi. Following Blench (1993a), it is tentatively divided into West Benue–Congo (corresponding to the former eastern Kwa languages) and East Benue–Congo, corresponding to the original Benue–Congo of Greenberg (1963). Ohiri-Aniche (forthcoming) has placed Ukaan–Akpes in an intermediate position. The current view is that the various groups of Benue–Congo fanned out around the Niger–Benue confluence. North of the confluence the Kainji and Platoid branches in particular have interacted intensively with Chadic; it is likely they replaced Chadic languages in part of their range before the more recent spread of Hausa. The groups of Benue–Congo will be discussed here in a roughly west-to-east order, in line with the previous treatment of families.

4.2.10.1. Ọkọ-Nupoid-Idomoid (Table 23.11)   Ọkọ and Nupoid (apart from the Ebira cluster) share a root #-pise (or #-bise), formula PS, which is specific to them and for which there is no obvious etymology. Ọkọ and Nupe Tako are separated from the main body of Nupoid by Yoruba and Ebira speakers but share this root, suggesting that it predates the spread of Yoruba and Ebira which isolated them from the other languages. It has not been suggested that Ọkọ and Nupoid share a common proto-language distinct from the Ebira cluster and Idomoid, so that this root is likely to have spread later than the proto-language of Ọkọ-Nupoid-Idomoid, and, if Ebira is correctly classified within Nupoid, later than proto-Nupoid. Ebira has a reduplicated root, probably #-kwʊkwɛ (KKR), with the first consonant weakened to -w- in à-wúhwɛ́ (Kropp Dakubu 1977:280) before being lost. Idomoid either uses the old #-nʊ root which originally meant "bird" (NN), or -gwʊ/gbi, which is probably a reduced form of the #-kwʊkwɛ root of Ebira. For "cock", the rather incomplete information shows compounds of "fowl" for most languages, but #-kokor- roots in a few of the more southern languages.

4.2.10.2. Yoruboid–Edoid–Akokoid–Igboid   (Table   23.12)   Westermann (1911:128–9) cites Yoruba as adire and connects it with Kunama dir-uwa, Dinka adjid, and Nuba dir-bad, suggesting that the second element in Kunama and Nuba may mean "male" and reconstructing proto-Sudanic dụilị. It appears highly unlikely that Yoruba is connected with the other forms, as there are no intermediate links. More plausibly, Akinkugbe (1978:430, 712) has reconstructed proto-Yoruboid *ā-dìwē̄ "chicken", which she suggests may be a compound with the second element *ēwē̄ "bird". But a construction "bird of ..." would have the qualifying word in second, not first, position; nor does this etymology account for the older forms with -r-, recorded for Yoruba by Crowther (1843:34) as adieh, adireh, and for Igala by Koelle ([1854] 1963) as adʒurɛ. Perhaps "fowl" should be reconstructed as *ā-dìrē̄ and derived from the same root as Ufe ɔ̀-yèlè̀. The Ufe form is a survival from the extinct language of Ufe (Ade Obayemi, pers.comm.), whose genetic classification is unknown. Akinkugbe (1978) also reconstructs *à-kìkɔ̄ "cock", suggesting that it is derived from *kɔ̄ "crow"; but given the many cases of similar words for "cock" or "fowl", such as the #a-kɔkɔ noted in Kwa, it is more likely that this is a loan at the proto-Yoruboid level. The Itsẹkiri form recalls the Igboid ègbélé, ègbénú "cock", with the meaning generalized to "fowl"; Degema (Edoid) has an apparently metathesized form èlègbé "cock". These seem to be isolated remnants of a once more widely spread root, formula QL (where Q stands for a labial-velar sound).

The #-kɔkɔ (KKR) root recurs in Edoid, where Elugbe (1989) has reconstructed *O-khɔkhɔ "chicken (domestic fowl)" (kh represents a lenis or weak stop and O- a prefix harmonizing with the stem). The reflexes of *kh correspond fairly well with the regular sound correspondences given by Elugbe; this suggests that the chicken was known to the speakers of proto-Edoid by this name, which immediately to the west, in Yoruboid, denotes the cock. Because of the differences in vowels and tone pattern, it is unlikely that the Edoid form was borrowed directly from the Yoruboid "cock". The form *O-kpa for "cock" (formula QK) was apparently introduced into Edoid later; see Table 23.15 (Platoid). Interestingly, Elugbe (1989:114) reconstructs a third word *O-kokodhoko "cock (onomatopoeic)", which in most northwest Edoid languages has, in shortened forms, become the normal word. This root resembles the #-kokor- forms noted in Nupoid and could thus be a borrowing at the proto-Edoid level from further north.

The Ahan–Ayere group of Akokoid shares the Yoruba root for "fowl", either by inheritance or more likely, considering the time scale, by borrowing, while the Arigidi group has a quite distinct word ɛ-hɛ. The words for "cock" are quite complex; some look like nasalized forms of the Yoruba word, some look like compounds with ɛ-hɛ, and the Ahan–Ayere word, a form of the #-kɔkɔ root, looks like a plausible source for the Edoid word for "fowl".

Igboid has as its normal word for "fowl" a form of the #-kɔkɔ root in which the first vowel has been raised to -ʊ-; between consonants the change -ɔ- to -ʊ- is more likely than the reverse. This would suggest that proto-Igboid borrowed from (proto-) Edoid. The only counter-evidence is the ɔ-tʃítʃì form in Ndele and closely related lects, which suggests a front vowel in the stem that is unattested in Edoid. Possibly the loan was not from Edoid but from some other language, now lost, which had a form such as #-kɪkɔ; this could yield #-kʊkɔ on one hand, further progressing to #-kʊkʊ, and #-kɪkɛ progressing to #-kɪkɪ on the other, from which -tʃítʃì could easily develop. Although one would not like to speculate too wildly about lost languages, it should be recalled that this is an area where language shift is well attested (the original language of Ufe, mentioned above, was spoken by grandparents of those now in middle age). Furthermore, to the immediate north of Igbo lies the Yoruboid language Igala, which shows very slight dialectal variation compared to Igbo. The natural conclusion is that Igala expanded at a period when Igboid was already well differentiated, and that speakers of other languages, Igboid or other, in what are now Igala-speaking areas might have shifted to Igala; thus before the period of Igala expansion there may well have been other languages immediately to the north of Igboid from which it could borrow. For "cock" the Igboid word is usually ɔ̀kpà in western areas in contact with Edoid, while in other areas it is qualified by dí "husband, master" or óké "male"; there is also a combination "male fowl". In Edoid, on the other hand, ɔ̀kpà is never qualified. This suggests that Igboid has almost certainly borrowed the word from Edoid rather than the reverse. The ègbélé root, mentioned above, occurs in northern areas and is probably an older one which has been largely replaced by ɔ̀kpà.

4.2.10.3. Akpes-Ukaan (Table 23.13)   Akpes and Ukaan consistently show the #-kɔkɔ root for "fowl", like Edoid and Igboid. The first vowel is -ɔ- in Ukaan, as in Edoid, and apart from the tone it could be seen as a possible source for Edoid; alternatively, both groups may have borrowed from a common source. The -o- vowel of Akpes suggests a possible borrowing from the "cock (onomatopoeic)" root of Edoid in one of its reduced forms (see Table 23.12). The words for "cock" seem to be specific to the group.

4.2.10.4. Kainji (Table 23.14)   The Kainji roots are highly diverse, suggesting that this area may have been a crossroads for different introductions of the fowl with different names. The #t-k- root of Kambari might be the same as the TKR root of Kurama–Jere–Sheni, with the #t-l-k- of Reshe being a metathesis of TKR. Other roots are KRM (Laru), KYT (northern group and Pongu), and YLM (southeastern, Basa, and Kamuku groups). The groups of superficially similar forms, which have here been represented by formulae, do not seem to correspond to any genetic groupings. We may therefore conclude that chickens were unknown to the speakers of proto-Kainji and entered the daughter languages at later periods from various directions; thus, even if some of the roots are originally related, they diverged not within Kainji but outside, and were then borrowed in different forms at the various periods when fowls were introduced.

Very few words for "cock" have been recorded; the -pɛ́ of Kambari recalls the #pan- "cock" of Atlantic, but this may well be an accidental similarity.

4.2.10.5. Platoid (Table 23.15)   Northern Plateau, the northwestern subgroup of western Plateau, Eggon, Yashi and Migili all use the NN "bird" root for "fowl"; the na/ne of Nungu and Ake also perhaps belongs here, as do the forms of the south–central subgroup of the central group. Because the semantic shift from "bird" to "fowl" is so common, it is not possible to decide whether this represents a single or a multiple shift.

The part of the southwestern subgroup of western Plateau which does not use NN uses #-ko or #-kiko/kuko, forms of KKR, obviously to be connected to the #-kiko/kuko of the Yoruboid "cock" and Edoid and Igboid "fowl". Possibly, then, #-kiko/kuko has a northern origin and has moved south. This is supported by the occurrence of #a-gbak "cock" in the same subgroup, a plausible source for Edoid #ɔ-kpa "cock" (cf. 4.1.10.2).

The west–central subgroup and southeastern group of Plateau, and Cara, have a root which could be either KKR or KT. Ayu yanaŋ looks like a loan from Kainji YLM, although it is a considerable distance away.

Aten tsɔ̀ɔ̀rɔ̄ looks like the TKR root of eastern Kainji, either a loan or the remains of an old root which spread over the area before being partly replaced by more recent roots.

In Benue, the -de root of Toro–Alumu seems a unique innovation. Tarok -rugu could be a form of TKR. Proto-Jukunoid *kúǹ (k)i-/i- (Shimizu 1980.2:143) recalls the KN of Adamawa and is a possible loan or retention from an Adamawa substrate.

4.2.10.6. Cross River (Table 23.16)   Cross River is currently believed to have spread from the Niger–Benue confluence area eastwards along the Benue and then southwards. Bendi is found in a relatively small area adjacent to North Bantoid. The branches of Delta-Cross lie to the southwest of Bendi; they are assumed to have expanded from north to south and west, and this is supported by the greater diversity of Upper Cross as compared to Lower Cross to its south, and by the lesser diversity of Kegboid (Ogoni) and Central Delta, which both lie west of Lower Cross and extend into the Niger Delta.

Bendi has a single root #-kua that probably originates from one of the K- roots already discussed. "Cock" is "male fowl" in Bekwarra, the only language where data is available.

Upper Cross has forms of the root #-kɔkɔ (KKR), found in all languages of the first major division and overlapping into Ikom–Olulumo. Dimmendaal (1978:317) reconstructs proto-Upper Cross *ì-kɔ́k and Sterk (n.d.) `-kʷɔ́kɔ̀. The root cannot, however, be proto-Upper Cross, for it does not occur in all the main groups of Upper Cross. It occurs in the more northern languages and so is apparently introduced from the north. The ancient "bird" root, NN, is general in Upper Cross (it is listed for comparison under Notes in Table 23.16), and in the languages which do not have #-kɔkɔ its meaning has been extended to "fowl"; in languages (starting with Lokukoli) which show a compound form it appears likely that the gloss "chicken" has been understood as "young fowl", since the second element is "child". In at least Mbembe (Akam) "bird" is qualified as "of the bush". A few languages have DN instead of NN for both "bird" and "fowl", confirming that the semantic extension has been made within Upper Cross. In the few languages where "cock" has been recorded, it is a compound "male fowl".

In Lower Cross the word for "fowl" also resembles the "bird" root, but is never identical with it. This suggests that the "fowl" word was borrowed from an Upper Cross language, either once with subsequent developments within Lower Cross, or several times for the differing forms. In addition, Lower Cross has #-kikɔ for "cock", often followed by "fowl", resulting in a combination "cock-fowl". I suggest that this longer form is the original, being reduced in some languages to "cock". It is not clear from which language this root entered Lower Cross, since although both its Igboid and Upper Cross neighbours have the root, it is always with the meaning "fowl" and not "cock".

Ikoro (1989:77) has reconstructed proto-Kegboid (Ogoni) *ɔ-kɔ̝ɔ̝ (KN) and "cock" is "male fowl".

It does not seem possible to reconstruct a Central Delta word for "fowl". The apparently collective Abuan form ìsù-ɣúnà looks like the Defaka and Nkoro root -kuna (KN), which suggests it may have been an old root spread in the coastal areas and carried into the heart of the Delta as Central Delta languages expanded there. This same root, possibly originating from Ubangi KTN, is the source of proto-Kegboid *ɔ-kɔ̝ɔ̝. The modern Abuan–Oḍual form is possibly a compound from this root, while the Kụgbọ-Ọgbịa forms are from "bird"; East Ọgbịa "bird of town" is a calque on Nembe or Ịzọn.

To summarize, there is no evidence that chickens were known to the speakers of proto-Cross River. They were probably known to the speakers of proto-Bendi. They were not known to the speakers of proto-Upper Cross, but were probably introduced from somewhere to the north, though not from Bendi, after Upper Cross speakers had begun to diversify. Lower Cross speakers probably acquired fowls from Upper Cross speakers, and perhaps another source as well, after they had begun to spread out. Proto-Kegboid speakers probably had fowls, perhaps acquired from speakers of Defaka (or other, now extinct, Defakoid languages). Just possibly Central Delta speakers also acquired them from the same source, but the linguistic evidence is not conclusive.

4.2.10.7. North Bantoid (Table 23.17)   Proto-Dakoid has a root #kpàá, probably a form of QK; Tiba, a little known language, has kùnkḗrā, perhaps KKR although it has nasalization. Ndoola and Nizaa have a root ʃo or ʃu, here tentatively identified with KT. Nizaa has another root sìm, which does not seem to be cognate with ʃo/ʃu. All the remaining languages appear to share a root #kuandu, probably a form of KTN, spreading from Ubangi via Adamawa. Data for "cock" is limited, but is either "fowl + male" or a special root: ŋgɔ̀ɔ̀ could be a form of KKR, kwaa (QK) was apparently originally a modifier ("cock-fowl" in Somié). koro is taken to be a shortened form of TKR; tákòrò is identified with TKR although it is some distance from the other forms. Alternatively, Bruce Connell (pers.comm.) has proposed that #ta-, which is a diminutive prefix in Mambiloid, might have been added to koro, which would then be a KT root. None of the roots seem to be derived from "bird" in North Bantoid, which is usually #nunuŋ (NN).

4.2.10.8. South Bantoid (Table 23.18)   South Bantoid is the most complex sub-branch of Niger–Congo, and its classification is constantly under review. Because the Bantu languages proper cover a vast area, mainly outside West Africa, they have not been listed here; the citations are limited to the "Bantu borderland" area where relationships are complex.

Buru is an isolated language and has an unusual root -dɔ (D). Tikar is another isolated language and has a root ɲìì. Ekoid has forms of the KKR root. Mbe has -kùɔ, apparently borrowed from the #-kua of neighbouring Bendi. Tivoid has a variety of roots: -sagi, which may originate from a "bird" root (cf. "bird" recorded under Dakoid in Table 17); -hù, which might be a version of a K- root; -kèɣ́, which seems to be a reduction of -KKR, and -ðɔ, which may be the D root of Buru. Beboid has a single root #ʃie, which seems to reconstruct to proto-Beboid. Kenyang has a form of KKR. In Jarawan, Blench (1995:206) suggests that the presence of the Hausa root argues against the fowl being old in Jarawan, but Koelle in 1854 recorded a different root for Jarawa, cognate with the other languages, showing that the Hausa root is due to modern influence; from the evidence here, a proto-Jarawan root #gubu is reconstructible. The remaining languages in Table 23.18 are grouped as Grassfields; all of them show the same root as Jarawan. A series of sound changes in the initial consonant appears quite clearly: g > gw (before a labial vowel) > gv > bv. This root is identical with the Common Bantu root *kúbà (Guthrie 1967–71i:137); the voiced initial g in Jarawan and Grassfields results from the nasal prefix. Guthrie (1967–71iii:321) proposes that the fact that the reflexes of this root cover an unbroken area in the northwest suggests that the ancestors of these people acquired the domestic fowl from a distinct source from the rest of the Bantu area, which have the -KKR root. Unfortunately the root *kúbà has not been traced outside northwest Bantu unless its source is the plural kùùbá "fowls" of Wom in the Leko group of Adamawa, which seems unlikely. Whatever its source, it must have been adopted while proto-Jarawan speakers were still in contact with Grassfields speakers, which would be at an early stage of South Bantoid differentiation.

Apart from *kúbà, forms of the KKR root are the most widespread in Bantu. The #kɔkɔ root in Bantu was discussed by Johnston (1886:483–6). He argued that since in nearly every Bantu language the fowl is known by variants of the same root, it must have been known to the Bantu before they began their expansion. He placed their homeland in west–central Africa. Later Johnston (1919i:22–3) gave the same argument, but revised their homeland to East Africa. He proposed that the fowl was introduced to Africa from its original place of domestication in India through Egypt and moved up the Nile valley until it reached the speakers of proto-Bantu "in the Nile valley north of the Albert Nyanza". He believed that the fowl was not known in Egypt "till after the Persian invasion of 525 B.C.", and that "even supposing it spread rapidly up the Nile valley as a domestic bird, it could hardly have reached Central Africa for another hundred years – if so soon". He proposed to allow 300 years for it to arrive and be fully accepted among the Bantu before their expansion began. This is his argument for dating the Bantu expansion at approximately 2500–2000 BP.

There are three major problems with adopting Johnston's argument today. First, as already noted, the fowl is recorded once in Egypt from an earlier period. Secondly, it is now generally accepted that Bantu originated in West and not East Africa, as a branch of the West-African-based Benue–Congo languages, part of the Niger–Congo family. Thirdly, our study of Niger–Congo has already shown that it is wrong to view the root as an essentially Bantu phenomenon.

Guthrie (1967–71i:123–4) reconstructs proto-Bantu *-KOKO 9/10, but notes that the large number of "skewed reflexes and apparent mutations" suggest that there was no proto-Bantu-X item because the bird was unknown; chickens were introduced "round about the Bantu threshold", and that "extensive loaning and cross-contamination accompanied the introduction of the bird at a time when the Bantu dispersion was about to begin". Like Johnston, he placed the homeland of the Bantu in East Africa; his conclusion differs only in that he places the introduction in the early stages of the expansion rather than before it.

Vansina (1990:290) regards the root as an "early loanword" in Bantu and believes that: "The animal was introduced from the east. Hence the greater diversity of terms in western Bantu is not surprising". The rationale of this argument is not clear.

An alternative interpretation of the two roots is as follows. Henrici's lexicostatistic classification of Bantu (1973) shows Zones A, B and C splitting off first, the remaining zones constituting central Bantu. Now it is precisely Zones A-C, together with Jarawan and Grassfields Bantu, where KB is found, KKR occurring as the basic root in central Bantu only; KKR is sometimes found as "cock" outside central Bantu, but this is assumed to be a later borrowing supplementing KB. I therefore conclude that KB, and presumably the fowl itself, entered the family tree at a point where Jarawan, Grassfields and Narrow Bantu had not yet diverged, suggesting an early introduction of the fowl into their common homeland around the Nigeria–Cameroon border; KKR, perhaps reflecting a new introduction of the fowl, entered at a later period when proto-central Bantu was spoken somewhere in Central Africa, and replaced KB in its daughter languages.

KKR has often been regarded as an onomatopoeic root derived from the crowing of the cock or the clucking of the hen, originated independently not only in Niger–Congo but also in other language families, including Indo–European. There are, however, arguments against this. First, the root is by no means universal; Table 23.19 shows that it is entirely absent in Berber, and Tables 23.1–18 show many alternative roots with more restricted distribution. Secondly, some languages contrast it with an obviously onomatopoeic "cock-a-doodle-doo" root, such as proto-Edoid (Table 23.12). Thirdly, some languages show evidence of KKR in the proto-language, but it then undergoes normal weakening changes; see Degema in Edoid (Table 23.12). I therefore conclude that while this root may have been onomatopoeic in origin at some remote period, and may have been favoured as a loanword by onomatopoeic associations, it is unlikely to have been invented independently many times. We should rather treat it as a normal root and study its distribution for historical evidence.

The wide distribution of KKR within Niger–Congo led Mukarovsky (1976–7ii:176–7) to reconstruct proto-West Nigritic #-KUKI "fowl, chicken", with forms cited from Atlantic, Gur, Kru, Kwa, and Benue–Congo. There is, however, a dating problem with this reconstruction. Proto-Atlantic–Congo (roughly equivalent to proto-West Nigritic) is not yet dated, but may well be older than 8,000 years, the period when the fowl was apparently first domesticated in Asia; and it would take some time for it to reach West Africa. It is therefore not plausible that the name of the bird occurred in the lexicon of proto-Niger–Congo. Consequently, the widespread occurrence of the root is due not to common origin from proto-Niger–Congo but to contact.

It is also observed that while many groups show evidence of the root, it cannot usually be found consistently throughout some of the major branches and therefore cannot be reconstructed to the proto-language of such major branches. It has, however, been reconstructed to the proto-language of some of the lower-level sub-groups, as shown in the preceding discussion, which indicates that it was known by the time these more recent proto-languages were spoken.

Tables 23.1–18 show that the root had a consonantal shape KKR, where K stands for a velar stop, probably [k], and R stands for an alveolar, [r], [l], or perhaps [d] or [ɗ]. This is rather long for a Niger–Congo root, which most commonly has two consonants; not surprisingly, therefore, many languages have shortened the root by dropping either the first or second K or the R. A clear example of the dropping of the first K is in Igala, compared with proto-Yoruboid (Table 23.12); of the dropping of the second K, in Konkomba compared with Gamgam (Oti-Volta, Table 23.7); of the dropping of the R, in the alternation in Busa (Mande) reported by Mukarovsky (1987:122) between koo "chicken" and kod-sa "rooster", where the compound preserves a [d] which has been lost in the simple word. It is tentatively suggested that the sequence of vowels (-uɔ-) in the Ekoid Bantu languages may have been attributable to metathesis between K_2 and L, followed by loss of -L-, giving developments like #-kugɔl- > #-kulɔg- > -kuɔg-.

The vowels are generally back rounded [u], [ʊ], [o], or [ɔ], but in a few languages, notably Bobo (Mande) and some of the Oti-Volta (Gur) languages, the final vowel is [o] in the singular, [i] in the plural. This yields singular #-kokol-o, plural #-kokol-i for suffixing languages (some Niger–Congo languages use suffixes rather than prefixes).

The KKR root is the most widespread in Niger–Congo with the specific meaning of "domestic fowl" or sometimes "cock". There are, however, languages which have a KTN root, where nasalization of the vowel(s) and -T-, with loss of the final -N, yields KN forms. In many cases, this seems to be a specialization of an old root for "bird" (cf. the specialization of *fowl* in English, compared with the retention of the original meaning in German *Vogel*); Mukarovsky (1976–7) proposes proto-West Nigritic #-GWÜN- (-GWYÜN-), to which we can relate Defaka -kuna and proto-Jukunoid -kúṅ "fowl", Manding kɔ̀rɔ̀ "bird". This root, unlike KKR, clearly included nasality, which could affect the initial consonant, as in Mɛnde ŋɔ̀ní corresponding to Vai kòndé. In Bantu *ŋ is lost, resulting in the proto-Bantu form *-(j)ùní. Proto-West Nigritic #-NÙNI "bird" is perhaps another form of the same root in which the velar [N] has assimilated in place to the alveolar [n]; Meeussen (1980) by implication and Guthrie explicitly treat them as the same root in Bantu.

They have, however, been kept apart here; one has been listed as KTN (yielding KT or KN) and the other as NN.

4.3. Afroasiatic

Table 23.19 lists forms in Afroasiatic.

4.3.1. Omotic

In Omotic the most widely spread root is #baak-, found in all branches except Ometo North and South; it has undergone different sound changes in the various groups and is thus likely to be quite old.

4.3.2. Cushitic

Agaw (central) Cushitic has a root #diruwa that looks as if it could go back to proto-central Cushitic. It is obviously the same word as Nubian #dirbad; one has clearly borrowed from the other. Sasse (1979) has reconstructed *lukkʊ "chicken" and *kormV "cock" for proto-East Cushitic; Roger Blench (pers.comm.) suggests that the latter probably originally meant "male". #lukkV "chicken" has been borrowed into at least two Omotic languages, Dache and Zayse–Zergula, which are relatively close to East Cushitic.

Ehret (1980:367) has reconstructed *koko, *kokomo "chicken" for the proto-Rift branch of South Cushitic; the forms he gives all actually contain a nasal and are rather of the shape #konki "fowl/chicken", #konkomo "cock"; formula KNK. The Bantu languages in the area, although they have borrowed other words from South Cushitic, appear to have the normal Bantu root *kúkú (Meeussen 1980), so there is no evidence of borrowing either way.

4.3.3. Semitic

Semitic has not been fully compiled, but it is important to note Classical Arabic dadʒaadʒa, Shuwa dʒidaad "fowl/chicken" and particularly diik "cock" as the source of borrowings in many languages. South Arabian has a form #degooget that does not seem to be found in Africa unless North Ethiosemitic #derho is a much compressed form. South Ethiosemitic has a root #kutto, noted by Roger Blench (pers.comm.), which also occurs sporadically in Cushitic and Omotic as a loanword.

4.3.4. Berber

Basset's studies of poultry names (1959a,b) show that the basic root in this family means "cock"; the feminine form "hen" is derived by prefixation and suffixation. Basset (1959a:117) reconstructs the root as *y-z-ḍ, noting that the first consonant also occurs as g, k or w and that therefore the original first consonant is unclear. The feminine form is basically t-yaziḍ-t (Basset 1959a:119). In the Saharan group, he reconstructs *ikazī "cock", plural *ikăzān, *tikazīt "hen", pl. *tikăzātīn (1959a:122). The "hen" form, shortened to two syllables (cf. 1959b:132), is a very likely origin for the TK root of Niger–Congo noted in Mande and elsewhere. Basset (1959b:135–6) also cites Iqraien (Rif) aɛtuq "fowl good for eating", where the ɛ indicates a loan from Arabic; he compares Jbala ɛ̣ttūqa "pullet", already noted in Arabic dialects of the Maghreb.

Finally, Ntifa afullust "hen" is clearly derived by semantic shift from fullus "chick" in dialects such as Metmata and Figuig (cf. the shift in English of chicken from "chick" to "fowl"); Basset (1959a:124) derives this from Latin pullus. This root is much more limited in distribution than the *y-z-d root, and obviously was borrowed later into Berber, in the Roman period of North African history.

4.3.5. Chadic   A number of suggestions have been made about Chadic recon-structions for "fowl" and "cock". Newman & Ma (1966:233) reconstruct proto-Chadic *k-z- "chicken", which corresponds to Jungraithmayr & Ibriszimov's (1994i:33) root kwz noted for West Chadic. In turn, this would appear to be a loan from the common Berber root for "cock" (4.2.4); k-z-ḍ, one of the alternants suggested by Basset, with loss of ḍ, would correspond well to the scattered Chadic forms. Even closer to forms like Hausa kàazáa are forms from the Saharan group of Berber such as *ikazi, pl. *ikazaan (Basset 1959a:122). It is therefore reasonably certain that these forms come from Berber with a meaning shift from "cock" to "fowl", and not from Kanuri kaji "guinea-fowl", as suggested by Skinner (1977:175) and Blench (1995:205, 207). While a shift from "guinea-fowl" to "fowl" is con-ceivable in the same language, it is difficult to envisage from one language to another; there is no evidence of such a shift in Kanuri itself (Table 23.20). While fowls newly introduced from Kanuriland might be described as "Kanuri guinea-fowl", such an expression would surely make use of the Hausa word for "guinea-fowl" and not an otherwise unknown Kanuri word. In any case, the k-z- (formula KZ) forms are not widespread and do not appear to be convincingly reconstructed to proto-Chadic or even proto-West Chadic. It seems more likely that the root was borrowed from Berber when West Chadic had already undergone internal differen-tiation, suggesting an introduction of the fowl across the Sahara.

Newman & Ma also reconstruct *g-z- "cock" (1966:233) (formula GZ). The reflexes they cite are quite varied and may not all be cognate. I suggest that forms like #gaza, #gadʒa might be borrowed from Berber *g-z-ḍ, another of the variants meaning "cock", again with loss of the third consonant but without the semantic shift to "fowl".

Thirdly, Newman & Ma (1966) reconstruct *( )k-r- "fowl". Under this they include highly varying forms of the shape #zakara, #kokor, #kuyo, #kara, #takur, #tsakala. Jungraithmayr & Ibriszimov (1994i:33) include most of these within their root kwd or kwz, which also includes KZ forms. They suggest that the great variety of reflexes "points towards a considerable age for the root". An alternative sugges-tion is that these represent differing borrowings from a set of variants. The set of Berber forms which have already been suggested as the source of KZ and GZ could also yield #zakara "cock" (formula ZKR) from *k-z-ḍ with metathesis of the first two consonants. The feminine forms with initial t- and with -z- lost (as in some of Basset's Saharan forms, #teka(h)it) have been suggested as the source of TK; alternatively, with -z- rhotacized to -r-, they could be the source of the #takur and #tsakala forms (formula TKR), which we have already observed in Kainji and possibly Plateau. On the other hand, forms such as #kokor appear to belong to KKR, already noted widely in Niger–Congo. Forms such as #kuyo, #kara might be shortenings of either TKR or KKR; probably detailed study of the various groups will be necessary to determine which.

In central Chadic, the Bura and Higi groups show forms of the TK root, usually with a nasal prefix. The Tera and Bata groups show forms like #ɗek-, where I suggest the initial implosive derives from the nasal + t-. Like the TK roots of Niger–Congo, this is probably derived from shortened forms of the Berber feminine #tekazit. An alternative source would be the Berber form from Arabic aɛtuq "fowl good for eating", but this could not be earlier than the spread of Arabic in North Africa, whereas the forms in the Bura and Higi groups show various developments which suggest they may have existed in the proto-language of those groups and therefore be an earlier loan from Berber. The same argument applies to Blench's suggestion (1995:234) that they derive from Arabic diik; languages like Mubi which actually have this word have it in a very obvious form. Probably the #ɬek forms of Masa are also TK roots; they might alternatively be forms of East Cushitic *lukkʊ, but this is geographically distant and connecting forms have not been found.

To summarize, I propose that Berber forms of the approximate shapes below are the origin of the Chadic words of the formulae given:

Berber	#k-z-ḍ m.	#g-z-ḍ m.	#t-k-z-ḍ-t f.
Chadic	K Z	G Z	T K
	Z K R *		T K R

*with metathesis

A Berber origin helps to explain the bewildering variety of Chadic forms that have caused problems for all who have previously discussed them. It is also likely that with so many varying forms borrowed there has been contamination of one form by the other, or by competing KKR forms. There are also apparent prefixes which have been added to these roots.

If these proposals are accepted, they suggest that chickens were adopted by various Chadic groups (as well as by Mande speakers) from a variety of Berber groups, not a single source; the variety of the sources is reflected in the variety of the Chadic forms. A more detailed study trying to relate the different forms to different trans-Saharan trade routes might be of interest.

Skinner (1977:182–3) notes that "there may be more than one etymon involved" and "there is certainly more than one morpheme", but he gives a composite reconstruction *nD-(r)k-r, with the explanation:

The core unit seems to be *D-, presumably meaning "chicken, fowl, bird". Then possibly a *k-r either prefixed or suffixed meaning "female". A *g- "male" prefix is more dubious. The *D- may not be unconnected with C of "guinea-fowl".

Skinner (1984) develops further an idea of compounds based on an old root for "bird". The hypothesis of multiple loaning adopted here seems preferable to this rather complex explanation.

Returning to our old friend KKR, we note first that it is definitely not of Berber origin, for there is no trace of it there. Secondly, we find that, just as in Niger–Congo, it occurs in a scattered way in Chadic, rarely reconstructible to the proto-language of a group. Again as in Niger–Congo, many forms have lost one consonant, a process called "root-thinning" by Jungraithmayr & Shimizu (1981:22–3). Some languages, however, show four consonants, e.g. Pero kòkkúròk "cock". Jungraithmayr (1971:288–90) has argued that some final consonants in animal names, including [k], represent old suffixes; Ebert (1978:41–5) argues that there is a masculine suffix -kí in Kera, as in gòglókí "cock", and in Mofu–Gudur Barreteau (1978:109) com-pares the generic gwágwàr "poule, poulet" with the masculine gwágwàlák "coq". It therefore seems safe to conclude that in other Chadic languages the final [k] is the remnant of a masculine suffix; the roots then correspond very well to KKR.

A root meaning "cock" of the shape #pukum-/fugum/vugum occurs in the Bura and Bata groups (formula PKM). This appears to be a loan from Kanuri or a closely-related language. Modern Kanuri has gùdòwúm, but older sources show a PKM form: Koelle ([1854] 1963) gubo:gum, Barth ([1862] 1971) gábugum, Benton (1911) gubogəm, Lukas (1937) gəvagəm ~ godogúm. The modern form is thus apparently of fairly recent origin, and the older one is the source of PKM in Chadic. There is perhaps a trace of the ga-, which does not appear to be a prefix in modern Kanuri, but may have been so originally, in Putai ʔávùgùm.

Jungraithmayr & Ibriszimow (1994ii:33) identify #y-b in the Bole and Zaar (southern Bauchi) groups, commenting that it must be of recent introduction; Blench (1995:233) identifies this as #yab (formula YB). YB has already been identified in Adamawa (Table 23.8) in languages adjacent to Chadic (Map 23.2). The movement of the KKR root from east to west, indicating that the fowl has moved in the same direction, suggests that Adamawa might be the source, with movement in the same direction from Adamawa to Chadic.

Various other roots have been identified in Chadic. Jungraithmayr & Ibriszimow (1994ii:33) note *z-m (ZM) in the Warji (northern Bauchi) group. In Table 19, I have been able to relate most of the forms to one of the major roots already discussed. This has involved postulating a number of prefixes, represented by lower-case letters in the formulae in Table 19. I have tentatively identified a few other roots of restricted distribution, such as KRN in the Ron and Sura–Gerka groups, which might be a development of the KTN that spread westwards from Ubangi.

4.4. Nilo–Saharan

Table 23.20 presents a selection of Nilo–Saharan languages; there is no attempt at comprehensive coverage. Songhay has a root which Nicolaï (1981:278) recon-structs as *gòròNgò; this might be a KTN root, which has been suggested as possibly developing to KRN in some West Chadic groups, with the -k masculine suffix yielding KRNK; the second K has, as so often, become voiced by assimila-tion to the preceding nasal, and the first has also been voiced.

KKR forms are still apparent elsewhere. The Kanuri form koki: "fowl, hen" recorded by Barth (1862:200) appears to be the oldest one in the sense that it retains intervocalic -k-. Koelle ([1854] 1963:126) records kugui with two vowels at the end, which has weakened to modern kùwî. From these forms taken together an early Kanuri form #kòkúì can be reconstructed, a good fit with KKR. Kanuri "cock" has already been discussed (4.2.5).

The most dramatic evidence from Nilo–Saharan languages is the proto-eastern Nilotic *-kɔr- and proto-Teso–Lotuko–Maa kɔ-kɔr- (Voßen 1982:345–6), supported by proto-Karamojong–Teso *kɔkɔrɔ (Ehret 1971:174). I follow Voßen's segmentation of the forms; he states that the sV/xɔ/ɣɔ syllable cannot yet be claimed as a reduplicating morpheme. Here are forms that, if borrowed as a whole, are a good match with West African KKR, coming from roughly the part of Africa where Johnston suggested the Bantu adopted the fowl. Perhaps it was the eastern Nilotes who first adopted it from further north or east. Voßen (1982:345–6), characterizes his reconstructions as "safe" or regular, but does not speculate on the date when proto-eastern Nilotic was spoken. If the fowl was known to the speakers, it could have spread westwards through other language groups; forms from geographically intermediate Daju and Maba are cited in Table 20. They show KRK forms that could possibly be metathesized forms of KKR.

The Central African distribution of KKR needs fuller study, but the evidence so far suggests that the word moved with the fowl from East to West Africa through Nilo–Saharan to reach the Niger–Congo languages. Exact routes are not yet clear because of the various shortened forms that often allow alternative interpretations. A form shortened to KK appears to have given rise to central Bantu somewhere in Central Africa. Another KK form, #kikɔ/kukɔ, occurs in Platoid languages and appears to have spread south to Yoruboid, Akokoid, Edoid, Igboid, and Upper and Lower Cross. Meanwhile, a full KKR form survived in East Chadic (Dangla group) and spread sporadically to West Chadic; although there is no clear direct route, a KKR form reached Gur and spread to Mande in the west and to Kru and Kwa to the south.

5. KKR outside Africa

The resemblance of English cock to KKR is not likely to be accidental. The OED compares Old English cocc, coc, kok with Old Norse kokkr (rare) and French coq, observing that the k-spelling points to a loanword; further, that the proto-Germanic root is *hanon- and the proto-Romance one Latin *gallus; French coq is from late Latin *coccus. The fact that cock is a loanword is also shown by its [k], which would have become [h] in Germanic if it had been inherited from an earlier stage of Indo–European. Chicken is believed to be derived from a palatalized form of the same root. Most European languages have borrowed forms related to this root. Grierson ([1928] 1994:118–19) shows large numbers of Indo–European languages in India with forms such as #kukkur "cock". This suggests that the KKR root originated in Asia. As in Africa, other roots developed in Germanic and Romance, while KKR spread westwards during the Roman empire (possibly with a particular variety). Possibly the kakara root of pre-Columbian America represents a spread of the same root eastwards. KKR also spread into eastern Africa and thence across the continent to the west, as shown above. The point of introduction into

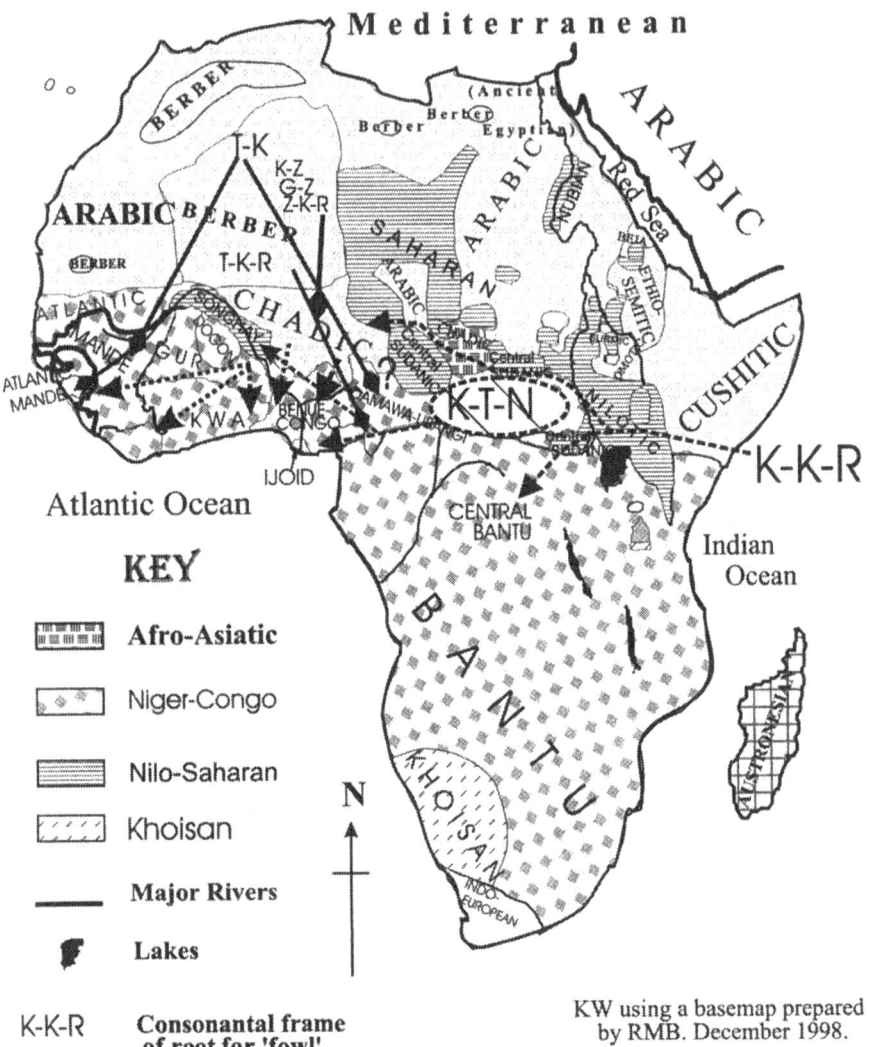

Figure 23.1 Distribution of three major roots for "fowl" in west–central Africa.

Africa is not yet clear. There is no evidence that KKR moved down the Nile valley, as different roots are found in Nubian and Cushitic; an early introduction to the East African coast seems to best fit the geographical spread (Fig. 23.1).

6. Ethnographic evidence for the domestic fowl in Africa

Carter (1971:194, 212) and Crawford (1984:302) briefly refer to chickens in Africa, and, quoting Sauer (1969), claim that a type with black feathers, flesh and bones was described from Mozambique in 1635; Carter adds that these melanotic strains, together with dark brown eggs, the use of cock-fighting, and linguistic evidence (mentioned above) all suggest an introduction from India. Black flesh, however, is unknown in West Africa, and so perhaps in East Africa represents a geographically limited and relatively late introduction from India.

Cock-fighting in West Africa has been reported only for use in divination, for example, by Barth on his way to Adamawa:

> When two are litigating ... , each of them takes a cock which he thinks the best for fighting: and they go to Kobshi. Having arrived at the holy rock, they set their birds a-fighting, and he whose cock prevails in the combat is also the winner in the point of litigation. (Kirk-Greene 1958:222, quoting Barth)

In the same general area cocks are also used for divination in a different way, which suggests that cock-fighting is not a deeply ingrained practice:

> By the Humbitode hill the spirit dwells, in a baobab tree near a spring. Both litigants set forth with a cock, and the one whose cock crows first is declared to have won his suit. The cocks may crow anywhere along the bush path, whereupon it is said that the liar's cock's crest will at once turn white and droop in shame. Should neither cock crow before they reach Humbitode, both parties are credited with telling the truth and the oath is declared void. (Kirk-Greene 1958:222)

There are occasional clear if unexpected statements in the ethnographic literature, such as the following:

> The only domestic animals which the Tiv had at the beginning were the dog, the cow, and the chicken. (East [1939] 1965)

For the fowl to be regarded as earlier than, for example, the goat, whose name appears to be reconstructible almost to proto-Niger–Congo, is surprising.

Fowls are very much taken for granted in West Africa, and the summary of their traditional use by the Igbo is probably valid for much of West Africa:

> Meat is preferred to fowl and there is a certain prejudice against eating the fowls reared in one's own house, except in an emergency as when an unexpected guest arrives. Fowls are looked upon "a little as one's own children" though no care is taken of them beyond shutting them up for the night, and they will be sold in the market without compunction. Eggs are rarely eaten and then only hard boiled. (Leith-Ross [1939] 1965:63)

Another way of testing the degree of antiquity of fowls in the culture is by investigating the richness of the vocabulary associated with them. Most of the

Table 23.21   Varieties of fowl listed in Jula (Keita 1989/90).

sìsè fį	"la poule noire"	<	sìsê – fį	poule-noir
sìsègwê	"la poule blanche"	<	sìsê – gwé	poule-blanc
sìsèwúlȩ̂	"la poule rouge"	<	sìsê – wúlȩ̂	poule-rouge
sàgàsísê	"la poule qui a de longues pattes"	<	sàgâ – sìsê	mouton-poule
dùgàsísê	"la poule dont le cou est dépourvu de plume"	<	dùgâ – sìsê	vautour-poule
ká̧wúlánî	(same as preceding)	<	ká̧ – wúlá – ni	cou-écorché-dérif.
kòlò̧sísê	"la poule dont le plumage est tâcheté de blanc"	<	kòlò̧ – kísé – sį́sê	cauris-graine-poule
férétósísê	"la poule sans plumage"	<	férétó – sìsê	vêtements-privatif-
kúnátósísê	"la poule dont les pattes sont rongées naturellement"	<	kúnátó – sìsê	lépreux-poule poule
dùsùsùmàsísê	"la poule dont le plumage de la poitrine est blanc et qu'importe le reste"	<	dùsù – sùmà – sìsê	coeur-frais-poule
bùgùrìgwèsísê	"la poule au plumage couleur cendre"	<	bùgùrìgwè – sìsê	cendre-poule
sá̧kábásísê	"la poule au plumage couleur des nuages (grise)"	<	sá̧ – kábá – sìsê	ciel-nuage-poule
tùbàbùsísê	"la poule des blancs"	<	tùbàbû – sìsê	toubab-poule
fàràfįsísê	"la poule locale"	<	fàràfį – sìsê	noir-poule
sìsêtùrúkèmè	"la poule à la crête épaisse"	<	sìsê – tùrú – kèmè	poule-crête-cent

better-described languages seem to have at least a few words for special types of fowls. No special search has been made, but a few examples have been noted. In Jula (Manding) Keita (1989/90) distinguishes the varieties listed in Table 23.21.

Bobo has at least two special names for breeds of fowls: gbìgbìrì *poules d'une race spéciale, qui ont ont les plumes recourbées et ne peuvent voler* and siân *espèce de poule aux plumes déchiquetées. En bambara saga sise* (Le Bris & Prost 1981:198). Note the cross-reference to Bambara, like Jula a Mandekan dialect.

In Hausa (Bargery 1934:323), notes fìngíi "fowl with permanently ruffled feathers ... in great demand for magical purposes" (also known as bíngi, bùr̃tʃ'áttʃ'àakíi, tʃíkírkítà, gùzgús, kúdúgúu, kúdúkkúu, kùrkúr, ʃikírkítà). Obviously, some of the other roots recorded in Chadic recur here with a specialized meaning.

Bekwarra has ìtʃákátʃáká "kind of fowl (bird?) with all feathers curving outwards" (Stanford n.d.:26).

Echie (Igboid) distinguishes two special kinds of cock valued by native doctors: ákpùká "smallish in appearance" and áyáyárá "with dishevelled feathers" (Sylvester Nwala, pers. comm.). Onitsha Igbo has òkúkù ábùké "small tough fowl used for sacrifice" (Williamson 1972:413).

For Nupe, Banfield (1914) gives the varieties listed in Table 23.22.

Since the morphology of African fowls remains undescribed and there are believed to be no significant African breeds (Blench & MacDonald, in press), it would be interesting to pursue the types distinguished in the various languages and find out their status. They might offer interesting genetic clues to origins and introduction routes.

Table 23.22   Varieties of fowl listed in Nupe (Banfield 1914).

biʃe àlăʒì	"a pure white fowl"
biʃe bibi	"a red, black, and white fowl"
biʃe èdzwǒ	"a red, black, and white fowl"
biʃe gbàtá	"a dark red and white fowl"
biʃe gbədụ	"a large jet black fowl"
biʃe gǔgoró	"a light brown fowl"
biʃe kparò	"a partridge plumaged fowl"
biʃe kpàyì	"a silver grey fowl"
biʃe kpìkpì	"a curly feathered fowl, has no wing feathers"
biʃe lìàlìăgi	"a very short-legged fowl"
biʃe lǔkpa	"a red and black fowl"
biʃe sákpa	"cf. biʃe gbàtá"
biʃe sòkógùtʃi	"cf. biʃe kpìkpì, but has wing quills"
biʃe ʃèlanla	"any fowl with red bill and red legs"
biʃe ʃèlǔ	"a guinea-fowl plumaged fowl"
biʃe tutumpèrè	"an ash coloured fowl"

Note the resemblance between Nupe biʃe kpìkpì and Bobo gbìgbìrì in both sound and meaning.

7. Conclusion

There is a basic conflict between the relatively recent attestation of chickens in archaeological contexts and their apparently deep embedding in West African culture. The most important traditional uses of the fowl in West Africa are ritual, for sacrifice and divination, and not for cock-fighting, egg production, or eating; they give the impression of a considerable time depth. The linguistic evidence presented here supports the deeper embedding suggested by ethnography rather than the shallower one attested by archaeology.

It is true that in West Africa words for "fowl" and "cock" cannot be reconstructed to proto-languages of the highest level. They can, however, be reconstructed to proto-languages of intermediate and lower levels. Unfortunately there is no method for dating proto-languages in absolute terms; it is only possible to say that proto-languages at a higher level are obviously older than those at a lower level. Table 23.23 summarizes.

Apart from languages that have specialized their word for "bird" to "fowl", three major and widespread roots have been observed. The first is TK. This has here been derived from the feminine form of the Berber word. Alternatively, it could be regarded as a form of the ancient #tax(V) noted by Blench & MacDonald (in press) as spread "from Korea across central Asia to the Near East, North Africa and south to Lake Chad". If TK belongs to this root, this would accord with its apparent great age in Mande. On the other hand, it is hard to explain how it reached both Mande and Chadic without passing through Berber or the Saharan branch of Nilo–Saharan. The Berber forms are clearly analyzable within Berber (Basset 1959a) and therefore cannot be loans from outside Berber, and the TK root has not been discovered

Table 23.23 Summary of reconstructions in Niger-Congo, Berber and Chadic.

Proto-language	"Fowl"	"Cock"	Remarks
Proto Niger-Congo	none	none	
Proto-Kordofanian	?	?	
Proto-Mande	TK	?	<Berber
Proto-Atlantic-Congo	none	none	
Proto-Atlantic	none	none	
Proto-North Atlantic	none	none	
Proto-South Atlantic	none	none	
Proto-Mel	TK	none	<Mande
Proto-Ijoid	none	none	
Proto-Dogon	NY-?	?	
Proto-Kru	?	?	
Proto-Gur	KKR?	?	
Proto-Central Gur	KKR	?	Manessy 1979
Proto-Oti-Volta	KKR	?	Manessy 1975
Proto-Adamawa	none	none	
Proto-Ubangi	KTN	none	
Proto-Kwa	?	?	KKR <Gur?
Proto-Benue-Congo	none	none	
Proto-Ọkọ-Nupoid-Idomoid	none	none	
Proto-Nupoid	?	none	PS not in Ebira
Proto-Nupe Group	PS	none	
Proto-Yoruboid-Edoid-Akokoid Igboid	?	?	KKR loaned?
Proto-Yoruboid	DR	KKR	Akinkugbe 1978
Proto-Edoid	KKR	QK	Elugbe 1989
Proto-Akokoid	?	?	
Proto-Igboid	KKR	QL?	QK <Edoid
Proto-Ukaan-Akpes	KKR	?	
Proto-Kainji	none	none	
Proto-Platoid	none	none	
Proto-Northwestern subgrp	NN	none	Gerhardt 1983
Proto-South-Central subgrp	NN	QK	Gerhardt 1983
Proto-Southwestern subgrp 1	KKR	QK	Gerhardt 1983
Proto-Jukunoid	KN <KTN		Shimizu 1980
Proto-Cross River	none	none	
Proto-Bendi	KKR/KT?	?	
Proto-Upper Cross	none	none	
Proto-Lower Cross	?*	KKR	*loaned <UC
Proto-Kegboid	KN <KTN	none	Ikoro 1989
Proto-Central Delta	?	none	
Proto-North Bantoid	none	none	
Proto-Dakoid	QK	none	
Proto-Mambiloid	KT?	none	
Proto-South Bantoid	none	none	
Proto-Ekoid	KKR	?	
Proto-Beboid	SY	?	
Proto-Jarawan-Grassfields-Bantu	KB	?	
Proto-Central Bantu	KKR	?	
Proto-Afroasiatic	none	none	
Proto-Berber	T-KZD-T *	Y/K/GZD	*feminine "hen"

within Saharan. I therefore take TK to be derived from a shortened form of the Berber feminine, loaned into Mande at an early date and into some groups of central Chadic at a somewhat later date. The feminine was also borrowed in a longer form, TKR, and spread in a somewhat erratic fashion, surfacing today in such various groups as Kainji, Chadic, and North Bantoid. At a later date still, forms of the masculine were also borrowed. A question not answered here is the origin of the Berber forms themselves. The root YZD does not resemble either the tax(V) or the ka(C)i roots mentioned by Blench & MacDonald (in press) as widespread in Eurasia, though the alternative form KZD could possibly be related to ka(C)i. The limited spread in Berber of a Latin loan, compared with YZD/KZD, shows that YZD/KZD must be considerably older.

The second major root is KKR. This has been traced from the east across Central Africa into West Africa. It is believed to be of Asian origin and to have reached the East African coast. Originally it meant "cock" but in a number of groups, probably those for whom this was the first introduction of the fowl, it was generalized to mean "fowl/chicken".

The third major root is KTN. It is suggested that this is an ancient Niger–Congo "bird" root which was specialized to mean "fowl" in Ubangi when the chicken reached speakers of either proto-Ubangi or its immediate daughter languages. Thence it spread westwards with the meaning "fowl" into some groups of Adamawa and Chadic, finally reaching into Songhay, and also south through Jukunoid to the Atlantic coast, where it can be traced in Defaka, Kegboid, and central Delta.

The different distribution of these three major roots presumably represents at least three separate introductions of the domestic fowl to West Africa; two across Central Africa from the east, and one across the Sahara from the north. Later adoptions from the Sahara are probably reflected in the more limited KZ, GZ, ZKR, and TKR forms from Berber into Chadic and some northern branches of Benue–Congo.

Even so, the time depth implied by some of the reconstructions is far beyond that currently supported by archaeology. Given the very scanty archaeological coverage of West Africa to date, it would not be surprising if earlier remains were to be found in the future.

Acknowledgements

I am most grateful to Roger Blench, Anneke Breedveld, Vaclav Blažek, Bruce Connell, Aron Dolgopolsky, David Dwyer, Stefan Elders, Rolf Endresen, Suanu Ikoro, Kevin MacDonald and Thilo Schadeberg for discussion and use of unpublished data.

Note

An earlier version of this paper, restricted to the KKR root, was presented at the Third World Archaeological Congress at New Delhi in 1994. A second version, adding the TK root, was presented at the Conference on African Livestock in London, September 1995. This version has been expanded to include all known roots in West Africa.

Appendix: Tables 23.1 to 23.20. Comparative tables of names for "fowl (chicken)" and "cock (rooster)" in African languages. The tables are arranged by genetic groups following current classifications. The major sources are Grimes 1992, Bendor-Samuel 1989 (for Niger-Congo) and Crozier and Blench 1992 (for Nigerian languages). Abbreviations for sources are explained at the end of each table. Numbers under sources refer to the number in a word list or to the page of the work. Capital letters under "Notes" refer to a proposed root; affixes are occasionally added in lower-case. The transcription is IPA except that y is used for [j] and the tilde marking nasalization is placed below the vowel in order to avoid interference with tonemarks.

Table 23.1 "Fowl" and "cock" in Kordofanian.

Kordofanian	Language	Fowl	PL.	Cock/Rooster	PL.	Notes	Source
Rashad	Orig	kú - dúr -ik	súdúrgin			DR	SE68
Koalib	Ngíeere	k - a gwuro	y-			?KR	S
	Nguwurang	k - ag uro	c-			?KR	S
Heiban	Heiban	k - ag aro	j-			?KR	S
Shwai	Cerumba	a urɔ	y-			?KR	S
	Ndano	g - ag arɔ	y-			?KR	S
Laro	[Nobbs]	th ag irɔ				?KR	S
	[MacD]	dh ag arɔ				?KR	S
Otoro	Kwara	g - ag arɔ	j-			?KR	S
	Kwijur	k - ag arɔ	j-			?KR	S
	Orombe	g - ag arɔ	j-			?KR	S
Tira	Tira	orɔ				?KR	S
Moro	UmmDorein	w - arɔ	l-			?KR	S
	Lebu	w arɔ				?KR	S
Fungor	Fungor	k ak arɔ				?KR	S

Sources: S = Stevenson MS; SE = Schadeberg & Elias 1979.

Table 23.2 "Fowl" and "cock" in Mande.

Mande	Language	Fowl/hen	PL.	Cock/Rooster	PL.	Notes	Source
Western: NW	Sambla (= Sembla)	t ɛ̃ g ɛ̃	tégɛ̃	kūkɛ̄ɛ̄		TK KKR	P 146, 103
	Jalunka (= Yalunka)	t ɔ y ɛ - n a		dontoːna		TK DNT	K II.5.a
	Soso (= Susu)	t ɔ y oː e i		koŋkoːre		TK KKR	K II.5.b
	Soso (= Susu)	t ɔ y ɛ		teː tɔyɛ		TK	K II.6
	Ligbi	tùgɔ́				TK	PP 38
	Vai	ti ɛ		tie kaima		TK	K II.4
	Kono	t ɛ		te koːŋe		TK -KKR	K II.3
	Maninka	sìsɛ́				TK?	G 196
	Bambara			dónónkɔ̀rɔ́		DNT-KKR	T 166
	Mandinka (Kankanka)	s i s ɛː		doːndoːŋ		TK? DNT	K II.1.e
	Bambara	s i s ɛ		dundoŋ		TK? DNT	K II.2
SW	Kpelle	t ɛ́ ɛ́		tɛː ʃiːre		TK	W 1; K II.10
	Loma	t ɛ ˞		tɛː sina		TK	K II.11
	Bandi	t ɛ ˞		tɛː hina		TK	K II.7
	Mende	t ɛ́		tɛ́ hina		TK	I 139
	Loko	t ɛ ˞		tɛː hina		TK	K II.8
Eastern	Bobo	nā - nɔ̄ n		kòkór-ó/-í		NN KKR	LBP 208
SE: Sthn	Tura	t ɔ̄ ɔ̄				TK	M 122
	Tura (Nao)	m ã̀ à				M	KD TOU

Table 23.2 (cont'd)

Mande	Language	Fowl/hen	PL.	Cock/Rooster	PL.	Notes	Source
	Mano	tɔ ˥		tɔːgʊ		TK	K II.13
	Yakuba (= Dan/Gio)	t ɔ̂		tɔːgʊ		TK	KD, K II.14
	Yaoure	m ɛ̄ ĩ̄				M	H 122
	Wan (Mwan)	m ã̰				M	H 112
	Wan	s ɛ	sɛmṵ			?	KD WAN
SE: Eastn	Samo Kouy	k ɔ̀ ɔ̀ r ɔ́				KKR	M 122
	Wowara	k ɔ̀ t ɔ̀				KKR	M 122
	Nyankoro	k ɔ̀ r ɔ̀				KKR	M 122
	Laakwe	k ɔ̂ l ɔ̂				KKR	M 122
	Toeni	k ɔ̂				KKR	M 122
	Bisa	k ẁ r				KKR	M 122
	Bisa Lebir	k ŵ f́				KKR	M 122
	Busa	k ɔ̄				KKR	W 256
	Busa	k o o		kod-sa		KKR	M 122

Sources: G = Grégoire 1986; H = Halaoui 1983; I = Innes 1969; K = Koelle 1854; KD = Kropp Dakubu 1980; LBP = Le Bris et Prost 1981; M = Mukarovsky 1987; P = Prost 1971; PP = Persson & Persson 1980; T = Touré 1996; W = Wedekind 1972; Wl = Welmers n.d.

Table 23.3 "Fowl" and "cock" in Atlantic.

Atlantic	Language	Fowl/Chicken	PL.	Cock/Rooster	PL.	Notes	Source
Northern: Cangin	Ndut	p aː n				PN	W 65
	Sili (Palor)	p aː n				PN	W 65
	Safi (Safen)	pʰ a m b i				PN-BT	W 65
	Lala (Lehar)	ˈp a b ɛ t				PN-BT	W 65
	Non	ˈp a b ɛ ʔ				PN-BT	W 65
Senegambian	Fulfulde	g e r t o gal	gertoode	donton	-al/-e	? DNTN	S 69, 52
	Sine (Serere-S.)	ɔ - ç ɛ kʰ	[g class]	séq	[g cl.]	?	W 65
	Wolof	g a n aa r		nyâtyɔ̃ngwɔ́n	ɔ-	?	MG 47,116
Tenda	Basari	ɛ - ty ã r ɛ́	ɔ-syãrɛ́	ɛ́-tyér ɛ́-tyán		TYR	F 247–8
	Bedik (Budik)	ɛ - ty é r	ɔ-syér		...ɔ-sy..	TYR	F 247–8
	Biafada	a - dʒ uː a	maː-s	ŋaːna		?	K I.C.1
	Pajade (Badyara)	pa - dʒ a f e		pa-ŋane		?	K I.C.2
Bak	Ganja	Ø - ɲ ɛ̄ g	g-			?	Nd 82
	Fulup (Ejamat)	ɛ - x u l ɔː l	su-	gɛin		?	K I.A.1
	Filham (Diola)	ɛ - r aː s a	si-lɔːl	keːn		?	K I.A.2
	Diola (Fogny)	ɛ - l ɔ ɔ l	si-			?	KD1 144

Table 23.3 (cont'd)

Atlantic	Language	Fowl/Chicken	PL.	Cock/Rooster	PL.	Notes	Source
	Bola (Mankanya)	o - g oː k	ŋgu-	ogihaːɪn ~ -dʒ-		KKR	K I.B.1
	Sarar (Mandyak)	u - g oː k	ŋgu-	woyuaːɪn ~ -dʒ-		KKR	K I.B.2
	Pepel (Papel)	ɔ - g o k a	ŋgeː-	oːdʒahã		KKR	K I.B.3
Southern:							
Mel	Baga	a - ts ɔ g ɔ ~ -ɣ-	ɛ-	ketseːpi		TK	K I.D.1
	Timne (Temne)	a - t ɔ k ɔ	ɛ-	ketopiː		TK	K I.D.2
	Bulom (Bullom)	i - s ɔ k	ʃi-	isɔk ipuɣan		TK	K I.D.3
	Mmani	i - s ɔ k	si-			TK	KD2 MMA
	Mampa (Sherbro)	i - s ɔ k		sɔk puɣan		TK	K I.D.4
	Sherbro	si - s ɔ̀ k		sɔk pùkan		TK	P 50
	Krim	- s ɔ g	-si			TK	KD1 375
	Kisi (Kissi)	- s ɔː	-a	sɔ kɔŋgiːli		TK	K I.D.5
	Gola	- t ɔ				TK	T 38
Limba	Limba	- t ɛ				TK	T 17
	Limba Warawara	- t ɛ n i				TK	T 17

Sources: F = Ferry 1972; K = Koelle 1854; KD1 = Kropp Dakubu 1977; KD2 = Kropp Dakubu 1980; MG = Munro & Gaye 1991; Nd = N'diaye-Correard 1970;
P = Pichl 1963; S = Sow 1971; T = Thomas 1916; W = Williams 1994; Wn = Wintz 1909.

Table 23.4 "Fowl" and "cock" in Ijoid.

Ijoid	Language	Fowl/Chicken	PL.	Cock/Rooster	PL.	Notes	Source
Defaka	Defaka	ò - k ù n à		ɔ́-ɓáí ò-kùnà *		KN * "male fowl"	J 146–7
East Ijo	Nkọrọ	ò - k(ùn)à		ó-wói ɓɔ̀kɔ̀ *		KN * "male fowl"	KW
	Kalabarị	ɔ̀ - ɓɔ́ᵗkɔ́		ó-wí ɔ̀-ɓɔ́ᵗkɔ *		BK * "male fowl"	J 146–7
	Ịbanị	ɔ̀ - b í ᵗ ɔ́		ó-wí ɔ̀-ɓíᵗɔ́ *		BK * "male fowl"	KW
	Kịrịkẹ (Okrika)	ɔ̀ - ɓɔ́ᵗkɔ́		ó-wú ɔ̀-ɓɔ̀kɔ *		BK * "male fowl"	KW
	Nembe	ò - f ò n ì		ò-wèi òfóni **		= "bird"; * "ground-bird"; ** "male bird"	K 127
		kịrí ò - fóni *					
West Ijo	Izọn (Kolokuma)	ò - f ó n í		ò-wèi òfóni **		= "bird"; * "town-bird"; ** "male bird"	WT
		àmá ó - fóni *					
	Biseni	à - w ọ̀ m ɛ́ ɛ̀		ó-wéi à-wọ̀mɛ́ɛ̀ *		= "bird"; * "male bird"	KW
	Akịta (Okordia)	à - wị̀ y ɛ̀		é-wéi àwị̀yɛ̀ *		= "bird"; * "male bird"	KW
	Oruma	ámá féni *		ò-wéi féni **		* "town bird"; ** "male bird"	KW

Sources: J = Jenewari 1983; K = Kaliai 1966; KW = personal data; WT = Williamson & Timitimi 1983.

Table 23.5 "Fowl" and "cock" in Dogon.

Dogon	Language	Fowl/Chicken	PL.	Cock/Rooster	PL.	Notes	Source
Dogon	Tɔrɔ Sɔɔ	é - ɲ e		éɲe ána		NY	CG 83
	Kamba Sɔɔ	é - nj e				NY	CG 83
	Jamsay	é - n e				NY	CG 83
	Toro Nomo	é - s a				?	CG 83
	Donno Sɔ	e - n dy é		endje-na		NY	K 132
	Piñay	s - ˈí d ˈà				?	KD 39

Sources: CG = Calame-Griaule 1968; K = Kervran 1982; KD = Kropp-Dakubu 1991.

Table 23.6 "Fowl" and "cock" in Kru.

Kru	Language	Fowl/Chicken	PL	Cock/Rooster	PL	Notes	Source
Western	Grebo	haˡbɛ́	haˡbé	kòɔ̀	kòɛ̀	HP	I 38,54
	Tépo	haˡpɛ́		haˡpɛ̀bju˧		HP	M 341,337
	Jrewe	haˡbɛˡgbá		haˡbɛ̀bɪˡʊˡ		HP-	M 341,337
	Klao	sɔ̃̀				TK?	M2 195
	Guéré	sɔ̃̀ˡ		sɔ̃̀bɛ̄ɔ		TK?	M 341,337
	Wobé	sɔ̃̀ˡ		sɔ̃̀gbã̀ɔ̃̀		TK?	M 341,337
	Niaboua	sɔ̃̀ɔ̃̀ˡ		sɔ̃̀ɔ̃̀gbã́á		TK?	M 341,337
Eastern	Bété (Daloa)	núnú		kɔ̀kɔ́bɛ̀rʊ̃̄		NN	M 341,337
	Bété Guibéroua	kɔ̀kɔˡ		kɔ̀kɔ́ɓɛˡlu˧		KKR	M 341,337
	Néyo	kɔ̃́kwɛ̄ɛ́		kɔ̃́kɔ̃́ɓɛ̄lʊ̄		KKR	M 341,337
	Godié	kɔ̃́kʷɛ̄		kʊkɔɓɛlu		KKR	M 341,337
	Koyo	kɔ̀kʊ̄jɔ̄		kɔ̀kʊ̄ɓɛ̄lʊ̄		KKR	M 341,337
	Dida	gʊ̀gɔ̄jɔ̃ŋwnɔ̄		gʊ̀gɔ̃ɓɛ̄lʊ̄		KKR?	M 341,337
	Aïzi	kɔsɔ		kɔgɔzɪsɪ		KS	M 341,337

Sources: I = Innes 1967; M = Marchese 1983; M2 = Marchese 1986.

Table 23.7 "Fowl" and "cock" in Gur.

Gur	Language	Fowl/Chicken	PL.	Cock/Rooster	PL.	Notes	Source
Bariba	*Baatonum (Bargu)	g oː a		goːɔ		KKR?	K XII.B.2
Central:	Buli	kp e a k ~ -ia-	kpesa	cɔ		QK	M 289
Oti-Volta:	Buli (Guresha)	gb e aː y a		gbaleaːy		QK	K IV.A.3
Buli-Koma	Konni	kp e e -n	-si	kpa-raa	-si	QK	Na 103
Eastern	Biali (Bieri)	k o g e ~-u-, kux	kusi			KKR	M 289
	Tayari	k o t a/e t/r a	kohir			KKR?	M 289
	Tamberma (Tamari)	k o t e	kuo			KKR?	M 289
	Waama (Wama)	k o k a	kosu			KKR?	M 289
Gurma	Moba	k ò l ī g ~ -e-				KKR	M 289
	Akaselem (Kasele)	k ɔ c	kɔlɛ			KKR	M 289
	Ntcham (Basari)	k ɔ l				KKR	M 289
	Ntcham (Tobote)	k ɔ l	kɔl			KKR	M 289
	Gourmanchéma	k o k o l o	kokoli	ko -tongu	ko-tondi	KKR	P 448
	G. (Gurma)	k o k uː r o	kokuːri	koːtuŋ u		KKR	K IV.A.4
	Konkomba	u k uu l	ikuul			KKR	KD1 359
	Konkomba	k ŭ ꞉ l	kò꞉/ɔ̄l			KKR	M 289
	Ngangam (Gangam)	k u k o l	kokole			KKR	M 289
Western	Boulba (Bulba)	n u ã̄	nuose			NN	M 289
	Birifor	n ũ ã̄				NN	M 289
	Dagaari	n ɔ̃-				NN	M 289
	Dagara	n w õː	noːr			NN	M 289
	Dagara	n u ȭ	nor	nu-ra	nure	NN	P 448

Loberu	l u l e	luli			NN	M 289
Dagbane	n o:	nóhi			NN	M 289
Gyoore	n ʊ̃: ã	nʊ́hí			NN	M 289
Gbanyan	n w ã				NN	M 289
Moore (Mo:re)	n oa ɣ a	noɛ́sɪ	no:ra:go		NN	K IV.A.1
Moore (Mɔ:se)	n ɔ: ɣ a	nose	no-raogo		NN	P 448
Moore	n oa ɣ a	nɔ:se		norado	NN	M 289
Kusaal (Kusa:l)	n ɔ: g a	noosi			NN	KD2 HAN
Hanga	n ó ̗o	nó:sé			NN	M 289
Gurenne	n ó á	nɔɔhe			NN	KD2 FRA
Frafra	n o w a	noosi			NN	KD2 MAM
Mampruli (West'n)	n oo w a	nuosi			NN	KD2 MAM
Mampr. (Walewale)	n o o	nobisi			NN	M 289
Mampelle	n o bila				NN	M 289
Nabte	n o bil				NN	M 289
Talne	n u bil				NN	M 289
Dagare	n o bila	nobile			NN	M 289
Yom- Nawdm — Nawdm (Naudem)	kw ɔ/a r g a	kwari			KKR	M 289
Nawdm	k ɔ̀ r g	kɔ̀rh			KKR	Ni 267
Yom	k o ːɣ a	kɔːsa			KKR	M 289
Grusi, Eastern — Yom (Djelaŋa)	k u:a r x	kuːarʃ	kurax	kurai	KKR	K IV.A.2
Kabiye (Kaure)	k a le m u: r ɛ	kaleːmɛ	kalemɔːɣu		KKR	K IV.B.2
Lukpa (Le:gba)	k a (r u) mb i r ɛ		kambaːɣo		KKR	K IV.B.1
Tem (Ki:amba)	k a li mb i r ɛ		kalombaːo		KKR	K IV.B.3
Grusi, Northern — Kasem (Kasm)	tʃ i:u r o:	kiemu	tʃiaːbɛːɔ		KKR?	K IV.D.1
Kasem (Yu:la)	k/tʃ io r o		kiːabɛːa	ki:abɛ	KKR?	K IV.D.2

Table 23.7 (cont'd)

Gur	Language	Fowl/Chicken	PL.	Cock/Rooster	PL.	Notes	Source
	Kasem (Kasséna)	ky é		kyébio	kyébèro	KKR?	Cr 15
	Kasem	tʃ ǒ r o	tʃěeni	kibya	kibir	KKR?	KD1 319
	Lyélé (L'élé)	ky o l o	kyélné	giba:		KKR?	P 448
Grusi, Western	?Sisaala (Koa:ma)	g i m e				GM	K IV.C.1
	Sisaala	dz ɩ m -ɩŋ				"hen"	Rw 24
	Sisaala (Bagba:laŋ)	g iːme n	giŋŋa	giːebaːl	-eŋa	GM	K IV.C.2
	Vagla	z a l -a				?	C 338
Kirma-Tyurama	Cerma (Kirma)	k o n a ŋ o	konamba	ko-haluŋo	ko-halamba	KTN?	P 448
	Turka (Tyurama)	k u l u	kulaba	kwol	kwalaba	KKR	P 448
SENUFO	Senufo, Mamara	x/h u	x/huu	hu-po	hu-pee	KKR	P 448
	Senufo, Shenara	g o	gobe	go-to	go-tɛbɛ	KKR	P 448
	Senufo, Tyembara	g ò-piàl		go-páò		KKR	R 53
	Senufo, Karaboro	nk ù l ò		ngo-plɔ	ngo-plɛb	KKR	P 448
	Senufo, Sucite	nk ù l ò	nkùu	nkù-pólô	nkù-pée	KKR	G 344
	Téén (Lorhon)	li mì	liɩu			?	KD2 LOR
	Toussian, S.	s è m	sɛnɛ	sem-pɛ	sempɛla	?	P 448
	Toussian, N.	s e/i nnin	sennii/sini	sim-pa	simpɛ	?	P 448

Sources: C = Crouch & Herbert n.d.; Cr = Cremer 1924; G = Garber 1987; K = Koelle 1854; KD1 = Kropp Dakubu 1977; KD2 = Kropp Dakubu 1980; M = Manessy 1975; Na = Naden 1986; Ni = Nicole 1980; P = Prost 1964; R = Roulon 1972; Rw = Rowland 1966.

Table 23.8 "Fowl" and "cock" in Adamawa.

Adamawa	Language	Fowl/Chicken	PL.	Cock/Rooster	PL.	Notes	Source
Duru:Kutin	Patapori (Peere)	k ɔ n				KN < KTN	BCCW 1.170
Leko	Pere (Peere)	k o n i				KN < KTN	B2 108
	Samba Leko	k ọ̀ ọ̀				KN < KTN	B2 108
	Dong	k o á				KN < KTN	M 1.374
	Wom (=Perema)	k ù á	kùùbá			KN < KTN	RB
	Mumbake (Nyong)	k ò á	kòòpá			KN < KTN	RB
Mumuye	Mumuye Zing	k ì ǹ		bàntí kìǹ		KN < KTN	S 201, 187
	Pugu	k ị̃				KN < KTN	M 1.507
	Yakọkọ	k ị̃				KN < KTN	M 1.516
	Gomla	k ìn				KN < KTN	M 1.519
Longuda	Longuda	s ụ y a w a				?	M 2.362
	Hill	tʃ ĩ tʃ ı w ɛ				?	M 2.368
Vere-Dowayo	Momi	k ɔ̃ z (az)	kəgi (ai)			KZ < Hau.	BE 39
Yungur	Yungur (Bena)	g o				KKR?	M 2.473
	Mboi	g o				KKR?	M 2.482
	Libo (Kaan)	i - y u a				YB	M 2.489
	Lala	y a - a	-aza			YB	RB 288
	Roba	y a a	-a			YB	RB 288
	Rọba (Lala)	y a a				YB	M 2.479
Waja	Waja	cw ī y ō	cwīyē-ī	kɔ̄l-ɔ̀	-íì	YB? KKR?	K
	Tula	y ı b -ɛ́				YB	J MS
Jen	Jen (Dza)	i - y e				YB	M 2.532
Bikwin	Munga (Leelau)	y i e				YB	M 2.538
Mbum	Mbum	k a x a				KKR?	M 2.499

Table 23.8 (cont'd)

Adamawa	Language	Fowl/Chicken	PL.	Cock/Rooster	PL.	Notes	Source
	Mbum	k á k á		téúkáká		KKR?	F
	Tupuri	k ā g ḗ		bɔ̄lɔ̄ kāk		KKR?	R
	Mundang	k á̀ à̀		tɔ́cúo		KKR?	El
	Pam/Mono	k ā̃ k				KKR?	B1
	Mono	k á̃ k		ŋwàà tōō kạ̀k		KKR?	SE
	Mambai	k á̃ g à		kàɡ syá ' ..?'		KKR?	Eg
	Dama	k ā̃ k ī				KKR?	B1
	Galke (Ndai)/Pormi	k à̀ k ā̃ i				KKR?	B1
	Kali	k ā̃ kā̃				KKR?	B1
	Karang	nd ò̃ y		túú ndóy		= "bird"	U
	Kuo	k ā̃ y		túó/tóó kā̃y		KKR?	U
	Tuburo, Pandjama, Kuo	k ā̃ ĩ̃				KKR?	B1
	Njak Mbai	nj ā̃ kā				KKR?	B1
	Ngumi	k ā̃ kā̃				KKR?	B1
Bua	Lua	t wàā̃r				?	Bo
	Kulaal	h à à l	-é			?	P
Day	Day	b ū̃ ù̃				?	N
Kam	Kam	k u m ɛ				KM	M 2.549
Fali	Fali	ɟ à: b á l				YB	KD1 228
(Unknown)	Yingilum	d ú ŋg ò		sàlām būū		DNG	BCCW 1.170

Sources: B1 = Boyd 1974; B2 = Boyd 1994; BCCW = Williamson & Shimizu 1968; BE = Blench & Edwards 1988; Bo = Boyeldieu 1985; Eg = Eguchi 1971; El = Elders forthcoming; F = Fløttum 1974; K = Kleinewillinghöfer 1991; KD1 = Kropp Dakubu 1977; M = Meek 1931; N = Nougayrol 1980; P = Pairault 1969; R = Ruelland 1988; RB = Roger Blench 1994; S = Shimizu 1983; SE = Stefan Elders MS; U = Ubels 1981.

Table 23.9 "Fowl" and "cock" in Ubangi.

Ubangi	Language	Fowl/Chicken	PL.	Cock/Rooster	PL.	Notes	Source
Gbaya	Proto-Gbaya	* k ò r á				KTN	M 132
	Gbaya 'Bodoe	k ò r á				KTN	M 132
	Gbaya 'Biyanda	g ò l έ				KTN	M 132
	Gbeya	k ọ̀ r á̧				KTN	M 132
	Manza	k ò r ā				KTN	M 132
	Mbodomo	k ò r á				KTN	M 132
	Bangando	k wà l á				KTN	M 132
	'Bofi	k à r ā				KTN	M 132
Sere-Ngbaka-Mba	Ngbaka Ma'bo	ng ō̧				KTN	M 132
	Monzombo	ng ō̧				KTN	M 132
Ngbaka	Gbanzili	k ɔ́ k ɔ́				KKR < Bantu	M 132
	'Baka	k ɔ̄ k ɔ̄				KKR < Bantu	M 132
	Mayogo	- ng ū̄				KTN	M 132
	Mundu	ng ō̧		móko ngò		KTN	M 132, V 215
Mba	Ndunga-le	ng ō̧ -				KTN 9/6	M 132
	Mba-ne	ng ō̧ -				KTN 9/6	M 132
	'Dongo-ko	ɓ έ̄ -				?	M 132
	Ama-lo	- nz ò -				KTN 14/10	M 132

Table 23.9 (cont'd)

Ubangi	Language	Fowl/Chicken	PL.	Cock/Rooster	PL.	Notes	Source
Sere	Sere	ng ù r ù				KTN	M 132
	Bare	ng ù r ù				KTN	M 132
Banda	Linda	ng ā t ɔ̄		oko-ngato		KTN	M 132, Ti 26
	Yangere	ng ā t ō				KTN	M 132
	Ngao	ng ā t ū				KTN	M 132
	Vara	ng ā t ɔ̄				KTN	M 132
	Wojo	ng ō t ō				KTN	M 132
	Dakpa	ng ɔ t ɔ̄				KTN	M 132
	Langbasi	ng ō t ō				KTN	M 132
	Mbanza	ng ɔ̄ t ɔ̄				KTN	M 132
Ngbandi	Sango	k ɔ́ nd ɔ̀		kɔ́lí-kɔ́ndɔ̀		KTN	B 552, 450
	Yakoma	k ɔ́ nd ɔ̀				KTN	M 132
	Kpatiri	k ɔ́ nd ɔ̀				KTN	M 132
Zande	Zande	k ó nd ó		bá-kondo		"father-"	M 132, T 250
	Nzakara	k ɔ̄ nd ɔ́		bá-kondo		"father-"	M 132, T 250
	Geme	k ɔ́ nd ɔ̀				KTN	M 132

Sources: B = Bouquiaux 1978; M = Moñino 1988; T = Tucker 1959; Ti = Tisserant 1930; V = Vallaeys 1991.

Table 23.10 "Fowl" and "cock" in Kwa.

Kwa	Language	Fowl/Chicken	PL.	Cock/Rooster	PL.	Notes	Source
Left Bank: Avati.-Nyangbo Gbe	Avatime	ó - (kɔ) k ɔ	-ɛ			KKR	Mk 176
	Proto-Gbe	kòkóló				KKR	Capo 153
	Gen lects	kòk lɔ̃				KKR	Capo 153
Kebu-Animere	Animere	-fu r u	a-			KKR?	H 220
Kposo-Ahlo-Bowili	Akposo (= Kposo)	ó - w o l o				KKR	H 220
	Igo (= Ahlo)	o - l o				KKR	H 220
NYO: Agneby	Abé (= Abbey)	y ɔ̀sɔ̀ - ʃɩ		yɔ̀sɔ̀-cà		KS	Hé 148, 48
	Abidji	ká r ò - jí		kárò-gbòdú ?		KKR	Hé 148, 48
	Adioukrou	ɦ - gɔ̀s		ɦ-gɔ̀s-ɔ̀rɔ́tʃ ?		KS	Hé 148, 48
Attié	Attié	k w ɑ̰̀ - ʃɩ		kwɑ̰̀-sɑ̰̀		KKR?	Hé 148, 48
Avikam-Alladian	Aladian	ɛ́ - kɔ̀sɔ̀		ɛ́-kɔ̀sɔ̀-kɔ̰ɩ		KS	Hé 148, 48
	Avikam	ɛ́ - s ɔ̀		ɛ́-sɔ̀-sɑ̰̀		KS	Hé 148, 48
Ga-Dangme	Dangme (= Ad.)	k ʊ ŋ ɔ́	-í	kuŋɔ-ku *		KTN?	KD1 128 * A 17
	Gã	w ù ɔ́		wúɔ́-nũu		KKR?	KD 240
Potou-Tano: Ega	Ega	ū - w l á	ɛ̄-	ɔ́-wlá-gbi		KKR	BR 60 Hé 48
Lelemi	Lelemi	à - kɔ́kɔ̀ *	bà-	à-kɔ́kɔ̀-ɔ̀dzàni		KKR * < Twi	Hö 106, 124
	Siwu (= Akpafu)	kɔ́kɔ́	mà-			KKR	KD2 SIW
	Sele (= Santrokofi)	k ɔ́ɔ́ k ɔ́ɔ́	bà-			KKR	KD2 SEL
	Sekpele	ù - k ù s ɛ	bɛ̀-			KS	KD2 LIK
Potou	Ébrié	kɔ̀sɔ̀ - ɓjè		kɔ̀sɔ̀-sɛ̀		KS	Hé 148, 48
	Mbato	ɠ ɔ̃ sɔ̃		à - ɠ ɔ́ ɠ ɔ́		KS	Hé 148, 48

Table 23.10 (cont'd)

Kwa	Language	Fowl/Chicken	PL.	Cock/Rooster	PL.	Notes	Source
Tano: Central	Akan	à - kú̀ k ɔ̀	ŋ-	à-kú̀kɔ̀-nínì		KKR	Ch 1933:243
	Abron	à - kù̀ k ɔ̀ -b(ɛ)rɛ̀ɛ̀		à-kù̀kɔ̀-ɲíɲɩ̀		KKR	Hé 148, 48
	Anufo (= Chokosi)	a - ˊ k ɔ̀		á-kɔ̀-ɲúmá		KKR	Kr
	Anyin (= Agni)	a - ˊ k ɔ̀		á-kɔ̀-ɲì		KKR	Hé 148, 48
	Baule (= Baoulé)	à - k ɔ̀ - blà^		à-kɔ̀-ɲì-má̀ ?		KKR	Hé 148, 48
	Baule	a - n ṵ m ạ̀				?	KD2 BAU
	Nzema	à - k ɔ́ l ɛ́-blìjɛ̀		à-kúlɛ́-ɲìɲɩ̀		KKR	Hé 148, 48
Guang	Chumburung	kì-tʃáŋî	à-			?	S 262
	Krache	kɔ̀-bʷàtɛ̀ːdʒ ̀ì	m̀-			?	S 262
	Nawuri	-tʃásɛ́ ̀	ɩ̀-			KS	S 262
	Gikyode	-tʃáːsɛ́ ̀	ɩ̀-			KS	S 262
	Nchumbulu(Banda)	-tʃàsì	ì-			KS	KD2 NCH
	Genyanga	gà-bwɛ́				?	KD1 261
	Gonja	-kòʃí	ŋ-			KS	S 262
	Gua (= Hill Guang)	à-k lɛ́nì	ŋ-			KKR	KD2 HIL
Krobu	Krobu (= Krobou)	kɔ́kɔ́-brīsí		kɔ́kɔ́-ndä^		KKR	Hé 148, 48
Western	Abure	à - ˊ k ɔ̀		á-kɔ̀-vènì		KKR	Hé 148, 48
	Éotilé	è -jíkɔ̀		è-jíkɔ̀-mìà		KKR?	Hé 148, 48

Sources: A = Accam 1966; BR = Bole-Richard 1983; Capo = Capo 1991; Ch = Christaller 1933 (retranscribed); H = Heine 1968; Hé = Hérault 1983;
Hö = Höftmann 1971; Kr = Krass 1970, 1973; KD = Kropp Dakubu 1973; KD1 = Kropp Dakubu 1977; KD2 = Kropp Dakubu 1980; Mk = Mukarovsky 1976–7;
S = Snider 1990.

Table 23.11 "Fowl" and "cock" in Oko-Nupoid-Idomoid.

Oko-Nup-Idom	Language	Fowl/Chicken	PL.	Cock/Rooster	PL.	Notes			Source
Oko	Ọkọ (= Ogori)	á - b é s ɛ́		ɔ́kɔ́kɔ́rịkɔ̀		PS	ɔ̃nɛ́nɛ̃ "bird"		I 155–6
	Magongo	á - b é s é				PS	ɔ̃nĩñɛ̃̄ "bird"		I 155–6
Nupoid	Ebira	ù - ú hʷ ɛ̃̄				KKR	KKR		Ad 14
	Koto	ò - ó hʸ ɛ̃̄		ɔ̀kɔ̀kɔ̄rɔ̀		KKR	KKR	KKR	S 258
	Gade	gù - t e n ɛ̀	à-	kókórēki		?		KKR	S 258
	Nupe	bī ʃ ē̄		bīʃē 'bá		PS			Ba 48
	Nupe Tako (= Basa Nge)	a - bi ʃ e		akukorɔ		PS			K 126
	Basa Kuje	sèyó ʃiʃànàmà		ò-kóyé		?			S 258
	Dibo (= Esitako)	a - bí ʃ é		biʃɛ-ba		PS			B, K
	Dibo (= Ganagana)	à - bí s é		àbisé róg^wói		PS			S 258
	Gupa	ā̄ - b ī ʃ ē̄				PS			B
	Kakanda	à - b ī ʃ ɛ́		àbīʃé òrùnî *		PS	* "husband"		S 258
	Kupa	ā̄ - b ī ʃ ē̄		abiʃe-nu		PS			B, K
	Asu (= Ebe)	bi ʃ i e		biʃiɛludʒi		PS			K 126
	Gbari-Sumakpna	p ī s é				PS			B
	Gbagyi (= Kuta)	p i s é				PS			B?
Idomoid	Eloyi	ɛ̀ - nū	ɛ̄-			NN			A 71
	Igede	ū - nū	ā-			NN			A 71
	Idoma	ù - gʷ ū		ò-búg^wū		KKR			Ab 114
	Yala (Ogoja)	ù - g ū				KKR			Bu 12
	Etulo	òl - gb î				KKR?			A2 60

Sources: A = Armstrong 1983; A2 = Armstrong 1964; Ab = Abraham 1951; Ad = Adive 1989; B = R.M. Blench 1994; Ba = Banfield 1914; Bu = Bunkowske 1976; I = Femi Ibrahim MS; K = Koelle 1854; S = Sterk 1977.

Table 23.12 "Fowl" and "cock" in Yoruboid–Edoid–Akokoid–Igboid.

Yor-Ed-Ak-Igb	Language	Fowl/Chicken		PL.	Cock/Rooster	PL.	Notes			Source
Yoruboid	Ife (Togo)	ɛ̄	d ɛ̀ ɛ̄		à-kìkɔ̄		DR	KKR		Ar
	Itsẹkiri	è	- gbé l é		è-gbélé ɔ̄kɛ̄rɛ̃̄ *		QL	*	"fowl + man"	Ak 430, 712
	Yoruba	ā	- d ì(ì)ɛ̄		à-kùkɔ̄		DR	KKR		Ar
	Yoruba (Ufe)	ɔ̀	- y ɛ̀ l ɛ̀		à-kìkɔ̄		YL	KKR		I 155–6
	Igala	á	- dʒù w ɛ̄		à- 1kɔ̄		DR	KKR		Ar
Edoid: Delta	Degema	ɔ́	- h ɔ́ h ɔ̀	í-	èlègbé *		KKR	QL * + ɔ́. mɔ̀sì		E 192, 196
	Egenẹ	à	- f ɛ̀ n ì		àfɛ̀nì mɔ̀sì		<Ijo?			E 192, 196
	Epie (Yenizue)	ɔ̀	- fìɛ̀ n ì		ò-kókòrókò		<Ijo?	KKR +		KW
	Epie	ɔ	- w ɔ ɔ		ɔwɔɔ mɔ̀sì		KKR?			E 192, 196
Edoid: Southwest	Ẹruwa	à	- r ɪ̀ f ɛ́		ò-kòkòróókò		?	KKR +		E 192, 196, 188
	Isoko	à	- r ì f ɛ̀	i-	ɔ-kpà		?	Q		E 192, 196, 188
	Okpẹ	ɔ̀	- r ì l è l è		ò-kòkòɔ̀rókò *		?	* also ò-gboɽuale		E 192, 196, 188
	Urhobo	ɔ́	- h ɔ̀	é-	ò-kòkòɔ̀lókò		KKR	KKR +		E 192, 196, 188
	Uvbiẹ	ɔ̀	- ɔ́ ɔ̀		ɔ-kpà		KKR	Q		E 192, 196, 188
Edoid: North-central	Ẹdo	ɔ̀	- x ɔ́ x ɔ̀		ɔ-kpà		KKR	Q		E 192, 196, 188
	Auchi	ɔ́	- kh ɔ̀	é-	ɔ-kpà	é-	KKR	Q		E 192, 196, 188
	Avbianwu	ɔ́	- kh ɔ̀	é-	ɔ-kpà	è-	KKR	Q		E 192, 196, 188
	Unẹmẹ	ɔ́	- khɔ̀khɔ̀	é-	ɔ-kpà	é-	KKR	Q		E 192, 196, 188
	Ghotuọ	ɔ̄	- h ɔ̀	ē-	ɔ-kpà *	è-	* also ò-kòkòyókò			E 192, 196
Edoid: Northwest	Ọloma	ó	- kh ò	í-	ɔ-kpà		KKR	Q		E 192, 196, 188
	Ẹmhalhẹ	ò	- kòókò		ɔ-kpà	ɛ̀-	KKR	Q		E 192, 196, 188
	Ibilo	ò	- k ò	ì-	ɔ-kpà	è-	KKR	Q		E 192, 196, 188

Group	Language						Source
	Uhami	ò - kòkò		ɔ̀-kpà	KKR	Q	E 192, 196, 188
	Ehuẹun	ò - hō		ɛ́-kpà	KKR	Q	E 192, 196, 188
	Ukue	ò - kókò		ɛ́-kpà	KKR	Q	E 192, 196, 188
	Iyayu	ú - sʷásʷé		ò-kōkò	?	KKR	I 155–6
	Ekpimi	ɔ̀ m ɔ̀ - hɔ̄		ɛ̀-kpàhō	KKR	Q-	I 155–6
	Ukue	ū m ɔ̄ - kōkò		ɛ̄-kpà	KKR	Q	I 155–6
Akokoid	Arigidi	ɛ́ - hɛ̄		à-kíkɔ̀	H	KKR	I 155–6
	Ajaşi	ūwɛ̄-ɛ̄	- hɛ̀	ɔ̄-kʷɛ̄hɛ̀	H	KKR-	I 155–6
	Ojọ	ūwɛ̰̄n-ɛ̄	- hɛ̄	ū-kpɔ̄ɛhɛ̄	H	KKR-?	I 155–6
	Oge	ɛ̄	- hɛ̄	ɔ̄-kɔ̄hɛ̀	H	KKR-	I 155–6
	Uro	ɛ́	- hɛ̄	ɔ̄-kúɛhɛ̄	H	KKR	KW O
	Ahan	ɔ̄mā	- dīɛ̄	ākɔ̄kɔ̄	-DR	KKR	I 155–6
	Ayere	ā	- dʒiɛ̄	ākɔ̄kɔ̄	DR	KKR	I 155–6
Igboid	Ekpeye	ú	- 'n ú úɗɔ̀		NN	KKR "bird of house"	WO
	Ndele	ɔ̀	- tʃí tʃí		KKR		WO
	Ogba	ò	- ríɛ̀bà		?		WO
	Onicha	ɔ̀	- kúkɔ̀	ɔ̀kpà, ɛ̀gbénú	KKR	Q, QL	W 413, 414, 106
	Nsụka	ɔ̀	- kɨkɔ̀	ɛ̀gbélé	KKR	QL	KW
	Ẹhugbọ	ɔ̀	- kúkù		KKR		WO
	Izii	ɔ̀	- kù		KKR		WO
	Obolo	ò	- kúkò		KKR		WO
	Ọkọcha	ɔ̀	- ɔ̄kɔ̀		KKR		WO

Sources: Ar = Armstrong 1965; Ak = Akinkugbe 1978; E = Elugbe 1989; I = Femi Ibrahim MS wordlists; KW(O) = MS wordlist; KW O = Williamson (orthographic); WO = Williamson & Ohiri-Aniche MS; W = Williamson 1972.

Table 23.13 "Fowl" and "cock" in Ukaan–Akpes.

Ukaan-Akpes	Language	Fowl/Chicken	PL.	Cock/Rooster	PL.	Notes		Source
Ukaan	Ikan (= Kakumo)	ɔ̄-ɲɛ̄ - k ɔ̀ k ɔ̀		ò-kòràkò		KKR	KKR +	I 155–6
	Aiyegbe (= Ishe)	ɛ̀ - k ɔ̀ k ɔ̀		ù-gbɔ̀dɔ̀		KKR	QT	I 155–6
Akpes	Akpes (= Akunnu)	ì - k ō k ò		ɔ̀-fɔ̀tɔ̀		KKR	FRT	I 155–6
	Ibaram	ì - k ō k ò		o-furɔtɔ		KKR	FRT	I 155–6
	Esuku (Ajowa)	ì - k ō k ò		ɔ̀-fùrɔ̀dɔ̀		KKR	FRT	I 155–6

Source: I = Femi Ibrahim MS.

Table 23.14 "Fowl" and "cock" in Kainji.

Kainji		Language	Fowl/Chicken	PL.	Cock/Rooster	PL.	Notes	Source
Lopa		Lopa	amɔ̀akiarɛ̀		kugu	?	KKR?	A 10, 37
West	Laru	Laru A	kurum				KRM	RB 93
		Laru	kúm				KRM	RB 93
	Kambari	Kambari (Central)	mɔ́-tɔ́ɔ́	ń-	ú-pɛ̂	p-	TKR?	BCCW, H 14
		Kambari (S'thwest)	mɔ̀-tòkù	ǹ-			TKR?	BCCW
	Northern	Hun (= Duka)	ù-kyɨ̀t	0-			KYT	BCCW
		Hun (West)	u-kyɛt				KYT	CR
		Hun (East)	kit				KYT	CR
		Hun (Dukku)	o-kɛɛt				KYT	CR
		Hun (Darangi)	o-kiitʰ				KYT	CR
		Ror	u-kuːt				KYT	CR
		Lela (= Dakarkari)	kɔ́ci	-ni			KYT	BCCW
		Gwamhi (= Lyase)	əv-kəti				KYT	CR
	Southeastern	Pongu (Azhiga)	kuya				KYT	D1
		Pongu (Casu)	kwia				KYT	D1
		Pongu (Cagere)	u-kwiya				KYT	D1
		Fungwa (= Ura)	i-yanəma	ca-			YLM	BCCW
		Bauchi-1	o-waroma				YLM	CR
		Bauchi-2	u-alomo				YLM	CR
		Bauchi-3	waiyam				YLM	BCCW
	Basa	Basa	alma	ʃa-			YLM	BCCW
		Basa-Benue	a-loma	ʃa-			YLM	RB ined
	Kamuku	Kamuku (Cinda)	æ-æɾom	ʃ-			YLM	D4
		Kamuku (Regi)	aɾɔm				YLM	D4
		Kamuku (Kuki)	aɾɔm				YLM	D4
		Shama	äɾɔm				YLM	D4
		Rogo	wa-ɾəm				YLM	CR

Table 23.14 (cont'd)

Kainji		Language	Fowl/Chicken	PL.	Cock/Rooster	PL.	Notes	Source
		Hungworo	ʔw-arəm			YLM		D4
		Sagamuk	ka-rəm			YLM		CR
		Cepu (Randeggi)	ka-romi			YLM		D4
		Cepu (Bobi)	ka-rom			YLM		D4
		Cipu (Kumbashi)	kuto			KTK?		D4
		Cepu (Kakihum)	kutowu			KTK?		D4
		Cepu (Karisen)	kutɔ			KTK?		D4
Reshe		Reshe (Pisabu)	hi-talokwa			TKR?		D3
		Reshe (Shabanda)	hi-tal kɔ			TKR?		D3
		Reshe (Harris)	hi-taluko			TKR?		D3
East	A	Piti	o-kur			KKR		BCCW
		Chawai	kɔr	a-		KKR		BCCW
	B	Amo	ko-koro	a-		KKR		BCCW
	C:	Kuda (= Kudu)	ka-kwio			KKR		IM
Nth'n Jos Group		Chamo, Butanci	kóró			KKR		IM
		Butu, Ningi	kò-hɔ́ró			KKR		IM
		Gyem	kwo-kwoɱ			KKR		BCCW
		Takaya (Taura)	ku-kuyu			KYT?		BCCW
		Gbiri-Niragu*	ku ru	a-		* = Gure-Kahugu		BCCW
		Surubu	kukuru	a-		KKR		BCCW
		Kurama	bə-tó óró	e-		TKR		BCCW
		Jere (Ibunu)	bè-tókóró			TKR		BCCW
		Jere (Buji)	bi-tókóró	i-		TKR		BCCW
		Sheni	u-tɔkɔrɔ			TKR		M

Sources: A = Adekunle 1986; BCCW = Williamson & Shimizu 1968; CR = unpublished field materials of †Clark Regnier (SIL); D1 = Dettweiler 1992; D2 = Dettweiler 1995a; D3 = Dettweiler 1995b; D4 = Dettweiler 1996; H = Hoffmann 1965; IM = Ian Maddieson MS; M = Meek MS; RB = Blench ined.

Table 23.15 "Fowl" and "cock" in Platoid.

Platoid		Language	Fowl/Chicken	PL.	Cock/Rooster	PL.	Notes	Source
Plateau	A. Nthn	Kadara	a -n o				NN	BCCW
		Doka	a -n o ŋ				NN	BCCW
		Kuturmi	o -n u ŋ				NN	BCCW
		Ikulu	gí -n ü fí	bè-nùŋ			NN	Sh2 11
B. Western	1. NWa.	Ashe (= Koro)	i -n ɔ̀				NN	G 102
		Idun (= Lungu)	i -n ù				NN	G 102
		Yeskwa	e -n u	e:-	i-fɔk	i-	NN FK	BCCW, RB O
	b.	Jaban Ketare	nyú				NN	G 102
		Jaban Kwoi	n ú n ú				NN	G 102
	c.	Kagoma	n ú				NN	G 102
		Kamantan	tʃab nyon				NN	BCCW
	2. SW a.	Ya (= Boj)	nakurre				?	BCCW
		Che (= Rukuba)	i -k ɔ́	ī-			KKR	CH
		Ninzam	ù -vrirīkō *	i-vikò	à-gbà	ànà-gbá	? *also á-mármíkō	H 97–8
		Nindem	k i k ɔ̀		à-gwàk		KKR	G 127
		Kaningkom			gwàk		QK	G 127
		Mada West	k ɔ́		gbà		KKR	G 127
		Mada North	k ɔ̀		gbà		KKR	G 127
		Mada	myekye		gba		?	RB O
		Gwantu	mun? k u k o				KKR mun = child?	KW O

Table 23.15 (cont'd)

Platoid	Language	Fowl/Chicken	PL.	Cock/Rooster	PL.	Notes	Source
	Kanufi(= Karshi)	k i k o	i-			KKR	KW O
	Ayu	y a n a ŋ				?	BCCW
	Shall	g i r i		gwaŋ giri		KKR QK	RB O
b.	Eggon	è -nú	é-			NN	KW
	Yashi	n u n				NN	RB O
	Nungu	mē -nà				NN	BCCW
	Ake	e -ne				NN	RB O
Central 1. N. Cent.a.	Berom	c ò ŋ ō t f u r	cóŋōt	rùŋàt *	bi-	? * + zàkàrà < H.	LB
	Cara	f u r	a-kur			KKR/KT?	RB O
b.	Aten	ts ɔ̀ ɔ̀ r ɔ̄	i-	rɔ̄m	i-	TKR	B 240–1
2. S. Central a.	Irigwe	ŋ̀wiæ̀	rè-			NN?	BCCW
	Irigwe	ŋwyέ		gbá		NN? QK	G 76
b.	Izere (= Afusare)	i -ɲɔ̃n	ī-	á-tsòn		NN	BCCW, Kr
c.	Jju (= Kaje)	ɲ ʉ̀ ɛ̀ n	ɲ ʉ έ n	à-ŋ-gbwàk		NN QK	G, BCCW
d.	Tyap, Gworok	ɲ ʉ à n		ɔ̀-gbàk		NN QK	G 76, 72
	Sholio	n -nwãn	Ø-	ɔ̀-gbàk		NN QK	KW
3. W. Central	Tari	a -kut		-gbak		KT/KKR?	RB O
South-Eastern	Horom	k u n z o		-gagak		KN?	RB O
	Horom	g o n g o				KN?	BCCW
	Barkul	k o r	a-			KT/KKR?	RB O

Southern	Language						NN	Source
	Migili (= Koro)	ko	-no	a-	ikókú-			BCCW
BENUE 1. Tarokoid	Tarok	i	-rùgú	ï-rúgú	ikòkúlɔ́k		TK? KKR	DS
	Bashar		gwo n g a	dembogaga			KN?	BCCW
2. Toro-Alumu	Toro (Turkwam)	a	-de				D	RB O
	Alumu (Arum)	a	-de				D	
	Chesu	a	-de	-bo			D	RB O
3. Jukunoid a.	Kutep	mbà	+k ù n	à-			KN	S 2.143
b. i.	Icen	i	k ŋ́				KN	S 2.143
	Kpan	i	-f ù n			ifún nã ùzùn	KN ùzùn = husband	Sh 12
ii. a.	Mbembe	i	f w ὲ			fwὲndɛ	KN?	RB O
	Nama		f w ɑ̃̀				KN?	RB O
b.	Jibu		kwɨn				KN	VD
	Jukun Takum		k u n a				KN	D
	Wase		k ὺ				KN	S 2.143
c.	Abinsi		k ὺ				KN	S 2.143
	Wapan	ā	-kwɨ̀				KN	S 2.143
	Kona		kwu n i				KN	S 2.143
d.	Chomo (Dhu)		kwɨ i				KN	S 2.143
	Bandawa		z a u				?	M
e.	Tita		k ọ				KN	M

Sources: B = Bouquiaux 1964; BCCW = Williamson & Shimizu 1968; CH = Carl Hoffmann MS; D = Margaret Dykstra MS; DS = Danjuma Siman p.c.; G = Gerhardt 1983; H = Hoerner 1980; Kr = Kraft 1981; KW (O) = personal (orthographic); LB = Bouquiaux MS; M = Meek MS; RB O = Roger Blench, orthographic wordlist; S = Shimizu 1980; Sh = Shimizu 1972; Sh2 = Shimizu n.d.; TD = Tyap dic.; VD = Van Dyken 1986.

Table 23.16 "Fowl" and "cock" in Cross River.

Cross River	Language	Fowl/Chicken	PL.	Cock/Rooster	PL.	Notes "bird"	Source
Bendi	Alege	o - k uːo	i-			KKR/KN?	BCCW
	Bete	u - k u o	i-			KKR/KN?	TLC
	Bayobiri, Ukpe	ɔ̀ - kw ɔ̃				KKR/KN?	CC
	Utugwong, Okorotung	ù - k ɔ̀	ì-			KKR/KN?	CC
	Okorogung, Obe	ò - k ɔ̀	ì-			KKR/KN?	CC
	Bekwarra	ù - f à ā		ù-yām ù-fāā		?	RS 78, 89
	Bokyi	ɛ - kw a				?	BCCW
Upper Cross	Koring (Otonkon)	wáá - k ɔ̄	6uákɔ̄ɔ̄			kí-nɔ̀/6ú-	S 55,178
	(Effiom)	i - k ɔ́ ʔ ɔ́	vù-	ɔ́rú⁺kɔ́ʔɔ́/í-		kí-nùɔ̀/vú-	S 55,178
	(Ntezi)	kik ɔ́;				ké-nɔ̀ŋ	S 55,178
	(Amuda)	kik ɔ́				ki-nɔ̀	S 55,178
	Kukele (Wanikade)	bi - k ɔ́ k ɔ́	bù-			bi-nɔ̀n/bù-	S 55,178
	(Waniheem)	bi - k ɔ́ k ɔ́	bù-			bi-nɔ̀n/bù-	S 55,178
	(Ntrigom)	bi - k ɔ̀ k ɔ̀	bù-			bi-nùɔ̀n/bù-	S 55,178
	Uzekwe (Ikke)	i - kwɔ́ kwɔ́	bù-	à-ràmkwɔ̀kwɔ̀	bà-	iniɔ́ŋ/bù-	JF
	Uzekwe	àyà - kʷ ɔ́ k ɔ̀	i-bʷɔ́kʷɔ̀			i-niɔ́ŋ/bú-	S 55,178
	KoHumono	βi - h ɔ h ɔ̀	βù-			βi-yɔ̀ŋ/βù-	Ck 24,19
	Ubaghara (Biakpan)	i - tʃ ɔ́ ɣ ɔ̀	bi-			i-nɔ̀n-	S 55,178
	(Ikun)	i - g ɔ́ ɣ ɔ̀	i-			i-nɔ̀n-	S 55,178
	Agwaagune	i - g ɔ́ ɣ ɔ̀				ɛ́-'nɔ̀n	S 55,178
	(Abini)	i - ɣwūɣwā				ɛ́-'nɔ̀n	S 55,178
	(Etuno II)	i - k ɔ́ ɣ ɔ̀				ɛ́-nɔ̀n	S 55,178
	(Adim)	i - ɣ ɔ́ ɣ ɔ̀				ɛ́-'nɔ̀n/i-	S 55,178
	Umɔn	β ì - g ɔ̃				bisá	S 55,178
	Ikom (Ajijikpo)	βì⁺ - k ɔ́ ɣ í				ɛ́-ninɔ̀n/mí-	S 55,178
	(Ikom)	bi - k ɔ̀ ɣ i	bi-	òrúmá bíkɔ̀ɣì *		* "male fowl"	KW
	Olulumo (Okuni)	ɛ̀ - k ɔ̃ k	i-			ɛ́-ninɔ̀n/í-	S 55,178
	Lokạạ	yi - n ɔ̀ n	yɔ̀-			yì-nɔ̀n li-kólá	S 55,178

Group	Language							Form	Source
	Mbembe (Ogada)	i	- nɔ̀ n	ɔ̀-				i-nɔ̀n/ɔ̀-	S 55,178
	(Ofonokpa)	i	- nɔ̀ɔ̀n	ɔ̀-				i-nɔ́ɔn rɛ́ráŋ	S 55,178
	(Obabene, Ahaha)	iyi	- nɔ̀ɔ̀n	ɔ̀-				i-nɔ̀ɔ̀n/ɔ̀-	S 55,178
	(Akam)	ɛ̀	- nɔ̀ n kwà	ɔ̀-				i-nɔ̀n tʃɛ́jám	S 55,178
	(Apiapum)	i	- nɔ̀ɔ̀n	ɔ̀-				i-nɔ̀nɔ̀ná	S 55,178
	Legbo (Itigidi)	i	- nɔ̀ɔ̀n					i-zó	S 55,178
	(Lemabana)	i	- nɔ̀ɔ̀					i-yòò	S 55,178
	Leyigha (Assigha)	i	- nɔ̀ ŋ					i-hó/ó-	S 55,178
	Lenyima	i	- nɔ̀	ɔ̀-				i-yó	S 55,178
	Lokukoli	ɔ̀	- nɔ̀ n -g-wɔ́	là- -b-				ɔ̀-nɔ̀n/là-	S 55,178
	Lubilo	ɔ̀	- nɔ̀ n -wɔ̀ŋ	ɔ̀-nɔ̀n-b-				kɔ̀-nɔ̀n/là-	S 55,178
	Agoi (Agoi Ibami)	ɔ̀	- nɔ̀ n -ɔ́iŋ	wà- -biŋ				kɔ̀-nɔ̀n/là-	S 55,178
	Doko (Iko Ekperem)	di	- dɔ̀ ŋ -ɸiŋ	dà-				ɔ̀-dɔ̀n/di-	S 55,178
	(Iko Ekperem)	ɔ	- dɔ̀ n -goi					ɔ-dɔn	JTB
	(Uyanga Okposung)	ɔ̀	- dɔ̀ n		ɔ-dɔn-dum			ɔ̀-dɔ̀n nɛ̀ tàmdɛ̀pù	Si 128
	Iyoŋiyoŋ (Uwet)	e	- nu n -pin	i- -piŋ	o-nun-dum			ko-nuŋ	OAU O
	(Akwa Ibami)	ɔ̀	- nɔ̀ n -ɛ́ŋ					gɔ̀-nɔ̀n gá gùdì	S 55,178
	(Akwa Ibami)	o	- nɔ n					ko-nɔngabudum	OAU O
	Ukpet (Akpet)	i	- ɣɔ̀ ɔ̀	ù-				i-nɔ́ʔ/ú-	S 55,178
	(Betem)	ɔ̀	- xɔ̀ -gbɛ́ŋ	ù- -biɛ̀ŋ				i-nɔ̀ŋ/ú-	S 55,178
Lower Cross	Kioŋ		- bɛ̀n-ú-'-nɔ̀n					í-tɔ̀t	S 55,178
	Korop	ú	- nɔ́n	nɔ́-	ó-dòmà ú-nɔ̀n *	bú-bà-		* "male fowl"	K 109
	Usakade	ú	- nɔ́ n	ì-	ɔ̀-kɔ̀kɔ́ ɔ́-nɔ̀n			éí-nɔ̀n/ni-	C 296
	Ibino	ú	- nɔ̀ n		ɔ̀-kpɔ̀n ú-nɔ̀n			í-'nùún	C 296
	Iko	ú	- nɔ̀ n		à-kpɔ̀n ú-nɔ̀n			í-'nùɛ́n	C 296
	Obolo	ú	- nɔ̀ n		ɔ̀-rí ú-nɔ̀n			ò-fùt	C 296
	Ebughu	ú	- nié		áɛ́-gò ú-nìè			ú-núŋɔ̂	C 296
	Efai	ú	- nié		à-dúkó ú-nìè			í-'núŋú	C 296
	ItuMbuso	ú	- níín		ɔ̀-kíkó ú-nɛ́n			í-núɛ́n	C 296
	Ekit	ú	- ni		á-kíkó ú-ni			í-'núgò/í-'nògò	C 296
	Enwang	ú	- ni		á-kíkó ú-ni			ú-núŋù	C 296
	Etebi	ú	- ni		á-kíkó ú-ni			í-núŋó	C 296

Table 23.16 (cont'd)

Cross River	Language	Fowl/Chicken	PL.	Cock/Rooster	PL.	Notes "bird"	Source
	Ilue	ú - n ì		à-kúkɔ́ ú-nî		ú-núŋù	C 296
	Oro	ú - n ì		ɔ̀-gɔ́kɔ̀ ú-nî		ú-núŋù/n	C 296
	Uda	ú - n ì		à- íkɔ́ ú-nî		ú-núŋù	C 296
	Anaang	ú - n ɛ̀ n		à-kíkò		í-núɛ́n	C 296
	Efik	ú - n ɛ̀ n		é-kíkó ú-nɛ̀n		í-'núɛ́n	C 296
	Ibibio	ú - n ɛ̀ n		à-kíkò		í-núɛ́n/í-'núɛ́n	C 296
	Ibuoro	ú - n ɛ̀ n		à-kíkò		í-núɛ́n	C 296
	Ukwa	ú - n ɛ̀ n		à-kíkò		í-núɛ́n	C 296
	Okobo	ú - n è		ó-kó ú-nè		í-núŋù	C 296
Kegboid	Gokana	k ɔ̀ ɔ̀		dōō kɔ̀ɔ̀ *		* "male fowl"	B 247-8
(Ogoni)	Kana	k ɔ̀ ɔ̀		àdóó kɔ̀ɔ̀ *		* "male fowl"	SI
	Tẹẹ	k ɔ̀ ɔ̀		ādōō kɔ̀ɔ̀ *		* "male fowl"	TD
	Baan (Ogoi)	k ɔ̀ ɔ̀		dóó kɔ̀ɔ̀ *		* "male fowl"	NW
	Eleme	ɔ̀ - k ɔ̀				è-nò "bird"	KW
Central	Abuan	(à) - k í n ì ɣ	(à)sí-	ò-léβiri à-kíniɣ *		* "male fowl"	W 44
Delta		(à) - k ɛ̀ n ɛ́ ɣ	(à)sí-				
		isừ - y ú n à				"fowls"	G 28
	Odual	(à) - k ɛ̀ n ɛ́ ɣ	(à)sí-	òléβiri àkiniɣ *		* "male fowl"	W 44, 49
	Kugbo	ɔ̀-ɲɛ̀ - n ú r *	à-wìy ì-	ò-lòβíri ɛ́-nɔ́r		* "child + bird"	W 44, 49
	West Ọgbịạ	ɔ̀-ɲɛ̀ - n ú r ú *	à-wíɛn ú-	ò-nòβíri ɛ̀-nùrù		* "child + bird"	W 44, 49
				ò-kpókpòrókpò			W 49
	East Ọgbịạ	ɛ̀ - n ú r ɛ́-ma*	ì- ... ì-	ò-lòβìr ɛ́-nùr		* "bird + town"	W 44, 49

Sources: B = Brosnahan 1967; BCCW = Williamson & Shimizu 1968; C = Connell 1991; CC = Bruce Connell (from D.W. Crabb MS); Ck = Cook 1969; G = Gardner 1980; JF = John Fajen MS wordlist; JTB = J.T. Breetvelt (MS wordlist); K = Kastelein 1994; KW (O) = Kay Williamson MS wordlist (orthographic); NW = Nwinee Williamson MS wordlist; OAU = O.A. Ubi MS wordlist; RS = Ron Stanford MS dictionary; S = Sterk MS wordlist; SI = Suanu Ikoro (pc); Si = Simmons 1976; TD = Tẹẹ dictionary (fc); TLC = T.L. Cook (MS); W = Wolff 1969.

Table 23.17 "Fowl" and "cock" in North Bantoid.

North Bantoid	Language	Fowl/Chicken	PL.	Cock/Rooster	Notes	Source
Dakoid	Chamba Daka	kp à á			QK	Bd 108
	Lamja	kp à á w ɛ̃ m			sãāwɛ̃m "bird"	RB 306
	Samba of Mapeo	kp à á			QK sãā "bird"	RB 306
	Samba Jangani	kp ɑ ɑ			QK saa "bird"	RB 306
	Samba Nnakenyare	kp ằ m ī			QK sằmī "bird"	RB 306
	Taram	kp a			QK sa "bird"	RB 306
Tiba	Tiba	k ù n k ɛ̃ r ã			KKR?	RB 306
Mambiloid	Ndoola (Ndoro)	ʃ ú r ã		ʃumá	ʃutʃuŋ "hen"	P 3
Ndoola	Ndoro: Ánwe	ʃ o' r ã			KT?	K 1
	Ndoro: Baissa	ʃ ó r a	-bú	ʃotʃóŋ	KT?	K 1, 2
	Ndoro: Tatindoro	ʃ õ r a'			KT?	K 1
Nizaa	Nizaa (Nyan-Nyan)	a - ʃ u ʃ ú			KT?	MM 5
	Nizaa (Nyam-Nyam)	s ì m			SM	BM 10
	Nizaa	s ì w̃		ŋ-gɔ̃	SM KKR	E
Kwanja	Kwanja	c ɨ́ nd è	cɨ́ndé	twàmbɨ	KTN ?	WW 12, 83
Mambila cluster	M. Lemele (Mbamnga)	k ɔ̃ r		kúlé	KTN	C 358-9
	Mambila	c üä r		kwaá	KTN QK	PD 109
	Mambila Mbar	c ɔ r			KTN	RB 7
	Mam. Tungba (Gembu)	tʃ ɔ̃ r		tákùl	KTN TKR	C 358-9
	Mambila Len (Saam)	tʃ ɔ̃ r		kwà tʃɔr	KTN QK	C 358-9
	Mambila Ba (Somié)	tʃ üä r		kwá tʃüär	KTN QK	C 358-9
	Mambila: Warwar	c ɔ r - m e			KTN	RB 7

Table 23.17 (cont'd)

North Bantoid	Language	Fowl/Chicken	PL.	Cock/Rooster	Notes	Source
	Maberem (Kuma)	tʃ ɔ nd o			KTN	M 568
	M. Cambap (Camba)	tʃ ɔ́ nd ɔ̄		kɔ̀rɔ́	KTN TKR	C 358–9
	M. Langa (Mabundu)	tʃ ɔ nd ɔ		tahɔrɔ	KTN TKR	C 358–9
	Mam. Karbap (Kara)	tʃ ō d ō		tɔ́kɔ̄r	KTN TKR	C 358–9
	Mam. Gelep (Titong)	tʃ ɔ nd ɔ		takoro	KTN TKR	C 358–9
	M. Torbi (Hore Taram)	tʃ ɔ̄ nd ɔ̄		tákɔ̀rɔ́	KTN TKR	C 358–9
	Mvanɔ (Kamkam)	s ɔ nd u			KTN	M 580
	Mambila: Kabri	k ɔ d u			KTN	RB 7
	Twendi	tʃ ɔ́ nd ɔ̄		kɔ̀rɔ́	KTN TKR	C 358–9
	Mam. Njerep (Somié)	tʃ ō ō			KTN	C 358–9
	Luo (now extinct)	tʃ ú nd ʌ̀ ŋ		kōrō	KTN TKR	C 358–9
Tep	Tep	nuŋ - c ɔ n i			KTN	RB 7
	Tep (Tep Kwar)	tʃ ō		kōū tʃōnì	KTN	C 358–9
Somyev	Somyev (Kila Yang)	tʃ ūɔ̄ r ɔ̄		tūtōgō	KTN ?	C 358–9
Wawa	Wawa (Oumiari)	tʃ ɔ̄ nd ā ì		tʃɔ̄ndāi tóŋgō	KTN	C 358–9
Vute	Vute (Bute)	tʃ ɔ̀ː n ɛ			KTN	BCCW 171
	Vute (Buti)	ʃ ua n e			KTN	BCCW 171

Sources: Bd = Boyd 1994; BCCW = Williamson & Shimizu 1968; BM = Raymond Boyd MS; C = Bruce Connell MS; E = Rolf Endresen p.c.; K = Rob Koops MS; M = Meek 1931a; MM = Meek MS (Linguistic Notes); P = Mona Perrin MS; PD = Perrin Dictionary (MS); RB = Roger Blench 1994; WW = Weber & Weber MS. The classification is that of Blench 1993b, expanded by Connell 1995 and p.c.

Table 23.18 "Fowl" and "cock" in South Bantoid.

South Bantoid	Language	Fowl/Chicken	PL.	Cock/Rooster	PL.	Notes	Source
Buru	Buru	è -d ɔ̄	ē-dɔ̄	kúdɔ̄	bā-	D	KC 155-6
Tikar	Tikar	ɲíì				?	BCCW 170
Ekoid	Ekparabong/Balep	ŋ -k ɔ̃ g	bɔ̀-			KKR	C 69
	Ejagham (= FGH)	ŋ -k ɔ̃ g	ɔ̀-			KKR	C 69
	Ejagham (Cameroon)	-k ò g				KKR	W 18
	Bakor (Efutop)	ŋ -k u ɔ̃ g	ò-	ǹlómàkuɔ̀k	òló⌄m-ókuɔ̀k	KKR	C 69 OA
	Bakor (Nde/Nselle)	ŋ -k ɔ̃ g	ɛ̀-	ǹnóàŋkɔ̀k	ɛ̀nôkɔ̀k	KKR	C 69 OA
	Bakor (Nta/Abanyum)	ŋ -k ɔ̃ g	ɔ̀-			KKR	C 69
	Bakor (Nkim/Nkumm)	ŋ -k ɔ̃ g	ù-			KKR, Nkim also -kɔ̀gí	C 69
	Bakor (Nnam)	ŋ -kɔ̀ɔg	ɛ̀-			KKR	C 69
	Bakor (Ekajuk)	ŋ -kɛ̀ɔg	ɛ̀-			KKR	C 69
Mbe	Mbe	b ù -kùɔ	bè-			KKR/KT	BP
Tivoid	Angwe (= "Batu")	ɛ̃ -z ā ᵘ	é-zāᵘ	o-lúɔŋ-kúɔ		SK	K
	Bitare (= Njwande)	è -z à kɨ	é-zàkɨ	kokó (zāᵘ)		SK	K
	Afi/Amanda (= Batu)	s à g ī	sági	kóngwé di-	kóngōi	SK	KC 155-6
	Kamino	s à	s à	kɔ̃gō bisaā	tí sà	SK	KC 155-6
	Esimbi	è -h ù	e -h ú	ke-rúŋgu	be	? 9/10, 7/8	H 31, 32
	Tiv	i -kè ý	í-ké⌄ý	nôm ì-kè ý	á--ái.	KKR	BCCW 171
	Abo	ɛ -ð ɔ				D	BCCW 171
Beboid	Mairago-Mashi	ʃ ē				SY	KC 155
	Noni	ʃ è è	ʃ é é	ɲèm̄ ʃéè	ɲēm̄ʃéé	SY 9/10	Hy 81:113
	Noni	ʃ ɔ̀ ɔ̀	ʃ ɔ́ ɔ́			SY	Ho 91
	Mekaf	ʃ y è°	ʃ y é			SY	Ho 91
	Koshin	ʃ i ə	ʃ í ə̀			SY	Ho 91
	Missong	i -ʃyɛ̃a	íʃyɛ̃à			SY	Ho 91
	Bu	ʃ ī	ʃ ī			SY	Ho 91

Table 23.18 (cont'd)

South Bantoid	Language	Fowl/Chicken	PL.	Cock/Rooster	PL.	Notes	Source
	Nchanti	ʃ ì ì̄	ʃ í ī			SY	Ho 91
	Akweto	ʃy è	ʃy é			SY	Ho 91
	Bebe-Jatto	ʃ ì ò	ʃ í ò			SY	Ho 91
	Dumbo	ʃ i ò	ʃ i ə			SY	CK 29
Nyang	Kenyang	ŋ́ - k ɔ́ k				KKR	BCCW 170
Jarawan	Jaku	ń - g ù b ú		túgwòŋ		KB QK	Kr 196–7
	Mama	g u				KB	BCCW 170
	Bankala	k à t *		gwàŋ		* < Hausa QK	Kr 196–7
	Jarawa	k a z a				< Hausa (mod.)	BCCW 170
	Jarawa	ŋ - g u b		bi:tŋ gub		old root (Koelle)	Ko XII.E.2
	Mbula	ɩŋ - g u k u lɛk				< Chadic?	BCCW 170
	Wurkum	n - g u b u				KB	BCCW 171
Ring: West	Aghem	m - bv ɨ̵				KB 9/10	Hy 79:212
	Aghem (Nyos)	a m - v ɔ				KB	HJ 95
	Aghem (Isu)	m̄ - bv ì				KB also mbvū	HJ 95
	Aghem (Weh)	m̄ - bv ɔ̄				KB	HJ 95
	Aghem (Weh (= Wɛ))	m - v ɔ́	tə-m-			KB	CK 51
Central	Mmen (= Mme)	m̀ - v e	à-sè-			KB	CK 52
	Mmen (= Bafmeng)	ʌ̄ m - bv ū̄				KB	HJ 95
	Bum	wán-ŋ - gw ò ʔ	wá-sù̧-ŋ-			KB	CK 42
	Bum	à ŋ - gw û				KB	HJ 95
	Babanki (= Kidzɵm)	ŋ - gf ú	mvú-si			KB	CK 48
	Babanki	m̀ - bv ú'				KB	HJ 95
	Kuɔ (= Ukfwɔ)	ŋ - gv ɔ̌	-fse			KB	CK 48
	Kom	ǹ - g u	-sĩ			KB	CK 48
	Kom	ɔ̄ ŋ - gv ū̄	-sĩ			KB 9/10	HJ 95

Group	Language	Form	(prefix)	Variant		Reference
North	Vəŋo (= Ngo)	ŋ - g ú	es-		KB	CK 54
	Bamunka (= Muka)	ŋ - ʌ ú	ŋ-wə́hə		KB	CK 54
	Kenswei Nsei	ŋ̂ - g ú			KB	BCCW 170
	Wushi (= Babessi)	m - pf ɔ́ ʔ			KB	HJ 95
East	Lamnsɔ'	ɲ́- gv ɔ̄ f	ŋ́-		KB	BCCW 170
	Meta' (= Uta)	ŋ̂ - g ú p	ɲ́-		KB	BCCW 170
	Menka (= Bamenka)	n' - g a' p			KB	CK 79
East Grassfields	Kaka	m̄ - v ɛ̄ p		mbò	KB	BCCW 171
Northern	Adere	m ɛ̄ ŋ - v ū b	vɔ̀-	lɔ́-ngú	KB	ELV 129, 32
(= Nkambe)	Lus	n̂ - g ù '°		ntó-ngó'	KB KKR	ELV 129, 32
	Kofa	n̂ - g ó '		ñtô-mvɔ̄p	KB KKR	ELV 129, 32
	Mbat	m - v ɔ́ p		ñtô-ngvɔ̄p	KB	ELV 129, 32
	Limbum	máaŋ - gv ɔ̄ b		tɔ́-mvɔ̄p	KB	ELV 129, 32
	Kwak	ḿ - v ɔ́ p		ntɔ́-mvɔ́b	KB	ELV 129, 32
	Mfe	m - v ɔ́ b		ñdɔ̄-mvɔ̄p	KB	ELV 129, 32
	Nkot	m - v ɔ̄ p		ñtɔ́-mvɔ́p	KB	ELV 129, 32
	Ntong	m̄ - v ɔ́ p'			KB	ELV 129, 32
Nun (= Noun)	Mungaka (= Bali)	ŋ̂ - g ɔ́ p		ŋ-kə̀''	KB KKR	ELV 129, 32
	Mungaka (Bati)	ŋ̂ - g w ɔ́ b			KB	BCCW 170
	Shü Paməm (Bamun)	m ʌ ŋ - g ɔ́ p		ŋ-kə̀ʔ	KB KKR	ELV 129, 32
	Shü Paməm (Bafanji)	m ú ŋ - g ú ò		mûŋ-kə̀ʔə́	KB KKR	ELV 129, 32
	Shü Paməm (Bapi)	ŋ̂ - g w̃ ɔ́ b			KB	ELV 129, 32
Bamileke	Mədumba (Banganté)	ŋ - g ā p		khə̄ə̄'	KB KKR	ELV 129, 32
	Ngombale (Babadjou)	m ɔ̄ ŋ - g ó b		máŋ-kwɔ̄'	KB KKR	ELV 129, 32
	Bamenyam (= Bagam)	m ɔ́ ŋ - g ú b			KB	BCCW 170
	Yemba (Bafu)	ŋ̂'- g á p		ŋ-kə̀ ɔ'''	KB KKR	ELV 129, 32
	Ngwe	ŋ - g ɛ́ p	mὲ-ŋ-		KB	BCCW 170
	Ghomala' (Baleng)	m ə̀ - g á p		ŋ-kə̀ ɔ̀'''	KB KKR	ELV 129, 32
	Ghomala' (Bafoussam)	n - g w ɔ́ p		kò'''	KB KKR	ELV 129, 32
	Ghomala' (Bamendjou)	m ə̀ - g á p		mə̀ -kə̀'''	KB KKR	ELV 129, 32
	Ghomala' (Bandjoun)	g ɔ́ p		kò'''	KB KKR	ELV 129, 32

Table 23.18 (cont'd)

South Bantoid	Language	Fowl/Chicken	PL.	Cock/Rooster	PL.	Notes	Source
	Ghomala' (Bayangam)	g ɔ́ p		kò‘ᵒ		KB KKR	ELV 129, 32
	Ghomala' (Batie)	g ɔ́ p		kɔ̀‘ᵒ		KB KKR	ELV 129, 32
	Fe'fe' (Balafi)	n - g ŧ p		khɯɔ̀‘		KB KKR	ELV 129, 32
	Fe'fe' (Fotouni)	n - g ŧ p		khɯʌ̀‘		KB KKR	ELV 129, 32
	Fe'fe' (Bangam)	n - g ɔ́ p		kà̰ʾ		KB KKR	ELV 129, 32
	Fe'fe'	n - g ɔ̄ p		khɯ̀à‘ᵒ		KB KKR	ELV 129, 32
	Fe'fe'	ŋ̀ - g ɔ́ p	ŋ-			KB	BCCW 170
	Nda'nda' (Bangou)	g ɔ̄ p		kúɔ̀‘		KB KKR	ELV 129, 32
	Nda'nda' (Batoufam)	ŋ - g ī p		khɔ̀‘		KB KKR	ELV 129, 32
	Nda'nda' (Bangwa)	ŋ - g ī p		kɛʌ̀‘		also ŋgɯ̄p KB KKR	ELV 129, 32
	Nda'nda' (Balengou)	ŋ - g ɔ̄ p		khɯɔ̀‘		KB KKR	ELV 129, 32
	Nda'nda' (Bazou)	ŋ - g ū p		khɯɔ̀‘		KB KKR	ELV 129, 32
	Ngyɛmbɔɔŋ			mên-kɔ̀ʔ	ŋ-k ɔ̀ʔᵒ		A 38
	Ngyɛmbɔɔŋ (Batcha)	ŋ̀ - g ú	à	xɨɔ̀‘		KB KKR	ELV 129, 32
Ngemba	Bafut	ŋ̀ - g à		àŋ-kàgà		KB KKR	ELV 129, 32
	Bambili	m̀ - p˺ ɛ́		àŋ-kɔ̄		KB KKR	ELV 129, 32
	Nkwen	ŋ̀ - g ū				KB	ELV 32
	Mankon	ŋ̀ - g ú b à		àŋ-kɨ̀ʔɔ̄		KB KKR	ELV 129, 32
	Mendankwe (= Manda-)	ŋ - g w ú		àŋ-kɨ̀ʔɔ̄		KB KKR	ELV 129, 32
	Awing	ŋ̀ - g ɔ́ b à		àŋ-kɔ̀ʔɔ̄		KB KKR	ELV 129, 32
	Awing (Bamukumbit)	m ú ŋ - g ù p		mĭŋ-kì‘		KB KKR	ELV 129, 32
	Pinyin	ŋ̀ - g ú b à		àŋ-kɔ̄ʔɔ̄		KB KKR	ELV 129, 32

Sources: A = Anderson 1980; BCCW = Williamson & Shimizu 1968; BP = Ayɔ Bamgbose MS transcribed by Pascale Piron; C = Crabb 1965; CK = Chilver & Kaberry 1974; ELV = Elias, Leroy & Voorhoeve 1984; H = J.-M. Hombert MS; HJ = Hyman & Jisa 1978; Ho = Hombert 1980; Hy 79 = Hyman 1979; Hy 81 = Hyman 1981; K = Robert Koops MS; KC = Robert Koops MS typed on computer by Bruce Connell; Ko = Koelle 1854; Kr = Kraft 1981; OA = Osbert Asinya, p.c.; W = Watters MS.

Table 23.19 "Fowl" and "cock" in Afroasiatic.

Afroasiatic	Language	Fowl/Chicken	PL.	Cock/Rooster	PL.	Notes	Source
Omotic: North	Wolaitta	kútto				KT	B 106
Ometo North	Kullo-Konta	ku tu				KT	B 106
	Dache	lukko				LK	B 106
	Maale	koydo				?	B 106
	Zala	kuttaa				KT	B 106
Ometo South	Zayse-Zergula	lúkko				tʃʼuutʃ-é "chick"	B 106
Gimira	Benc Non	g y ã m LM		gyām bā´ LM LM-M		?	W 146–7
Janjero	Yemsa	buta				?	B 106
Kefoid	Kefa-Mocha	baakk-e				BK	B 106
	Shinasha	baa ká (la)				BK	B 106
Dizoid	Nayi			kʷobʷu		?	B 106
Mao	Mao	waakɛ				BK	B 106
	Hozo	waayɛ				BK	B 106
	Sezo	waay				BK	B 106
Aroid	Hamer	ba c-a				BK	B 106
	Aari	baatʃ-é				baatʃitán "hen"	B 106
	Galila	baac-				BK	B 106
Ongota	Ongota	baaʃa				BK	B 106
Cushitic: Beja	Beja	seüwi				?	B 107
		endirhoo*				*cf. guineafowl	
		jeddáad**				**< Arabic	
Agaw	Bilin	dir wa				DRBD	B 107
= Central	Xamtanga	dʒir wa				DRBD	B 107
		g iru wa					
	Kemant	d ir wa				DRBD	B 107
	Awngi	d úr a				DRBD	B 107

Table 23.19 (cont'd)

Afroasiatic	Language	Fowl/Chicken	PL.	Cock/Rooster	PL.	Notes	Source
East: Highland	Burji	lukk-ancoo				LK + koróomi	B 107
	Sidamo	lukko				LK	B 107
		kutto				KT	
	Gedeo	lukko				LK	B 107
	Kambata	anta beẽʔu				?	B 107
	Qabena	(h)anta baaqu-ta				cuwatʃeeta "ch."	B 107
	Haddiya	anta bā ʔa				?	B 107
Dullay	Yaaku	kɔkɔ	-n			KKR?	B 107
	Harso	lukkal-akkó	-le			c'aac'ut-e "ch."	B 107
	Dobase, Gollango	lukkal-akkó	-le			LK	B 107
	Gawwada	lukkal-e				LK	B 107
	Dullay	lukkal-akkó				LK	B 107
	Tsamay	luukal-e				LK	B 107
Saho-Afar	Afar	dorho				DR	B 107
	Saho	kukkunay				KKR	B 107
Western Omo-Tano	Daasanec	lug				LK	B 107
	Arbore	lúkku				LK	B 107
	Ba'iso	luk al-e				LK	B 107
Rendille-Boni	Rendille	luk-u				LK	B 107
	Somali	dígàag*		diig*	diigag	*< Ar. collect. "chicken, hen"	Ab 59, 60, 66
		dìgàagád	digáagádó-yyín			"hen"	
		dóoró					
	Somali	digag		dig*		*< Arabic	L 35, 39, 101
		doro**				**collective	
		luki***				***"hen"	
	Somali	digaag	diyaad	diiq*	diiqaq	*< Arabic	N76 62
		dooro	-yin				
	Jabarti	doora		diig*	-agi	*< Arabic	B 107

Group	Language	Form	Alt. form	Related term	Notes / source	Ref.
Oromo	Oromo	hindaanqoo / henda ko*			tʃutʃoo "chick" *; also lukku'	B 107
	Borana	lu ku			LK	B 107
	Munyo	lû ku'			LK	B 107
Konso-Gidole	D'iraassh (= Gidole)	lukka l-itt			c'aac'ut-ét "ch."	B 107–8
	Mossiya	luhhal-e			LK	B 108
	Konso	lukka l-itta			LK	B 108
South	Dahalo	gúúk-o(u)			? < Swahili?	B 108
	Kw'adza	ko kombo			KNK	B 108
	Iraqw	koonki		konkomaaye	tsii(am)o "hen"	B 108
	Gorwaa	ko nki		kookumó	tsii'mó "pullet"	B 108
	Alagwa	ko nki		konkomo	KNK	B 108
	Burunge	ko nkiya		konkoku	conciya "pullet"	B 108
Semitic: Central	Mandaic	ʃoprina		zaaya	?	B 114
Arabic	Arabic (Classical)	dadʒaadʒa		diik	katkuut "chick"	B 115
	Arabic (Egyptian)	firaax*			* also farxa	B 115
	Shuwa Arabic	dʒidaad			JD	B 115
South: South Arabian	Mehri	degooget		dik	DGGT DK	B 115
	Jibbāli (= Shahri)	dg g t			DGGT	B 115
	Sheri	degŝget		diik	DGGT DK	B 115
	Harsūsi	deyaayeh		dik	DGGT? DK	B 115
	Soqotri	degégeh	dawarrih		DGGT DK	B 115
Ethio-Semitic: North	Ge'ez	dorho	ʔadya:k	di:k	DR	N82 114
	Tigre (Beni Amer)	diːrho			siwsiwaːy "ch."	B 115
	Tigre	derho			('am)tʃatʃu "ch."	B 115
	Tigrinya	dãrho			tʃatʃut "chick"	B 115
South	Amharic, Argobba	dor o			tʃatʃut "chick"	B 115
	Harari	atãwaaq			tʃaatʃu "chick"	B 115
	Gafat	kuttä			< Cushitic	B 115
	Caha	kut ara			tʃutʃu "chick"	B 115
	Ezha	kuttära			tʃutʃuyä "chick"	B 115

Table 23.19 (cont'd)

Afroasiatic	Language	Fowl/Chicken	PL.	Cock/Rooster	PL.	Notes	Source
	Ennemor	kut ara*		aḇiyä		*also aŋgʷäroʻä "ch."	B 115
	Endegen	uŋgoroʻ				siisiitʃfä "ch."	B 115
	Gyeto	kut ara		aḇiyä		tʃawatʃweyä "c"	B 115
	Muher	kuttäna				tʃutʃeyyä "ch."	B 116
	Masqan	kuttäna		werwɔro		tʃatʃewä "ch."	B 116
	Gogot	kuttäna*				*also ǧärä	B 116
	Soddo	ǧärä				tʃutʃiyyä "ch."	B 116
	Selti	intʃaaqo		weruuro		tʃatʃewä, sasut	B 116
	Wolane	entʃaaqo				tʃutʃiyye "ch."	B 116
	Zwaay	əntaaqˑ				entʃatʃut "ch."	B 116
Berber	Siwa	t -yäzẹ̈t̩*	-yäzit-ín	yazẹt	yazitin	*"hen"; /tẹkazẹt	Ba.a 126
	Djebel Nefousa	tẹ -gazẹt*	-gazit-in	gazet	igazitẹn	*"hen"	Ba.a 126
	Ghdamès	t - azit̩*	-in	azit	zitẹn	*"hen"	Ba.a 126
	Zenaga	t -aɣ ðuḍ*	tuɣðaḍan	awaʒuḍ	uʒaḍan	*"hen"	Ba.a 126
	Mzab	t -yazit̩*	-yaziḍin	ayaziḍ	yaziḍan	*"hen"	Ba.a 126
	Ouargla	t -yazit̩*	tyazitin	yaziḍ	yaziḍen	*"hen"	Ba.a 126
	Figuig	t -yazit̩*		yazit	iyzdan	*"hen"; fullus'c'	Ba.a 126
	Bougie	φa -yazit̩*	φyuzaṭ	ayezit̩		*"hen"	Ba.a 126
	B. Salah	(φ)a -yazẹ̈t̩*	φyazeḍin	aɛ̄qqūq	iɛāqāq	*"hen"	Ba.a 126
	Metmata	φa -yazẹ̈t̩*	φyázeḍin	ʒiɛāðer	iʒiɛāðren	*"hen";füllūs'c'	Ba.a 127
	Ntifa	ta -fullus-t*	tifullūsin	afullus		*"hen"	Ba.a 127
Saharan	Ghat	tʃi -ka i-t*	-kaiatʃin	ẹkähi, ikai	ikahan	*"hen"	Ba.a 128
	Ahaggar	tẹ -kahi -t*	tikähain	ikahi	ikähän	*"hen"	Ba.a 128
	Kel Oui			ẹkähi	ikahaten	*"hen"	Ba.a 128
	Ioulimmiden			äkes	ikasen	*"hen"	Ba.a 128

Group	Language			kàadʒii	zàkárà	zàkáríu	(dán) tsàakóo'c-	NN
Chadic: West / Hausa	Hausa		k àà z áa		zàkárà	KZ	ZKR	M 288–9
	Gwandara (Karshi)		k à dʒ á		dʒàkára	KZ	ZKR	M 288–9
	Gwandara (Arabishi)		k a dʒ a		dʒàkára	KZ	ZKR	M 288–9
	Gwandara (Garaku)		k à z á		zàkála	KZ	ZKR	M 288–9
	Gwandara (Gitata)		k à z á		zàká a[1]	KZ	ZKR	M 288–9
	Gwandara (Gwagwa)		k à z á		zàkára	KZ	ZKR	M 288–9
	Gwandara (Nimbia)		k a h a		zàgə́re	KZ	ZKR	M 288–9
Sura-Gerka	Mwaghavul		kw é é è		teel	KKR	TKR?	Kr 1.29
	Miship (Chip)		k ò		dil-kò	KKR	m.-KKR	Kr 1.41
	Goemai		k è		dəl-kè	KKR	m.-KKR	Kr 1.53
	Angas		k ì		dtl	KKR	m.-KKR	Kr 1.20
	Kofyar		n ə g ə n k ó *			* "hen"		B ined
	Montol		k i y ě					B ined
	Yiwom				kɔ ŋ	KKR	KRN	B ined
Ron	Fyer	k ù	kw è			KKR?		JI70
	Bokkos	ʃ i	k ó ò r		koroŋ	KKR	KRN	JI70,S182
	Daffo		c à à n		koroŋ	KKR?	KRN	JI70,S182
	Kulere		k ô d			KKR?		JI70
Bole	Karekare		k é z ì		gə̀dʒà	KZ	GZ	Kr 1.62
	Gera		y ìì b í	yiibà	zèkírà yiibí	YB	ZKR YB	B ined
	Gera		ì b í		mizíbí	YB	? YB	Kr 1.74
	Geruma		ì bb í	ibbá		YB		B ined
	Giiwo		y ìì b í	ibbà	yàawó	YB		B ined
	Galambu		y ìì w ú		yúwá	YB		B ined
	Bole		y àa w í	yàbbi	gàdʒà	YB	GZ	IG 136,Kr
	Ngamo		y á b í		gàdʒà	YB	GZ	Kr 1.95
	Beele		y àa w í			YB		B ined
	Kwaami		y áa b é	yàbbi	gádʒà	YB	GZ	B ined
	Pero		p óo dʒ è		kòkkúròk	?	KKR-k	F 62, 84
	Tangale		y á ɓ è		kʷ ɔ̀lók	YB	KKR-k	Kr 1.122
	Dera (= Kanakuru)		y áa ɓ è ~ -w-	yáapiyén	kó lə̀k	YB	KKR-k	N 2

Table 23.19 (cont'd)

Afroasiatic	Language	Fowl/Chicken	PL.	Cock/Rooster	PL.	Notes		Source
Bade	Ngizim	gâzá		gàs-kém		GZ	GZ-?	Kr 1.248
	Bade	káazɔ́ɗaakôn		gàs-kámáan		KZ	GZ-?	Kr 1.258
Warji	Warji	dlàrɓéyài				LRKY		Jl 2.70
	Tsagu	cóòmái		cukuran *		? *	Sk 182	Jl 2.70
	Kariya	dlìrkí				LRKY		Jl 2.70
	Pa'a	dlurkíyà				LRKY		Jl 2.70
	Siri	dlàrkáyi				LRKY		Jl 2.70
	Mburku	zámá				ZM		Jl 2.70
	Jimbin	zámá				ZM		Jl 2.70
	Diri	àtûrà				?		Jl 2.70
	Miya	ðírki		ʃóʃó		LRKY		Kr 1.147
Zaar	Boghom (= Burma)	kùk		pégwàŋkùk		KKR p-QN-		Kr 1.173
	Boghom	yàap		gwàŋdi		YB	QN-	Sh 28
	Jimi	yabo				YB		Jl 2.70
	Guruntum	yàa				YB		Jl 2.70
	Geji	kówùl		gúndà		KKR QN-?		Kr 1.185
	Polci (= Pəlci)	kó rò		tóxsì		KKR TK-S		Kr 1.237
	Polci (Buli)	kó r		gwàkórókór		KKR Q-KKR-		Kr 1.198
	Dir (= Dirawa)	kwórí		tús kwɔ́ri		KKR TS-		Kr 1.160
	Saya cluster (= Seya)	gyèrí		tór wigyèri		KKR TS-		Kr 1.210
	Zaar	kwàar		gúŋ kwàar		KKR QN-		Sh 28
	Dot (= Dwot)	ʔiðì		dùs kùrók		?	TS-KKR	Kr 1.224
Central: Tera	Tera (Pidlimdi)	kúdʒà		dʒírti		KZ	LRKY	Kr 2.6
	Tera	kúʒà		gátʃàk		KZ	g-TKR?	Nl 42
	Hwona	dʼiyárá		kìlŋàrá		?	?	Kr 2.16
	Ga'anda	cémsà		cìtkéftá		KMS CT-KRT		Kr 2.26
	Ga'anda (Gaɓin)	cìmsé		cìtgìrté		KMS CT-KRT		Kr 2.35
	Boga (= Boka)	dèktá		cìkárá		TK-t CKR		Kr 2.45

Group	Language					
Bura	Bura	m̀ -t ɨ k á	fùgùm	TK	PKM	Kr 2.54
	Cibak	ń -t ɨ k á	vùkùm	TK	PKM	Kr 2.64
	Putai (= West Margi)	m̀ -d ì g à	ʔávùgùm	TK	PKM	Kr 2.74
	Nggwahyi	ń -t ɨ k á	ávùgùm	TK	PKM	Kr 2.84
	Huba (= Kilba)	t ɨ g à	vùkùm	TK	PKM	Kr 2.94
	Margi South (Hildi)	t ɨ k à	vígùm	TK	PKM	Kr 2.104
	Margi S. (Wamdiu)	t ù k à	vùgùmù	TK	PKM	Kr 2.114
	Margi	m̀ -t ɨ k à	vúgùm	TK	PKM	Kr 2.123
Higi	Kamwe * (Nkafa)	kʷà-n-t ɨ kʷá	ŋgúli	TK	KL	Kr 2.133
	Kamwe (Baza)	wà-n -t ɨ x á	ŋgùlù	TK	KL	Kr 2.143
	Kamwe (Kamale)	ká-ŋ - k à	ɲɨli	TK	KL	Kr 2.153
	Kamwe (Ghye)	kó-ŋ - k à	ŋɨli	TK	KL	Kr 2.163
	Kamwe (Futu)	ká-ŋ - k à	ɲúli	TK	KL	Kr 2.173
	Kamwe (Fali Kiria)	kʷɔ́ n -t ɨ k á	ŋgɨɬkú	TK	KL	Kr 2.183
	Kamwe (Fali Gili)	kwó - kw ù l á	ŋgùlùkú	KKR, kɨ-k	KL	Kr 2.193
Bata	Zizilivakan (F Jilbu)	ʔ y ù k ɨ y	gɨ̀msá		KMS	Kr 3.8
	Fali (Madzarin) *	b ù dʒ è y í	gɨ̀msá	*(= Mucella)		Kr 3.18
	Fali ('Bwagira)	ʔ i k ú n	gɨ̀msɨ̀n		KMS	Kr 3.28
	Gude	gy á gy à	ŋgúr gyágyà	GY-GY?		Kr 3.38
	Nzanyi (= Njanye)	d è k ì c í	dékɨ	TK-	TK-	Kr 3.48
	Bacama (Mwulyen)	d è ɔ́ k ò	pùgùmrí	TK-	PKM	Kr 3.58
	Bata (Bacama)	d ɛ̀ y k t ó	dɛ́ykɔ̀y	TK?		Kr 3.68
	Gudu	d ɔ́ y ù	dɔ́yùd ímìrà tsàkálá	TK?		Kr 3.78
Mandara	Wandala (= Mandara)	kw ù l á	ywà-tíkálá	KKR	TKL	Kr 3.88
	Glavda (= Gələvda)	yw à - c í k á	yá-túkúlú	TK	TKL	Kr 3.100
	Dghwede Zəyvana	g ú - s k è	yɔ̀-tɔkwúlá	TK	TKL	Kr 3.113
	Guduf-Gav.(Gava)	w ó - c í k à	wá -tókálá	TK	TKL	Kr 3.125
	Guduf-Gav.(Nakatsa)	w à - tʃ i k á		TK	TKL	Kr 3.137
Daba	Daba (= Musgoy)	gɔ̀-m - d á k	bisól	TK	?	Kr 3.158

Table 23.19 (cont'd)

Afroasiatic	Language	Fowl/Chicken	PL.	Cock/Rooster	PL.	Notes	Source
Mafa	Mafa (= Matakam)	wɔ́-tsáx		ŋglák wɔ́tsáx		TK	Kr 3.147
	Matal	gò-tsák		àgòɖák		TK	B 112
	Ndreme	dzùgùrák		águrɖá		?	B 112
	Hurza	dzùgùrá				?	B 112
South	Wuzlam	wà̰-cə̀kár		ágwáɬ		TKR	B 112
	Muyang	mì-c kwir		ágwáɬ		TKR	B 112
	Mada	mì-c kwér		gù̃ŋgwáɮ		TKR	B 112
	Zəlgwa-Minew	mè-n-dzí kwir		bókóy		TKR	B 112
	Giziga North	kə-ts kar				TKR	B 113
	Giziga South	ku-cukur				TKR	B 113
	Mofu-Gudur	gwágwàr		gwágwàlák		KKL-KKL?	Br 109
	Sukur	takur		iʒakʷ		TKR	B 113
E.-Central	Yedina	kògwí				< Kanuri	B 113
Mandage	Kotoko	kù-skú				TKR?	B 113
	Afadə	ku-sku				TKR?	B 113
Munjuk	Muzuk	mɨ-skɨr				TKR	B 113
	Mpus	yuguriy	yogoray			DGR?	B 113
	Vulum	yùgúr		yugur		DGR?	B 113
	Mbara	'dùgúr				DGR?	B 113
Kada	Kada	gəmda				KM?	B 113
Masa: North	Masa	ɬek-ka		ɬek-ŋa		TK?	B 113
	Banana	ɬíyèkó		ɬikúlɔ́ká		TK? -KKL-k	Kr 3.180
	Banana (Mouseye)	hlek-ka		bɔlɔkhlekŋa		TK?	Kr 3.188
	Musey	ɬek				TK?	B 113
	Koumo-Tagal	a-ɬek				TK?	B 113
South	Lame	yà-ʃ̣íỹà		gúlwà		SYK KKL-k	Kr 3.199
	Lame (Peve)	ʃínyék		gúlɔ́k		SYK KKL-k	Kr 3.207

Group	Language					Notes	Source
	Peve	ʃe kne	goylok singue			SYK	B 113
	Misme (= Zime)	singue	gùlôk			SYK KKL-k	Kr 3.216
	Dari (= Zime)	ʃe knè				SYK KKL-k	B 113
	Batna (= Zime)	sé kné				SYK	JI 2.71
	Sorga-Ngete	si nye				SYK	B 114
	Hede-Rong	sekné				SYK	B 114
East: Sibine	Sibine (= Sumray)	dùree				?	JI 2.71
	Tumak	kɔ́ɲ		kiyáɲ		< Barma (CSu)	Ca 154
	Mawer	káɲ				< Barma (CSu)	B 114
	Ndam	kiiŋii				?	JI 2.71
	Lele	tò rò		kòrà		TKR? ?	JI 2.71
	Kabalai	tùwàrrɛ̂				TKR	JI 2.71
A.3	Kera	gàgláw	gòglóki			dɛ̀bàrgɔ̂ "hen" gɛ̀bgúr "collec."	E 43
	Kwang	bógoortó *				* also bògɔ̀rɔ́	JI 2.71
B.1	Dangla	kókìrà*	bokinûmam	bóròròm		KKR * JI 2.71	B 114
	Migama	kúkkìrà*	bòkòɽkòɽò			*also dùlpú	B 114
	Jegu	kókóré				KKR	JI 2.71
	Bidiya	kókir-a	kókir-o	kókir-i	kókir-i	kókir "poultry"	AJ 91
	Mubi	kúrri	diɫk *	kúrôorú k	dàyàkà	KKR *<Arab.	B 114
	Birgit	kòkóréy				KKR	JI 2.71
	Toorom	kòkór-e				KKR	B 114
B.2	Mokilko	dʒìdât	kòkór-Ø	bàkòrkó		?	B 114
	Mokulu(= Mokilko)	ʔòssó				?	JI 2.71
B.3	Sokoro	wàárō				?	JI 2.71

Sources: Ab = Abraham 1962; AJ = Alio & Jungraithmayr 1989; B = Blench ined.; Ba.a = Basset 1959a; Br = Barreteau 1978; Ca = Caprile 1978; E = Ebert 1978; F = Frajzyngier 1985; IG = Ibriszimow & Gimba 1994; JI = Jungraithmayr & Ibriszimow 1994; Kr = Kraft 1981; L = Larajasse 1897; M = Matsushita 1974; N76 = Nakano 1976; N82 = Nakano 1982; N1 = Newman 1964; N2 = Newman 1974; NN = Newman & Newman 1977; S = Skinner 1977; Sh = Shimizu 1975; W = Wedekind 1990. "ch." or "c" in Notes column indicates "chick".

Table 23.20 "Fowl" and "cock" in Nilo-Saharan (partial).

Nilo-Saharan	Language	Fowl/Chicken	PL.	Cock/Rooster	PL.	Notes	Source
Songhay	Dendi	gɔ̀rɔ̀ŋgɔ̂		gɔ̀rɔ̀ŋgɔ́ hàriyá		"fowl male"	Z 94
	Kaado	gɔ̀rɔ̀ŋ-g ɔ̀rŋà				GRNG <KRN-k?	N 278
	Eastern	gɔ̀r gɔ́				GRNG <KRN-k?	N 278
	Songhay (Gao)	gʷar ga				GRNG <KRN-k?	W 13
	Western	gor go				GRNG <KRN-k?	N 278
	Dendi	gɔ̀lɔ̀ŋɔ̂ gɔ̀lɔ̀ŋɔ̂				GRNG <KRN-k?	N 278
	Tihishit	garaŋ'go				GRNG <KRN-k?	N 278
	Tasawaq	gɔ̀rɔ̀ŋgɔ́				GRNG <KRN-k?	N 278
	Songhay	gɔ̀rùŋgú		gòrùŋgú har		"fowl male"	B 201
	Zarma	gɔ̀r ŋ ɔ̀				GRNG <KRN-k?	N 278
	Zarma			k ɔ̀lóntò		?	T 166
Saharan	Kanuri	kokī		gábugum		KKR g-PKM	B 200
	Kanuri	kugui		gubo:gum		KKR g-PKM	K 126
	K. (Munio,Nguru,Kanem)	kuɣui		geba:ɣem		KKR g-PKM	K 126
	Kanuri	kugui		gubogəm		KKR g-PKM	Be 55, 46
	Kanuri	kùw î		gùdowúm		KKR ?	C 30, 156
	Teda	kókora, kókoya		kóki aŋkir		KKR	B 200

						KRK *"large bird"	
Maba	Maba	k ā r i k		koʃ émba *		KRK	B 201
	Aiki	k è è d é	kédé-t	kédé már		KKR?	No 150,209
Kado	Deiga			dʒɔ̀fik	gi-	?	R 12
Fur	Fur	d ó g à	dóngà	zòngà		DG?	J 23, 40, 30
Daju	Kas	u - k u r ge	ukurunge			KRK	
	Sila	u - k u r ge	ukurunge			KRK	
	Mongo	ku - g u r ge	kuguriŋge			KRK	
Central Sudanic	Bagirmi (= Bagrimma)	k i n dʒ a		kindʒa kala*		*"fowl male"	B 201
	Barma (Goundi)	k ɔ̄ ɲ dʒ á				KND?	Ca 154
East Sudanic:	Nobiin (= Nubian)	d ì r b á d	dirbàdii	dirbàdnóndì		DRBD	We 346
East Nilotic	Kakwa, Kuku, Mondari ...	só - k ú r - ì	só-kór-ò			KKR	V 345
	Teso	á - k ò r				KKR	V 345
	Turkana	έ - k â r ɔ́i t				KKR	V 345
	Turkana	a - k o k o r o - i t	nga-	e-kokoroi-t	ngi-	KKR	Ba 34–5
	Dongotono	x ɔ k ɔ́ r ɔ̀				KKR	V 345
	Lokoya	á - ɣ ɔ́ ɣ ɔ́ r ɔ́ ŋ				KKR	V 346
	Lopit	x ɔ x ɔ r ɔ				KKR	V 346
	Lotuko	ɔ - x ɔ x ɔ r ɔ				KKR	V 346
Komuz	Uduk	à - ŋ w á		dhàhó *		? "big rooster"	BC 18, 56

Sources: B = Barth 1862; Ba = Barrett 1990: BC = Beam & Cridland 1970; Be = Benton 1911; C = Cyffer 1994; Ca = Caprile 1978; J = Jakobi 1990; K = Koelle 1854; N = Nicolaï 1981; No = Nougayrol 1989; R = Reh 1994; T = Tersis 1972; V = Voßen 1987; We = Werner 1987; W = Williamson 1967; Z = Zima 1994.

References

In the case of collections of data from many sources, I have for economy cited only the collection and not the individual contributors: e.g. Kropp Dakubu 1977, 1980.

Abraham, R.C. 1967. *The Idoma language; Idoma wordlists, Idoma chrestomathy, Idoma proverbs*, 2nd edn. London: University of London Press.

Abraham, R.C. 1962. *Somali–English dictionary*. London: University of London Press.

Accam, T.N. 1966. *Adangme vocabularies*. Legon, Ghana: Institute of African Studies.

Adekunle, A.A. 1986. *Aspects of the phonology of Lopanchi*. BA project, University of Ilorin.

Adive, John R. 1989. *The verbal piece in Ebira*. Dallas: Summer Institute of Linguistics *and* University of Texas at Arlington.

Akinkugbe, O.O. 1978. *A comparative phonology of Yoruba dialects, Itsẹkiri, and Igala*. Unpublished doctoral dissertation, University of Ibadan.

Alio, K. & H. Jungraithmayr 1989. *Lexique bidiya*. Frankfurt am Main: Vittorio Klostermann.

Anderson, S.C. 1980. The noun classes of Ngyembɔɔn-Bamileke. In *Noun classes in the Grassfields Bantu borderland*, L.M. Hyman (ed.), 37–56. Los Angeles: Department of Linguistics, University of Southern California.

Armstrong, R.G. 1964. Notes on Etulo. *Journal of West African Languages* 1(2), 57–60.

Armstrong, R.G. 1965. Comparative wordlists of two dialects of Yoruba with Igala. *Journal of West African Languages* 11(2), 51–78.

Armstrong, R.G. 1983. The Idomoid languages of the Benue and Cross River valleys. *Journal of West African Languages* 13(1), 91–149.

Banfield, A.W. 1914. *Dictionary of the Nupe language*. Shonga: Niger Press.

Bargery, G.P. 1934. *A Hausa–English dictionary*. London: Oxford University Press.

Barrett, A.J. 1990. *Turkana–English dictionary*. London: Macmillan.

Barreteau, D. 1978. Aspects de la morphologie du mofu-gudur. See Caprile & Jungraithmayr (1978), 115–42.

Barth, H. [1862] 1971. *Collection of vocabularies of Central–African languages* [2 volumes], 2nd edn. London: Frank Cass.

Basset, A. 1959a. Le nom du coq en berbère. In *Articles de dialectologie berbère*, 117–30. Paris: Klincksieck.

Basset, A. 1959b. Sur quelques termes berbères concernant la basse-cour. In *Articles de dialectologie berbère*, 131–54. Paris: Klincksieck.

Beam, M. & E. Cridland. 1970. *Uduk dictionary*. Khartoum: Sudan Research Unit.

Bendor-Samuel, J.T. (ed.) 1989. *The Niger–Congo languages*. Lanham, Md: University Press of America.

Bennett, P.R. n.d. Tentative lexical reconstructions for the Mundu-Gbanzili languages. Unpublished paper, Department of Linguistics, University of Wisconsin.

Bennett, P.R. & J.P. Sterk 1977. South–central Niger–Congo: a reclassification. *Studies in African Linguistics* 8, 241–73.

Benton, P.A. 1911. *Kanuri readings*. Oxford: Oxford University Press.

Blench, R.M. 1993a. Recent developments in African language classification and their implications for African prehistory. In *The archaeology of Africa. Food, metals and towns*, T. Shaw, P. Sinclair, B. Andah, A. Okpoko (eds), 126–38. London: Routledge.

Blench, R.M. 1993b. An outline classification of the Mambiloid languages. *Journal of West African Languages* 23(1), 105–18.

Blench, R.M. 1995. A history of domestic animals in northeastern Nigeria. ORSTOM *Cahiers des Sciences Humaines* 31(1), 181–237.

Blench, R.M. 1994. Domestic animals in Afro–Asiatic. Paper presented at the VIth Hamito–Semitic Congress, Moscow.

Blench, R.M., in press. The languages of Africa: macrophyla proposals and implications for archaeological interpretation. In *Archaeology and Language III*. R.M. Blench & M. Spriggs (eds). London: Routledge.

Blench, R.M. & A. Edwards 1988. A dictionary of the Momi language. Cambridge: the editors.

Blench, R.M. & K.C. MacDonald, in press. Domestic fowl. *Cambridge Encyclopaedia of Nutrition, Vol. I.* Cambridge: Cambridge University Press.

Bole-Richard, R. 1983. La classification nominale en Ega. *Journal of West African Languages* 13(1), 51–62.

Bouquiaux, L. 1964. A word list of Aten (Ganawuri). *Journal of West African Languages* 1(2), 5–25.

Bouquiaux, L. 1978. *Dictionnaire sango–français.* Paris: SELAF.

Boyd, R. 1974. *Etude comparative dans le groupe adamawa.* Bibliothèque de la SELAF, 46. Paris: SELAF.

Boyd, R. 1994. *Historical perspectives on Chamba Daka.* Köln: Rüdiger Köppe.

Boyd, R. & R. Fardon 1992. *Chamba English lexicon.* MS. School of Oriental and African Studies.

Boyeldieu, P. 1985. *La langue lua ("niellim"). Phonologie – morphologie – dérivation verbale.* Descriptions de langues et monographies ethno-linguistiques, 1. Paris: SELAF.

Brosnahan, L.F. 1967. A word list of the Gokana dialect of Ogoni. *Journal of West African Languages* 4(2), 43–52.

Bunkowske, E.W. 1976. *Topics in Yala grammar.* Unpublished PhD dissertation, UCLA.

Calame-Griaule, G. 1968. *Dictionnaire Dogon: dialecte tɔrɔ.* Paris: Klincksieck.

Capo, H.B.C. 1991. *A comparative phonology of Gbe.* Dordrecht: Foris.

Caprile, J.-P. 1978. Les mots voyageurs dans l'interfluve Bahr-Erguig/Chari/Logone. See Caprile & Jungraithmayr (1978), 145–56.

Caprile, J.-P. & H. Jungraithmayr (eds) 1978. *Préalables à la reconstruction du proto-Tchadique.* Paris: SELAF.

Carter, G.F. 1971. Pre-Columbian chickens in America. In *Man across the sea: problems of pre-Columbian contacts,* C.L. Riley, J.C. Kelley, C.W. Pennington, R.L. Rands (eds), 178–218. Austin: University of Texas Press.

Chilver, E.M. & P.M. Kaberry 1974. *Western Grassfields: linguistic notes.* Ibadan: Institute of African Studies, University of Ibadan.

Christaller, J.G. 1933. *Dictionary of the Asante and Fante language called Tshi (Twi),* 2nd edn. Basel: Basel Evangelical Missionary Society.

Clarke, J. [1848] 1971. *Specimens of dialects,* 2nd edn. Farnborough: Gregg International.

Clutton-Brock, J. 1993. The spread of domestic animals in Africa. In *The archaeology of Africa. Food, metals and towns,* T. Shaw, P. Sinclair, B. Andah, A. Okpoko (eds), 61–70. London: Routledge.

Connell, B. 1991. *Phonetic aspects of the Lower Cross languages and their implications for sound change.* Unpublished PhD thesis, University of Edinburgh.

Connell, B. 1995. Dying languages and the complexity of the Mambiloid group. Paper presented at the 25th Colloquium on African Languages and Linguistics, Leiden.

Cook, T.L. 1969. Some tentative notes on the Kòhúmónò language Pt.I. *Research Notes from the Department of Linguistics and Nigerian Languages,* University of Ibadan 2(3).

Crabb, D.W. 1965. *Ekoid Bantu languages of Ogoja, eastern Nigeria.* Cambridge: Cambridge University Press.

Crawford, R.D. 1984. Domestic fowl. In *Evolution of domesticated animals,* I.L. Mason (ed.), 298–311. London: Longman.

Cremer, J. 1924. *Grammaire de la langue kasséna ou kassené.* Paris: Paul Geuthner.

Crouch, M. & P. Herbert n.d. *Vagla–English, English–Vagla dictionary.* Tamale: Ghana Institute of Linguistics, Literacy and Bible Translation.

Crowther, S. 1843. *Vocabulary of the Yoruba language.* London: Church Missionary Society.

Crozier, D.H. & R.M. Blench (eds) 1992. *An index of Nigerian languages,* 2nd edn. Abuja: Language Development Centre, NERDC, *and* Ilorin: Department of Linguistics and Nigerian Languages, University of Ilorin, *with* Dallas, Texas: SIL.

Cyffer, N. 1994. *English–Kanuri dictionary.* Köln: Rüdiger Köppe.

Dalby, D. 1965. The Mel languages: a reclassification of southern "West Atlantic". *African Language Studies* 6, 1–17.

Dettweiler, S.H. 1992. Level one sociolinguistic survey of the Pongu people. Unpublished paper, University of Ilorin, Nigeria.

Dettweiler, S.H. & S.G. Dettweiler 1995a. Sociolinguistic survey (level one) of the Lela people. Unpublished paper, University of Ilorin, Nigeria.

Dettweiler, S.H. & S.G. Dettweiler 1995b. Sociolinguistic survey (level one) of the Reshe people. Unpublished paper, University of Ilorin, Nigeria.

Dettweiler, S.H. & S.G. Dettweiler 1996. Sociolinguistic survey (level one) of Kamuku language cluster. Unpublished paper, University of Ilorin, Nigeria.

Dimmendaal, G.J. 1978. *The consonants of proto-Upper Cross and their implications for the classification of the Upper Cross languages*. Leiden: Unpublished doctorandus scriptie, University of Leiden.

Dwyer, D.J. 1987/88. Towards proto-Mande morphology. *Mandekan: Bulletin semestriel d'études linguistiques* **14/15**, 139–52.

Dwyer, D.J. 1989. Mande. In *The Niger–Congo languages*, J. Bendor-Samuel (ed.), 46–65. Lanham, Md.: University Press of America.

East, R. (ed.) [1939] 1965. *Akiga's story*. London: Oxford University Press.

Ebert, K.H. 1978. Lexical root and affixes in Kera. See Caprile & Jungraithmayr (1978), 41–50.

Eguchi, P.K. 1971. Esquisse de la langue mambaï. *Kyoto University African Studies* **6**, 139–94.

Ehret, C. 1971. *Southern Nilotic history: linguistic approaches to the study of the past*. Evanston, Illinois: Northwestern University Press.

Ehret, C. 1980. The historical reconstruction of southern Cushitic phonology and vocabulary. Berlin: Dietrich Reimer.

Elias, P., J. Leroy, J. Voorhoeve 1984. Mbam-Nkam or eastern Grassfields. *Afrika und Übersee* **67**, 31–107.

Elders, S. (forthcoming). *Grammaire mundang. Grammaire, textes, vocabulaire*. Leiden: Research School CNWS.

Elugbe, B.O. 1989. *Comparative Edoid: phonology and lexicon*. Port Harcourt: University of Port Harcourt Press.

Ferry, M.-P. 1972. Deux langues tenda du Sénégal oriental. *Bulletin de la* SELAF 7.

Ferry, M.-P. 1991. *Thesaurus Tenda: dictionnaire ethnolinguistique de langues sénégalo-guinéennes -nóyàn (bassari), -nók (bedik) -mǎy (konyagi)*. [2 volumes]. (SELAF No. 325.) Paris: Peeters.

Fløttum, S. 1974. *Dictionnaire mbum–français*. Oslo: [l'auteur].

Frajzyngier, Z.A. 1985. *Pero–English and English–Pero vocabulary*. Berlin: Dietrich Reimer.

Freemann, R.A. 1972. *Okoyai aumgbọmọ: seeds of poetry*. Ibadan: Institute of African Studies.

Fumihito, A., T. Miyake, S.-I. Sumi, M. Takada, S. Ohno, N. Kondo 1994. One subspecies of the red junglefowl (*Gallus gallus gallus*) suffices as the matriarchic ancestor of all domestic breeds. *Proceedings of the National Academy of the Sciences, USA* **91**, 12505–9.

Garber, A.E. 1987. *A tonal analysis of Senufo: Sucite dialect*. PhD dissertation, University of Illinois.

Gardner, I. 1980. *Abuan–English, English–Abuan dictionary*. Port Harcourt and Jos: University of Port Harcourt Press and Nigeria Bible Translation Trust.

Gerhardt, L. 1983. *Beiträge zur Kenntnis der Sprachen des nigerianischen Plateaus*. Glückstadt: J.J. Augustin.

Greenberg, J.H. 1963. *The languages of Africa*. Bloomington, Ind: University of Indiana.

Grégoire, C. 1986. *Le maninka de Kankan*. Tervuren: ACCT.

Grierson, G.A. [1928] 1994. *Linguistic survey of India*. Delhi: Government of India.

Grimes, J.E. & B.F. Grimes (eds) 1993. *Ethnologue language family index*. Dallas: Summer Institute of Linguistics.

Guthrie, M. 1967–71. *Comparative Bantu* [4 volumes]. Farnborough: Gregg.

Halaoui, N., K. Tera, M. Trabi (eds) 1983. *Atlas des langues mandé-sud de Côte d'Ivoire*. Abidjan: ACCT–ILA.

Heine, B. 1968. *Die Verbreitung und Gliederung der Togorestsprachen*. Berlin: Dietrich Reimer.

Henrici, A. 1973. Numerical classification of Bantu languages. *African Language Studies* **14**, 82–104.

Hérault, G. (ed.) 1983. *Atlas des langues kwa de Côte d'Ivoire*. Tome 2. Abidjan: ACCT–ILA.

Hoerner, E.F. 1980. *Ninzam*. Hamburg: Universität Hamburg.

Hoffmann, C. 1965. A word list of central Kambari. *Journal of West African Languages* **2**(2), 7–31.

Höftmann, H. 1971. *The structure of Lelemi language*. Leipzig: Veb Verlag Enzyklopädie.

Hombert, J.-M. 1980. Noun classes of the Beboid languages. See Hyman (1980), 83–137.

Hyman, L.M. (ed.) 1979. *Aghem grammatical structure*. Los Angeles: Department of Linguistics, University of Southern California.

Hyman, L.M. 1980. *Noun classes in the Grassfields Bantu borderland*. Los Angeles: Department of Linguistics, University of Southern California.

Hyman, L.M. 1981. *Noni grammatical structure*. Los Angeles: Department of Linguistics, University of Southern California.

Hyman, L.M. & H. Jisa (compiled) 1978. *Word list of Comparative Ring*. MS.

Ibriszimov, D. & M. Gimba 1994. *Bole language and documentation unit (BOLDU): Report 1*. Köln: Rüdiger Köppe.

Ikoro, S.M. 1989. *Segmental phonology and lexicon of proto-Keggoid*. Unpublished Master's thesis, University of Port Harcourt.

Innes, G. 1967. *A Grebo–English dictionary*. London: Cambridge University Press.

Innes, G. 1969. *A Mende–English dictionary*. Cambridge: Cambridge University Press.

Jakobi, A. 1990. *A Fur grammar*. Hamburg: Buske.

Jenewari, C.E.W. 1983. *Defaka: Ịjọ's closest linguistic relative*. Port Harcourt: University of Port Harcourt Press.

Johnston, H.H. 1886. *The Kilima-Njaro expedition: a record of scientific exploration in eastern equatorial Africa*. London: Kegan Paul, Trench.

Johnston, H.H. 1919–22. *A comparative study of the Bantu and Semi-Bantu languages* [2 volumes]. Oxford: Clarendon Press.

Jungraithmayr, H. 1971. Reflections on the root structure in Chadohamitic (Chadic). *Actes du 8ᵉ Congrès International de Linguistique Africaine*, 1, 285–92. Abidjan: Université d'Abidjan.

Jungraithmayr, H. & D. Ibriszimow 1994. *Chadic lexical roots* [2 volumes]. Berlin: Dietrich Reimer.

Jungraithmayr, H. & K. Shimizu 1981. *Chadic lexical roots*. (Marburger Studien zur Afrika- und Asienkunde, Serie A: Afrika, Band 26.) Berlin: Dietrich Reimer.

Kaliai, M.H.I. 1966. *Nembe–English dictionary, vol. 2*. Ibadan: Institute of African Studies, University of Ibadan.

Kastelein, B. 1994. *A phonological and grammatical sketch of DuRop*. Doctorandus scriptie, University of Leiden.

Keita, A. 1989/90. *Esquisse d'une analyse ethno-sémiologique du jula vernacularisé de Bobo-Dioulasso*. Thèse pour le doctorat (nouveau régime), Université de Nice-Sophia Antipolis.

Kervan, M. 1982. *Dictionnaire dogon: donno sɔ*. Bandiagara, Mali: Paroisse catholique.

Kirk-Greene, A.M.H. 1958. *Adamawa past and present*. London: Oxford University Press *for* International African Institute.

Kleinewillinghöfer, U. 1991. *Die Sprache der Waja (nyan wiyàụ)*. (Europäische Hochschulschriften, Reihe XXI Linguistik, vol. 108). Frankfurt: Peter Lang.

Koelle, S.W. [1854] 1963. *Polyglotta Africana*. London: Church Missionary Society.

Kraft, C.H. 1981. *Chadic wordlists* [3 volumes]. Berlin: Dietrich Reimer.

Krass, A.C. 1970. *A dictionary of the Chokosi language*. Local Studies Series No. 4. Legon: Institute of African Studies, University of Ghana.

Krass, A.C. 1973. *A dictionary of the Chokosi language: English–Chokosi*. Local Studies Series No. 5. Legon: Institute of African Studies, University of Ghana.

Kropp Dakubu, M.E. 1973. *Ga–English dictionary*. Legon: Institute of African Studies, University of Ghana.

Kropp Dakubu, M.E. (ed.) 1977. *West African language data sheets, vol. 1*. Legon: West African Linguistic Society.

Kropp Dakubu, M.E. (ed.) 1980. *West African language data sheets, vol. 2*. Leiden: West African Linguistic Society and African Studies Centre.

Kropp Dakubu, M.B. 1991. A note on Dogon in Accra. *Journal of West African Languages* 21(2), 35–40.

Langdon, R. 1990. When the blue-egg chickens come home to roost. *Journal of Pacific History* 25, 164–92.

Le Bris, P. & A. Prost 1981. *Dictionnaire bobo–français*. Paris: CNRS and ACCT.

Leith-Ross, S. [1939] 1965. *African women*. London: Routledge & Kegan Paul.

Lukas, J. 1937. *A study of the Kanuri language.* Oxford University Press *for* International African Institute.

MacDonald, K.C. 1992. The domestic chicken (*Gallus gallus*) in sub-Saharan Africa: a background to its introduction and its osteological differentiation from indigenous fowls (Numidinae and *Francolinus* sp.). *Journal of Archaeological Science* 19, 303–18.

MacDonald, K.C. 1995. Why chickens?: the centrality of the domestic fowl in West African ritual and magic. In *Animal symbolism and archaeology*, K. Ryan & P.J. Crabtree (eds). Philadelphia: MASCA/University of Pennsylvania Press.

MacDonald, K.C. & D.N. Edwards 1993. Chickens in Africa: the importance of Qasr Ibrim. *Antiquity* 67, 584–90.

Manessy, G. 1975. *Les langues Oti-Volta.* Paris: SELAF.

Manessy, G. 1979. *Contribution à la classification généalogique des langues voltaïques.* Paris: SELAF.

Marchese, L. 1983. *Atlas linguistique kru*, 2nd edn. Abidjan: ACCT-ILA.

Marchese, L. 1986. *Tense/aspect and the development of auxiliaries in Kru languages.* Dallas: Summer Institute of Linguistics *and* University of Texas at Arlington.

Matsushita, S. 1974. *A comparative vocabulary of Gwandara dialects.* Tokyo: ILCAA.

Meek, C.K. 1931. *Tribal studies in Northern Nigeria* [2 volumes]. London: Kegan Paul.

Meeussen, A.E. 1980. *Bantu lexical reconstructions* [Reprt W. Index]. Tervuren: Musée Royal de l'Afrique Centrale.

Moñino, Y. (ed.) 1988. *Lexique comparative des langues oubangiennes.* Paris: Geuthner.

Mukarovsky, H.G. 1976-7. *A study of western Nigritic* [2 volumes]. Wien: Afro-Pub.

Mukarovsky, H.G. 1987. *Mande–Chadic common stock.* Wien: Afro-Pub.

Munro, P. & D. Gaye 1991. *Ay baati Wolof: a Wolof dictionary.* Los Angeles: Department of Linguistics, University of California.

Naden, T. 1986. Première note sur le kɔnni. *Journal of West African Languages* 16(2), 76–112.

Nakano, A. 1976. *Basic vocabulary in Standard Somali.* Tokyo: Institute for the Study of Languages and Cultures of Asia and Africa.

Nakano, A. 1982. *A vocabulary of Beni Amer dialect of Tigre.* Tokyo: Institute for the Study of Languages and Cultures of Asia and Africa.

N'diaye-Correard, G. 1970. *Études fca ou balante (dialecte ganja).* Paris: SELAF.

Newman, P. 1964. A word list of Tera. *Journal of West African Languages* 1(2), 33–50.

Newman, P. 1974. *The Kanakuru language.* Leeds: Institute of Modern English Language Studies, University of Leeds.

Newman, P. & M. 1966. Comparative Chadic: phonology and lexicon. *Journal of African Languages* 5(3), 218–51.

Newman, P. & R.M. Newman 1977. *Modern Hausa-English dictionary.* Ibadan: Oxford University Press.

Nicolaï, R. 1981. *Les dialectes du Songhay.* Paris: SELAF.

Nicole, J. 1980. *Phonologie et morphologie du nawdm.* Horsleys Green: Summer Institute of Linguistics.

Nougayrol, P. 1980. *Le day de Bouna (Tchad). I Phonologie, syntagmatique nominale, synthématique.* Bibliothèque de la SELAF, 71–2. Paris: SELAF.

Nougayrol, P. 1989. *La langue des Aiki dits Rounga.* Paris: Geuthner.

Ohiri-Aniche, C. 1999. Language diversification in the Akoko area of Western Nigeria. In *Archaeology and Language IV,* R.M. Blench & M. Spriggs (eds). London: Routledge.

Pairault, C. 1969. *Documents du parler d'Iro, kùláál du Tchad.* Langues et littératures de l'Afrique noire, 5. Paris: Editions Klincksieck.

Persson, A. & J. Persson 1980. *Collected field reports on aspects of Ligbi grammar.* Legon: Institute of African Studies, University of Ghana.

Pichl, W.J. 1963. *Sherbro–English and English–Sherbro vocabulary* [2 volumes]. Freetown: Fourah Bay College.

Plug, I. 1996. Domestic animals during the early Iron Age in southern Africa. *Aspects of African Archaeology: Papers from the 10th Congress of the PanAfrican Association for Prehistory and related studies,* G. Pwiti & R. Soper (eds), 515–20. Harare: University of Zimbabwe Publications.

Prost, A. 1964. *Contribution à l'étude des langues voltaïques*. Dakar: IFAN.

Prost, A. 1971. *Éléments de sembla*. Lyon: Afrique et langage.

Reh, M. 1994. A grammatical sketch of Deiga. *Afrika und Übersee* **77**, 197–261.

Roulon, P. 1972. Essai de phonologie du tyembara. *Bulletin de la Selaf* **9**, 1–55.

Rowland, R. 1966. Sissala noun groups. *Journal of West African Languages* **3**(1), 23–8.

Ruelland, S. 1988. *Dictionnaire tupuri-français-anglais*. Langues et cultures africaines, 10. Louvain: Peeters/Paris: SELAF.

Sapir, J.D. 1971. West Atlantic: an inventory of the languages, their noun class systems and consonant alternation. In *Linguistics in sub-Saharan Africa*, vol. 7 of *Current trends in linguistics*, T.A. Sebeok (series ed.), 45–112. The Hague: Mouton.

Sasse, H.J. 1979. The consonant phonemes of proto-East-Cushitic (PEC): a first approximation. *Afroasiatic Linguistics* **7**, 1–67.

Sauer, C.O. 1969. *Agricultural origins and dispersals. The domestication of animals and foodstuffs*, 2nd edn. Cambridge, Mass.: MIT Press.

Schadeberg, T.C. & P. Elias 1979. *A description of the Orig language*. Tervuren: Musée Royal de l'Afrique Centrale.

Shimizu, K. 1972. The Kente dialect of Kpan, Pt. 2. *Research Notes from the Department of Linguistics and Nigerian Languages, University of Ibadan* **5**(1).

Shimizu, K. 1975. *Boghom and Zaar: vocabulary and notes*. Kano: Abdullahi Bayero College. [Mimeo.].

Shimizu, K. 1980. *Comparative Jukunoid* [3 volumes]. Wien: Afro-Pub.

Shimizu, K. 1983. *The Zing dialect of Mumuye*. Hamburg: Buske.

Simmons, D.C. 1976. Notes and a word list of Basanga: the language of Uyanga Okposung, Nigeria. *Research Notes from the Department of Linguistics and Nigerian Languages, University of Ibadan* **7**(3), 1–17.

Skinner, A.N. 1977. Domestic animals in Chadic. In *Papers in Chadic linguistics*, P. Newman & R.M. Newman (eds), 175–98. Leiden: Afrika-Studiecentrum.

Skinner, A.N. 1984. Afroasiatic vocabulary: evidence for some culturally important items. *Africana Marburgensia* Special Issue 7.

Snider, K. 1990. *Studies in Guang phonology*. Doctoral thesis, University of Leiden.

Sow, A.I. (ed.) 1971. *Dictionnaire élémentaire fulfulde-français-English elementary dictionary*. Niamey: Centre de Recherche.

Stanford, R. n.d. *Bekwarra-English dictionary*. MS.

Sterk, J.P. 1977. *Elements of Gade grammar*. Unpublished doctoral dissertation, University of Wisconsin.

Sterk, J.P. n.d. *Unpublished wordlists of Upper Cross*. Typed on computer under the supervision of Bruce Connell.

Stevens, L. 1991. *Genetics and evolution of the domestic fowl*. Cambridge: Cambridge University Press.

Swadesh, M., E. Arana, J.T. Bendor-Samuel, W.A.A. Wilson 1966. A preliminary glottochronology of the Gur languages. *Journal of West African Languages* **3**(2), 27–65.

Tersis, N. 1972. *Le zarma*. Paris: SELAF.

Thomas, N.W. 1916. *Specimens of languages from Sierra Leone*. London: Harrison.

Tisserant, C. 1930. *Essai sur la grammaire Banda*. Paris: Institut d'Éthnologie.

Tucker, A.N. 1959. *Le groupe linguistique zande*. Tervuren: Koninklijk Museum van Midden-Afrika.

Ubels, E. 1981. *A comparative study of Koh and Karang*. Unpublished paper. Summer Institute of Linguistics Yaoundé.

Vallaeys, A. 1991. *La langue mondo*. Tervuren: Koninklijk Museum van Midden-Afrika.

Van Dyken, J. 1986. The genetic linguistic postulates – some applications. *Notes on Linguistics* **36**, 25–41.

Vansina, J. 1990. *Paths in the rainforests*. London: James Currey.

Voßen, R. 1982. *The eastern Nilotes*. Berlin: Dietrich Reimer.

Weber, M. & J. Weber n.d. *A dictionary of the Kwanja language*. Electronic MS. [The Kwanja word list is supplied by Martin and Joan Weber, who work with the Kwanja people in Cameroon and are sponsored by Lutheran Bible Translators of Canada and the Evangelical Lutheran Church of Canada.]

Wedekind, K. 1972. *An outline of the grammar of Busa (Nigeria)*. Doctoral dissertation, Faculty of Philosophy, University of Kiel.

Wedekind, K. 1990. Gimojan or Benyemo: Beng-Yemsa phonemes, tones and words. In *Omotic language studies*, R.J. Hayward (ed.), 68–184. London: School of Oriental and African Studies.

Welmers, W. E. n.d. Notes on Kpelle. Unpublished paper.

Werner, R. 1987. *Grammatik des Nobiin (Nilnubisch)*. Hamburg: Buske.

West, B. & B.-X. Zhou 1988. Did chickens go north? New evidence for domestication. *Journal of Archaeological Science* 15, 515–33.

Westermann, D. 1911. *Die Sudansprachen*. Hamburg: L. Friedrichsen.

Westermann, D. 1927. *Die westlichen Sudansprachen und ihre Beziehungen zum Bantu*. Berlin: Walter de Gruyter.

Williams, G. 1994. Intelligibility and language boundaries among the Cangin peoples of Senegal. *Journal of West African Languages* 24(1), 47–67.

Williamson, K. 1967. Songhai word list (Gao dialect). *Research Notes from the Department of Linguistics and Nigerian Languages, University of Ibadan* 3.

Williamson, K. (ed.) 1972. *Igbo–English dictionary*. Benin: Ethiope.

Williamson, K. & C. Ohiri-Aniche in prep. Comparative Igboid.

Williamson, K. & K. Shimizu (eds) 1968. *Benue–Congo comparative wordlist, vol. 1*. Ibadan: West African Linguistic Society.

Williamson, K. & A.O. Timitimi (eds) 1983. *Short Ịzọn-English dictionary*. Port Harcourt: University of Port Harcourt Press.

Wolf, P.P. de 1995. *English-Fula dictionary* [3 volumes]. Berlin: Dietrich Reimer.

Wolff, H. 1969. *A comparative vocabulary of Abuan dialects*. Evanston: Northwestern University Press.

Zima, P. 1994. *Lexique dendi (songhay)*. Köln: Rüdiger Köppe.

CHAPTER TWENTY-FOUR

Linguistic evidence for the prehistory of livestock in Sudan

Marianne Bechhaus-Gerst

1. Introduction

Trying to reconstruct the prehistory of livestock in the Sudan is by no means an easy venture for a historical linguist. Owing to the dramatic climatic, socio-political and socio-linguistic changes that took place over an extended period, we are confronted with a vast area either more or less devoid of recent settlement or inhabited by various ethnic groups who no longer speak any genuine African languages. The desiccation of the region west of the Nile valley led to a gradual migration of its inhabitants to more fertile areas. Although excavations of recent years, for example along the Wadi Howar, have brought to light the material remains of prolonged settlements by different peoples, we still know next to nothing about their ethnic or linguistic identities. The middle Nile region, on the other hand, is inhabited by various ethnic groups which during the last centuries have shifted to Arabic as their first and later only language, giving up their original mother tongues.

At the same time, the territory of present-day Sudan, as the presumed homeland of two of the language phyla of Africa, Afroasiatic and the Nilo–Saharan, is of great interest for historical linguistics. The assumed time depth of about 10,000 years for the proto-stages of both families implies that speakers of various proto-languages of their sub-groups lived in and migrated through precisely this area for most of prehistory. These different groups were not isolated, but came into contact at different stages of their development at different times and influenced each other culturally, socio-politically, economically, and, of course, linguistically. Most groups involved in these processes have disappeared from their former homelands. Some were absorbed by other ethnic groups, while others can be found through most of northeast Africa where many Nilo–Saharan and Afroasiatic languages are spoken today.

There are two exceptions to this general development which can help us to reconstruct at least part of the settlement and migratory as well as the economic history of the Sudan area: Old Egyptian and Nobiin. Old Egyptian, an isolated branch of Afroasiatic, is known to have been spoken in northern Africa before 3000 BC when it was written down in hieroglyphic characters. Investigation of the language might bring to light different sets of loanwords which could help us to determine which languages were spoken west and south of Egypt between 3000 BC and the first centuries AD. Research of this kind has, however, only just begun. Just the opposite is the case of Nobiin, a Nubian language of the eastern-Sudanic sub-group of Nilo–Saharan spoken in the Nile Valley. Nobiin has been the subject of historical studies during the last ten years (Behrens 1981, Thelwall 1982, Bechhaus-Gerst 1984/85, 1989, 1994), with the result that we are now able to reconstruct most of the history of the language and its speakers.

Linguistic and non-linguistic evidence shows that the pre-Nobiin speakers were the first to split off the Nubian proto-group. They migrated towards the Nile Valley where they arrived at about 1500 BC. This date is supported by Old Egyptian loanwords in Nobiin which, on the basis of their specific phonological structure, could not have been borrowed later in the development of the Egyptian language. This solid dating allows the reconstruction of the history of various other languages once spoken in the Sudan either genetically related to or in contact with Nobiin.

2. Nubians and livestock

It has been generally assumed that the proto-Nubian people were pastoralists before the initial split of the group, i.e. before 1500 BC, and that the Nobiin as well as the Dongolawi/Kenzi adopted a sedentary form of existence with agriculture as the main economic basis only when they reached the Nile valley. Up to this day animal husbandry, with breeding of cattle, sheep, and goats, is only slightly less important than agriculture in Sudanese Nubian society. However, there is hardly any livestock terminology that can be identified as common or proto-Nubian.

2.1. Proto-Nubian

The only term that can be found in all branches of Nubian except Nobiin is "milk", which shows that the proto-Nubians made use of secondary products.

2.2. milk

Dongolawi/Kenzi	icci < *ej-ti
Dilling	ɛj
Dair	ij
Dulman	ijji
Ghulfan	ij
Kadaro	eju
Birgid	eʃʃi
Midob	iccidi

In the Nubian languages of the Nile Valley and in Hill Nubian there are common terms denoting "small ruminant livestock, possession" and "lamb". They might go back to proto-Nubian, in which case we have, however, to assume a complete loss in the Darfur-Nubian languages Birgid and Midob.

2.3. small livestock, possession

proto-Nubian   *orti

Nobiin/Dongolawi/Kenzi	urti	"small livestock"
Dilling	ɔrti	"sheep"
Dair	orti	"cast. ram"
Ghulfan	ɔrti	"lamb"
Koldegi	ordi	"small livestock"

2.4. lamb

Nobiin/Dongolawi/Kenzi	katti
Dilling	kwote
Dair	kote

The Darfur-Nubian languages Birgid and Midob on the other hand share a common designation for "sheep", although there is no evidence that these two ever constituted a linguistic unit within Nubian.

2.5. sheep

Birgid	tii
Midob	tii

Restricted to Darfur-Nubian Midob and Hill Nubian Dilling we find a probable correspondence for "he-goat".

2.6. he-goat

Midob	uwar
Dilling	war

On the basis of this rather scant evidence we may conclude that by 1500 BC the proto-Nubians were familiar with the herding of livestock and the use of secondary products. It seems to be quite obvious, however, that the introduction of livestock goes back to an earlier period and that there never has been what might be called a proto-Nubian stage in the development of the keeping of livestock.

2.7. Nubian–Tamare–(Nera)

This becomes even more evident, if we take a look at the closest relatives of the Nubian language family, the languages of the Tama group, spoken in eastern Chad and western Sudan, and – less certain – the Nera language, spoken in Eritrea. All three have been classified recently by Ehret (e.g. Ehret 1993:105) as constituting the Astaboran sub-group of eastern Sudanic. Here, we find the following correspondences:

2.8. cow

Nobiin/Dongolawi/Kenzi	ti	Tama	tɛɛ
Hill Nubian	tɛ	Ereŋa	te
Birgid	tei	Sungor	te
Midob	təə	Miisiirii	tɛɛ
		Ibiri	tee

This is the one term present in all branches of Nubian. There are related forms not only in the Tama group, but also in many other languages within eastern Sudanic so that it must be dated back to the proto-eastern Sudanic stage. Nowhere, however, are forms as near to identical as between Nubian and Tama.

2.9. goat

Dongolawi/Kenzi	ber-ti	Tama	bil	Nera	bile
Haraza	fiyla	Ereŋa	bil		
Dilling	hel-ti	Sungor	bil		
Midob	bile	Miisiirii	bil		

This is the most convincing correspondence, because of the presence of more or less identical roots in all three groups. The term cannot be found elsewhere in Nilo–Saharan with the exception of central Sudanic Moru-Madi where it has to be regarded as a loanword (Ehret 1993:114). One rather interesting piece of evidence seems to be that among the peoples of Eritrea and Ethiopia only the Nera and very few surrounding ethnic groups are breeding the so-called "Nubian goat" (Ayalew & Peacock, pers. comm.). The assumption that this has some historical significance seems justified.

2.10. bull

Tama	wɛr	Nera	bero
Ereŋa	wuer		
Sungor	wer		

No regular sound correspondences between Tama and Nera have yet been established, but both the similar forms and the identical meaning speak for themselves.

Of some importance in the comparison of Nubian, Tama and Nera is the common term for "goat" not found in other Nilo–Saharan languages. It suggests the introduction of the goat, or at least a new type of goat, to the proto-Nubian–Tama–Nera society. As there seems to be no evidence for a very early acquaintance with small stock in East Africa this in itself is not unlikely. If we assume the existence of a former proto-Nubian–Tama–Nera group of pastoralists, the question remains open as to where and when it might have existed. The present-day geographical distribution, with most of the languages spoken between Chad and Kordofan, points to the Darfur–Kordofan region. This localization would coincide with the assumed homeland of the proto-Nubians (Behrens 1981, Bechhaus-Gerst 1984/85). Ehret

(1993:117) proposed a date of 4000 BC for the initial split of "proto-Astaboran", consisting of what was to become present-day Nera. With regard to the few correspondences between Nera and the other members of the group the dating seems to be probable. Owing to a total lack of supporting evidence it has, however, to be regarded as highly speculative.

3. Livestock in the Nile valley

Around 1500 BC the ancestors of the Nobiin reached the Nile Valley probably around or south of the third cataract where they met with two different groups. Archaeologists distinguish between the Nubian C-Group and the Kerma culture. It has been hypothesized by Behrens (1981) that the people of the C-Group culture were speakers of Berber languages. Part of the evidence comes from loanwords in the Nobiin language, of which the most important one is *aman* "water, Nile". Cattle was important for the C-Group people, but excavation at the C-Group settlement at Sayala revealed that sheep was the predominant domestic animal and the main provider of meat (Bietak 1968:38). The term for "sheep" was borrowed from Berber into Nobiin from where it later diffused into other languages of the Nubian family.

Kabyl	igid	Nobiin	eged	"sheep"
Siwi	egaite	Dongolawi/Kenzi	eged	
Dilling	ɔgud	"sheep, goat"		
Dair	ogud	"goat"		
Ghulfan	ogot	"goat"		
Birgid	εgidi	"goat"		

Considering the limited number of Berber loans in Nobiin (Bechhaus-Gerst 1984/85:109ff.) the influence of the C-Group people and culture was probably a restricted one. The distinctive traits of the C-Group disappeared shortly after the Egyptian conquest of Nubia and the arrival of the Nubians. It can be assumed that the Berber people were absorbed and very soon gave up their original language.

The question remains open whether we can in a similar way identify the Kerma culture, which was contemporary with the Nubian C-Group. In the Nobiin lexicon there are many loanwords which originate in the Cushitic languages now spoken in Ethiopia. The various semantic fields covered by Cushitic loans suggest that at an early time in the independent development of Nobiin its speakers came into contact with speakers of one or more Cushitic languages. In Egyptian sources of the Old Kingdom there is indeed some linguistic evidence (Bechhaus-Gerst 1989:95ff.) that allow us to tentatively connect the Kerma culture with speakers of Cushitic. Upon their arrival at the Nile Valley, it seems therefore that Nobiin speakers not only met with people speaking Berber but also with speakers of Cushitic. The latter, however, were much more influential on Nobiin culture and lexicon, as can be judged from the great number of Cushitic loans in the field of livestock terminology.

3.1. sheep/goat

Nobiin	fag	"goat"	Bilin	bagaa	"sheep"
	faag	"wool"	Chamir	begaa	
			Chamta	bigaa	(v. Kefa)
			Kemant	bagaa	
			Iraqw	beeʕi	
			Alagwa	beʕi	
			Burungwe	beʔi	

Correspondences can be found in other branches of Afroasiatic (Behrens 1984/85:167f.), e.g.

Omotic:	Kefa	begee	"sheep"	
	Moca	bago		
	Shinasha	baggoo	(v. Kefa)	
Berber:	Tuareg-Niger	á-beggug	"ram"	
Chadic:	Mwulyen	mbágà -ti	"sheep"	
	Bacama	mbàgà -to		
	Gudu	mbóek		

Nobiin *fag* was most probably borrowed as **pag*, because the initial *f* is a result of a recent regular sound shift **p > f*. The meaning of *faag* "wool" suggests that it was also borrowed from a Cushitic or Omotic source. *fag* "goat" replaced the original term, which can be reconstructed for proto-Nubian and is also found in the Tama languages and Nera (s.a.).

3.2. calf

Nobiin	gor	Agaw	gär	(v. Burji)	
Midob	kor	Bilin	gar		
		Kemant	gar		
		Somali	agor	"m.-calf - 2 y."	(Heine 1978:74)
		Rendille	ogor	"gazelle"	
		Oromo	ajoro	"weaned calf"	(Haberland 1963:63)
		Burji	gare	"calf"	

On the basis of Semitic correspondences Behrens (1984/85) reconstructed proto-Afroasiatic **ʕigali* "calf". The initial vowel might then be original for proto-Cushitic, as there are other examples of the deletion of initial vowel in Cushitic loanwords in Nobiin.

3.3. young (he-)goat

Nobiin	mogor	"young he-goat"	*Sam	maqal	"young sheep and goat"
			Afar	baʔal	"kid he-goat"

The similarity in form and meaning suggests that we are dealing with a Cushitic loan in Nobiin. Because of the irregularity of the sound correspondences within East Cushitic, it is by no means clear whether one of the two (proto-)languages was the donor language of the term.

3.4. young (of animals)

Nobiin   kalissi   "young of animals" (M)

Burji   k'al-a   "son, male child; young of animal"
          k'al-   "to give birth"
Sidamo   qallo   "offspring"
Chamir   gulaʃa   "young boy?"

Nobiin *kalissi* "young of animal" is found in the compound noun *kaj-in kalissi* "young of donkey". Because of the initial g, the position of Chamir *gulaʃa* is doubtful.

3.5. ?

Nobiin   korj-(awi)   "big white sheep with brown spots"   (M)
Sidamo   karso   "mixture, half-caste"

The final *-awi* might be a nominal suffix borrowed from Sidamo (compare *diddawe* skin of sheep or goat).

3.6. sheep/goatskin

Nobiin   kurki   "sheepskin that covers the camel saddle"   (M)
Afar   kuxi   "small goatskin water-bag"

Because of the lack of additional Cushitic correspondences, the direction of borrowing cannot be determined.

3.7. hen/cock

Nobiin   dirbad
Kenzi   darbaad
Dongolawi   durmade

Bilin   dirwa   (Appleyard 1977:25)
Kemant   dirwa   (s.a.)
Awngi   duraʃi
Chamir   jirwa
Somali   doora   (s.a.)
Afar   dorrahe, sgt. dorrahiyta
Beja   andirhe

The final *-ad* of the Nubian forms might represent the borrowing of a singulative form like the one documented for Afar. There are correspondences in

Ethio–Semitic which can be traced back to an East Cushitic origin (Appleyard 1977:25). The Nubian forms must have been borrowed from central Cushitic.

3.8. animal enclosure

Nobiin	angi
Kemant	angii
Bilin	ingii
Chamir	inqii

The eastern Sudanic correspondences Maasai *(enk)aŋ* "home, kraal" and Bari *aŋ* "enclosure" have to be regarded as independent borrowing from an East Cushitic source.

3.9. stable, enclosure

Nobiin/Dongolawi/Kenzi	kudee/kuude	"stable, enclosure"
Bilin	koodaa	"resting-place for cattle during nighttime"

Because of the singular occurrence both in Nubian and in Cushitic the direction of borrowing cannot be determined.

3.10. manure, droppings

Nobiin	fuudee	Oromo	faando	
		Somali	faanto	(Zaborski, n.d.)
		Hadiya	finda	"horse manure"
		Chamir	fandiyaa	

There are more examples which document the loss of a nasal consonant in Nobiin in loanwords from Cushitic. Chamir *fandiyaa* has to be regarded as a secondary borrowing from Ethio–Semitic, e.g. Tigre *fandəya* (Zaborski, n.d.). In Nobiin the original Nubian terms for the two most important secondary products have been replaced by loanwords from Cushitic.

3.11. butter

Nobiin	furuu	HEC	buuru		
		Somali	bur	"fat"	(Cerulli 1938:195)
		(Shinasha	fuura)	(s.a.)	

3.12. milk

Nobiin	suu	Timbaaro	az-u
		Kambaata	az-uta
		Alaba	az-uta
		Hadiya	ad-o
		Libido	ad-o
		Sidamo	ad-o
		Burji	ad-a

The Cushitic forms are most probably derivations of proto-(highland-East-) Cushitic *'az- "white" (H.-J. Sasse, pers. comm.). There is a regular sound correspondence HEC z/d : Nobiin s and the loss of the initial vowel is documented in several examples. This would be an important example for the borrowing of a term including a Cushitic derivational suffix.

There is one loanword in the field of livestock terminology that apparently goes back to a non-Cushitic source. The Nobiin name for the "pig" was borrowed from Omotic, another sub-family of Afroasiatic consisting of languages now spoken in southwest and south Ethiopia. This supports Fleming's hypothesis (1965:378) of the original domestication of the pig by Omotic-speaking peoples and its later spread from their area of settlement.

3.13. pig

Old Nobiin	kutun	"pig"	Kefa	gudinoo	"pig, wild boar"
			Moca	gudino
			Ometo	gudunza

4. Conclusion

The data given above allows only a partial reconstruction of the rather complex prehistory of livestock in the Sudan. With the Nobiin language and the available information on the history of the Nobiin speakers, some solid dating becomes possible that goes beyond the sphere of mere speculation. It has been shown that the Nubians were familiar with herding cattle from proto-eastern-Sudanic times. The introduction of the goat might go back to the level of proto-Nubian–Tama–Nera. There is a common term for all three members of the sub-group. Owing to the scarcity of correspondences between Nera and the other languages, we must proceed from a time depth of at least 6,000 years. The area of settlement of the proto-Nubian–Tama–Nera group must have been in the far west of the Sudan, somewhere between Darfur and Kordofan. At about 4000 bp the proto-Nubian community was in existence, perhaps more to the east of the former homeland. Livestock by then seems to have become more important for the Nubian economy. The proto-Nubians had a common term for small livestock with the meaning of "possession" in general. At this period, secondary products were used as can be seen by the term for milk which can be reconstructed for proto-Nubian.

Before 1500 BC the pre-Nobiin separated from the proto-Nubian community and migrated towards the Nile Valley. Within the lexicon of the Nobiin language we find evidence for the presence of two different languages which at that time were spoken in the Nile Valley in the region between the first and fourth cataract. The first, a Berber language, can be related to the Nubian C-Group. Osteological evidence shows that sheep were the basis of the economy of the C-Group people (Bietak 1987:118), and the term for "sheep" was borrowed from Berber into Nobiin.

The second and more important one was a Cushitic language, the identification of which makes it possible to link the Kerma culture with a Cushitic people. The hypothesis that these "Nile–Cushites" had been settling in the Nile Valley since about 5300 bp was – on the basis of archaeological evidence – suggested by Randi Haaland (1992:59). She further assumed that the "Nile–Cushites" were specialized pastoralists who later expanded into other parts of northeast and East Africa, thus introducing cattle to these areas. The great number of Cushitic loanwords in the field of livestock terminology in Nobiin actually bears witness to the importance of animal husbandry for the Cushites of the Nile Valley. By the time of the arrival of the Nobiin, around 1500 BC, the economic basis of the Cushitic society had already undergone considerable changes. The great bulk of Cushitic loans in Nobiin refer-ring to economic activity do not come from the field of animal husbandry but from the field of agriculture. This clearly indicates that by 1500 BC the Nile–Cushites were well adapted to their Nile environment and practised some sort of mixed economy with emphasis on agriculture.

Under the influence of the Cushites, Nubian pastoralists adopted a sedentary form of existence. They took up agriculture, went on keeping small ruminants, adopted Cushitic techniques and the corresponding Cushitic lexicon. Thus we not only find Cushitic loanwords referring to (small) livestock, but also in the field of agriculture and – from a much later period – iron-working. The Nobiin terms for "house" and "village" can also be traced back to a Cushitic origin. There is some indication as to which sub-group of Cushitic the Nile–Cushites can be attributed to. Of the branches of Cushitic represented in Nobiin–Cushitic correspondences, High-land East Cushitic – a sub-group nowadays consisting of several languages spoken in the southern part of Ethiopia – appears most often, exclusively in 27 sets and altogether in 47 out of 83 sets. This is, of course, a purely statistical approach that will have to be substantiated by further investigations.

At least one other group of Afroasiatic speakers played a role in the spreading of livestock in the Sudan. The term for the "pig" clearly originated in Omotic lan-guages. There is some more linguistic evidence which makes it plausible that speakers of Omotic languages played an important role within the Sudan as an area of contact (e.g. Böhm 1986, Bechhaus-Gerst 1989).

Still, there are large gaps in our knowledge of the prehistory of livestock in the Sudan. The more we learn about the presence of speakers of Afroasiatic languages, like Berbers, Cushites, Omotic and probably Chadic people, in the neolithic period, the less certain becomes the role the Nilo–Saharans played in the process of the development of animal husbandry, use of secondary products and agriculture. It is far from clear whether it was really the Nilo–Saharans who, as Haaland (1992:61) pointed out, occupied the Khartoum area before the assumed arrival of the Cushites at about 5300 bp. In the same way it cannot be decided whether it was the Cushites who adopted milking and bleeding of cattle from eastern Sudanic people as Ehret (1974:56) proposed, or the other way round, as Haaland (s.a.) would have it. Con-cerning this area only one factor is evident: there is at the moment no indication that Eastern Sudanic speakers played a role in the domestication of grain and the development of agriculture.

Sources of linguistic data

1. Nubian languages

Nobiin	Werner 1987
Nobiin (M)	Murray 1923
Dongolawi	Armbruster 1965
Kenzi	v. Massenbach 1962
Dilling	Kauczor 1929/30
Dair	Junker & Czermak 1913
Hill Nubian (other)	Meinhof 1918/19
Birgid	Thelwall 1978
Midob	Werner 1993

2. Other Nilo-Saharan languages

Tama group	Edgar 1991
Nera	Reinisch 1874

3. Berber languages

Kabyl	Murray 1923
Siwi	Basset 1890
Tuareg-Niger	Barth 1858

4. Cushitic languages

Bilin	Reinisch 1887
Awngi	Hetzron 1978
Chamir	Reinisch 1883
Kemant	Conti Rossini 1912
Rendille	Schlee 1978
Oromo	Gragg 1982
Burji	Sasse 1982
proto-Sam	Heine 1978
Afar	Parker & Hayward 1985
Sidamo	Gasparini 1983
Hadiya	Plazikowsky-Brauner 1964
Highland-East-Cushitic	Korhonen et al. 1986
Iraqw	Whiteley 1958
Alagwa	Whiteley 1958
Burunge	Whiteley 1958
Beja	(field material) Bechhaus-Gerst

5. Omotic languages

Kefa	Cerulli 1951
Moca	Leslau 1959
Ometo	Moreno 1938

6. Chadic languages

Chadic   Kraft 1981

References

Appleyard, D.L. 1977. A comparative approach to the Amharic lexicon. *Afroasiatic Linguistics* **5**(2), 43–109.

Armbruster, C.A. 1965. *Dongolese Nubian. A lexicon.* Cambridge: Cambridge University Press.

Barth, H. 1858. *Reisen und Entdeckungen in Nord- und Central-Afrika.* (Bd. 5) Gotha.

Basset, R. 1890. *Le dialecte de Siouah.* Paris.

Bechhaus-Gerst, M. 1984/85. Sprachliche und historische rekonstruktionen im Bereich des Nubischen unter besonderer Berücksichtigung des Nilnubischen. *Sprache und Geschichte in Afrika* **6**, 7–134.

Bechhaus-Gerst, M. 1989. *Nubier und Kuschiten im Niltal. Sprach- und Kulturkontakte im "no-man's land".* Afrikanistische Arbeitspapiere–Sondernummer 1989.

Bechhaus-Gerst, M. 1994. *Möglichkeiten und Grenzen einer diachronen Soziolinguistik. Sprachwandel durch Sprachkontakt am Beispiel des Nubischen im Niltal.* ms.

Behrens, P. 1981. C-group-Sprache–Nubisch–Tu Bedawiye. Ein sprachliches sequenzmodell und seine geschichtlichen implikationen. *Sprache und Geschichte in Afrika* **3**, 17–49.

Behrens, P. 1984/85. Wanderungsbewegungen und Sprache der frühen saharanischen Viehzüchter. *Sprache und Geschichte in Afrika* **6**, 135–216.

Bietak, M. 1968. *Studien zur Chronologie der Nubischen C-Gruppe.* Vienna: Österreichische Akademie der Wissenschaften.

Bietak, M. 1987. The C-Group and the Pan-Grave Culture in Nubia. In *Nubian Culture Past and Present.* Main Papers presented at the Sixth International Conference for Nubian Studies in Uppsala, 11–16 August 1986. T. Hägg (ed.), 113–28. Stockholm.

Böhm, G. 1986. Beobachtungen zur Frage meroitisch-omotischer Wortbeziehungen. *Beiträge zur Sudanforschung* **1**, 115–20.

Cerulli, E. 1938. *La Lingua e la Storia dei Sidamo.* (Studi Etiopici II). Roma.

Cerulli, E. 1951. *La Lingua Caffina.* (Studi Etiopici IV). Roma.

Conti Rossini, C. 1912. *La langue des Kemant en Abyssinie.* Wien.

Edgar, J. 1991. First steps towards proto-Tama. In *Proceedings of the Fourth Nilo–Saharan Linguistics Colloquium,* M.L. Bender (ed.), 111–31. Hamburg: Buske.

Ehret, C. 1974. *Ethiopians and East Africans. The problem of contacts.* Nairobi Historical Studies, 3. Nairobi: East African Publishing House.

Ehret, C. 1993. Nilo–Saharans and the Saharo–Sudanese neolithic. In *The archaeology of Africa. Food, metals and towns,* T. Shaw, P. Sinclair, B. Andah, A. Okpoko (eds), 104–25. London: Routledge.

Fleming, H.C. 1965. *The age-grading cultures of East Africa: an historical inquiry.* Michigan: Ann Arbor.

Gasparini, A. 1983. *Sidamo–English Dictionary.* Bologna: EMI.

Gragg, G. 1982. *Oromo Dictionary.* Monograph No. 12. Ann Arbor: Michigan State University Press.

Haaland, R. 1992. Fish, pots and grain: early and mid-Holocene adaptations in the central Sudan. *The African Archaeological Review* **10**, 43–64.

Haberland, E. 1963. *Die Galla Süd-Äthiopiens.* Stuttgart: Kohlhammer.

Heine, B. 1978. *The Sam Languages.* Afroasiatic Linguistics, 6. Malibu: Undena Press.

Hetzron, R. 1978. The nominal system of Awngi. *Bulletin of the School of Oriental and African Studies* **41**, 121–41.

Junker, H. & W. Czermak 1913. *Kordofan-Texte im Dialekt von Gebel Dair.* Wien: Hölder.

Kauczor, P.D. 1929/30. Bergnubisches Wörterverzeichnis. *Bibliotheca Africana* **3**(4), 342–83.

Korhonen, E., M. Saksa, R.J. Sim 1986. A dialect study of Kambaata–Hadiya (Etiopia). Part 2: Appendices. *Afrikanistische Arbeitspapiere* **6**, 71–121.

Kraft, C.H. 1981. *Chadic wordlists* [3 volumes]. Berlin: Reimer.

Leslau, W. 1959. *A Dictionary of Moča*. University of California Publications in Linguistics, 10. Berkeley, Los Angeles: University of California.

Massenbach, G. v. 1962. *Nubische texte im dialekt der Kunuzi und Dongolawi*. (Abhandlungen für die Kunde des Morgenlandes, 34.4). Wiesbaden: Steiner.

Meinhof, C. 1918/19. Sprachstudien im egyptischen Sudan. *Zeitschrift für Kolonialsprachen* 9, 43–64, 89–117, 167–204.

Moreno, M.M. 1938. *Introduzione alla Lingua Ometo*. Roma: Mondadori.

Murray, G.W. 1923. *An English–Nubian Comparative Dictionary*. Oxford: Oxford University Press.

Parker, E.M. & R.J. Hayward 1985. *An Afar–English–French Dictionary*. London: SOAS.

Plazikowsky-Brauner, H. 1964. Wörterbuch der Hadiya-Sprache. *Rassegna di Studi Etiopici* 20, 133–82.

Reinisch, L. 1874. *Die Barea-Sprache. Grammatik, Texte und Wörterbuch*. Wien: Hölder.

Reinisch, L. 1883. *Die Chamirsprache in Abessinien, I*. Sitzungsberichte der Phil.-Hist. Classe der Kaiserlichen Akademie der Wissenschaften. Wien: Akademie der Wissenschaften.

Reinisch, L. 1887. *Wörterbuch der Bilin-Sprache*. Wien: Hölder.

Sasse, H.-J. 1982. *An Etymological Dictionary of Burji*. (Kuschitische Sprachstudien, Bd. 1). Hamburg: Buske.

Schlee, G. 1978. *Sprachliche studien zum Rendille*. Hamburg: Buske.

Thelwall, R. 1978. A Birgid vocabulary list and its links with Daju. In *Gedenkschrift Gustav Nachtigall 1874–1974*. H. Ganslmayr & H. Jungraithmayr (eds), 197–210. Bremen: Übersee-Museum.

Thelwall, R. 1982. Linguistic aspects of Greater Nubian history. In *The archaeological and linguistic reconstruction of African history*, C. Ehret & M. Posnansky (eds), 39–56. Berkeley: University of California Press.

Werner, R. 1987. *Grammatik des Nobiin (Nilnubisch)*. (Nilo-Saharan, 1). Hamburg: Buske.

Werner, R. 1993. *Tidn-áal: a study of Midob (Darfur-Nubian)*. Berlin: Reimer.

Whiteley, W.H. 1958. *A short description of item categories in Iraqw*. Kampala: Institute of Social Research.

Zaborski, A. n.d. *Beja-Nubian contact and interference*. ms. Institut für Afrikanistik, Köln.

CHAPTER TWENTY-FIVE

Ethnographic perspectives on cattle management in semi-arid environments: a case study from Maasailand

Kathleen Ryan, Karega Munene,
Samuel M. Kahinju, Paul N. Kunoni

1. Introduction

In the course of their evolution, humans have developed many strategies to enhance their chances of survival. In his discussion of one such strategy, nomadic pastoralism, Spooner (1973:3) states that "all populations live in a particular relationship with their environment and this relationship is mediated by their subsistence pattern". There are two common subsistence adaptations to grasslands. Since humans cannot digest grass directly, energy has to be filtered through grass-eating animals by either hunting and eating wild animals or managing domesticates in a grassland environment. The difference between the two is the degree of control over animals and environment. A pastoral ecosystem has three main elements: people, animals and environment, that under certain conditions combine in a dynamic relationship to achieve a balance among all three (Swift 1977:273). How this balance is achieved is largely dependent on decisions taken by the herd-owners who, in coming to these decisions, are constrained by the environment in which they live and the biology of the animals on which they depend.

The emphasis in this chapter is on the interactions of humans and herd animals, and how social relations become part of the ecology of the animals. We re-examine the hypothesis that cattle in many African societies are part of both the biological and social environments and that social relationships, cemented through cattle, are an essential part of the survival strategy of humans in pastoral ecosystems. Husbandry of a particular species of herd animals imposes certain constraints on a society, and these are reflected in the behaviour, social organization, and socio-political relationships of pastoral groups. Many researchers have discussed these linkages, notably: Barfield 1993, Bernus 1981, Blench 1994, Dahl & Hjort 1976,

Deshler 1965, Downs & Ekvall 1965, Dupire 1970, Dyson-Hudson 1980, Dyson-Hudson & Dyson-Hudson 1980, Evans-Pritchard 1940, Fratkin et al. 1994, Galaty 1982, 1989, 1994, Galaty & Bonte 1991, Galaty & Johnson 1990, Goldschmidt 1977, 1986, Homewood & Rodgers 1991, Hopen 1958, Ingold 1980, Kinahan 1991, Lamprey & Waller 1990, Leeds 1965, Lefébure 1977, Lincoln 1981, Massey 1987, Oxby 1986, Rigby 1969, 1985, 1992, Spear & Waller 1993, Spooner 1973, Stenning 1959, Swift 1977, Sylla 1995, Vayda 1965, Vivello 1977, to name but a few. That similar human–animal interactions occurred in the past is suggested by the work of Robertshaw & Collett 1983, Smith 1983, 1984, 1992, Parkington et al. 1986, Marshall 1990a,b, 1994, Robertshaw 1990, MacDonald & MacDonald, Ch. 8 in this volume.

In this paper, we focus on one, until recently semi-nomadic and now largely settled, pastoral group, the Maasai of Kenya. We illustrate links between the special needs of cattle and some features of Maasai social relations, concentrating on three aspects of cattle management: how cattle are acquired; how they are exchanged or redistributed; and how, and under what circumstances, they die.

2. Research design

The research area is Kenya Maasailand (Fig. 25.1), an area of semi-arid grasslands with low and unreliable rainfall and limited permanent sources of surface water, generally unsuitable for crop cultivation. Until recently, through use of a variety of

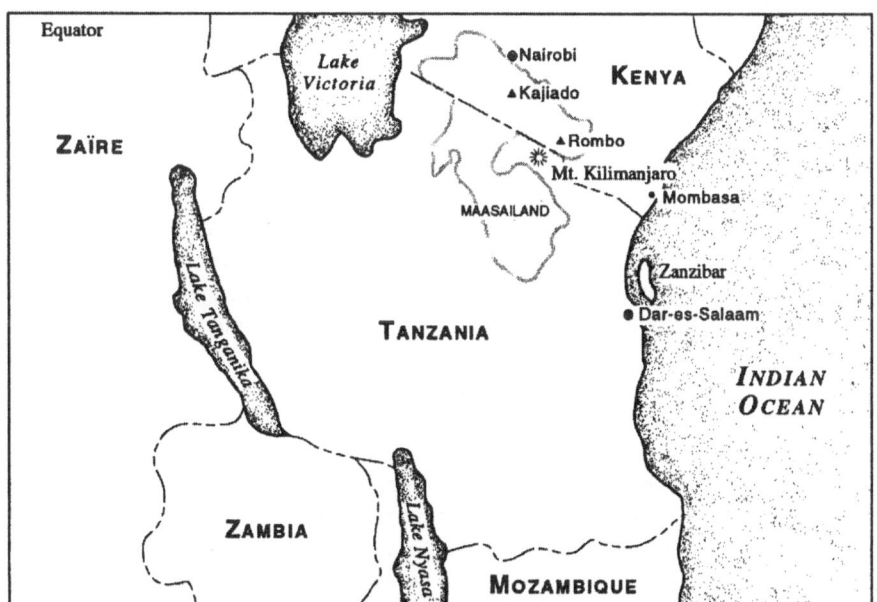

Figure 25.1 Map of Maasailand showing research locations.

livestock (cattle, goats, and sheep) and extensive ranging and rotational grazing, Maasai lived relatively well off these marginal lands (Sindiga 1994). Our study area is Kajiado district, a large Maasai Section, southeast of Nairobi. There we have selected seven settlements or *bomas*: three near Kajiado town, and four in the Rombo area.

The project is designed to answer questions about demography, exchange, redistribution, slaughter, and disease of cattle in Maasai society. Our primary informants are elders (from various age-sets) and one young man from the current warrior age-set, together with their extended families. Elders were chosen because they exercise the ultimate control over the family herds but wives and older children were also consulted, particularly when we discussed milk yields, animal health, or behaviour of individual animals. We adopted an informal interview method, modified from those used by Galaty (1989) and Grandin (1991), in order to construct cattle genealogies of animals owned by our informants. Most of our data were gathered during these discussions.

The classificatory system embedded in the naming of individual cows is discussed in detail by Galaty (1989:218–230). He identifies distinct modes of signification, using symbols of *social* transactions (such as "in exchange", "gift from a friend", or "bridewealth") and forms of visual description (such as "speckled", or "without horns") or a combination of both (such as "exchanged for reddish-brown ox"). All cattle in a particular lineage are named after a bovine matriarch. Her name tends to commemorate the event or transaction by which she was acquired and her descendants continue to take that name. For further discussion of cattle naming systems see Rigby (1969) and Schoenbrun (1993:46). In Table 25.1 we give examples of names given to cattle in our informants' herds.

A typical family herd of 50–60 animals may be subdivided into 10 matrilineages, comprising 80 per cent females and 20 per cent males (including castrates) although there is quite a lot of variation in these ratios. Cattle are allotted to specific matri-centric households in the polygynous Maasai family. Each animal can therefore be classified genealogically (from the matriarch), spatially (to a household), and by individual description (gender, age, colour, and behaviour) (Galaty 1989). A detailed examination of naming and genealogical data, undertaken in 1991 (Ryan et al. 1991) in an attempt to record strategies of herd management, led us to examine more closely the social significance of cattle transactions in Maasai society and to identify those actions that we suggest have a direct causal link to animal welfare.

3. Acquisition of cattle

Cattle are acquired through inheritance, bridewealth, gift-giving, raiding and purchase. Avenues of acquisition are different for men and women. Acquisition here does not imply ownership but simply rights in an animal (Oboler 1985 discusses this differentiation). Although both sexes eventually acquire herds, they do so in a different sequence and they maintain different levels of authority over their allotted animals.

Table 25.1 Examples of names given to informants' animals (heifers)*.

	Name	Meaning	How or why obtained
Named for transaction	Noontoyie	for girls	bridewealth received for sister
	Entotua	of love	gift from age-mate
	Pokurot	after a struggle	from a cattle raid
	Kereti	blessing	brought in by informant's mother
	Naoli	exchanged one	adopted into family matriarch line
Received from	Nolmongi	reminder of an ox	exchanged for steer
	Mong'o Sampu	exchanged for a brown-striped ox	received in exchange for steer
	Nesingo	from Singo in Kambaland	bought for disease resistance
Descriptive/ behavioral names	Meshuri**	no defence/no cover	cattle raid; first cow received from father
	Noomaroro**	named after a tree; long ears	first cow received from father
	Kilera**	named after acacia tree	paid for by selling dead elephant tusk found in an acacia bush in Amboseli; first cow received from father
	Nkeyi	speckled, black and white	descendant of main family matriarch
	Naimai	no horns and small hump	descendant of main family matriarch
	Merurai	troublesome, cried at night when a calf, doesn't allow people to sleep	descendant of main family matriarch

* following Galaty 1989.
** main family matriarch descended from "first cows" given to informants.

In constructing our genealogies we started with the first cow given to each man when he was a small boy, usually by his father. We asked what her name was and how she was acquired, how many female and male calves were born to her, how many of those females produced offspring of their own, how many males were kept whole and how many castrated, and how and under what circumstances the castrates were disposed of. Our oldest informant clearly remembered the first cow given to him by his father, over 80 years ago. That cow was an offspring of one raided by our informant's father, about 90 years ago, from the Laikipia Maasai, and renamed *Meshuri* (meaning "no defence" or "no cover" from driving rain) in memory of the hardship endured during the arduous but successful raid. This original *Meshuri* matriarch had seven female and three bull calves (Table 25.2). All but one of the female calves were given to our informant and his brothers; the remaining female, brighter coloured than her siblings, remained in the father's herd and was given a

Table 25.2 Sample of "first cow" genealogy (first generation) given by father when informant was a child.

Cow name	How acquired	Number of offspring	
Meshuri	from cattle raid	females 7	males 3
		1 remained in father's herd; 6 given to informant & brothers:	
		1 given to informant had 7 male calves →	7
		3 "exch. heifers" adopted into **Meshuri** line and called **Meshuri** 1 given separate name **Mparasi**	← 4 exch. for heifers 2 killed for meat by *morans* 1 died

separate name, *Natapatu*, meaning "one who stands apart" and formed a new matriarch line. She and her progeny passed to the eldest son at the father's death.

The calf shown in Figure 25.2 follows in the line of the original *Meshuri* matriarch and carries the *Meshuri* line name, although she will not be officially recognized until she produces her first calf. This line remains dominant in the herd today. Males do not get named for the line as females do; they are given individual descriptive names, although it is known from which matriarch's line they descend. In some cases where the direct blood line of a matriarch (particularly the "first cow" line) dies out, one of the male offspring of that line is exchanged for a heifer from another herd and that heifer is "adopted" into the line and given the matriarch's name, although she has no direct blood link to that line. In fact, all of our informant's *Meshuri* line derive from this kind of exchange because his second generation *Meshuri* heifer produced only bull calves (Table 25.2). Other direct blood line *Meshuri* cattle continue in his brothers' children's herds.

Everyone remembered his "first cow" and at least one or two generations of offspring. Between the third or fourth generation and the recent past, as a man's herds increased and cattle were allocated to wives' households, redistributed to sons, sent out as bridewealth or dowry, or given or received as gifts, memories were less exact. However, because our informants came from different age-sets, the time of confused memories for one often coincided with the clear "first cow" stage of another and so we were able to collect cattle histories spanning the past 90 years.

More animals can be given by a father at any time; the number given depends on the love between father and child, and a father's wealth. Other avenues of acquisition include animals given from a mother's herd, starting from when a boy is old enough to herd; and sons eventually inherit a share of their mother's herd. At the time of marriage, parents and relatives try to give gifts to build up a son's herd. Some of these are reallocated to his bride's herd but he retains a residual unallotted

Figure 25.2   Young calf from the *Meshuri* matriarch line.

portion to which he has sole rights. Animals acquired through special gift or purchase generally are added to his personal herd.

Daughters are also given a heifer from the family line and it is descendants of
this animal, usually a cow and calf, that go out with the girl when she marries.
Unlike a boy, she does not inherit all of the progeny. However, she will be allotted
cattle by her husband, and her husband's relatives and her co-wives are expected to
give her gifts according to their means. These gifts are intended to cement a special
bond with the new wife. They are always remembered by donor and recipient and
are reflected in the reciprocal terms of respect they use when addressing each other
(see Spencer 1973:31). Our oldest informant still remembers the names and colours
of the heifers he gave to his eldest son's two wives at marriage; they were both
heifers descended from his "first cow" lineage. The new wife's herd then will be
made up from the cow and calf from her own family; the allotment from her
husband; and other gift animals (from relatives and co-wives). From her herd, her
sons will build up herds of their own. She cannot give any cattle outside the family.

Table 25.3 *Noontoyie* (bridewealth) lines.

What	How acquired	Disposition of bridewealth animals or their offspring		
		females	bulls	castrates (steers)
Noontoyie 1 1 mature cow	from sister's bridewealth	4 heifers 1 "exch. heifer" "	Offspring: 1 bull: remained in herd	3 steers: 1st exch. for 1 heifer*; 2nd exch. for 2 steers*; 3rd sold

*1st "exch. steer" exch. for 6 sheep & 4 goats; 2nd "exch. steer" exch. for 1 female donkey
Number from line still surviving in present herd: 3 mature females (no calves); 1 heifer

What	How acquired	females	bulls	castrates (steers)
Noontoyie 2 1 mature cow with male calf; 5 heifers; 1 steer	from 1st daughter	1 mature cow: died of drought 5 heifers: 2 given to informant's brothers	1 male calf: died 1 steer	

Number from line still surviving in present herd: 1 mature cow with calf; 3 calves; 2 heifers; 1 steer

What	How acquired	females	bulls	castrates (steers)
Noontoyie 3 1 mature cow with male calf; 3 heifers; 1 steer	from 2nd daughter	1 mature cow 3 heifers		1 male calf: castrated, slaughtered eve of daughter's initiation ceremony 1 steer: slaughtered during wedding ceremony

Number from line still surviving in present herd: 1 mature female with 1 heifer offspring

What	How acquired	females	bulls	castrates (steers)
Noontoyie 4 1 mature cow with male calf; 3 heifers; 2 steers	from 3rd daughter	1 mature cow 3 heifers: 2 given to informant's brothers 1 "exch. heifer" "		1 male calf: died of East Coast fever 2 steers: 1 exch. for heifer "

Number from line still surviving in present herd: 3 heifers; 4 steers; 1 female calf; 1 male calf

In addition, animals received by parents, as bridewealth for a daughter, can be reallocated to her brother's herd; or a father can keep an animal from his daughter's bridewealth in his own herd. Each of these animals is named after the bridewealth transaction as *Noontoyie* although each is identified and remembered as a separate line. Table 25.3 lists the *Noontoyie* transactions for one elder in the southern area, who had one for a sister and three for his daughters. All marriage contract animals are included here, not just the one cow and calf plus a heifer[1] that are given to the bride's parents when the bride officially goes to her husband's *boma*, but only the official cow and calf and their descendants are named *Noontoyie*. In our area, the extra animals are usually redistributed right away, rather than kept in the father's herd. In the present herd 23 animals derive from bridewealth transactions (Table 25.3).

A third category, *Sotua* (a gift from a friend), included gifts between age-mates, between brothers or half-brothers, from a man to his future father-in-law, or from a rich man to a poorer relative. These were often reciprocal with the original recipient returning the favour at some future time. Examples of *Sotua* transactions are given in Table 25.4.

Table 25.4 *Sotua* transactions.

---

1. Given by friend and age-mate when they were *morans*;
   informant gave back a steer after 2nd marriage.
   (offspring: 8 calves: 5 females, 3 males)
2. Given by friend one age-set older than informant;
   informant gave a daughter in return as 2nd wife.
   Just before the marriage, informant gave his future father-in-law a heifer
   (offspring: 6 calves: 3 females, 3 males)
3. Given by son of 1st wife's father
   (offspring: 6 calves: 3 females, 3 males)
4. Given by sister's husband when informant married 2nd wife
   (offspring: 10 calves: 7 female, 3 males)

Cattle raiding, although illegal since colonial times, was, until quite recently, an acceptable means of increasing one's herd. Cattle were raided from other adjoining tribal groups or from rival Maasai sections, as in the case of the raid on the Laikipia which garnered the matriarch *Meshuri*. However, today, purchase has largely replaced this option.

Some typical lines in a herd from the southern area are shown in Table 25.5. All the matriarch lines (except the descendants of the one purchased animal) derive from social transactions enacted over a considerable period of the herd owner's lifetime: from the first cow received from his father; bridewealth from his sister; a gift from an age-mate while they were both young men; gifts from the fathers of his two wives; gifts from a step-brother and son; and finally bridewealth from a daughter and a gift from a son-in-law. These transactions occur sequentially over a period of 40 to 50 years, but all are still remembered through the matriarch lines.

Table 25.5   Typical herd "lines" from southern area.

Kilera	first cow
Noontoyie	from sister
Noontoyie	from oldest daughter
Sotua	from age-mate during moranhood
Sotua	from father of first wife
Sotua	from father of second wife
Sotua	from older step-brother
Sotua	from son of same step-brother
Sotua	from son-in-law
Purchased	from Kambaland

Table 25.6   Avenues of cattle acquisition for men and women.

	Men	Women
Inheritance	from father and mother	from father
Bridewealth	from sister or daughter	from daughter
Gift	from friends, relatives	from husband, relatives, co-wives
Purchase	option open to men	rarely option for women
Cattle-raiding	men-only activity	may get reallocated cattle

An example here is the gift from an elder brother to his young step-brother (our informant) that forged a special bond between the two and was reinforced in the next generation by a gift from the older man's son.

Table 25.6 summarizes the differences in avenues of acquisition between men and women. While sons can inherit animals from both father and mother, daughters can only inherit from their fathers. A mother's herd is reserved for her sons and those sons, in particular the last-born, are responsible for their mother's welfare – especially in her widowhood or old age. A father is initially responsible for a daughter's acquisition in that he gives her an animal from the family herds, which she is expected to care for until she leaves in marriage, taking with her perhaps one or two animals from the progeny. The bulk of the herd she will care for over her lifetime is given to her by her future husband and his relatives and her co-wives and will pass on to her sons. Gifts probably account for a majority of exchanges, marking social transactions that form special bonds and hence the reciprocal ties necessary to a family's survival.

4. Redistribution and exchange

Animals are redistributed and exchanged through inheritance, bridewealth, gift, loan, purchase or sale and, until recently, cattle-raiding. The genealogical charts show that animals are seldom exchanged for purely economic reasons. Most exchanges also fulfil social obligations or mark social transactions, as providing bridewealth; making a gift to bind a friendship; exchanging steers for heifers to put in the care of a new wife entering the homestead; or exchanging heifers for steers to obtain the latter for ceremonial slaughter.

Even sales of animals often have an underlying social significance: one of our informants sold one steer to pay for initiation ceremony expenses and another for his father's burial expenses. Of course, most exchanges tend to combine social and economic rationale – such as the exchange of a mature steer for a heifer and a small steer, when one has no immediate ceremonial or monetary use for that mature steer.

Besides these exchanges, animals are redistributed in other ways. Parts of a herd are often lodged temporarily at other *boma*s in an attempt to spread risk from predators or cattle-raiders, while the home *boma* may have some additions from outside. Labour requirements may necessitate moving animals to other *boma*s when labour is scarce at the home *boma*. Another strategy is to move children to labour-scarce settlements. Animals are also moved temporarily (or, in some cases, semi-permanently) to other settlements in times of drought or when disease in one area reaches proportions to threaten the herds severely. Our oldest informant moved his whole herd for an extended period when East Coast fever hit his home range in 1990. Usually animals are moved to kin settlements or to those of age-mates; in future years the herder will reciprocate by accepting animals from his friends or relatives when they are in trouble.

In early September 1991, at the end of two years of drought, one of our southern informants had moved most of his herd to other settlements, keeping only a few lactating cows and calves to provide milk for the small children at his own *boma*. He also retained his female sheep and goats but by the end of the month he had to move them also. Regrettably, many of the lactating cows with calves held at the home *boma* died that year. Under severe drought conditions, elders recognize that some animals must be exposed to almost certain death for the good of the family.

Exchanges or purchases made outside of the home locality bring in new breeding stock into the herds. Boraana bulls are especially prized. In addition, purchases are often arranged for health reasons. Animals are especially preferred from areas known to produce animals with immunity to certain diseases, such as the Kamba area, where cattle are believed to be resistant to East Coast fever (currently the main cause of death among our informants' herds). One such animal (named *Nesingo* = from Singo), was purchased from the town of Singo in Kambaland and is now 18 years old. She has produced many healthy offspring and may have passed on her immunity to her progeny.

5. Herd Mortality

To maintain a herd balance of approximately 80 per cent female and 20 per cent male, culling strategies have to be employed. Throughout the lifetime of Maasai men or women, as they pass from one life stage to the next, each stage is marked by a distinctive ceremony, many involving the slaughter of an animal. The animal chosen is almost always a mature castrated male, preferably fattened. Because the beauty and health of the sacrificed animal enhances the status of the family, there is an incentive to keep male animals to maturity rather than cull them at a young age, as is done in many modern dairy economies. Retaining male animals to maturity also allows use of blood from the living animal when milk is scarce. Although cows may stay in the herd until they are well over 15 years of age, old or sick female animals are culled (slaughtered, or sold, for meat). These cullings represent deliberate decisions under unstressed conditions.

However, prevalence of drought and disease in most of Maasailand often tips the balance. Under conditions of stress, as during a prolonged drought, young animals are more vulnerable than mature ones since they depend on continued lactation of their dams. A whole generation can be lost. Strategies of care may favour young female calves over males by allowing them longer access to their mother's milk although Homewood & Rodgers (1991:168) found no such bias in mortality data for Maasai herds they studied in Tanzania. None of our informants admitted the slaughter of young male calves, even under stress conditions. Division of the adult herd into male and female grazing units favours the selection of better quality forage for the breeding cows but, weakened by drought and starvation, even mature cows succumb to disease. Great efforts are made to save animals by dispersing them to better watered and pastured areas or disease-free zones but, inevitably, under extreme stress, many of the herd will die; in some cases all may be lost.

It is under extreme conditions such as loss of the majority of the herd that other less palatable strategies are tried, such as selling of small stock to buy grain and resorting to wage labour or labour in exchange for food. Reverting to hunting and gathering, one of the options employed in the past, is hardly an option today in our area. However, the most successful strategies are those that utilize those social bonds already enacted during a man's lifetime. During the immediate crisis, these may include moving all or some of the family in with relatives; fostering children at other *bomas*; borrowing milking cows from relatives or friends. In order to rebuild a herd, a man will have to call on his reciprocal ties with his own relatives; or with age-mates and in-laws, contracted over the years in many *Sotua* and *Noontoyie* transactions. When Maasai herds were decimated by epizootics and drought in the early 1930s, our oldest informant lost the best part of his herd, so he called on his father's Kikuyu relatives who supplied him with the means to acquire the nucleus of a new herd. A son-in-law will be expected to come to the aid of his father-in-law, father-in-law to aid a son-in-law; brothers, half-brothers, brothers-in-law, and age-mates are all potential donors, provided they have not themselves suffered a similar loss. When asked why he so freely gave a cow to an age-mate in trouble, an elder said: "Today I am well off but who knows about tomorrow?" Delayed payments of those "extra" animals contracted as part of a bridewealth settlement can be called in many years later. One elder in our group delegated his senior wife to travel to a far off *boma* to negotiate the release of a contracted animal. Meantime he travelled many miles in the other direction to make arrangements to lodge his few remaining animals at an age-mate's *boma*.

6. Conclusions regarding archaeological significance

What is the possible archaeological significance of these data? The living herds we studied comprise a high percentage of mature lactating females but also a significant number of prime age castrates. These are reflected in our projected mortality profiles under "deliberate culling" or "died from old age" (ranging from 3 to 18 years of age, if all sexes are combined and we exclude normal calf mortality). They are consistent with Marshall's (1990a) findings from some neolithic sites in southwestern Kenya, which suggest similar management strategies to those of contemporary specialized herders under unstressed conditions. If we add in our deaths from "disease and disaster", however, we see increased mortality in all age classes but, significantly, many juveniles (1–3 yr) enter the death assemblage.

Although we are only at the preliminary stage of our analysis of mortality data, we notice that in several herds many more deaths resulted from disease or starvation than were deliberately culled. This was particularly noticeable when we constructed a herd profile from a recent time period (late 1960s to 1980s) of animals owned by a herder who is now in his mid to late thirties. We traced four of his matriarchs and two generations of their progeny to determine survivorship patterns and cause of death. Fifty per cent of the herd survived beyond maturity, many well over 10 years, of which 96 per cent were female with bulls making up the other 4 per cent. No castrates appear to have survived beyond 4 or 5 years. The females' ages were estimated by their age at first calving and numbers of progeny; one giving birth to as many as 10 live calves, which would put her age at 15+. The others ranged from 8 to 14 years. No females were recorded as deliberately culled; they seem rather to have died of old age. However, culling of old females and younger infertile females was recorded in some other herds.

We took note also of exchange and redistribution transactions: 6½ per cent were exchanged for an equivalent value. Most usually, a mature steer was exchanged for a young heifer and one or two small steers (or in some cases for four or five small stock or one or two donkeys). While these exchanges reflect equal value they alter the age or sex classes (and even the species) that might eventually enter the death assemblage and would be impossible to detect archaeologically. The same can be said for the "redistribution" category comprising 17 per cent of the outgoings of the herd, in this case all sent out as bridewealth or dowry to acquire wives. The assumption is that an equal number will eventually enter the herd but, in this case, since our informant's daughters are still young, none of the animals we focused on in this herd derived from daughters' bridewealth.

If we exclude the females and bulls who died of old age, the deliberate culling rate was about 6 per cent within the whole cohort. This includes animals slaughtered as part of a ceremony or, in rare cases, culling of an unsuccessful bull. In addition, steers were sometimes donated to the current warrior age-set for slaughter, or sold for cash. In either case, the carcasses would not become part of the death assemblage at the settlement. No animals were slaughtered just to supply the family with meat, although if death was due to starvation or disease, the animal was butchered and the meat was eaten. Significantly, 20 per cent of the animals tracked died as a result of disease or starvation. All were butchered in the same fashion, there being no differentiation between butchery of a ceremonial or diseased animal. There were some differences in cooking methods and food distribution among participants for different occasions but these would be difficult to detect in the faunal record.

Under ideal conditions, a faunal assemblage derived from a Maasai homestead should represent a very high percentage of female animals in the older age class, a lower percentage of castrate males/infertile females in the mature class, very few juveniles and relatively low mortality of young calves. Under conditions of stress, one would expect to see high mortality among calves and juveniles with little change in the mature class except slightly higher numbers in the oldest range, representing deaths from stress or deliberate culling of old females for food.

It is perhaps significant that the herd we analyzed lived during a time of extreme stress (late 1960s to late 1980s) and this stress is certainly reflected in the high mortality among young calves and juveniles. Detection of patterns such as these – reflecting fluctuations between stressed and unstressed conditions – may serve as indicators of specialized pastoralism in the archaeological record in semi-arid contexts and, in conjunction with other indicators of stress, such as hypoplastic banding on teeth (Gifford-Gonzales 1984, Marshall 1990), may help to track fluctuations in environmental crises over time.

7. General conclusions

We believe that the Maasai livestock management strategies as described here have a direct relationship with social transactions within the culture. It is through cattle genealogies and the remembered transactions in cattle that past and present social relations of humans are preserved. In addition, however, the strategies of acquisition, exchange, and culling carry implications for the welfare of the cattle. The acquisition of "first cow" from the father at a very young age facilitates bonding between child and animal; gradual increase in responsibility under adult supervision prepares a young man (or woman) for future cattle-rearing responsibilities. While exchanges are often couched in social terms, they usually have a sound management rationale. Cattle not only serve as biological, environmental filters for herding families but as a mnemonic for the social relations that enable herders to spread risk in a risky environment. We recognize that subsistence strategies employed in the present cannot act as exact templates for the past. However, patterns perceived during case studies of present-day herding practices may serve as guidelines for interpretation of the behaviour of prehistoric pastoralists under similar environmental conditions.

Note

1. There were slight variations reported by our Kaputiei Maasai informants in the north and the Kisongo in the south in numbers of animals transacted as part of bridewealth. In the south, "official" bridewealth

was reported as always one cow and calf, and one steer. In the north, it was one cow and calf and one heifer; in case the cow died, the heifer would take its place.

References

Barfield, T. 1993. *The nomadic alternative.* Englewood Cliffs, NJ: Prentice Hall.

Bernus, E. 1981. *Touaregs Nigériens. Unité culturelle et diversité régionale d'un peuple pasteur.* Mémoires ORSTOM No. 94. Paris: Editions de l'Office de la Recherche Scientifique et Techniques Outre-Mer.

Blench, R.M. 1994. The expansion and adaptation of Fulße Pastoralism to subhumid and humid conditions in Nigeria. *Cahiers d'Études africaines* **133–5**, 197–212.

Dahl, G. & A. Hjort 1976. *Having herds: pastoral herd growth and household economy.* Stockholm Studies in Social Anthropology, No. 2. Stockholm: University of Stockholm.

Deshler, W.W. 1965. Native cattle keeping in eastern Africa. See Leeds & Vayda (1965), 153–68.

Downs, J.F. & R.B. Ekvall 1965. Animals and social types in the exploitation of the Tibetan Plateau. See Leeds & Vayda (1965), 169–84.

Dupire, M. 1970. *Organisation Sociale des Peul. Études d'ethnographie comparée.* Paris: Librairie Plon.

Dyson-Hudson, N. 1980. Strategies of resource exploitation among East African pastoralists. In *Human ecology in savanna environments,* D. Harris (ed.), 171–84. London: Academic Press.

Dyson-Hudson, R. & N. Dyson-Hudson 1980. Nomadic Pastoralism. *Annual Review of Anthropology* **9**, 15–61.

Evans-Pritchard, E.E. 1940. *The Nuer. A description of the modes of livelihood and political institutions of a Nilotic people.* Oxford: Clarendon Press.

Fratkin, E., K.A. Galvin, E.A. Roth 1994. *African pastoralist systems. An integrated approach.* Boulder: Lynne Rienner.

Galaty, J.G. 1982. Being "Massai"; being "people-of-cattle": ethnic shifters in East Africa. *American Ethnologist* **9**, 1–20.

Galaty, J.G. 1989. Cattle and cognition: aspects of Maasai practical reasoning. In *The walking larder: patterns of domestication, pastoralism, and predation,* J. Clutton-Brock (ed.), 215–30. London: Unwin Hyman.

Galaty, J.G. 1994. Rangeland Tenure and Pastoralism in Africa. See Fratkin et al. (1994), 185–204.

Galaty, J.G. & P. Bonte 1991. *Herders, warriors and traders: pastoralism in Africa.* Boulder: Westview Press.

Galaty, J. & D. Johnson (eds) 1990. *The world of pastoralism: herding systems in comparative perspective.* New York: Guilford Press.

Gifford-Gonzales, D. 1984. Implications of a faunal assemblage from a Pastoral Neolithic site in Kenya: findings and a perspective on research. In *From Hunters to Farmers: the causes and consequences of food production in Africa,* J.D. Clark & S.A. Brandt (eds), 240–51. Berkeley: University of California Press.

Goldsmidt, W. 1977. A general model for pastoral social systems. In *Pastoral production and society/ Production pastoral et société,* Equipe écologie et anthropologie des sociétés pastorales (ed.), 15–27. Cambridge: Cambridge University Press.

Goldschmidt, W. 1986. *The Sebei. A study in adaptation.* New York: Holt, Rinehart & Winston.

Grandin, B.E. 1991. The Maasai: socio-historical context and group ranches. In *Maasai herding: an analysis of the livestock production system of Maasai pastoralism in eastern Kajiado District, Kenya,* S. Bekure, P.N. de Leeuw, B.E. Grandin, P.J.H. Neate (eds), 21–39. ILCA Systems Study 4. Addis Ababa: International Livestock Centre for Africa.

Homewood, K.M. & W.A. Rogers 1991. *Maasailand ecology. Pastoralist development and wildlife conservation in Ngorongoro, Tanzania.* Cambridge: Cambridge University Press.

Hopen, C.E. 1958. *The pastoral Fulbe Family in Gwandu.* London: Oxford University Press for IAI.

Ingold, T. 1980. *Hunters, pastoralists and ranchers. Reindeer economies and their transformations.* Cambridge: Cambridge University Press.

Kinahan, J. 1991. *Pastoral nomads of the central Namib desert.* Windhoek: New Namibia Books.

Lamprey, R. & R. Waller 1990. The Loita-Mara region in historical times: patterns of subsistence, settlement and ecological change. See Robertshaw (1990), 16–35.

Leeds, A. 1965. Reindeer herding and Chukchi social institutions. See Leeds & Vayda (1965), 87–128.

Leeds, A. & A.P. Vayda 1965. *Man, culture, and animals.* Washington: American Association for the Advancement of Science.

Lefébure, C. 1977. Accés aux ressources collectives et structure sociale: l'estivage chez les Ayt Atta (Maroc). In *Pastoral production and society/Production pastoral et société*, Equipe écologie et anthropologie des sociétés pastorales (ed.), 115–26. Cambridge: Cambridge University Press.

Lincoln, B. 1981. *Priests, warriors, and cattle. A study in the ecology of religions.* Berkeley: University of California Press.

Marshall, F. 1990a. Cattle herds and caprine flocks. See Robertshaw (1990), 205–60.

Marshall, F. 1990b. Origins of specialized pastoral production in East Africa. *American Anthropologist* **92**, 873–94.

Marshall, F. 1994. Archaeological perspectives on East African pastoralism. See Fratkin et al. (1994), 17–43.

Massey, G. 1987. *Subsistence and change. Lessons of agropastoralism in Somalia.* Boulder: Westview Press.

Oboler, R.S. 1985. *Women, power, and economic change: the Nandi of Kenya.* Stanford: Stanford University Press.

Oxby, C. 1986. Women and the allocation of herding labour in a pastoral society (Southern Kel Ferwan Twareg, Niger). In *Le Fils et Le Neveu. Jeux et enjeux de la parenté touarégue*, Publié sous la direction de S. Bernus, P. Bonte, L. Brock, H. Claudot, 99–127. Cambridge: Cambridge University Press.

Parkington, J., R. Yates, A. Manhire, D. Halkett 1986. The social impact of pastoralism in the south-western Cape. *Journal of Anthropological Archaeology* **5**, 313–29.

Rigby, P. 1969. *Cattle and kinship among the Gogo.* Ithaca: Cornell University Press.

Rigby, P. 1985. *Persistent pastoralists.* London: ZED Books.

Rigby, P. 1992. *Cattle, capitalism, and class: Ilparakuyo Maasai transformations.* Philadelphia: Temple University Press.

Robertshaw, P. (ed.) 1990. *Early pastoralists of south-western Kenya.* Nairobi: British Institute in eastern Africa.

Robertshaw, P.T. & D.P. Collett 1983. The identification of pastoral peoples in the archaeological record: an example from East Africa. In *Transhumance and pastoralism. World Archaeology* **15**(1), 67–78.

Ryan, K., Karega-Munene, S.M. Kahinju, P.N. Kunoni 1991. Cattle naming: the persistence of a traditional practice in modern Maasailand. In *Animal use and culture change*, P.J. Crabtree & K. Ryan (eds), 90–96. MASCA Research Papers, Supplement to **8**. Philadelphia: MASCA, University of Pennsylvania Museum of Archaeology and Anthropology.

Schoenbrun, D.L. 1993. Cattle herds and banana gardens: the historical geography of the western Great Lakes region, *c.* AD 800–1500. *The African Archaeological Review* **11**, 39–72.

Sindiga, I. 1994. Indigenous (medical) knowledge of the Maasai. *Indigenous Knowledge and Development Monitor* **2**(1), 16–18.

Smith, A.B. 1983. Prehistoric pastoralism in the southwestern Cape, South Africa. In *Transhumance and pastoralism. World Archaeology* **15**(1), 79–89.

Smith, A.B. 1984. Adaptive strategies of prehistoric pastoralism in the south-western Cape. In *Frontiers: southern African archaeology today*, M. Hall, G. Avery, D.M. Avery, M.L. Wilson, A.J.B. Humphreys (eds). Cambridge Monographs in African Archaeology 10. Cambridge: BAR International Series 207.

Smith, A.B. 1992. *Pastoralism in Africa. Origins and development ecology.* London: Hurst.

Spear, T. & R. Waller (eds). 1993. *Being Maasai: ethnicity and identity in East Africa.* London: James Currey.

Spencer, P. 1973. *Nomads in alliance.* London: Oxford University Press.

Spooner, B. 1973. *The cultural ecology of pastoral nomads.* An Addison-Wesley Module in Anthropology, No. 45. Reading, MA: Addison-Wesley.

Stenning, D. 1959. *Savannah nomads*. Oxford: Oxford University Press for International Africa Institute, London.

Swift, J. 1977. Pastoral development in Somalia: herding cooperatives as a strategy against desertification and famine. In *Desertification. Environmental degradation in and around arid lands*, M.H. Glantz (ed.), 275–305. Boulder: Westview Press.

Sylla, D. 1995. Pastoral organizations for uncertain environments. In *Living with uncertainty. New directions in pastoral development in Africa*, I. Scoones (ed.), 134–52. London: Intermediate Technology Publications.

Vayda, A.P. 1965. Anthropologists and ecological problems. See Leeds & Vayda (1965), 1–5.

Vivelo, F.R. 1977. *The herero of western Botswana*. St Paul: West Publishing.

CHAPTER TWENTY-SIX

The history of working animals in Africa

Paul Starkey

1. Introduction

The employment of domestic animals for tillage or transport is known as animal traction. The term is generally understood to include pack transport as well as the "pulling" work of animals. In various parts of the world cattle, buffaloes, yaks, horses, donkeys, mules, camels, llamas, elephants, reindeer, goats and dogs are used for transport, crop cultivation, water-raising, milling, logging and land excavation or levelling.

Cattle are the major work animals worldwide. It is most common to use male animals because they are stronger than females and cattle herds always produce a surplus of males. Castrated animals are more docile than intact males. Thus, the most common working cattle are castrated bulls, known as oxen or bullocks. In some texts, the word oxen has been used to describe any working cattle. Since working cattle are generally castrated males, the two uses of the word generally overlap. Nevertheless, confusion can occur in regions where cows or bulls are used for work. In this text, an ox is a castrated bull of any breed that is used for work. The term bovid encompasses the animals that are closely related to cattle, including water buffaloes and yaks.

After cattle, the main work animals worldwide are horses, donkeys (asses) and mules, known collectively as equids. In current English, a domestic ass is generally called a donkey. The word ass is now mainly confined to archaic, zoological or colloquial writings (and the statistics published by the Food and Agriculture Organisation FAO). Mules are non-breeding hybrid animals formed by crossing a female horse with a male donkey. They are stronger than donkeys and more hardy than horses. The other possible cross (female donkey and male horse) is known as a hinney. These are much less common, partly because the cross is biologically much more difficult to produce (Blench Ch. 21 in this volume).

This chapter explores the use and spread of animal traction in Africa from both chronological and geographical perspectives. However, since several different processes occurred in the continent, time sequences and country patterns will inevitably overlap. The study concludes by describing the present situation and current trends.

As far as possible, references relate to information that is readily accessible: these are often review studies rather than primary sources.

2. Origins and expansion

The recorded history of animal power in Africa starts in Egypt with the first drawings of oxen and ard plows occurring in the third Dynasty [2778–2723 BC] (Haudricourt & Delamarre 1955). These, together with the engravings of oxen and plows in early Mesopotamian civilizations, appear to constitute some of the earliest records of animal traction anywhere in the world. It is possible that the *maresha* animal-drawn plow was spreading in Ethiopia at the same time that animal traction was developing in ancient Egypt, but there are few artistic records from this period in Ethiopian history (but see Phillipson 1993).

The early Egyptian ard plows were clearly illustrated in wall paintings and on papyrus. Further evidence comes from the intact plows that have been found in some tombs, and also from some detailed models of plowing teams. The plows comprised a long wooden beam that pulled a horizontal wooden plow body fitted with a metal share. The plow was controlled by two handles. The ard plows widely used in Egypt to this day are not dissimilar to the ancient designs. The early drawings and models show animals plowing in yoked pairs. Some early illustrations suggest that the yokes were tied to the horns (head–horn yokes). Other illustrations and models suggest the use of withers yokes similar to those used in present-day Egypt.

Several ancient Egyptian illustrations clearly show that the animals used for plowing were cows. One possible historical and cultural explanation is that the ox was considered more sacred than the cow. In most parts of the world, oxen are the first bovids to be used for work. Cows only start to be employed when smallholder farming becomes intensive, when animal feed resources are limited and when the work operations are light or highly seasonal. These conditions may well have developed early in Egypt, particularly as equids were available for certain transport tasks (year-round work). In present-day Egypt, almost all animals used for plowing are cows.

Some ancient Egyptian illustrations show cattle pulling sledges. In the Papyrus of Ani (about 3,300 years old), oxen are seen pulling the funeral sledge (Rossiter 1984). Wheeled ox-carts (as opposed to horse-drawn chariots) do not appear to have been common in Ancient Egypt. Drawings and models of ox carts in nearby Mesopotamian civilisations date back about 5,000 years ago (Haudricourt & Delamarre 1955).

Domestic donkeys were recorded at Maadi in Egypt 3500–4000 bp (Midant-Reynes 1992). Paintings of pack donkeys appear in Egyptian tombs about 5,000 years ago (Clutton-Brock 1992). The main use of donkeys in Ancient Egypt appears to have been for pack transport. Biblical evidence suggests donkeys were also important in the region for riding. Donkeys were not only maintained for work, as Cleopatra is said to have bathed in donkeys' milk.

The employment of horses in Egypt appears to have followed the use of donkeys. They do not appear until the about the thirteenth Dynasty [*c.* 1800–1750 BC].

Many Egyptian illustrations, 3000–3500 bp, show horses hitched in pairs for pulling two-wheeled chariots. A ceremonial chariot with lightweight spoked wheels was found in the tomb of Tutankhamun (1342 BC). Drawings dating from the same era show animals that appear to be mules or hinnies pulling two-wheel carts or chariots. However, there is no evidence that donkeys were used to pull carts in Ancient Egypt (Clutton-Brock 1992).

Water buffaloes were probably introduced into Egypt around 1,300 years ago (FAO 1977). They have thrived and their numbers have grown so that their population in Egypt now equals that of cattle. Buffaloes are used for some plowing and water raising, but cows (with their superior tolerance of heat) remain the main plowing animals in Egypt. Water buffaloes have not spread in significant numbers from Egypt to other parts of Africa.

3. Animal traction in North Africa

Ethnographic evidence suggests that animal traction technology spread from ancient Egypt southwards into Sudan. The technology also spread around the Mediterranean, and it was probably widespread along the coast of North Africa by 500 BC (Laoust 1918, 1930). Animal traction has been an integral part of farming and transportation systems in North Africa for over 2,000 years, and a wide range of species and technologies are now used. Camels are herded for meat, used for riding and pack transport and may assist with tillage operations or irrigation equipment. Horses are owned for riding, recreation and pulling carts, and some transport horses also assist with cultivation. Mules are mainly used for pulling carts and wagons in both urban and rural areas. They may also assist with soil tillage. Donkeys are mainly used as pack animals, and for riding. Donkey carts are found in urban and rural areas, and donkeys may also be expected to perform some soil cultivation work. While motorized transport has been increasing rapidly in the twentieth century, the use of animal power for local transport has not experienced a proportionate decline. However, long-distance transport of goods and humans using animals has been largely replaced by cars and lorries.

For the cultivation of heavy soils, notably in the irrigated fields along the Nile, cows remain the main type of working animal. Although ox-carts were used in North Africa in historic times, they are very rarely seen today. Animal-drawn carts in North Africa are almost invariably pulled by equids. Prior to the introduction of tractors in Morocco, some oxen were used for plowing. However, nowadays there is little economic justification for maintaining oxen that walk quite slowly and are relatively expensive to own. Plowing is now either performed by tractors or by other types of animals that have more uses during the rest of the year (horses, donkeys, mules, camels and cows).

In the Maghreb region (Morocco, Algeria, Tunisia), pairs of animals tilling the land are almost invariably harnessed with an unusual belly-yoke system characteristic of this region. The belly yoke is a wooden bar suspended from straps around the withers (the beginning of the back) and which hangs under the chests of the two animals without touching them. The tillage implement is attached to the yoke. Historically, farmers plowed with long-beamed wooden ards, and these are still very widely used in the Maghreb. Short-beamed metal plows and cultivators have been slowly spreading during the twentieth century. Since the yoke is suspended a set height above the ground, it can be used to harness animals of different sizes and types. The same system is used to plow with two mules, two donkeys, a horse with a donkey, a cow with a donkey, a camel with a mule, or other unusual combinations.

Elephants were employed for military purposes in North Africa about 2,300 years ago. They were used for ceremonial purposes prior to this, but their adoption as war animals around the Mediterranean followed the Asian battles of Alexander the Great. Hannibal of Carthage (in what is now Tunisia) remains world-famous for his attempts to deploy elephants against Rome. However, elephants were neither easy to manage during battles nor were they invincible. They were only deployed in military campaigns around the Mediterranean for about 300 years (Delort 1992) and thereafter reverted to a minor display role.

4. Ethiopia and the Horn of Africa

The single-handled Ethiopian scratch plow is very different in design from the two-handled plow used in Egypt. The *maresha* is more like a spear, pulled through the soil using a long beam. Goe (1987) reviewed several theories concerning its origins or introduction. Stiehler (1948) suggests the ard plow was introduced 2,600–3,000 years ago by Semitic-speaking peoples invading from South Arabia. Another view is that the plow was already in use at this time, having spread from Cushitic-speaking peoples of Nubia in northeast Sudan (Simoons 1965). Linguistic evidence suggests that the ard was in use "several millennia" before the South Arabian invasion, which might make the Ethiopian plow the oldest in Africa (Ehret 1979). Drew (1954) illustrates rock paintings from Eritrea that clearly show a *maresha* plow, but their dating is uncertain.

The Ethiopian *maresha* is not only old, it is highly persistent. To the present day, it is almost universally used by smallholder farmers for the cultivation of the tef grain crop. Although a variety of development programmes has attempted to introduce short-beamed steel plows for the past fifty years, there has been almost no adoption of these. The plowing animals are generally oxen, yoked in pairs with withers yokes, and controlled by a single person. Where oxen are in short supply, horses, donkeys or cows may be used, but oxen are the work animals of choice. Camels are occasionally used for cultivation. Ox-carts do not appear to have been part of Ethiopian traditional systems, and they remain extremely uncommon.

Transport of goods in Ethiopia has long been based on pack donkeys. Little is known about when donkeys started to be used in Ethiopia, and when they became common. Under the Pharaoh Pepi II (*c.* 2270 BC) caravans with pack donkeys were trading with Punt (Ethiopia) (Kitchen 1993). Donkeys have long been important in the history of the salt trade in northern Ethiopia (Wilson 1976, 1991) and are represented in traditional Ethiopian art. The population of donkeys has been rising in recent history. With about five million donkeys, Ethiopia now has the second largest population of donkeys in the world. Pack donkeys are extremely important in both rural and urban economies. The success of military campaigns in Ethiopia and Eritrea in the late twentieth century owed much to the use of pack donkeys. Despite the large numbers of pack animals employed in Ethiopia, transport of loads by humans (mainly women) is still common.

Horses and mules are mainly used for riding. Simple passenger-carrying, two-wheel, horse-drawn carts became common in Ethiopian cities around the middle of the twentieth century. They were banned by the authorities from central Addis Ababa around 1963, but remain common in other towns. They are almost invariably used as passenger taxis for hire, and there is negligible use of horse carts for freight purposes.

Donkeys have not been traditionally used for pulling carts. However, an innovative design of low-cost donkey cart started to be seen in the Rift Valley of Ethiopia in the 1970s. In the rest of the world, most carts pulled by one donkey have parallel shafts that pass on either side of the animal. The donkey pulls from a collar or breast band, and the weight is taken by a band between the two shafts that passes over a back saddle. In contrast, the Ethiopian carts are pulled from converging shafts attached to a simple pack saddle. The carts made from wooden poles, with steel wheels, appear of recent, indigenous design, and have evolved in a country where donkeys have always been used to carry on their backs rather than pull from harnesses. At the end of the twentieth century, these carts have been spreading rapidly in the Rift Valley where they are used for the transport of water, straw and other materials. The development and rapid spread of these carts has been within the informal sector. It contrasts with the low uptake of the more expensive steel carts promoted by government agencies in the 1980s and 1990s.

5. Stationary applications of animal power

In Egypt, there has been a long history of using work animals to raise water for irrigation. The ingenious *sakia* irrigation wheels appear to have been developed during the Ptolemaic period, about 2,200 years ago (Stead 1986). The traditional *sakia* wheels have internal spirals, allowing them to efficiently raise water that is within two metres of the surface (Löwe 1986). They remain in use in present-day Egypt and may be turned by cows, buffaloes or donkeys or less commonly, horses or camels. Some *sakias* are found in Sudan (Nicholls 1918).

Animals may also be used to pull water from wells. In North Africa, mote systems are employed, where an animal walks down a slope and pulls on the rope attached to a leather water bag (Löwe 1986). Some motes have self-emptying systems. Descending the slope makes it easier for the animal to raise the water. All types of work animals may be used. Elsewhere, notably in circum-Saharan Africa and the Horn of Africa, animals are also used to pull water from simple wells. Such systems are most common in pastoral areas, where large numbers of animals must be watered at the same time.

6. Threshing

Lagercrantz (1950) has reviewed the use of animals in Egypt, North Africa and Ethiopia and Northern Somalia for threshing. In this operation, the animals walk round in circles over beans or cereals, separating the husks from the grain. There is a strong geographical distinction between systems using a central tethering post and those which simply make use of random trampling movements. In Egypt, a special threshing sledge, *nōrag*, may be pulled by the animals to accelerate the process. Pigs were used for work in Ancient Egypt, both treading and threshing seed in the eighteenth Dynasty, a practice also confirmed by Herodotus (Zeuner 1963:262). The first iconographic evidence for animal threshing is in the Old Kingdom (i.e. prior to 2300 BC) and this use of animal power has continued in Egypt up to the present.

Threshing is a seasonal operation and the species used are those that are readily available because they are maintained for other work. Iconographic evidence from Egypt suggests that donkeys were used in the Old Kingdom and were supplanted by oxen in the New Kingdom. Exceptionally, camels are used to thresh grain in Tunisia. In the Ethiopian highlands, oxen are used to thresh the cereal, *teff*. In the Ethiopian Rift Valley, a similar technology has more recently been adopted for decorticating maize.

Animal threshing occurs in both the Canaries and the Azores, probably as a result of European introductions. It was also brought to South Africa by European settlers, and Lagercrantz (1950:23) reproduces an early eighteenth century engraving of horse-threshing in the Cape Colony. Through the agency of missions, animal threshing spread to Namibia and to the Sotho-speaking areas.

7. Milling

Animal power is used for milling in a band stretching from Somalia to Chad. Oilseeds such as sesame or groundnuts are placed in a large wooden pestle, carved out of the trunk of a large tree. The animal walks around pulling a counter-balanced frame attached to a large wooden mortar. This grinds the seeds, extracting the oil. The animals employed are often oxen but camels may be used in Sudan and Somalia. This grinding technology is pre-colonial but its exact origin is unknown. The pattern of transfer of this animal-powered grinding technology has not been investigated. Similar mills are found in the Seychelles and on the Indian subcontinent, but they have not spread elsewhere in sub-Saharan Africa.

8. Animal power for riding and pack transport in sub-Saharan Africa

Horses, donkeys, camels and cattle have been used for riding and pack transport in parts of sub-Saharan Africa for centuries, if not millennia. However, relatively little is known about the history of these applications of animal power. Certain pastoralist groups in the continent, including several in West Africa, ride cattle and use them as pack animals. There are historical observations of fifteenth-century-European seafarers concerning the Khoi-Khoi of South Africa. These rode cattle and used them for pack transport. Some cattle were trained for use in battles (Burman 1988, Joubert 1995).

Equid remains have been found in archaeological sites in West Africa that date back to between 2,000 and 1,500 years ago (Blench Ch. 21 in this volume, MacDonald & MacDonald Ch. 8 in this volume). Several authors, including Muzzolini (Ch. 6 in this volume), have noted that references to donkeys in traditional art (including rock-paintings) and literature are surprisingly few in sub-Saharan Africa (including Ethiopian and Arabic sources). This might be because the spread of donkeys was slow and scattered, or it could be because donkeys had low status, compared with horses and camels. Donkeys were used mainly as pack animals (as they are worldwide) although there is little evidence to support this. Certainly, by the time of recorded European exploration, donkeys were used as pack animals in parts of West Africa (Sahelian zone) and East Africa (in some coastal ports and among the Maasai). A Portuguese report of 1758 suggests that the Shona in Zimbabwe were using pack donkeys, and this may have been associated with the gold trade route to and from Sofala in Mozambique (Ellert 1993).

Horses became important for riding and prestige in West African civilizations across a wide zone of Muslim-influenced cultures from Senegambia to Sudan. Their high social status meant that they were seldom used for transporting goods (Law 1980a). In eastern and southern Africa, horses were introduced from ports in the past five hundred years by European settlers and traders.

Camels, used for riding, transporting and meat–milk production spread in circum-Saharan countries between 1,000 and 3,000 years ago, first in the Horn of Africa and later in West Africa (Banti 1993, Blench 1995, and Ch. 21 in this volume, MacDonald & MacDonald Ch. 8 in this volume, Muzzolini Ch. 6 in this volume). They are shown extensively in late period Saharan rock-paintings, but only as riding animals. Camels have probably been used to pull water from wells since their introduction, but their use for ploughing in West Africa is recent (RIM ii 1992).

9. Animal power for cultivation and wheeled transport in sub-Saharan Africa

Ethiopia, together with a few neighbouring parts of the Horn of Africa, is exceptional in sub-Saharan Africa, since farmers have been using animal power for tillage for thousands of years. However, in most sub-Saharan African countries, animal traction for tillage and wheeled transport was introduced during the colonial period. The process of introduction and adaptation is still continuing.

There are various factors that may be responsible for the late adoption of ploughs in sub-Saharan Africa. In much of the continent, different tribal groups have specialized in animal-rearing and in crop production. Thus many crop-growing farmers did not own potential work animals. Moreover, many traditional farming systems have been based on bush-fallow rotations. The bush is cut down and burned, and seeds or tubers planted in the cleared area. There is no need to till the land with a plough. In any case this would be difficult since the soil is full of roots. Seeds can be scattered or planted in small pockets, for which a simple digging implement is appropriate. In farming systems with long periods of bush fallow, weeds do not present major problems. Provided the fallow periods are long, such systems can be quite productive in terms of yield per unit of human labour. It is only when human population pressures necessitate short fallow periods, that it becomes justified to clear the land of roots and stumps and to plough. Thus, in much of sub-Saharan Africa, the necessary social, environmental and agricultural conditions to favour the use of ploughs have not really existed. Indeed, there are still parts of Africa where the plough is not really economically justified. The failure of animal traction to spread into some semi-humid areas in recent decades, is partly explained by the lack of the appropriate preconditions (Starkey 1986a, 1992, Pingali et al. 1987).

Another important constraint on the spread of the plough in pre-colonial times was the presence of tsetse flies and trypanosomiasis in virtually all lowland areas. The relatively low human populations that obtained almost everywhere in sub-Saharan Africa meant that hunting pressure on wild animal vectors was insufficient to eliminate reservoirs of trypanosomiases. Pastoral cattle that can survive when well-fed or moved regularly by expert herders have a much accelerated death rate from disease when subjected to work-stress (Blench 1997). It is possible that both wheeled vehicles and ploughs were introduced experimentally in prehistory, but failed owing to disease constraints. Increased human population in the colonial era following improved health-care, both allowed major clearance of regions of bush and eliminated large populations of tsetse vectors. This helped to make animal traction a viable proposition in many areas.

10. African ports and islands

Traders and colonial powers had contact with Africa's offshore islands and ports, before the hinterland was colonized. In most countries, the use of animal-powered wheeled transport was first introduced in coastal and river ports in the seventeenth, eighteenth or nineteenth centuries (Law 1980b). In a few cases where social, economic and ecological conditions proved favourable, the use of animal-powered transport gradually spread from the coastal region, through the activities of traders, settlers, missionaries and the administering authorities. Animal-drawn cart technology spread inland in South Africa (and neighbouring territories), French West Africa (from Saint Louis) and in East Africa. However, with the notable exception of South Africa (and nearby countries), the introduction of animal power for agriculture was largely a twentieth century phenomenon.

When the Dutch settled in the Cape in 1652, they bartered goods for oxen and used these to pull carts. Ox carts transported goods to and from ships, and building materials for the new settlement. Horses, mules and donkeys were imported within a few years and were also used for transporting people and goods. Although there are reports of settler farmers ploughing with oxen as early as 1657, crop production was not a major activity of the early colonists. Hunting, trading and stock-farming were more important. For these activities, the colonialists built four-wheel wooden wagons, pulled by large teams of oxen (10 was quite normal and 16 were sometimes used). The semi-nomadic *trekboers* travelled widely in *kakebeen* wagons into the hinterland of the Cape in the seventeen and eighteenth centuries. They were followed by traders, missionaries and settlers, who also used ox wagons. In the nineteenth century, there were some major treks of settlers, including the Great Trek from 1836 to 1852, from the Cape Colony through the Orange Free State to what became known as Transvaal and Natal. In 1874, some settlers took part in the long Dorsland Trek passing into what is now known as Botswana, Namibia and Angola (Burman 1988, Joubert 1995).

Behind the trekking settlers in southern Africa, an infrastructure based on animal-powered transport was developed by miners, traders, missionaries and the administrative and military authorities. Horses were used for riding and military purposes. Prior to the development of railways, freight was generally moved by ox wagon. It has been estimated that in the first ten years of its development 18,500 ox-drawn wagons entered Kimberley carrying supplies for the mines and its peoples (Joubert 1995). In the nineteenth century, inter-town coach services were operated using horses and mules. One exceptional service in northern Transvaal used a mixed team of zebras and mules to pull its coach. Donkeys worked in the mines, and donkey carts were common for local transport. Following the outbreak of rinderpest in 1896, donkeys became important for a time for long-distance transport. Caravans of wagons, each pulled by 16–20 donkeys carried produce and goods in what is now Zimbabwe.

At the end of the nineteenth century, and in the early years of the twentieth century, animal power played a crucial role in the agricultural, mining and transport sectors of the growing economy of South Africa. Animals were used for urban and rural transport, they worked in the mines and they ploughed for both large-scale (white) and small-scale farmers.

During the first half of the eighteenth century, horses were imported into Freetown (Sierra Leone) for riding, racing and wheeled transport (Dorward & Payne 1975). In 1811, a Horse Tax was imposed, and modified to a Horse and Carriage Tax in 1828. However, an outbreak of a disease (thought to have been trypanosomiasis) in 1856, severely restricted the subsequent use of horses. Although ox carts never fully replaced horses, they were used by the colonial authorities for port transport and some refuse collection (Starkey 1981).

Horses had been used for riding in the semi-arid zone of West Africa for centuries before the European colonists arrived. The French established an important base in the port of St Louis (Senegal) and used horses and donkeys for transport in the port and town. Horses were also used by the military for expeditions into the hinterland, both for riding and for pulling artillery. However, horses did not thrive in the humid zone, and attempts to use horses for transport around Conakry (Guinea) had little success (Bigot 1989).

Madagascar has long been influenced by Asian cultures, where animal power has been long-established. One report (van Nhieu 1982) suggested that animal traction was introduced into Mantasoa region during the regency of Queen Ranavalona 1 (1828–61). This is broadly in line with the period of 1850–80 mentioned by Bigot (1985). It is not clear whether animal power was used before this time. The animal-powered soil tillage implements in use in Madagascar today are steel ploughs, based on European designs of the industrial era. Had there been a long-standing use of animal power for soil tillage prior to the nineteenth century, one would expect to see some evidence of traditional long-beamed ploughs, such as those still widely used in North Africa, Ethiopia, the Middle East and South Asia. Although such implements do not appear to have been used, animal power was used for soil preparation in the rice fields. In the traditional *piétinage* system, cattle are made to walk round and round in rice fields, thereby creating a puddled soil suitable for rice cultivation. Wooden ox carts with large, spoked wheels became important for trade within the island in the nineteenth century, and such carts are still widely used today. The technology for making wooden cart wheels was also established in Zanzibar and Pemba islands, although it does not seem to have transferred to the East African mainland.

Lamu island, off the Kenyan mainland, was for a long time an important Arab trading post, and pack donkeys were widely used for local transport. In the nineteenth century, European traders, explorers, settlers, military expeditions and missionaries set off to explore the mainland from a variety of bases on the east coast including Lamu, Pemba, Zanzibar, Tanga and Beira. Animal-drawn carts (generally ox-carts) were used for establishing forts, trading posts, settlements and mission stations. For example, staff of the London Missionary Society travelled through Tanzania by ox wagon in 1876–7 (Koponen 1994, cited by Sosovele 1997). Roads were constructed suitable for animal-drawn carts and wagons. By 1901, there were 1,800 km of road suitable for ox wagons in and around Zambia (Müller 1986). About the same time, ox wagons were moving from the coast (Tanga) to Arusha, and as far as Uganda (Sosovele 1997). It seems probable that there were individual examples of early European traders, settlers and missionaries using animals for crop cultivation in East and Central Africa in the nineteenth century. However, the impact of such examples was small, for by the beginning of the twentieth century, animal traction was still largely absent from the agricultural systems in these countries.

11. Sledges

The origin and spread of the animal-pulled sledge in not clear. Simple triangular sledges are now very widely used throughout eastern and southern Africa and also in Madagascar. However, they are seldom, if ever, seen in West Africa, where they should be similarly useful as basic and cheap means of transporting materials. There is a wide range of sledge types and this suggests there has been considerable farmer innovation in the design of sledges. Three main types have been identified in

Zambia, the dug-out log (*umulangu*), and the flatter carved board (*mula*) and the simple Y-branch (Müller 1986). All of these, but particularly the Y-branch, may be modified with superstructures made from poles and/or basket work. There is some uncertainty as to whether or not the *umulangu* was in use in Zambia prior to the arrival of the traders and missionaries in the nineteenth century (Müller 1986). However, despite the different sledge designs, they are almost always pulled in a similar way. Oxen are used in pairs (one or more pairs) to pull the sledges by means of a steel chain attached to a standard withers yoke. This suggests that the present system of pulling sledges has spread in post-colonial times.

While considering sledges, it is interesting to note that there are several independent twentieth century examples of sledges being developed into simple carts, using wheels made from tree-trunks. One such example in northeastern Zimbabwe appears to have been a farmer response to the banning of sledges by the colonial authorities (who feared that sledges accelerated erosion). Four-wheel, articulating carts were developed from Y-branch sledges using cross-sections of tree-trunks and simple wooden axles. Similar innovations have been seen in the Mbeya region of Tanzania (Starkey & Mutagubya 1992).

12. Colonial promotion of animal traction in agriculture

During the first half of the twentieth century, the colonial authorities or agricultural production companies in several parts of sub-Saharan Africa, attempted to introduce animal traction for cultivation. The aim was to increase agricultural production in the colonies by teaching the indigenous population how to use work animals for ploughing.

One of the early schemes took place in 1900 in the West African country of Togo, then under German control. In the hopes of increasing cotton production, a team of black American experts from Alabama were hired by the Berlin Colonial Economic Committee to introduce animal power for cultivation. Further attempts were made in Togo in 1908 (at Mango) and 1913 (at Tabligbo). Although the idea of animal traction was not totally rejected, there was little adoption in Togo at that time. Even with further attempts at introduction in the 1950s, there were probably fewer than 1,000 ploughs in use at the time of independence in 1960 (Westneat et al. 1988).

The French colonial authorities in Guinea discussed the idea of using animal power in local farms in 1910 (Bigot 1989). One aim was to increase agricultural production: at that time Guinea exported both groundnuts and rice. A pilot farm, *ferme indigène*, was established in 1914 in the Niger valley near Kankan. Indigenous N'dama oxen were trained and found to work well. By 1918, there were twelve farms, locally known as *fermes de chefs*. In 1919, a total of 105 ha on these farms were ploughed with animals, and several crops including groundnuts and rice were grown successfully using animal power. The pilot farm approach was changed to that of a major agricultural campaign to encourage smallholder farmers to grow crops using animal power. The campaign, which was started in 1924, was supported by adaptive research, including on-farm evaluation of a variety of implements. A wide range of support services was made available, including training schools, provision of equipment, credit and marketing. Ploughing competitions were held. The number of farmers using animal power increased rapidly and, by 1929, 15,000 oxen had been trained and 24,000 ha were cultivated with animal power. The area ploughed per team of animals increased in Haute Guinée from 3.55 ha to 5.05 ha. Guinea led the way in animal power in West Africa. In 1931, there were reported to be 5,700 ox ploughs in Guinea, compared with 2,000 ploughs and cultivators in the whole of the rest of Francophone West Africa, including what is now Benin, Burkina Faso, Côte d'Ivoire, Mali, Mauritania, Niger and Senegal. The very rapid growth in animal traction slowed in the 1930s as the policies and practices of the support services (notably credit and equipment provision) varied, and certain ecological limits to animal traction were noted (Bigot 1989).

The humpless taurine N'dama oxen in Guinea were worked in pairs with simple yokes tied to the horns. In France, both head–horn yokes and withers yokes are used in different areas, and it may be that the officers responsible for the initial training of oxen came from a region in France where head–horn yokes were common. The fact that the animals were humpless should not have affected the choice of yoke, since European animals were also humpless. These were worked with withers yokes in parts of France and most of northern Europe and with horn–head yokes elsewhere in France and in Spain and Portugal. Whatever the original reason for the choice, horn–head yokes subsequently became standard for use with N'dama cattle in the countries surrounding Guinea (Côte d'Ivoire, Guinea Bissau, The Gambia, southern Mali, southern Senegal and Sierra Leone). Elsewhere in the region, Zebu cattle were worked with withers yokes. Withers yokes are now the norm throughout Africa, except in those countries around Guinea where to this day the taurine cattle continue to be worked with horn–head yokes (Starkey 1981, 1986b, 1991).

Information concerning the success of the animal traction scheme in Guinea reached neighbouring Sierra Leone, then under British administration. In 1927, the British authorities banned a traditional form of domestic slavery practised by the Mandinka people. The Mandinka elders asked how they could cultivate their rice fields without their traditional labour supply. The Sierra Leone Department of Agriculture, based at Njala, had already experienced problems with the first few tractors introduced, and so suggested the use of animal power. This was a new and innovative technology in the farming systems of Sierra Leone, where human labour was the only power source. Three Sierra Leonean agricultural instructors were sent to Kankan in Guinea in 1928 to learn how to work with N'dama oxen. They returned and taught people to train animals and to plough. The Director of Agriculture "had great confidence the ox-plough had come to stay". In 1930, "all operations in connection with ox ploughing succeeded almost beyond expectations" and animal power training centres were established in four locations (Starkey 1981, 1982). The technology was adopted and slowly spread. Initially, French-manufactured ploughs and harrows were purchased from Guinea, but later Ransome ploughs from Britain were imported. During animal power surveys in the 1980s, the first introduction of animals for ploughing in 1928 was still fresh in the minds of the village elders. Moreover, some of the words still used in the training of the animals, appear to have derived from the original French-language training (e.g. "*Allez!*" for "Move!"). One of the original agricultural instructors sent to Guinea reported that his great-grandchildren were ploughing his farm using oxen (Starkey 1982).

Animal traction did not spread as fast in Sierra Leone as it did in Guinea. There were two main reasons for this. The area of introduction in Sierra Leone was on the southern margin of the main cattle zone in West Africa, where cattle were few and ecological conditions were not particularly favourable for the use of work oxen. Secondly, work oxen were being introduced for basic food production, and not for an export-orientated cash crop. Thus, there was little financial incentive on the part of the authorities or the private sector to provide the same level of back-up services (produce marketing, credit, supply of equipment, training, animal health care) that proved important in establishing animal traction in Guinea. Both the contemporary reports of the Department of Agriculture and interviews with farmers suggested that shortage of equipment and lack of a support programme restricted the speed of adoption. Although animal traction had been firmly established in Sierra Leone since 1928, its expansion was slow until there was further formal promotion. Promotional schemes in the 1950s and again in the 1980s led to further adoption, encouraged by support services such as equipment provision, credit and training (Starkey 1981, 1994a, Bangura 1990).

The Gambia provides a later example of formal colonial extension combined with informal farmer innovation and diffusion. Prior to 1955, there was virtually no use of draft animals for crop cultivation in The Gambia. The colonial authorities introduced a very structured extension programme based on ox ploughing schools (subsequently known as Mixed Farming or District Extension Centres). By 1960, there were 13 schools with 104 trainees, and by 1965 (the year of Independence), there were 24 schools, with 377 people receiving training. The programme continued into the post-independence era, so that between 1955 and 1975, animal traction was successfully introduced into most Gambian villages. However, while the extension services were only promoting the use of yoked pairs of oxen, an alternative draft animal technology based on single-harnessed equines (donkeys and horses) was diffusing informally from Senegal. By 1988, more donkeys than oxen were being used in The Gambia. For cultivation of sandy soils, low-power implements (scarifying tines and seeders) were being used more than the higher-draft ridgers and ploughs first promoted by the extension services. Donkeys were inexpensive animals that appeared to offer faster and more timely cultivation. The Gambia therefore illustrates the importance of two separate processes in the introduction of animal power. The formal introduction programme was deemed necessary since animal traction had not arisen spontaneously nor did it appear to have diffused in any significant way from Senegal or Guinea prior to 1955. There was also the major change in technology from oxen to donkeys that diffused at least as rapidly without any official promotion. This change involved very different systems of harnessing, implements and husbandry. Through the two processes of formal extension and informal diffusion, animal traction became a normal part of farming systems in

The Gambia in the period of about one generation (Mettrick 1978, Starkey 1986b, 1988, Haswell 1991).

In Uganda, colonial farmers and the British authorities introduced ox ploughs for cotton production in the Teso District (east) at the beginning of the twentieth century. Uptake was rapid, with a favourable combination of training centres, available animals and implements and a cash crop. A variety of implements were tested, including wooden implements from India. Eventually, the British Ransome plough was considered the most suitable. Following the success in Teso, the technology spread in neighbouring Lango district, where there were 68 ploughs in use in 1930. This increased to about 2,500 in 1940, 10,000 in 1950 and 15,000 in 1960 (Kinsey 1984).

In 1903, European farmers and traders started to settle in the Machakos District of Kenya. They used heavy ploughs that required teams of six animals. There was no formal promotion of animal traction, but ploughs were available from trading stores. Some local Kamba farmers apparently started using ox plows in 1910. By 1912, the District Commissioner had noted an increase in farm size and cash-crop production associated with the innovation. By 1933, there were 600 ploughs in use. The lighter Ransome Victory plough became available in the 1940s, and became the most popular implement. By the late 1950s, almost all farmers in the District were making use of animal power, through ownership or hire. This high rate of adoption had taken place as a result of private-sector sources of equipment and without any formal extension or credit programmes (Tiffen et al. 1994).

In Tanzania, some settler farmers around Arusha used oxen for ploughing around the beginning of the twentieth century, but the need was not considered great since human labour was readily available. The German colonial authorities attempted to introduce ox ploughs in 1910, with the aim of increasing food exports from the colony. Ploughs and harrows were sent to all District Offices, but there was little uptake. Reasons for the lack of adoption of the plough included poor and limited promotion and lack of suitable animals in certain areas. In the 1920s, the British colonial authorities also attempted to introduce ox ploughs imported from Britain. Uptake was very localized, with most interest in Sukumaland where cotton was grown. In the 1930s, gold production in Mbeya District (southwest) stimulated a local market for food crops, which in turn stimulated interest in the use of ox ploughs for rice production. By 1945, there were about 700 ploughs in use in that region. In Tamine District, it is said that legislation requiring people to grow maize stimulated interest in ox ploughs in the 1940s. Through the middle decades of the twentieth century, animal traction spread slowly in several separate areas of Tanzania mainly through informal diffusion rather than active promotion. The flow of information was assisted by migrant workers, who would see animal power in one area and try it in their own villages. Adoption was assisted by the widespread availability of ploughs which were commercially marketed by traders of Asian origin (Sosovele 1991, 1997, Koponen 1994 cited by Sosovele 1997).

Similar localized schemes to promote animal traction were initiated by the colonial authorities in many sub-Saharan African countries between 1900 and 1960. The success tended to be very localized, and associated with semi-arid areas where there was a clear market for produce. By the time of independence, there were still some countries and many administrative regions in which animal traction was very rare.

13. Post-independence promotion of animal traction

Most sub-Saharan African countries achieved independence from the colonial authorities in the decade 1957–66 (1974–5 for the Portuguese colonies). In some countries, it was assumed that the rapid tractorization seen in Europe in the previous decades would follow swiftly in Africa. For this reason, many Ministries of Agriculture placed initial emphasis on tractorization. Following the failure of most smallholder tractor hire schemes of the 1960s and 1970s, together with the increases in the price of oil in the 1980s, animal traction became a major feature of development strategies in many countries in sub-Saharan Africa. Donor-supported development projects promoting animal traction became increasingly common from the mid 1970s to late 1980s. However, in a few areas, animal traction spread rapidly in the immediate post-independence period.

Senegal provides one of the most dramatic increases in the period immediately after independence in 1960. In Senegal, horses and donkeys had been used in the ports and towns since the nineteenth century. In the early twentieth century, some of the transport horses were brought to work on pilot farms, pulling tine cultivators (*houes*) rather than ploughs. During the 1920s, the agricultural authorities concentrated on promoting the use of horses and tine cultivators to assist in the production of groundnuts as an export crop in the groundnut basin of Thiès. By 1930, there were 770 cultivators in use, compared with only 50 ploughs, 20 seeders and 10 agricultural carts (Bigot 1989). In the 1950s, the agricultural authorities carried out much research on the use of animal power, testing many seeders suitable for use with groundnuts (Havard 1986). The timeliness of planting after the first rains was critical, and horses could pull seeders rapidly to ensure maximal benefits. However, horses were common only in the north of the country, and were not readily available for farm work in the central Siné Saloum area nor in southern Casamance. There was, therefore, increasing research emphasis on the potential for using oxen, which were rarely worked in Senegal prior to the 1950s.

The *Programme Agricole* was launched in 1958 (the year Senegal achieved internal self-government). Several institutions were created to support farmers through credit, training, provision of animal traction implements and other farm inputs and the marketing of crop products. The programme ran until 1980, and led to the rapid adoption of animal-traction technology. This is well illustrated by the numbers of implements acquired during this time. The numbers of animal-drawn tine cultivators increased from about 3,000 in 1958 to 340,000 in 1980, a 10,000 per cent increase. The numbers of seeders increased from 40,000 in 1958 to 310,000 in 1980. Carts (horse-, ox- and donkey-drawn) increased from 5,000 to 140,000. Ploughs increased from 1,000 to 65,000 (Havard & Faye 1988). To help meet the demand for animal traction equipment a local factory was established by a group of French firms. Although the programme was highly successful, funds ran out in 1980, and the period of rapid growth ended. With the ending of the credit, sales dropped and the factory went bankrupt, although it was subsequently re-launched under a new name. However, the technology had been transferred to the rural population, and it remains in use to this day. The technology has been gradually spreading into new areas, for example, into the rice-production systems of the southern province of Casamance (Fall 1990).

The promotion of animal traction in Senegal was associated with major research programmes. Initially these were based on on-station trials, notably at Bambey (Nourrissat 1965). Subsequently, emphasis was placed on model farms and then on-farm investigations by farming systems research teams. Several of the technologies developed and promoted failed, including animal-drawn wheeled toolcarriers which were heavy multipurpose implements. Although credit was made available for these, they were not widely adopted (Bordet 1988, Starkey 1988). At the same time, some farmer-invented technologies spread, including the use of cows (females) for work in Siné Saloum (Lhoste 1983).

A very rapid expansion of animal traction took place in southern Mali in the 1970s and 1980s. There had been some promotion of animal power in the French colonial period, resulting in about 8,000 animal-traction implements in use at the time of independence. There was then a major campaign to increase cotton production using work oxen. The donor-supported programme was initially implemented by the French CFDT (*Compagnie Française pour le Développement des Fibres Textiles*) and then by the national CMDT (*Compagnie Malienne pour le Développement des Textiles*). These provided a comprehensive package of inputs including extension, implements, seeds and fertilizers, credit, marketing and farming systems research. The growth was rapid, rising to about 50,000 implements in use in 1974, and 150,000 in 1988. In thirty years, animal-traction technology had changed from being rare, to being the normal agricultural practice in southern Mali (Zerbo & Kantao 1988, Traoré 1989).

A similar post-independence pattern of very rapid adoption of animal traction based on a cotton crop was seen in Chad. Although there were long-standing traditional systems of using animal power (packing, riding, milling) there appear to be no reports of cultivation with animal power prior to the 1950s. Promotion by government agencies started just before independence in 1960. At this time there were estimated to be only about 100 ploughs in use in the country. By 1966 there were 7,000, and by 1990 this had risen to about 115,000. This rapid rate of adoption was assisted by credit (almost all ploughs were bought with credit), the provision of the necessary inputs and a market for cotton as a cash crop. The proximity of northern Cameroon may also have been important, for by the time the promotional scheme started, news of an earlier, successful campaign in that country had been widely circulated (BDPA 1966, Starkey 1993).

There were many other examples of rapid localized growth of animal traction in West Africa resulting from promotional schemes, for example, in the semi-arid cotton zones of Burkina Faso, northern Cameroon, Côte d'Ivoire and Guinea Bissau (Bonnet et al. 1989, Peltre-Wurtz & Steck 1991, Starkey 1991, Mahdavi 1992).

As a result of the various public-sector initiatives and farmer-to-farmer diffusion, tillage with animal power had become part of the normal farming systems in most of semi-arid West Africa by the 1990s. The use of animal-drawn carts, pulled by oxen or donkeys also increased dramatically. In many places where the use of animal power for cultivation or cart transport had been nonexistent or very limited prior to 1960, 60–95 per cent of the rural population were using draft animals by 1990.

Further south, in the more humid parts of West Africa, there were some other examples of promotional schemes in the 1970s and early 1980s that resulted in low adoption rates. Such disappointing schemes tended to be near the margin of the forested area, for example southern Guinea, Sierra Leone, Ghana, Togo and the Northwestern Province of Cameroon. This proved to be a marginal zone for animal traction at this stage. Some projects faced major problems with animal sickness, and mortality rates as high as 50 per cent were cited. The de-stumping of land to allow ploughing was a serious labour constraint, and one that did not always seem justified. Some projects blamed social constraints, such as lack of tradition of animal husbandry and unfamiliarity with cattle. This overlooked the fact that similar constraints were rapidly overcome elsewhere, provided that animal traction was economically profitable, assisted by supporting services, including credit and animal health, and a good market for the crops (Munzinger 1982, Starkey 1986a).

The situation in the forest–cropland interface zone was changing quite rapidly in the last two decades of the twentieth century. The boundary of the semi-arid zone appeared to be moving southwards in West Africa, and deforestation was increasing in all zones. The ecological and disease balances appeared to changing. While the overall adoption of draft animals in the semi-humid zone was still low, compared with the semi-arid zone, animal traction was often increasing rapidly from a low base. For example, in the 1980s, numbers of draft cattle increased from 2,000 to 38,000 in Côte d'Ivoire and 2,000 to 12,000 in Togo (Westneat et al. 1988, Starkey 1992).

In eastern and southern Africa, immediate post-colonial promotion of animal traction during the 1960s and 1970s involved mainly national Ministry of Agriculture extension services. At this time, while national agricultural engineering services were placing emphasis on the development of tractor hire services, some centres were established to develop new "appropriate technology" implements and carts. Some centres were established by national Ministries, while others were developed by non-governmental organizations: few worked closely with the end-users and few produced implements that were adopted by farmers. By the 1980s, national "top-down" extension programmes and services appeared to have limited impact, and there was increased emphasis on area-specific, donor-assisted development projects. Some integrated projects had specific animal traction components, some of which proved highly effective. By the 1990s, there was increasing emphasis on participatory and farming systems approaches and linking projects that were working on animal traction (networking).

In Tanzania, ox training centres were established throughout the country to promote animal traction in the 1960s, 1970s and 1980s. Little attempt was made to concentrate animal traction extension in areas of greatest potential. An appropriate technology centre was established and an ox-plough factory was opened in 1970. Animal traction continued to spread slowly in the country, mainly through farmer-to-farmer diffusion of knowledge. In the 1980s, donor-assisted projects with animal-traction teams worked in several parts of the country including Mbeya, Tanga and Maswa, and attempted to identify and alleviate some of the constraints to the spread of animal traction, including the provision of credit, and the supply of implements, carts and animals (donkeys in Tanga region). In 1991, a national-network was formed to link animal-traction programmes in the country (Sosovele 1991, Starkey & Mutagubya 1992, Starkey & Grimm 1994).

At the time of independence in 1964, animal traction in Zambia was mainly found in the Southern Province. During the colonial period, the spread of the use of ox ploughs had been assisted by several favourable conditions. The smallholder farmers who were cultivating crops in the good agricultural zone had a tradition of cattle ownership. Ox ploughs were well known, having been used on colonial farms since the beginning of the century. Ploughs and spare parts manufactured in what is now Zimbabwe were readily available through private-sector stores. There was a market for produce along the line of rail. The technology had penetrated a little into some other areas, notably to the west and the east, probably through farmer and worker migration and/or the influence of various missions and traders. In the 1970s and 1980s, donor-supported integrated development projects in several regions started to actively introduce animal traction, through extension, credit and assistance with the provision of inputs. Some appropriate technology organizations developed and promoted wooden-wheeled carts (which were largely rejected). In some areas, uptake of ox power was rapid, and one project in the northwest had particular success in introducing animal-powered transport by providing credit and a supply of ox carts. In the more humid areas, where few cattle were owned, uptake was low, and mortality among the cattle acquired from other regions was high. In 1985, a national animal draft power programme was launched. This helped co-ordinate animal traction work in the various projects, as well as related research, implement production and training activities (Starkey et al. 1991, Löffler 1994).

14. Current situation and trends

As the twentieth century ends and the twenty-first century begins, the historical processes continue, and animal traction in Africa continues to spread and to evolve. There are still many areas where animal traction is absent from farming systems. In certain mountainous regions, forest zones, very arid areas or where the animal disease challenge is high, animal power may never be considered appropriate. However, at the margins of these areas, animal traction is likely to spread. The adoption of the technology in new areas is likely to result from farmer migration, farmer-to-farmer contact and the activities of development programmes. Development programmes, whether governmental or non-governmental, have been shown to be particularly effective at introducing technology rapidly into new areas, through training (technical knowledge is often a limiting factor), credit, input supplies and marketing arrangements. The introduction of animal traction into new areas can also be influenced by farmers migrating from areas of animal traction use. These on-going processes can be clearly seen in many countries where there are some areas of animal traction adoption and some regions where animal traction has never been used. For example, in Guinea, Ghana, Tanzania, Zaïre and Zambia animal traction is spreading each year into villages where it has never been used before, through both project promotion and the movement of farmers.

While official extension services and projects have proved successful at introducing animal traction for the first time, they have been less successful at promoting "improvements". Many innovative implements and harnessing systems have been developed and promoted by government services in Africa with negligible acceptance and adoption by farmers. Although extension services are meant to train farmers how to adjust ploughs "properly", the vast majority of ploughs in Africa are not adjusted and used according to conventional agricultural engineering wisdom. In contrast, many innovations that have spread, such as use of donkeys or cows, have done so despite the disapproval of extension services.

In those parts of the world with a long history of animal traction, including North Africa and Ethiopia, one person works with a team of animals. In areas of introduction, animals may not be well-trained in the initial years and farmers may lack confidence. Thus, immediately after the adoption of animal power, between two and four people may work with a single pair of animals. In much of sub-Saharan Africa, it is still common for at least three people (one may be a child) to work with animals that are plowing. There is a tendency for the numbers to decrease, and in areas where animals have long been worked, two is the most common number of people. A further decrease to one person is only rarely seen. This may be associated with the fact that most implements are pulled by chains, whereas in areas of traditional use, long-beamed implements are more common.

When animal traction is first adopted, the animals are generally only used for primary soil tillage. The mechanization of subsequent operations such as weeding may not follow for many years. Farmers are generally reluctant to allow their animals into fields with crops, lest they damage them. However, weeding is a labour-intensive operation, and animal-powered weeding is becoming more common in most African countries. In countries and areas with long experience of animal power, including southern Mali, Senegal, South Africa and Zimbabwe, the majority of animal-using farmers now weed with animal power. In other areas, including much of Malawi, Tanzania and Zambia, only a minority (albeit an increasing one) use weeders. While the introduction of weeding technology has sometimes been associated with a specific extension programme, in several areas weeding appears to have developed as a result of farmer-innovation. For example, in the Machakos area of Kenya, farmers weed using their ploughs (Tiffen et al. 1994).

Oxen are generally used when animal traction is first adopted. This is logical when animals and feed resources are plentiful. Cattle herds produce surplus male animals and cows are mainly valued for their reproductive and milk-producing capacities. The use of cows may start when oxen are in short supply or expensive, when farming systems become more intensive, when animal feed resources are a limiting factor and/or when work operations are light or highly seasonal. Cows are used widely in Egypt and North Africa and an increasing use of work cows has been reported in Senegal, Zambia and Zimbabwe. Such innovations are generally farmer-led, and not the result of extension programmes (Lhoste 1983, Starkey et al. 1991, Starkey 1994b).

In most sub-Saharan African countries (with the notable exception of Senegal), oxen were promoted and used as the main work animals. Horses were of high status and expensive and in tropical Africa generally only thrived in highland areas. Donkeys were considered too weak to plough, and were mainly used for pack transport. However, the range of the donkey is spreading in much of Africa. This expansion has been associated with changing climatic and agro-ecological conditions, including droughts and deforestation. Donkeys are increasingly used for pulling carts, for which they are well-adapted. In many cases, they are taking over from oxen for cultivation. This is not generally because they are better than oxen, but they are cheaper and more able to survive in drought conditions. The donkey population in Sahelian West Africa increased three-fold in the last half of the twentieth century (ten-fold in Senegal and The Gambia). Similar high increases have also reported in some southern African countries such as Botswana and Lesotho. The trend to increasing donkey populations is seen almost everywhere in Africa, apart from some North African states. Donkeys have replaced oxen as the main work animals in The Gambia, and they are becoming increasingly important in Namibia, South Africa and Zimbabwe. The move from oxen to donkeys has generally been a farmer-led innovation, which has sometimes been ignored or actively discouraged by agricultural authorities (Starkey 1987, 1994c, 1995, Starkey & Starkey 1997).

Animals are not only being used increasingly for tillage in sub-Saharan Africa, the transport role of animals is becoming more important. In much of Africa, head-loading has been the main means to transport for domestic water, fuelwood and market produce. In the majority of sub-Saharan African countries, animal-drawn carts are being used increasingly for trade, farm and domestic transport and local hire. In addition to within-village work, animal-drawn carts often deliver and distribute the produce and goods that are transported to and from towns by motor vehicles. There is a clear trend towards increasing use of motor vehicles for transport of people and goods in both rural and urban areas. However, evidence from the more urbanized areas of Africa (and elsewhere in the world) suggests that the number of motor vehicles can rise a very long way before there is any reduction in the use of animal-powered transport. The two systems can co-exist for a long time (provided legislation does not lead to the banning of animal transport). The implication is that the present overall growth in the use of animal transport in Africa will continue for the foreseeable future.

In some countries, including Zambia and Tanzania, the number of carts is still low compared with rural populations. Cart adoption appears limited by the low availability of carts (or components), their high cost and lack of credit. Elsewhere, including Namibia, Senegal and southern Mali, animal-drawn carts (mainly donkey carts) have become quite common, and in some areas most rural households own carts. Mauritania provides an example of the rapid increase in animal-drawn carts. Carts were very rare at the time of independence in 1960, perhaps 1,000 in the whole country. By 1996, this had increased to about 75,000, mostly donkey carts. This growth resulted not from promotional schemes, but from direct investment by farmers and transporters, dealing with local traders and artisans. This represented a private investment in 1996 terms of about fifteen million US dollars. The proximity of Mali and Senegal, where carts are manufactured, has facilitated the adoption of carts (Starkey 1996).

Most work animals in Africa have been owned and controlled by men. Most of the promotional schemes of the twentieth century have been directed by and at men. Involvement of women with work animals has been small, with the notable exception of donkeys. In many countries, women use donkeys for pack transport, and sometimes for riding. In a few areas, including Tarime in Tanzania, it has become common for women to work with oxen. Elsewhere, it is still unusual, but the involvement of women with work animals is increasing in many countries. This is particularly so in those rural areas where men travel far to work in towns, leaving women as heads of household. The trend towards increasing direct involvement of women in animal traction is being assisted in some countries by promotional and credit schemes targeted on women.

In most of Africa, animal power is being used to replace or supplement human power. This trend is continuing in most of the continent. With so many agricultural and transport tasks performed by humans, there is much scope for continued expansion. Long-distance transport by animals is now rare. Although there are still examples of caravans in the circum-Sahara countries and Ethiopia (where salt is transported by donkeys in Tigre), the trend is for motor vehicles to replace animals for long-distance transport. While some nomadic pastoralists still use animal power for transport, there is a trend towards sedentarization. Donkeys have generally been replaced in the mining industry.

In countries where farmers can own large areas of land suitable for crop production, tractor power is tending to replace animal power for primary cultivation. This trend has been seen in the second half of the twentieth century in several countries with large-scale farmers, including Kenya, Morocco, South Africa and Zimbabwe. In those areas of Africa where agricultural land is privately owned, the pattern of tractor mechanization has been similar to that of other parts of the world. Tractorization has been associated with large and increasing farm sizes. In inhabited rural areas, as some farms grow increasingly large to ensure motorization is profitable, other farming families give up their right to the land (through legislation, sale, bankruptcy or other means). The situation in South Africa illustrates the point, although the mechanization processes were obviously influenced by the apartheid policies. At the end of the twentieth century, there were just 60,000 large-scale farming families (historically white farmers) that used tractor power, with economies of scale making tractor utilization profitable. On less land, there were 1.2 million smallholder farming families (historically black farmers) that used a combination of human power, animal power and tractor power. The provision of tractor power to the smallholder sector, whether by government or private enterprise, tended to be capital depleting and economically unsustainable (Starkey 1995).

The picture throughout the continent (and elsewhere in the world) appeared broadly similar. The use of tractors by smallholder farmers to grow normal, rainfed crops is economically unsustainable in many circumstances. It is not possible to predict the long-term future of mechanization, owing to unknown developments in the technology and changing socio-economic conditions. However, in the immediate, foreseeable future, animal power is unlikely to be rapidly replaced by tractor power in the smallholder sector. Animal power may only be replaced by tractors where large land-holdings are possible and where people consider the economic cost of tractors to be justified for social reasons.

15. Conclusions

Animal power for transport and cultivation has been used in Africa for over 5,000 years. Animal traction has been an integral part of farming systems in North Africa and Ethiopia for at least two millennia. The use of animals for riding and pack transport has a long history in many parts of Africa, although details of its origins and spread are not well understood. The use of animal power for wheeled transport and soil tillage does not appear to have spread in sub-Saharan Africa prior to the colonial period. In the past five hundred years, the use of animal power for wheeled transport spread slowly from ports and colonial bases. The animal-drawn plough was introduced into South Africa by colonists in the seventeenth century, and it spread slowly to neighbouring territories. In most other sub-Saharan countries, it was introduced in the twentieth century by colonial authorities wishing to increase agricultural production. In most villages in Africa, animal-drawn ploughs were introduced in living memory. Many successful colonial and post-independence promotional schemes involved the provision of implements, training, credit and marketing channels. In circumstances where implements were readily available through private-sector traders, animal traction sometimes spread without any promotional schemes. Following the introduction of animal traction technology, many innovations were developed and spread through informal farmer-to-farmer processes. At the end of the twentieth century, animal traction was still spreading in most of sub-Saharan Africa, and appeared likely to continue to increase in the foreseeable future.

References

BDPA 1966. *La culture attelée et la modernisation rurale dans le sud du Chad*. Bureau pour le Développement de la Production Agricole (BDPA), Paris, France.

Bangura, A.B. 1990. Constraints to the extension of draft animal technology in the farming systems of Sierra Leone. In *Animal traction for agricultural development*. P. Starkey & A. Faye (eds), 324–7. Proceedings of the Third Regional Workshop of the West Africa Animal Traction Network, 7–12 July 1988, Saly, Senegal. Technical Centre for Agricultural and Rural Co-operation (CTA), Ede-Wageningen, The Netherlands.

Banti, G. 1993. Ancora sull'origine del cammello nel Corno d'Africa: osservazioni di un linguista. In *Ethno, Lingua e Cultura*, A. Belardi (ed.), 183–223. Roma: Calamo.

Bigot, Y. 1985. Quelques aspects historiques des échecs et des succès de l'introduction et du développement de la traction animale en Afrique sub-Saharienne. *Machinisme Agricole Tropical* 91, 4–10.

Bigot, Y. 1989. Un siècle d'histoire d'une technologie agricole: la traction animale en Guinée. See Raymond et al. (1989), 36–52.

Blench, R.M. 1995. A history of domestic animals in northeastern Nigeria. *Cahiers de Science Humaine*, ORSTOM 31(1), 181–238.

Blench, R.M. 1997. *Animal traction in West Africa: categories, distribution and constraints on its adoption and further spread: a Nigerian case study.* ODI Working Paper 106, Overseas Development Institute, London.

Bonnet, B., B. Guibert, O. Robinet, P. Lhoste 1989. Conduite, gestion des carrières et valorisation des boeufs de trait in zones cotonnières (Burkina Faso, Côte d'Ivoire et Mali). See Raymond et al. (1989).

Bordet, D. 1988. *From research on animal-drawn implements to farmers' appropriation: successes and failures of development experience in Senegal.* Paper 88:384, International Conference on Agricultural Engineering, 2–5 March 1988, Paris, France.

Burman, J. 1988. *Towards the far horizon.* Cape Town, SA: Human & Rousseau.

Clutton-Brock, J. 1992. *Horse power: a history of the horse and the donkey in human societies.* Cambridge, Mass.: Harvard University Press.

Delort, R. 1992. *The life and lore of the elephant.* London: Thames & Hudson.

Dorward, D.C. & A.I. Payne 1975. Deforestation, the decline of the horse and the spread of the tsetse fly and trypanosomiasis (nagana) in nineteenth century Sierra Leone. *Journal of African History* 16(2), 239–56.

Drew, S.F. 1954. Notes from the Red Sea Hills. *South African Archaeological Bulletin* 9, 101–2.

Ehret, C. 1979. On the antiquity of agriculture in Ethiopia. *Journal of African History* 20, 161–77. .

Ellert, H. 1993. *Rivers of gold.* Gweru: Mambo Press.

FAO 1977. *The water buffalo.* Rome: FAO.

Fall, A. 1990. Adoption et principales contraintes à la diffusion des équipements de traction animale en Basse Casamance, Sénégal. In *Animal traction for agricultural development.* P. Starkey & A. Faye (eds), 267–75. Proceedings of the Third Regional Workshop of the West Africa Animal Traction Network, 7–12 July 1988, Saly, Senegal. Technical Centre for Agricultural and Rural Cooperation (CTA), Ede-Wageningen, The Netherlands.

Goe, M.R. 1987. Animal traction on smallholder farms in the Ethiopian highlands. PhD thesis, Cornell University, UMI Dissertation Information Service, Ann Arbor, Michigan, USA.

Havard, M. 1986. Les conclusions des expérimentations (1950–1985) sur les semis en culture attelée des principales espèces cultivées. Document de Travail 1986:10. Département Systèmes et Transfert, ISRA, Dakar, Sénégal.

Havard, M. & A. Faye 1988. Eléments d'analyse de la situation actuelle de la culture attelée au Sénégal: perspectives d'études et de recherches. See Starkey & Ndiamé (1988), 241–52.

Haudricourt, A.G. & M.J. Delamarre 1955. L'homme et la charrue à travers le monde. Géographie Humaine No. 25, Gallimard, Paris, France.

Haswell, M. 1991. Population and change in a Gambian Rural Community, 1947–1987. In *Rural households in emerging societies*, M. Haswell & D. Hunt (eds), 141–71. Oxford: Berg.

Joubert, B. 1995. An historical perspective on animal power use in South Africa. In *Animal power in South Africa: empowering rural communities*, P. Starkey (ed), 125–38. Gauteng, SA: Development Bank of Southern Africa.

Kinsey, B.H. 1984. Equipment innovations in cotton-millet farming systems in Uganda. In *Farm equipment innovations in Eastern and Central Southern Africa*, I. Ahmed & B.H. Kinsey (eds), 209–52. Aldershot, UK: Gower Publishing.

Kitchen, K.A. 1993. The land of Punt. See Shaw et al. (1993), 587–608.

Koponen, J. 1994. *Development for exploitation: German colonial policies in mainland Tanzania.* Helsinki: Finnish Historical Society *and* Hamburg: Lit Verlag.

Lagercrantz, S. 1950. *Contributions to the ethnology of Africa.* Studia Ethnographica Upsaliensia, I. Lund: Håkan Ohlssons.

Laoust, E. 1918. Le nom de la charrue et de ses accessoires chez les Berbères. *Archives berbères* **3**, 4–30.

Laoust, E. 1930. Au sujet de la charrue berbère. *Hespéris*, 10, 37–47.

Law R. 1980a. *The horse in West African history*. Oxford: Oxford University Press.

Lhoste, P. 1983. Développement de la traction animale et évolution des systèmes pastoraux au Siné Saloum, Sénégal (1970–1981). *Revue d'Elevage et de Médécine vétérinaire des Pays Tropicales* **36**(3), 291–300.

Löffler, C. 1994. Transfer of animal traction technology to farmers in the North Western Province of Zambia. See Starkey et al. (1994), 354–9.

Löwe, P. 1986. *Animal powered systems: an alternative approach to agricultural mechanization*. Eschborn Germany: Vieweg, for German Appropriate Technology Exchange (GATE), GTZ.

Mahdavi, G. 1992. Development of animal traction in cotton areas of French-speaking African countries. In *The role of draught animals in rural development*, G. den Hartog & J.A. van Huis (eds), 83–7. Proceedings of an international seminar held 2–12 April 1990, Edinburgh, Scotland. Wageningen, The Netherlands: Pudoc Scientific.

Mettrick, H. 1978. *Oxenisation in The Gambia*. London: Ministry of Overseas Development.

Midant-Reynes, B. 1992. *Préhistoire de l'Egypte: des premiers hommes aux premiers pharaons*. Paris: A. Colin.

Müller, H. 1986. *Oxpower in Zambian agriculture and rural transport*. Edition Herodot Socio-economic Studies in Rural Development No. 65. Aachen, Germany: Rader Verlag.

Munzinger, P. (ed.) 1982. *Animal traction in Africa*. Eschborn, Germany: Vieweg for German Appropriate Technology Exchange (GATE) GTZ.

Nicholls, W. 1918. The saqia in Dongola province. *Sudan Notes and Records* **1**, 21–4.

Nourrissat, P. 1965. La traction bovine au Sénégal. *L'Agronomie Tropicale* **9**, 823–53.

Peltre-Wurtz, J. & B. Steck 1991. *Les charrues de la Bagoué: gestion paysanne d'une opération cotonière en Côte d'Ivoire*. Paris: ORSTOM Editions.

Phillipson, D.W. 1993. The antiquity of cultivation and herding in Ethiopia. See Shaw et al. (1993), 344–57.

Pingali, P., Y. Bigot, & H. Binswanger 1987. *Agricultural mechanisation and the evolution of farming systems in sub-Saharan Africa*. Baltimore: World Bank with Johns Hopkins Press.

Raymond, G., Y. Bigot, D. Bordet (eds) 1989. *Economie de la mécanisation en région chaude*. Actes du IX séminaire d'économie rurale, 12–16 September 1988, Montpellier, France. Centre de Coopération Internationale en Recherche Agronomique pour le Développement (CIRAD), Montpellier, France.

RIM 1992. *National Livestock Resource Survey* [6 volumes]. Abuja: Final Report to Federal Department of Livestock and Pest Control Services, Federal Government of Nigeria.

Rossiter, E. 1984. *The book of the dead: papyri of Ani, Hunefer, Ahhai*. Geneva: Liber.

Shaw, T., P. Sinclair, B. Andah, A. Okpoko (eds) 1993. *The archaeology of Africa. Foods, metals, and towns*. London: Routledge.

Simoons, F.J. 1965. Some questions on the economic prehistory of Ethiopia. *Journal of African History* **6**(1), 1–13.

Sosovele, H. 1991. *The development of animal traction in Tanzania: 1900–1980s*. PhD dissertation, University of Bremen, Germany.

Sosovele, H. 1997. The challenges of animal traction in Tanzania. Paper for workshop of Animal Traction Network for eastern and Southern Africa (ATNESA) 4–8 December 1995, Karen, Kenya. (Proceedings in preparation).

Starkey, P. 1981. *Farming with work oxen in Sierra Leone*. Freetown, Sierra Leone: Ministry of Agriculture.

Starkey, P. 1982. N'dama cattle as draught animals. *World Animal Review* **42**, 19–26.

Starkey, P. 1986a. *Draught animal power in Africa: priorities for development, research and liaison*. Network Paper 14, Farming Systems Support Project, University of Florida, Gainesville, USA.

Starkey, P. 1986b. *Strengthening animal traction research and development in The Gambia through networking*. Consultancy Mission Report and Annotated Bibliography on Animal Traction. Gambia Agricultural Research and Diversification Project, Banjul, The Gambia.

Starkey, P. 1987. Brief donkey work. *Ceres* **20**(6), 37–40.

Starkey, P. 1988. *Perfected yet rejected: animal-drawn wheeled tool-carriers*. Eschborn, Germany: Vieweg for German Appropriate Technology Exchange (GATE) GTZ.

Starkey, P. 1991. *Animal traction in Guiné-Bissau: status, trends and survey priorities.* Report of consultancy mission 22 February to 5 March 1991 in association with Pan Livestock Services, Reading University and Gaptec, Lisbon Technical University. Reading, UK: Animal Traction Development.

Starkey, P. 1992. Changes in animal traction in Africa and Asia: implications for development. In *The role of draught animals in rural development*, G. den Hartog & J.A. van Huis (eds), 11–24. Proceedings of international seminar 2–12 April 1990, Edinburgh, Scotland. Wageningen, The Netherlands: Pudoc Scientific.

Starkey, P. 1993. *La traction animale au Chad: politiques et approches.* Oxford, UK: Oxfam.

Starkey, P. 1994a. The transfer of animal traction technology: some lessons from Sierra Leone. See Starkey et al. (1994), 306–17.

Starkey, P. 1994b. A worldwide view of animal traction highlighting some key issues in eastern and southern Africa. See Starkey et al. (1994), 66–81.

Starkey, P. 1994c. Donkey utilisation in sub-Saharan Africa: recent changes and apparent needs. In *Working equines*, M. Bakkoury & R.A. Prentis (eds), 289–302. Proceedings of second international colloquium, 20–22 April 1994, Rabat, Morocco. Actes Editions, Institut Agronomique et Veterinaire Hassan ll, Rabat, Morocco.

Starkey, P. (ed.) 1995 *Animal power in South Africa: empowering rural communities.* Gauteng, SA: Development Bank of Southern Africa.

Starkey, P. 1996. La traction animale en Mauritanie: situation et perspectives. Project SPFP/MAU/4051, Food and Agriculture Organization of the United Nations, Rome.

Starkey, P., H. Dibbits & E. Mwenya 1991. *Animal traction in Zambia: status, progress and trends.* Wageningen, The Netherlands: Ministry of Agriculture, Lusaka *with* IMAG–DLO.

Starkey, P. & J. Grimm 1994. The introduction of animal traction in the Tanga Region, Tanzania. GTZ, Eschborn, Germany.

Starkey, P. & W. Mutagubya 1992. *Animal traction in Tanzania: experience, trends and priorities.* Chatham, UK: Ministry of Agriculture, Dar es Salaam, Tanzania and Natural Resources Institute.

Starkey, P. E. Mwenya, J. Stares (eds) 1994. *Improving animal traction technology.* Technical Centre for Agricultural and Rural Cooperation (CTA), Wageningen, The Netherlands.

Starkey, P. & F. Ndiamé (eds) 1988. *Animal power in farming systems.* Proceedings of workshop 19–26 September 1986, Freetown, Sierra Leone. Eschborn, Germany: Vieweg for German Appropriate Technology Exchange (GATE), GTZ.

Starkey, P. & M. Starkey 1997. *Regional and world trends in donkey populations.* Paper for workshop on improving donkey utilisation and management, 4–9 May 1997, Debre Zeit, Ethiopia. Proceedings to be published by the Animal Traction Network for Eastern and Southern Africa (ATNESA), Harare, Zimbabwe.

Stead, M. 1986. *Egyptian life.* London, UK: British Museum Publications.

Stiehler, W. 1948. Studien zur Landwirtschafts und Siedlungsgeographie Aethipiens. *Erdkunde*, **2**, 257–82.

Tiffen, M., M. Mortimore, F. Gichuki 1994. *More people, less erosion: environmental recovery in Kenya.* Nairobi: African Centre for Technology Studies.

Traoré, M. 1989. Mécanisation agricole et integration agriculture-elevage dans le cadre du Projet Mali-Sud CMDT. See Raymond et al. (1989), 111–31.

van Nhieu, J.T. 1982. Animal traction in Madagascar. In *Animal traction in Africa*, P. Munzinger (ed.), 427–49. Eschborn, Germany: Vieweg for German Appropriate Technology Exchange (GATE) GTZ.

Westneat, A.S., A. Klutse, K.N. Amegbeto 1988. Features of animal traction adoption in Togo. See Starkey & Nadiamé (1988), 331–9.

Wilson, R.T. 1976. Some quantitative data on the Tigre salt trade from the early nineteenth century to the present day. *Annali Instituto University of the Orient, Napoli* **36**, 157–64.

Wilson, R.T. 1991. Equines in Ethiopia. In *Donkeys, mules and horses in tropical agricultural development*. D. Fielding & R.A. Pearson (eds), 33–47. Proceedings of colloquium held 3–6 September 1990, Edinburgh. Centre for Tropical Veterinary Medicine, University of Edinburgh, UK.

Zerbo, D. & A. Kantao 1988. Traction animale au Mali. See Starkey & Nadiamé (1988), 175–81.

Zeuner, F.E. 1963. *A history of domesticated animals.* London: Hutchinson.

CHAPTER TWENTY-SEVEN

Bees and bee-keeping in Africa

Andrew D. Kidd & Berthold Schrimpf

1. Introduction

While there are over 20,000 species in the bee family (Apoidea), most are solitary not social, and bee-keeping in Africa is concerned almost exclusively with *Apis mellifera* L., which is also the common honeybee of Europe. Bee-keeping is practised throughout much of sub-Saharan Africa and also in North Africa. It is locally important in the economy of many regions, and the technology and management practices of some areas are complex and highly diversified. It is this last aspect which much of the existing literature on bee-keeping in Africa has tended to be missed or distorted largely as a result of ill-considered speculation by Eurocentric bee scientists.

The first and only comprehensive survey of bee-keeping in Africa was that of Seyffert (1930) who studied a wide range of source documents from early travellers and colonial officials. Seyffert, a German, was able to map beehive types based on available documents and his study is particularly strong for those parts of Africa formerly colonized by Germany. His synthesis has, unfortunately, been under-utilized in subsequent studies of African bee-keeping. Irvine (1957) brought together a wide range of material when giving an overview of indigenous African methods of bee-keeping. He, as Seyffert before him, also studied the writings of early travel writers and ethnographers and was able to give a descriptive overview when presenting a large number of highlights without drawing many conclusions. The more recent study of Villières (1987a,b) also covers, in some detail, traditional bee technology and focuses mainly on francophone West Africa. There have also been a number of useful studies on bee-keeping in various African countries (e.g. Svensson 1985, for Guinea Bissau; Gnägi 1992, for Mali; Himsel 1991, for Niger; Jessen 1967, Boyles 1991, Mutsaers 1993, for Nigeria; Silberrad (1976), Clauss & Clauss 1991, Wainwright 1992, for Zambia; Nightingale 1976, Geider 1989 for Kenya; Smith 1958, for Tanzania; Fichtl & Adi 1994, for Ethiopia). Crane, in a general study of the archaeology of bee-keeping (1983) and in a survey of world bee-keeping practice (1990), does little justice to the available literature on African bee-keeping.

A clear understanding of the position of Africa in terms of world production of honey and beeswax is lacking owing to incomplete records. The African *A. mellifera* is a particularly high producer of wax and existing evidence suggests that around half of the beeswax that reaches the world market comes from Africa (Crane 1990). However, honeycomb is discarded in many areas. Curtin (1985) notes that international trade in African beeswax was already of regional importance along the West African coast prior to the eighteenth century. While a trade in beeswax continued into the eighteenth century its importance seemed to decline in competition with the more lucrative slave trade. However, with the revival of "legitimate trade", the price of beeswax increased three-fold between the 1780s and the 1830s in Senegambia, with the export volume increasing by a factor of ten over the same period. MacGregor & Oldfield (1837) mention beeswax being used in trade by the Ful6e people of Rabbah in west–central Nigeria. Earlier this century in Nigeria, the trade and export of beeswax was of some importance to local economies (Taylor 1942, Corby 1943, Forde & Scott 1946, Buchanan & Pugh 1955). This seems to have declined since independence and many older bee-keepers describe how they used to sell beeswax 20–30 years ago but find there is no longer a market.

Honey is mainly processed and consumed locally in Africa, though there is clearly a long tradition of honey marketing in some areas, as has been described by early travellers in northern Nigeria (Denham *et al.* 1828) and in traditional stories (Skinner 1969). Its widespread use in various cultural ceremonies and production of honey beer give added significance to its role in the life of many Africans (Seyffert 1930). Mungo Park (1817) notes the production of mead in parts of West Africa. Presently, little of the honey produced in Africa reaches the world market and that which does tends to be produced in eastern and southern Africa from non-traditional sources.

In summarizing the relative importance of bee-keeping and honey production throughout the world, Crane (1990) shows that Africa has a similar number of hives to Europe, at around 13.5 million. However, she picks on the statistic of greatest hive density to support her claim to Europe's "long and rich tradition of bee-keeping". While this is undoubtedly true it distorts the overall picture.

Existing evidence suggests that the number of hives *per capita* in Africa and Europe are similar. Given a lower average number of hives per bee-keeper in Africa and the likely underestimates of the prevalence of bee-keeping there, it is reasonable to suggest that Africa has the greatest number of bee-keepers *per capita* of any continent. Last century writers found the level of bee-keeping in some areas remarkable. Monteiro (1875) along the River Quanza [=Kwanza], Angola, describes how some families have 300–400 hives. According to Bridges (Banso Reassessment Report 1934, quoted in Kaberry 1952) more than 10 per cent of the Nsaw men in the North West Province of Cameroon are involved in bee-keeping. Present observations suggest that this is still true and that the percentage is even greater among the neighbouring Oku. Corby (1943) notes that 28 per cent of the male population keep bees in several districts of Kontagora Emirate, central Nigeria.

The origin and history of bee-keeping in Africa has been little studied and the last comprehensive survey on its distribution was nearly 70 years ago (Seyffert 1930). Most recent opinion regarding its origin is represented by the cultural evolutionist assumptions of Crane (1983, 1990), the most widely known and respected bee scientist. Crane gives an elaborate description of the multiple influences and origins of bee-keeping in Europe. Unfortunately, she postulates that the practice of bee-keeping originated in Ancient Greece or Rome and spread to sub-Saharan Africa via the Nile valley based on her speculations concerning the superiority of "rational" European bee-keeping and a lack of early iconographic materials from Africa. However, evidence of the diversity and sophistication of African bee-keeping argue for a very ancient establishment on the continent.

2. African honeybees and their distribution

The species of main concern in world bee-keeping is *Apis mellifera*. It is thought to have originally evolved in Africa and spread to Europe (Ruttner 1992). South and southeast Asia has its own species of honeybee, *Apis cerana*, which is smaller and less productive, and the Americas had no native honeybee species. However, the geographical separation is being broken down under man's influence through introduction of the more productive *A. mellifera* to new areas including the Americas. Given its extensive geographical distribution and its importance to the local and world economy through its products, honey, wax, propolis and royal jelly, it can be regarded as a highly successful, adaptive species. The bee population in Africa is relatively high, though in some areas its preferred forest and bush–savanna habitat is becoming increasingly threatened by deforestation. The abundance of bees in Africa was noted by some early European travellers (e.g. Denham et al. 1828).

Tropical Africa also has a number of species of stingless bee (Meliponinae). Prior to the introduction of *Apis mellifera* in the 1700s and 1800s, the long tradition of bee-keeping in South America was exclusively with Meliponinae. These stingless bees are also social bees of the family Apidae and several different species are known to exist in Africa. They may also be kept in some areas, though often only owing to their colonizing a hive designed for the more productive *A. mellifera*. However, Gutmann (1922) noted that they have been traditionally kept among the Chagga in Tanzania in small horizontal log hives. They are usually only exploited through honey hunting and the quality of their honey is often greatly appreciated. Among the Oku, North West Province of Cameroon, Meliponinae honey is highly valued in traditional medicine. However, their low production means that they are chased from hives that they seek to inhabit.

Around 25 races of *A. mellifera* have been described, with at least 11 of these being found in Africa (Fig. 27.1; Ruttner 1992). Each of these seems to be somewhat morphologically and behaviourally distinct, though often introgressively. This variation has implications for bee-keeping practice though this has been little studied.

There are several races which are only to be found in North Africa. A small and relatively defensive race, *A. m. lamarckii*, which is black with yellow abdominal bands, is found in the lower Nile valley. Further along the North African coast, from the Libyan desert to the Atlantic coast, *A. m. intermissa* is found. This race is black, produces much propolis and stings readily. It is well adapted as a race to the climatic extremes of the region and though 80 per cent of colonies may perish

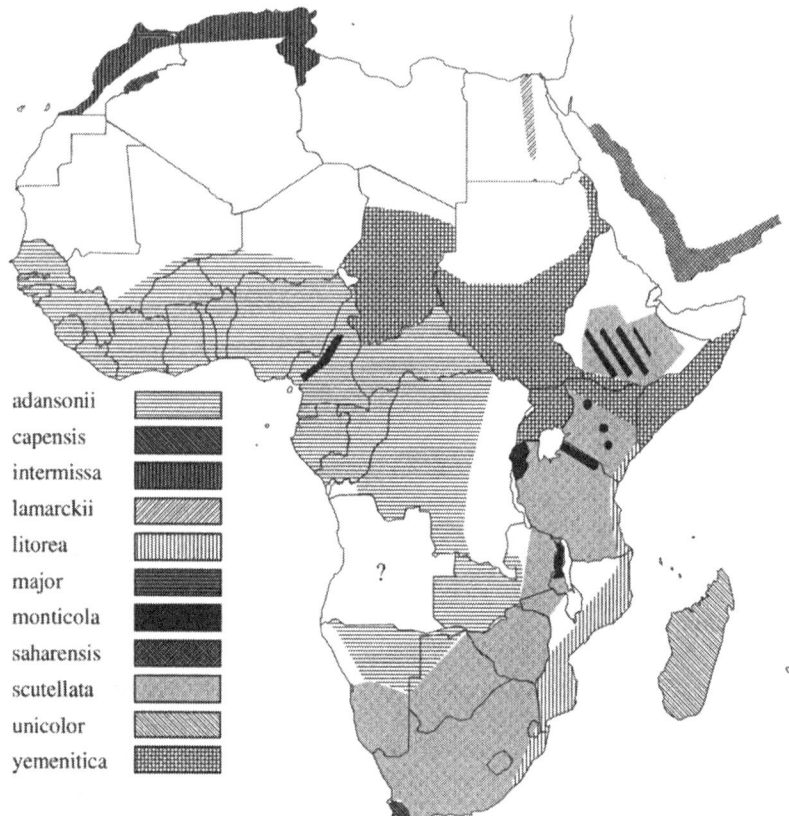

Figure 27.1   Distribution of races of *Apis mellifera* in Africa (adapted from Ruttner 1992).

during a drought it is able to recover rapidly. Those colonies that have been introduced to more temperate zones have perished. Another race that has become adapted
to a particular niche in harsh environments is *A. m. saharensis* which lives in oases
south of the Atlas mountains in Morocco and western Algeria. Crane (1990) states
that present populations of the race are likely to be relics of a much larger population that would have populated the area when the climate was more benign. Their
ability to survive under extreme conditions is a clear indication, however, that they
have undergone significant adaptation to the climatic change. The race has a tan/
yellow body colour and are also known for their ineffectiveness in defending their
nest. Ruttner (1992) also describes the race *A. m. major* found in the Rif mountains
in the north of Morocco, which is notable for its larger size.

A number of other races have been described in sub-Saharan Africa (Ruttner
1992). The known variability in morphological characteristics of bees from tropical
Africa is very high when compared to the total known variability of *A. mellifera*. In
general, tropical races of *A. mellifera* are smaller than temperate forms and have a
more slender abdomen. Colonies produce a greater number of swarms than the

European forms and whole colonies may abscond as a result of disturbance or food shortage. The bees are perhaps best known for their highly developed defensive character, which is often termed "aggressive" by bee-keepers used to the European bee. This is a characteristic which varies across regions and also diurnally and season-ally within a region. The defensive behaviour of African bees has been locally utilized in warfare by many African tribes (e.g. the Tiv in central Nigeria).

Crane (1990) rightly makes the point that the term race perhaps draws too clear a boundary between forms that are introgressive and whose geographical proxim-ity has allowed on-going hybridization. However, differences in morphology and behaviour can be observed with some of these according to climate (ecoclines) and distance (geoclines). The races will be briefly defined according to those described by Ruttner (1992).

The race commonly found in West Africa is *A. m. adansonii* and was named by Latreille in 1804 from bees collected in Senegal. For many years the name became used for tropical African bees in general. Little is known about these bees or indeed any honeybees west of the Rift valley (Crane 1990).

One other widely distributed race is *A. m. scutellata* which tends to inhabit open miombo woodland at about 500–1,500 m characterized by species of *Brachystegia* and *Julbernardia*. Though samples from Ethiopia to South Africa show it to be morphologically well defined and uniform, there appears to be considerable vari-ation in its defensive behaviour (Ruttner 1992). This is the most studied of tropical African honeybees and most bee-keeping development programmes have dealt with this race.

At higher altitudes the race *A. m. monticola* can be found. Ruttner (1992) notes its presence as far afield as the highlands of Ethiopia, Kenya, Tanzania, Rwanda, Burundi, Malawi and Cameroon. It is larger than *A. m. scutellata*, approaching European *A. mellifera* in size, and is dark with longer hair on the abdomen. Crane (1990) compared it to other tropical Africa races as being less "aggressive", though she also referred to Dietz & Krell (1986) who described most colonies sampled, particularly those found between 1,550 m and 2,100 m, as "aggressive to very aggressive".

The race found on the coast of Kenya, Tanzania and Mozambique is called *A. m. litorea*. It is quite yellow and, though small, has a relatively long tongue. It is most similar morphologically to *A. m. adansonii* of West Africa. Colonies of *A. m. litorea* are quite migratory in nature and move from one area to the next in search of new forage.

The honeybee native to the Arabian peninsula is named *A. m. jemenitica* and bees morphologically similar to this have been found further west in Chad and Sudan. Ruttner (1992) describes them as being particularly adapted to dry *Acacia*-dominated savanna. It is similar to *A. m. saharensis* in its ability to survive pro-longed drought.

The honeybees of Madagascar are considered to be an indigenous adaptation and have been named *A. m. unicolor*. They have been imported into Réunion and Mauritius which have no native honeybees. In Madagascar the bees in the coastal region show differences from those in the cooler highlands. Those in the highlands do not behave so defensively and do not abscond. Around the Cape Peninsula in South Africa there is an isolated population of *A. m. capensis*. It is similar in size to *A. m. scutellata*, but with a dark abdomen. Its reproductive behaviour is unique for the species in that its workers can develop into full females (thelytoky). Adult workers emerge relatively quickly after the cell is sealed, with the parasitic *Varroa* mites hardly able to reproduce on its worker brood.

Ruttner's mapping of the various races in Africa was based on incidental collections and on a very limited collection of specimens that were then extrapolated on a national and regional basis. His multivariate analysis of morphological characteristics was very sophisticated and ground breaking in terms of understanding the variation within the species. However, the narrow database from Africa, the continent where *A. mellifera* is said to have evolved (Ruttner 1992), can be expected to mask the full morphological and behavioural variation that is likely to exist. This is particularly true of West Africa, where studies have been particularly lacking.

Some bee races seem to exist in close proximity to each other, though often occupying distinct ecological niches. A recent study by Kassaye[1] described the presence of at least five distinct groups of honeybees in Ethiopia with their distributions correlating to ecological variation in the country. The races found were *A. m. litorea* in Gambella, *A. m. monticola* in the highland Bale Region, a yellow bee on the eastern escarpment resembling *A. m. jemenitica*, a new proposal of *A. m. abyssinica* in the western part of the country, and the bees in the lowland had a resemblance to *A. m. adansonii* (regarded as the West African bee) but not *A. m. scutellata* of East Africa. This study perhaps illustrates that the distribution of bee races is likely to be much more complex than as yet described.

African bee-keepers usually differentiate categories of bee. Most know how to distinguish the queen, workers and drones and have theories about their biology enabling them to carry out effective management practices. Size, colour, hairiness, defensiveness, the ability to sting, the orientation of comb placement and the quality of honey production may be used to describe different bee colonies and bee types, corresponding to different bee species and races. Taylor (1942) describes four types of honeybees that are differentiated by Hausa bee-keepers around Zaria, Nigeria.

The comparative advantage and adaptive nature of various genes from African races of *A. mellifera* is clearly demonstrated by the spread of the so-called "Africanized" *A. mellifera* in the Americas. In 1956, more than a hundred queens were taken from southern Africa to Brazil of which 49 survived the journey (Kerr 1967). In 1957, several colonies swarmed and hybridized with honeybees of European origin. The spread of these hybrid forms has been followed since then with growing interest as it has led to the introduction of dominant highly defensive and absconding characteristics into the honeybee population (Ruttner 1992). The potential threat to a European style of bee-keeping in the Americas has led to an intensive study of this phenomenon. Indeed, the fullest weight of the scientific community came to bear on the issue. As Crane (1990) has summarized, there have been many statistical analyses of its morphology, its DNA sequences and wing beating frequencies, whole bee hydrocarbon assays and high-resolution supercritical fluid chromatography of its beeswax, in order to understand further its spread and its impact on the more manageable European *A. mellifera* population. The African bees and their genes have never been so intensively studied as when they have proven more successful than was hoped in the New World. It is interesting that despite the Africanized bee being initially perceived as a major threat to bee-keeping in the Americas, some are of the opinion that they have forced a positive change in bee-keeping practice (De Jong 1996). However, as Chandler (1976) points out:

> It is one of the strange turns of history that scientists know least about the race of honeybees that is probably the most populous in the world, and certainly the one with the most potential for development as the major commercial honeybee of the tropics. It is an even stranger turn of history that more is known about a few *adansonii* genes halfway around the world from their home in the so-called Africanized bees of South America than about the varied complex of *adansonii* in their environment of evolution.

This complexity which exists among honeybees in tropical Africa has had great, but generally unappreciated, implications for the history and spread of bee-keeping in Africa. One of the key events in the evolution of modern European-style bee-keeping was the exploitation of the bee space in hive construction (Crane 1983). The bee space is the centre-to-centre distance by which bees naturally separate adjacent combs. When bars are placed in a hive with this spacing, the combs are attached one to each bar and become movable. Thus combs are more easily removed for inspection and replaced. However, the bee space among populations of *A. mellifera* in Africa varies. Clauss & Clauss (1991) record the bee space as 33 mm in Zambia, southern Africa; others have noted that in East Africa it is 32 mm; in Senegal, West Africa, 31 mm; and in southern Nigeria, West Africa, Mutsaers (1991) found it to be 30 mm. This physical manifestation of the heterogeneity in biology of the species in Africa is only one factor that illustrates how little we know about the species in Africa and its implications for current and future bee-keeping practice. It is to the complexity of traditional bee-keeping in Africa that we now turn.

3. Origins and diffusion of bee-keeping in Africa

There are several archaeological finds concerning the exploitation of bees for honey in Africa. However, these appear to be exclusively related to honey-hunting rather than bee-keeping, unlike some sites in Europe where bee artefacts have been found. Africa has the most known rock-paintings related to honey-hunting than any other continent (Crane 1983). This is attributable mainly to the work of Pager (1973, 1976) who describes a number of sites in southern Africa and north of the Sahara. Unfortunately, no rock-paintings depicting honey-hunting in Africa have been dated, though Pager (1976) claimed, probably erroneously, that rock-paintings suggest honey-gathering activities in the Ice Age. Similar honey-hunting practices to those depicted are still carried out in central and southern Africa.

Table 27.1 Location of rock-paintings in Africa related to bees or honey-hunting.

	Comb patterns	Catenary curves	Formlings	Ladders	"Swarms" of bees	Honey-hunting
North Africa						
Algeria		x				
Libya				x		
Tunisia			x	x		
western Sahara	x					
Morocco		xx				
East Africa						
Tanzania		x			?	
Southern Africa						
Lesotho		?		x		
Malawi		?	?	x		
Namibia	xxx	xx	x	x	x	
South Africa		xxx	xx	xxx	x	xx + ?
Zimbabwe		x	xxx	xxx	xxx	x + ?

x = one site, xx = several, xxx = many, ? = uncertain. After Crane (1983).

Table 27.1 shows the location of rock-paintings in Africa which relate, or seem to relate, to bees or honey hunting. Several of the rock-paintings show only depictions of honeycomb or forms characteristic of bee nests, while others have honey hunters in groups of up to five surrounded by bees as they take the honey. An alternative interpretation by Petie (1974) suggests, however, that some paintings depict the netting of birds. Ladders are often also depicted in the paintings and (more rarely) other types of honey collecting equipment. The use of fire or smoke to assist in warding off the bees is also depicted in one painting from Zimbabwe.

The earliest evidence of bee-keeping comes from Egypt (Kuény 1950, Crane & Graham 1985), where four scenes from temples and tombs have been dated at around 2400 BC, 1450 BC (two) and 600 BC. The oldest scene is a stone bas-relief found in the sun-temple of Neuserre at Abu Ghorab and is presently in the Egyptian Museum, Berlin. It depicts harvesting from stacked horizontal hives in a similar way to peasant bee-keepers in contemporary Egypt, though the ancient hives tended to be shorter. The honeybee native to Egypt is *A. m. lamarckii*, which builds smaller colonies and is more defensive than the European bee. Crane (1983) suggests that the style of beehive construction, horizontal hives built in stacks, is well adapted to the biology of the bee, an adaptation which first took place at least 5,000 years ago.

The other three scenes are found in tombs (73, 100 and 279) on the west bank opposite Luxor. The scene in tomb 279, where Pabesa of the Saite dynasty (660– 525 BC) is buried, again shows horizontal hives stacked on one another. The hives are painted blue-grey which Crane (1983) suggests indicates that they were made of clay or unbaked mud. The harvested honey is poured into containers or the honeycomb is sealed within round pottery containers that appear of the same type still in use in Kashmir (Crane 1983).

Crane (1994) states that though she believes that Roman knowledge of bees derived from ancient Greece, bee-keeping practices described by Roman authors were derived from the western Mediterranean region, particularly North Africa, including Carthage, rather than directly from Greece. Evidence for this is that the practice of managing a hive from the opposite end to the bee entrance was described by Roman authors but did not seem to be the practice of ancient Greece. The same practice is found in ancient Egypt and in contemporary North Africa, and indeed throughout much of Africa where conservative bee-keeping is practised.

The complex pattern of traditional bee-keeping in Spain was partially influenced by the importation of technologies from North Africa during the Arab occupation of the peninsula from about 1300–500 BP. In the foothills of the Pyrenees in Aragon, the word used for apiary (arnal) is a loanword from Arabic where it means "the bee" (Chevet & Chevet 1987). This can be seen in the walled apiaries or bee shelters resembling those of ancient and contemporary Egypt and present day Morocco by the Berber, as well as in the use of horizontal hives. In other parts of the Iberian peninsula and further north of the Pyrenees in southern France, upright hives are more characteristic, as they are in northern Europe (Crane 1983). Erup & Armbruster (1958) suggested that there may be a link to the bee boles common in parts of northern Europe and well described in the British Isles by Crane (1983).

Archaeological evidence for the *keeping* of bees in sub-Saharan Africa is missing. However, the rich diversity and complexity of bee-keeping suggests its ancient establishment. The complex historical cultural and economic links throughout the continent make it difficult to resolve the origins and diffusions of the many bee-keeping practices that exist. Research to address this complex issue is only beginning and the lack of early materials makes the task particularly difficult. Unlike many domesticated animals in Africa, the task is not likely to be assisted by a closer study of the genetics and biology of the bee population. The swarming and absconding behaviour of the African bee make for close introversion among the wild and exploited population.

Crane (1983) postulated a link between bee-keeping in Ethiopia and that found in Egypt and other parts of the Mediterranean. This is possible given the restricted distribution of horizontal, cylindrical clay or dung hives. However the direction of influence is at present unknown, though the first record of bee-keeping in Egypt dates from about 2400 BC. In terms of complexity, Ethiopia has the more complex traditional system compared to Egypt. As mentioned earlier, there are clearly links between Egyptian bee-keeping and that found in Morocco and the bee walls of Spain. This is likely to be the result of multiple influences, the major one of which would be the Moorish invasion of the Iberian peninsula. Other types of beehive are found in North Africa, such as bark and log hives. Given the more restricted use of such hives in southern Europe, it is possible that these originated in sub-Saharan Africa. Such hives are mentioned by early Roman writers (Crane 1994) and it is conceivable that they are of considerable antiquity, predating the desiccation of the

Sahara. The spread and present-day distribution of log and bark hives has some link to the vegetation. The miombo woodland of eastern and southern Africa, characterized by *Brachystegia* and *Julbernardia*, is where bark hives are most abundant and little traditional bee-keeping is found to its south (H. Hastings, pers. comm.).

Scott (1952) describes thatched conical hives that were usually placed singly in trees of Dracaenas in Ethiopia. He also mentions one occurrence of an apiary of such thatched hives on a low wooden platform which he states is a method used in southwest Arabia. Such an apiary has also been described more recently in Ethiopia by Fichtl & Adi (1994) and in southern Chad by Gadbin (1976). This illustrates again a bee-keeping link between Arabia and parts of east Sudanian Africa. Ethiopia seems to be a centre of diversity in bee-keeping, being influenced and perhaps influencing regions to its north and south. Ancient trade routes in various regions may well have had an influence on sharing bee-keeping technologies, particularly with lighter, more portable hives. Wainwright (1992) has noted possible ancient trade in bee products from Africa to the Phoenicians and the Chinese. Various trade routes across the Sahara would also seem to have included bee products (Curtin 1985, Lovejoy 1985) and may have driven further diffusion and innovation in bee-keeping practice.

It would seem that there may well be some links between the origin and diffusion of bee-keeping and of cereal cultivation. The distribution of cereal complexes approximates well with that of various different hive types and bee-keeping practice. Of the hives used, those made from trees in the form of logs or bark seem to have the widest distribution. Perhaps those made of logs are the oldest tree type, which were replaced in many areas owing to the weight advantage of the bark hive. The density of hives made of tree material has probably been reduced where woven hives have been introduced. Woven hives have a reduced weight and in some areas a lower labour requirement and impact on the environment. In systems where land ownership is related to user rights over trees, hives woven from plant material or crop residues have a particular advantage. The present distribution of log, bark, and woven hives is rather mixed and complex in many areas. However, their distributions suggest that log and bark hives were diffusing in much the same way as cereals such as sorghum and millet, and later in southern Africa also with finger millet. They suggest also a pre-Islamic diffusion through the continent. Woven hives seem to be more recent and their distribution and usage by different peoples has some relationship with the spread of Islam in West Africa and suggests a trans-Saharan influence.

In central Nigeria, log and bark hives have a provenance suggesting a spread linked more with the pre-jihad era prior to the early nineteenth century and are found used among those previously known as the "pagan peoples". Thus, while the Hausa term for the log hive, *kongo*, is used among some peoples in central Nigeria, many have there own term. The woven hive, of various sorts, is spreading in central Nigeria in the wake of other hive types and may be referred to by the Hausa loanword *kango*. The possible role of Fulɓe speakers in diffusing the woven hive in its region of distribution should not be underestimated. Indeed, a description of the local economies of Nigeria, published 50 years ago, notes that the pastoralist Fulani appear to be more concerned about bee-keeping than the mainly Hausa agricultural population in the north of Nigeria (Forde & Scott 1946). The woven hive is often particularly suited to seasonal bee-keeping which can be especially well integrated with pastoralist transhumance.

An intriguing form of woven hive is the upright woven skeps described among the Serer in Senegal (Ndiaye 1976). Their orientation makes them almost unique in Africa for such conical hives. The hives of closest resemblance are found in northern Europe, particularly in Denmark and the Netherlands. Given that Senegambia has been one of the oldest known sources of bee products exported to Europe, it is possible that a link exists between the two hive provenances. It is difficult perhaps to speculate on the direction of possible diffusion of this hive. However, it is worth noting that, in Europe, the Danish bee-keepers are unique for their tradition of hand-pressing the honey, a practice common in Africa.

The rarely studied conical mud-built hives correlate with the distribution of two ancient West African grains, *Digitaria exilis* and *D. iburua* (Portères 1946, 1976, Harris 1976). The main centres of cultivation of the latter are closely tied to more recent reports of mud-built hives, particularly for *D. iburua*. The most easterly cultivation of both cereals is close to the Jos Plateau, Nigeria, similar to the range of the mud-built hive. This would suggest a common development in the region of both the cereal culture and bee-keeping. In the same way that the *Digitaria* species are thought to be the most ancient of West African cereals, it may be postulated that the mud-built hive is the oldest form of beehive in the region. It has managed to stay prevalent in those areas where the people were able to avoid various forms of cultural imperialism over the centuries (Isichei 1983).

The double-chamber variants of mud-built hives are found close to the present northern limit of the mud-built hive type in central Nigeria. The area is in the hilly region of Kauru and on the western and southwestern fringes of the Jos Plateau. This is the present boundary of the Benue–Congo and Chadic language groups (Crozier & Blench 1992). It is also an area noted for its complex relationship of trade and intermarriage among the hill communities (Sharpe 1982), factors that are perhaps conducive for innovation.

4. Traditional bee-keeping in Africa

4.1. Bee-keeping techniques

Traditional bee-keeping in Africa is complex and highly diversified. In some areas, on a regional or local level, it is absent, particularly in the forest zone of West and Central Africa, and in others there are a large number of different hive types and management practices. Studies to date have only begun to unearth the rich traditional knowledge of African bee-keepers. All African bee-keepers have some knowledge of how to attract swarms to colonize hives. This may relate to the positioning of the hive and in many areas bee-keepers place their hives high in trees where swarms are often found to settle (Mutsaers 1991). Bee-keepers often prefer to place their hives in certain trees (e.g. *Daniella oliveri*, *Terminalia spp.*, *Parkia biglobosa*).

African methods of attracting bees to hives are numerous. This may include the practice of certain rites or the placement of bee attractants in the hive, often plants or mixtures of plants and frequently aromatic. Some bee-keepers are known for their ability to make potent attractants and one bee-keeper from Tahoss on the Jos Plateau, central Nigeria, travels about 20 km to get one such attractant from a bee-keeper in another village. Taylor (1942) mentions that these recipes are also a jealously guarded secret around the Zaria area, where he describes the ingredients as including fruits of *Swartzia madagascariensis*, *Bauhinia reticulata* (probably *Piliostigma reticulatum* or *P. thonningii*) pods, *Cassia goratensis* (*Senna singueana*) and inflorescences of *Cymbopogon giganteus*. These are pounded and dried and used as a fumigant. Cow dung is often used as a raw material in hive construction and is commonly regarded as a good attractant for bees. In Ethiopia, hives constructed only of cow dung are found (Fichtl & Adi 1994). More usually it is used as a plaster to smooth the sides of a hive woven from plant material or around the bee entrance holes.

Another interesting management practice, which seems rare in Africa, is migratory bee-keeping, where colonized hives are moved in response to changing honey flows in the vegetation. It has been practised in Egypt, where hives have been moved by raft along the Nile (Crane 1983), with basket hives carried as headloads around the Mount Oku area in Cameroon, and in Ethiopia (Fichtl & Adi 1994). A movable clay hive has also been used around Gashish on the Jos Plateau. This practice may well be more prevalent than present reports suggest.

Catcher boxes are placed high in a tree where bee swarms are normally found. When a swarm enters, the colony can then be transferred to the hive. Forms of catcher boxes have been seen in Nigeria, where a small pot is used, and in an area of the Bamenda Grassfields, North West Province of Cameroon where an innovative system was developed in response to increasing theft of honey and hives. Bee-keepers have transferred their log hives to the compound and placed them in a separate room of the kitchen hut or have even built new rooms for this purpose. The "beeroom" can contain up to 50 hives. Each is built into the wall with the bee entrances to the outside of the hut so the bees enter the hives through the wall. Having solved the problem of theft, the bee-keeper is now faced with colonizing the hive. As only few swarms would enter the hives as they are, small wooden boxes or small raffia or log hives of about 50 cm × 30 cm × 30 cm are used as "catcher boxes". They have an opening at the rear end and the bottom part is partially or totally movable. These are set up in places where it is known that swarms are likely to settle. When a swarm has colonized the catcher box and brood has been placed in newly built combs, the box is sealed and carried to the beeroom. There it is attached to one of the empty log hives. The log hive has a special opening for the catcher box at the top near the bee entrance. The bottom of the catcher box is partially removed and fitted to the log hive. The gaps are closed and sealed with mud. In this way, the catcher box becomes an integral part of the hive, sitting on top of it near the bee entrance. During harvest the catcher box is never touched. It is believed that the queen moves there during harvesting when smoke enters the hive and that this prevents absconding.

In Ethiopia, Fichtl & Adi (1994) describe the use of a queen cage made of bamboo which assists in moving a swarm into a hive. The queen is identified in a swarm and is caught and placed in the queen cage. Van der Burgt (1903, quoted in Irvine 1957) mentions Rundi bee-keepers catching a queen and transferring it to the hive after which the swarm will follow.

The occurrence of bee-keeping varies to some extent within a region, some peoples are bee-keepers while the neighbouring peoples may well not be. Again within the same ethnic group there may be some villages which have bee-keepers while others may not. Traditional bee-keeping is almost exclusively carried out by men, though Irvine (1957) noted that women may use ground hives.

4.2. Hive types

Most studies seem to assign a particular hive type to a particular ethnic group. However, other evidence shows that the story may not be so simple as a number of peoples have several different hive types used either perennially or seasonally in a complex system. Such evidence suggests that bee-keeping development in Africa does not follow a linear evolution of the type described by Crane (1983) in regard to European bee-keeping.

The literature has been able to show the innovative nature of much African bee-keeping and it also shows that hive types do not simply replace a more "modern" hive for a more "primitive" style. Indeed, the oversimplification of such substitution is further illustrated by the co-existence of both honey-hunting and bee-keeping, and also of conservative and non-conservative forms of hive management within the same village in various parts of Africa. Rather, it suggests an increasingly complex system of bee-keeping (which is also not isolated from the wider livelihood system) that is integrative and syncretistic. Thus, there is a "hybridization" of much bee-keeping technology and management practice. An improved understanding of both African bee biology and of this adaptive process helps grasp the history and future development prospects of bee-keeping in Africa.

4.3. Hives constructed from trees

The hive types of widest distribution in Africa are constructed from wood, usually in the form of cylindrical bark or log hives. Both hive types are found throughout the continent, in North, West, East and southern Africa. Neither have been recorded much south of 25°S, in Egypt or in much of Central Africa (though this may be partly due to lack of evidence).

Emin Pasha gives an early description of the use of bark hives when coming across them among the Dinka of East Sudan in 1888 (Irvine 1957). They are found in most African countries where bee-keeping is practised and are particularly pre-valent in the miombo woodlands which are characterized by *Brachystegia* and *Julbernardia*. Bark hives form horizontal cylinders, usually around 1–1.5 m long and 30–45 cm across, to which end pieces are added. They may also be sealed and insulated with grass and cow dung.

The trees from which the bark is taken vary. The Acholi of Uganda tend to use bark of *Terminalia*, whereas in much of the remainder of eastern and southern Africa bark of *Brachystegia* and *Julbernardia* is used. Harris (1940) also notes that bark hives are generally made from *Brachystegia* and *Julbernardia* trees. In Nigeria, though bark hives are present, they seem to be rare (Jessen 1967). In West Africa, the Yalunka of northeast Sierra Leone use the bark of *Daniellia oliveri* which is prevalent in the moister savanna zone. Pobéguin (1906) describes how the same tree is used in Guinea and in other parts of francophone West Africa. The wood of *Daniella oliveri* is resistant to termites, adding to the potential longevity of such hives. The tree extends from Senegal to Sudan, Uganda and Zaïre (Keay 1989).

Horizontal bark hives are also found in North Africa where they are made from *Quercus suber* (cork oak) which, when the bark is removed, is not killed as with many other species. The cork quality also helps maintain a more even hive temperature during both hot and cold weather. Similar hives are found in Spain and Portugal where they are used vertically. Hives made from cork bark are regarded as the best type of hive by the ancient Roman writer Varro (Crane 1994).

The use of log hives is not restricted to Africa. Log hives have been used in South America for stingless bees, throughout Asia with *A. cerana* and have been used upright in northern Europe for European *A. mellifera*. They have also been described by writers in ancient Rome (Crane 1994). Trees are often nesting sites for wild bees and it is not surprising that one of the preferred nesting sites of bees has been adapted for use as a hive.

An overview of several African log hive types is given by Irvine (1957), but he perhaps makes a false distinction between "hollow logs" and "specially hollowed wood". The view of Smith (1942) is that the log hives used around Buea, Cameroon, were not built specifically for the purpose but that the bee-keeper is merely exploiting rotted hollows in logs in which bees have nested. He also suggests that the bee-keepers were inspired to do so following the use of packing cases for bee-keeping by some German colonialists. While both of his suggestions would seem to be based more on his own prejudices, it is true that bee-keepers may exploit opportunities that bees provide in their natural nesting habit. For example, Villières (1987a) has described a hive in eastern Senegal formed by bricking up a hole in a baobab tree in which bees had nested. Preservation and exploitation of natural nesting sites has also been seen in central Nigeria, Cameroon and South Africa.

Log hives have been particularly well documented in East Africa, in Tanzania (e.g. Seyffert 1930, Ntenga & Mugongo 1991) and Kenya (e.g. Nightingale 1976). Various different tree species are used to construct hives. Ntenga & Mugongo (1991) found more than ten different species which bee-keepers preferred for the construction of log hives. There are different forms of log hive used in Africa. One of the most common types is made by hollowing a log from the ends to make a cylinder. This is then placed or hung horizontally in a tree. Harvesting is done either by lowering the hive or by climbing the tree and then removing an end-piece. Other types of hive are made by splitting a log prior to hollowing it out. The two pieces are then tied back together (Irvine 1957). Harvesting of such hives is often done by lifting the upper section to expose all the honeycombs. The upper section is sometimes purposely made bigger than the lower (Ntenga & Mugongo 1991). A third form of access is through a door placed mid-way along the length of the hive. Though all log hive types seem to be well described among different peoples and in different areas in East Africa, all three forms have also been seen in central Nigeria and elsewhere in Africa.

Barth (1857–9) describes the use of hollow logs, placed in trees, as beehives among the Hausa of northern Nigeria. A number of different tree species are used to make log hives in West Africa. One hive of wider distribution in West Africa is made from the stems of the fan-palm *Borassus aethiopium* which are hollowed out and used as a horizontal cylindrical hive. This is a common hive of the Hausa in northern Nigeria and is also used further west and in East Africa.

Horizontal wooden box hives made from boards have been described from Morocco and Algeria (Crane 1983). Such hives have also been found in the region sweeping from Slovenia through the Alps of northern Italy. Crane (1983) mentions that she also found a similar hive in Corsica. Somewhat further afield, Clauss & Clauss (1991) describe the use of wooden board hives in Zambia. One bee-keeper there has also used old dug-out canoes to build such hives. The Zambian wooden hives differ from the North African types in that one plank is removed during harvesting. Irvine (1957) mentions some other hive types that seem to be made from specially prepared boxes. Wooden sticks bound together into a cylinder and covered in grass and cow dung or mud have also been seen in Zambia. Another wooden form of hive is described from Guinea Bissau where old wine barrels used by the Portuguese are utilized (Svensson 1985).

4.4. Hives constructed from other plant material

Many types of hive are made of plant material other than from trees. These may be cylindrical or conical in shape and are usually placed in trees horizontally. The construction material varies both within and between areas. In Morocco, there are woven cane hives which Crane (1994) relates to descriptions in the writings of Varro, Virgil, Columella and Pliny from ancient Rome. Grasses are also often used to build straw hives. The species of grass varies and Mutsaers (1993) lists seven types of grass that have been used in Nigeria and Niger, including stems of the crop plants sorghum and millet (Himsel 1993). Depending on the form of construction, an inner woven frame is often used to provide the shape. Basket work hives constructed from spilt bamboo or other similar material are also found. The inside of grass and woven hives are often finished with a layer of cow dung and mud.

A typical example of a woven hive type can be found in the Bamenda area of Cameroon where the most intensive bee-keeping is practised around Mount Oku. The traditional hive is called the "Oku Hive", and it was recently reported that this hive constitutes 80 per cent of the 1,392 hives of the 58 members of the North West Beefarmers' Association. The other 20 per cent are top-bar hives, introduced within the last ten years, or pots, tins and boxes. The number of hives varies from about three to more than 100 per bee-keeper.

The Oku hive is a horizontal hive and is usually constructed mainly from raffia. It is cylindrical, has a diameter of about 30 cm to 55 cm and is about 1 m to 1.5 m long. Two or three woven cane rings form the inner structure to which split raffia is attached. It is sometimes sealed with mud, usually mixed with cow dung, and has another two or three rings of woven cane around the outside of the structure. The bee entrance is at one of the ends and harvesting is done from the opposite end by removing a door and cutting out the combs. Usually, there is more brood near the bee entrance with the honey towards the harvesting door. Most bee-keepers take this fact into consideration when harvesting by leaving the combs near the entrance untouched. Thus conservative bee-keeping can be practised as the bees are then less likely to abscond. This management procedure is made easier during construction by placing sticks in the shape of a star about 22–24 cm from the bee entrance so that they know where to stop cutting when harvesting at night with a machete. The hives are placed in trees about 3–7 m from the ground. However, some bee-keepers will place their hives only 1 m from the ground under certain conditions, for example where there are overhanging rocks or caves with wide openings.

Many bee-keepers around Mount Oku practice migratory bee-keeping to optimize cropping of the varying honey flows. They move their hives on headload up and down the mountain, following the flowering peaks of various vegetation and in this way obtain better yields. The distance is usually no further than what a man can walk in one night. The hives are completely plastered and sealed with mud at night, so that no bees can come out, and are then transferred the same night to their new position where the mud is removed before sunrise. Fichtl & Adi (1994) have also described migratory bee-keeping using basket hives from Ethiopia.

Gnägi (1992) describes six hive types used in the Ouélessébougou Arondissement of Mali, five of which are constructed from grass, bamboo or millet stalks. The sixth type is a cylindrical bark hive. There are two types of grass hive which he distinguishes, conical and cylindrical. The conical grass hive usually has only one entrance making it necessary to disturb the brood when harvesting the honey. This makes it likely that the colony will abscond which is not a problem when the hive is used seasonally. He notes the use of a local innovation whereby the bee entrance is placed at the apex of the horizontal cone. Thus conservative harvesting, leaving some comb in the hive and reducing the likelihood of the colony absconding, is made possible. It also allows the bees to be smoked out of the hive prior to harvesting, reducing the risk of the honey harvester being stung.

The use of cylindrical hives is more common where conservative bee-keeping is practised, as Gnägi notes from Mali. Cylindrical hives make the construction of two entrances easier; an entrance for the bees at one end and a harvesting door at the other. There are exceptions to this general rule and Schweinwurth (1873) described the use of a harvesting hole midway along the length of a long cylindrical hive made of basketry. In addition to the bark cylindrical hives, Gnägi describes three other types of cylindrical hive based on the construction material, grass, millet stalks and woven bamboo. Such variation in the type of hive and in the material used is common in many areas (Himsel 1991, Mutsaers 1993). Clauss & Clauss (1991) also describe the use of a rolled woven mat in forming a cylindrical hive.

The waterproofing of hives is also an important aspect. This is often done by plastering the hive with a mix of cow dung and mud. Gnägi suggests that plastering with the right mix is one of the critical points in maintaining the right environment for the bees to reduce the possibility of their absconding. It can also add to the potential longevity of the hive. Further east and south in Africa, banana leaves are also used to protect the hive (Irvine 1957, Clauss & Clauss 1991). The incidence of grass and woven hives seems to be much less in the region covered by the miombo woodland in east and southern Africa where bark and log hives are more common.

Though horizontal hives are the norm, examples of woven upright hives also exist. Conical upright basketry skeps plastered with cow dung have been described from Senegal where they hang from the apex of a tree. Such hives are remarkably similar to traditional skeps used in Denmark and The Netherlands (Crane 1983) and do not appear to exist in other parts of Africa. The only other upright basketry hive so far studied is found around Kontagora in western central Nigeria (Corby 1943). It would seem to be a mimicked copy of another local hive constructed from mud which is described below.

4.5. Container hives

Traditional household containers such as clay pots and calabashes are also used as beehives. Pots which have fallen out of use in the household are usually used for the hives; however, there are examples where pots are bought specifically for use as beehives.

Crane (1983) describes specially constructed fired-clay hives being used in Morocco. The hives are constructed in a tapered manner that allows the pottery cylinders to be put together end to end. This allows for the addition of special honey chambers, known as supers, after the establishment of the bees in the hive.

Clauss & Clauss (1991) describe the use of calabashes as beehives. Usually a single calabash is used, though they have seen one instance of a double calabash hive. In central Nigeria an upside down pot with another smaller pot as a brood chamber fixed on the side has been seen, along with other double pot types placed open end to open end horizontally. Usually such hives have one pot larger than the other, though sometimes they are the same size.

Other types of container are also brought into new use as beehives. These include old buckets, petrol cans, oil drums in francophone Africa (Villières 1987a; all three have also been seen in central Nigeria and western Cameroon), old wine barrels in Guinea Bissau (Svensson 1985) and cement barrels in Cameroon (Smith 1942).

4.6. Hives constructed from earthen material

The earliest recorded evidence of hives in Africa is of horizontal cylinders made of mud or clay which were placed in walled apiaries. These have been described from ancient Egypt and the same hive type is still in use (Crane 1983). This form of wall apiary is also found further west in North Africa and in northern Spain, and further east on the Arabian peninsula. Outside of Egypt the wall apiaries may be used as receptacles for horizontal woven cane hives covered with cow dung.

One key distinction that has been drawn between hive types is that between horizontal and upright hives. Crane (1983) has used this as the primary dichotomy in describing hives. The tendency seems to be that horizontal hives allow for smaller but more numerous combs. The placement of brood in relation to the honey is thus affected. There seems to be a greater likelihood of combs having both brood and honey in the upright hive.

Though upright and horizontal hives can be constructed from similar materials, Crane has noted that the hive orientation varies according to geographical locality. It was thought that upright hives were originally found in northern Europe, only later being used in Asia. However, evidence from central Nigeria (Corby 1943), northern Benin (Villières 1987a,b) and the travels of Frobenius in what is now Guinea and Mali (Seyffert 1930) suggest an ancient existence of upright hives constructed from mud and thatch in West Africa.

The German explorer Frobenius described *gemauerte Häuschen* (small house masonry hives) among the Malinke in Guinea and elsewhere in Mali (Seyffert 1930). Their exact form of construction is rather unclear. Corby (1943) gives a description of upright hives constructed from mud and thatched with grass among the Kambari in the Kontagora Emirate of west–central Nigeria. The hives may be constructed on a platform high in the trees. The Kambari have two other types of hive both constructed from basketry plastered with mud and cow dung. One of the basketry hive types is similar to that used further north among the Hausa, while the second basketry type is a "hybridization" of the two forms as its form is that of the hive type constructed from mud.

The hives used by the Kambari tend to have the bee entrance on one side and a harvesting door at the other. The harvesting door is removed at night and smoke is blown into the hive in order to drive the bees to the outside of the hive on the other side. The honey harvester does not remove all of the combs, reducing the probability of the bees absconding. Hives similar to those described by Corby can also be seen among the Somba people in the Atakora Mountains in the north of the Benin Republic (Villières 1987a,b). Our own research shows a much wider distribution of this type of hive than previously thought in Kaduna and Plateau States of central Nigeria, and particularly in the hilly areas on the fringes of the Jos Plateau. The approximate southern limit of traditional bee-keeping in Nigeria does not extend too far into the forest zone (RIM 1992).

The construction of hives built of mud in central Nigeria varies to some extent from village to village. Some hives are constructed in a similar manner to the local grain stores, especially in terms of the closure of the mud ceiling, while others used a method of closure specifically for the hive. This second type has been seen among the Berom at Tahoss and the Shagawu at Gashish on the southwest fringe of the Jos Plateau. At Tahoss the hive is conical with a flat ceiling onto which a thatched roof is placed. The closure of the hive ceiling is by placement of sticks across the top of the hive which is then sealed over with mud. The preferred stick is hollow and, when split and placed inside down, forms a ridged guide on which the bees can build their combs. The bee-keepers also say that these sticks have the advantage of having an aroma which attracts the bees. The sticks are placed in an orientation across the bee entrance and, of those ten measured, have an average width of 30 mm which corresponds exactly to measurements of bee space in Nigeria (Mutsaers

1991). When harvesting, the bee-keeper is only able to see one comb after the other as he removes them. The brood is mainly placed close to the bee entrance on the other side and some combs will be left, practising conservative bee-keeping.

Among the Shagawu in Gashish on the Jos Plateau there are three types of hive. One is conical in shape, similar to those found elsewhere in the region. However, the other two are straight-sided and, from above, have a more elliptical shape with the bee entrance and harvesting door on the longer sides. The ceiling is again formed with sticks covered with mud. However, the sticks are more roughly cut and the type seems of less importance. They fulfil the same function of guiding the bees in the orientation of comb building. The key difference is that the orientation is switched by laying the sticks between the two sides in which the entrance and harvesting door are placed. The bees, as in all hive types, tend to build their first and brood combs closest to the entrance and the combs of predominantly honey further from the entrance. The advantage expressed by a Gashish bee-keeper when comparing the comb orientation with that of the Tahoss hive was that he was able to see all of the combs when he opened the harvesting door and choose more carefully those with much honey at either end.

The difference between the two straight-sided hives in Gashish is mainly in size. However, the size then allows for two further interesting innovations. The larger hive may have a smaller chamber of similar style constructed on top of it which has the new bee entrance, while the entrance placed initially in the lower chamber is sealed. The two chambers are connected by holes pierced in the ceiling of the larger chamber. The upper chamber is not harvested and is left for the bees. Its position near the entrance means that the bees tend then to make this their brood chamber. During harvesting the bees are smoked into the upper chamber. The smaller straight-sided hive is constructed on two pieces of wood, allowing it to be moved according to varying honey flow of trees and shrubs in the area. The hive is therefore designed for migratory bee-keeping.

Double chamber hives constructed of mud are found in other parts of central Nigeria. However, those others seen have a main chamber conical in shape and the second chamber, attached on the side rather than the top, is a small, old pot with its mouth facing in and a bee entrance tapped through the pot floor – for example, as seen in a hive, found in the Kauru Hills between Kaduna and Jos in central Nigeria. Monber, in a letter to Irvine (1957), describes "bee rooms" among the Baju (Kaje) near Zaria, Nigeria. These seem to be something similar to the double chamber hives described above. It seems that in this case a swarm which has nested in a pot placed in a tree is brought down and fixed to "a much larger pot or small room". Thus the pot is used first as a "catcher box" and secondly as a brood chamber. Monber tells Irvine of having seen over 36 litres of honey being taken from such a hive in one season. The same use of a catcher pot has been seen further east close to Jos among the Bache (Rukuba).

Many of the peoples in central Nigeria in areas where hives built from mud are found also keep other forms of hive, particularly log hives and pot hives. While, as mentioned above, a small pot hive may be integrated into a mud-built hive, only one example of a hybridization of log and mud-built hive has yet been discovered.

This was found among the Tumi in the Kauru Hills between Jos and Kaduna, in central Nigeria. The mud-built section is similar in shape to two of the hives found in Gashish on the Jos Plateau and the ceiling is made from one half of a split log hive.

The present situation in central Nigeria shows that in one village several different hive types may be used. Thus mud-built hives, log hives and pot hives, and in some areas bark and woven hives, may all be used by the same bee-keepers, without excluding the practice of honey-hunting. The heavier mud-built and log hives are often used as perennial hives whereas the others may only be used on a seasonal basis. This type of system seems to be a relatively recent development and brings together advantages of both perennial and seasonal systems. This then enables the bee-keeper to exploit the swarming tendency of *A. mellifera adansonii*, with the perennial hives providing a colony source for the seasonal hives. The bee-keeper becomes the manager of a wider bee system rather than only focusing attention on individual hive management. Such a system of perennial "bee reservoirs" with seasonal hives was advocated for use in Kenya by Nightingale (1976) based on his long experience of African bee behaviour.

5. Bee-keeping development in Africa

There is a decline in bee-keeping in many parts of Africa owing to, *inter alia*, local competition with refined sugar, theft, environmental factors and the opportunity costs when compared with other enterprises. However, the demand for good quality, organic beeswax and other bee products is on the increase. The spread of *Varroa* mite and the widespread use of chemicals has meant that the demand for wax produced by the European bee is declining. The African bee is a particularly good producer of wax and the market potential seems vast. Wainwright (1992) states that colonialism was the worst event in the development of bee-keeping in Africa. It managed to disturb ancient trade in honey and beeswax that did not revive until well after independence. Presently, an excellent basis for the revitalization and expansion in the historic trade in bee products exists throughout Africa. However, the views of many bee scientists and developers do not seem to fit well with this challenge.

Crane (1992) follows Dzierzon (1862) in attempting to define "rational" bee-keeping. Both writers, Dzierzon for Europe and Crane for the tropics and sub-tropics, centre their concept of rationality on the idea of intensive management and, in particular, on the use of movable frame hives following the developments of Langstroth (1853) with European bees. The assumption that apiculture has spread to Africa (Crane 1983) is based on the fact that management in Africa is in many cases very limited, in some cases zero-management is practised; whereas in Europe, management is the basis for any type of successful bee-keeping. This assumption underlies a basic error: that intensive bee management is an indicator for "rational" bee-keeping for all bees races of the world. As have many other authors, Crane assumed that lack of intensive management is a result of ignorance and the use of inappropriate, "primitive" hives. However, non-intensive management of African bees is more a rational response to the behaviour of the bees (very defensive, absconding, swarming behaviour) and the complexities of African subsistence systems.

Furthermore, there is a great variability in the different populations of African bees, ranging from the more well known, highly defensive "mass attacking killer bee" (perhaps typified by *A. m. adansonii* and *A. m. scutellata*) to the relatively docile bee of the East African highlands, *A. m. monticola* (which can be kept around homesteads and managed without protective clothing during day time, cf. Ruttner 1992). One response to the difficulty in managing African bees was to import European bees, regarded as having better behavioural characteristics for bee-keeping. This has also proven unsuccessful.

Those management techniques that are practised, such as hive placement, have also been widely criticized. The high placement of hives is often only related to protection from thieves and bush fires by eurocentric bee experts. However, it is also a practice adapted to swarm-catching and increases the percentage colonization of hives. A ground apiary is necessary when intensive management is used. The use of a ground apiary has not proven attractive for both environmental (bush fires, pests) and social (thieves) reasons, as well as for their non-adaptation to African bee biology.

With many other authors, Smith (1958) noted in his observations of bee-keeping in Tanganyika (Tanzania) that traditional bee-keeping was "most inefficient". However, he then, in a curiously paradoxical juxtaposition, points out that the various attempts at using more "normal" European and American hives had been unsuccessful. Fichtl & Adi (1992) mention that traditional bee-keeping in Ethiopia, despite its antiquated appearance to eurocentric bee-keepers, is actually a very efficient undertaking. No capital investments are necessary to obtain sufficient honey yields. If the harvest must be increased, additional baskets will be prepared and hung up in trees. The biology of the bees does not require sophisticated management methods like nucleus making and queen rearing. The volume of baskets ensures that sufficient swarms will be produced to replace the bee colonies spoiled during traditional harvesting. Gnägi (1992) has also made this point for bee-keeping in Africa in general and for Mali in particular.

Since most colonial officers, development workers and scientists assumed wrongly that the management practices of African bee-keepers were less advanced for the African bee than the complicated management practised elsewhere, they tried to develop advanced and "modern" methods of bee-keeping in Africa. Almost all recent interventions such as introducing movable frame hives, or developing adapted hives like the highly promoted Kenyan Top-Bar Hive (KTBH) have been disappointing (Crane 1990). As hives they depend on intensive management, the antithesis of the rational exploitation of the African bee.

DeBold & Fondell (1996) follow Gentry (1984) in discussing stages in the development of the bee–human relationship as characterized by a continuum from "bee killing" through "bee-having" to "bee-keeping". This reflects the same linear evolutionary perspective of Crane and others. They suggest that there is no conservative bee-keeping in the Central African Republic. However, they also note the lack of success in introducing the KTBH. This is rightly put down in part to the prohibitive cost of the materials required, as well as such factors as the relative fragility of the KTBH and its need to be placed closer to ground level where it is accessible to bee-keepers and thieves alike. Unfortunately, they fall into the trap of also blaming "cultural factors" for the lack of impact of bee-keeping development programmes.

A better understanding of the biology of the African bee and the heterogeneity in its population is clearly necessary if we are to understand traditional bee-keeping management systems and their development. This then requires a development dialogue among bee-keepers and scientists, not only on hives and their management but also on bee-keeping systems and their relationship to livelihood systems in Africa. Working with bee-keepers, with their current technology and with their innovations, can begin to take the sector forward (Clauss & Clauss 1991, Gnägi 1992). However, until more bee-keeping development programmes free themselves from the hegemony of a false "modernization" and "rationality" based on an understanding of the European *A. mellifera*, the African bee-keeper will not be effectively served.

Acknowledgements

The authors would like to thank Ms Salma Zabaneh, the Librarian of the International Bee Research Association (IBRA), and Mr H. Geffcken of the Institute of Bee-keeping, Celle, Lower Saxony for their assistance with literature research. We are also both thankful for the patience of our families and friends.

Part of this study was carried out while one of us (ADK) was a Research Fellow on the EU-funded Jos Plateau Environmental Resources Development Programme, a linkage project between the University of Jos, Nigeria, and the University of Durham, UK. The views expressed are those of the authors alone.

Note

1. The survey covered 75 per cent of the country, but did not include the north owing to the instability in the area at that time. The information came from the *Ethiopian Bee-keeping Newsletter* 1(2), and was described briefly in *Bee-keeping and Development* 30, 8, March 1994.

References

Barth, H. 1857–59. *Travels and discoveries in North and Central Africa*. London: Longman.
Boyles, P. 1991. Possibilities for the development of honeybee, honey and wax resources on the Jos Plateau. *Jos Plateau Environmental Resources Development Programme Interim Report* No. 24. Durham: Department of Geography, University of Durham, UK.
Buchanan, K.M. & J.C. Pugh 1955. *Land and people in Nigeria*. London: University of London Press.
Chandler, M.T. 1976. The African honeybee–*Apis mellifera adansonii*: The biological basis of its management. In *Apiculture in Tropical Climates*, E. Crane (ed.), 61–8. London: IBRA.
Chevet, R. & B. Chevet 1987. *L'arna aragonaise*. Bordeaux: *Authors*.
Clauss, B. & R. Clauss 1991. *Zambian bee-keeping handbook*. Ndola, Zambia: Mission Press.

Corby, H.D.L. 1943. Bee-keeping and honey production in the Kontagora Emirate. *Farm and Forest* 4(1), 25–9.

Crane, E. 1983. *The archaeology of bee-keeping*. London: Duckworth.

Crane, E. 1990. *Bees and bee-keeping: Science, practice and world resources*. Oxford: Heinemann Newnes.

Crane, E. 1992. Traditional bee management: definitions and some examples in the tropics and sub-tropics. In *Traditional bee management as a basis for bee-keeping in the tropics*, J. Kaal, H.H.W. Velthuis, F. Jongeleen, J. Beetsma (eds), 13–24. Bennekom: NECTAR.

Crane, E. 1994. Bee-keeping in the world of Ancient Rome. *Bee World* 75(3), 118–34.

Crane, E. & A.J. Graham 1985. Bee hives of the Ancient World. *Bee World* 66, 23–41, 148–70.

Crozier, D.H. & R.M. Blench (eds) 1992. *An index of Nigerian languages*, 2nd edn. Dallas: SIL.

Curtin, P.D. 1985. The external trade of West Africa to 1800. In *History of West Africa, vol. I*, 3rd edn, J.F.A. Ajayi & M. Crowder (eds), 624–47. Harlow: Longman.

De Bold, K. & T. Fondell 1996. Bee-keeping in the Central African Republic. *Bee World* 77(2), 103–7.

De Jong, D. 1996. Africanized honey bees in Brazil, forty years of adaptation and success. *Bee World* 77(2), 67–70.

Denham, D., H. Clapperton, W. Oudney 1828. *Narrative of travels and discoveries in northern and Central Africa*. London: John Murray.

Dietz, A. & R. Krell 1986. Survey of honey bees at different altitudes in Kenya. *American Bee Journal* 126(12), 829–30.

Dzierzon, J. 1862. *Dzierzon's rational bee-keeping*. London: Houlston.

Erup, O. & L. Armbruster 1958. Nochmals spanische Bienenmauern. *Archiv Bienenkorben*, 35, 32–3.

Fichtl, R. & A. Adi 1994. *Honeybee flora of Ethiopia*. Weikersheim: Margraf.

Fischer, F.U. 1993. Bee-keeping in the subsistence economy of the Miombo savanna woodland of South-Central Africa. *ODI Rural Development Forestry Network Paper*, 15c, 1–12.

Forde, D. & R. Scott 1946. *The native economies of Nigeria*. London: Faber & Faber.

Gadbin, C. 1976. Aperçu sur l'apiculture traditionnelle dans le sud du Tchad. *Journal d'Agriculture Tropicale et de Botanique Appliquée* 23(4–6), 101–15.

Geider, T. 1989. Bee-keeping and honeywine in Pokomo: culture, history and lexicography. In *Transition and continuity of identity in East Africa and beyond: in memoriam David Miller*. E. Linnebuhr (ed.), 111–62. Bayreuth: Bayreuth African Studies.

Gentry, C. 1984. *Small scale bee-keeping*. Washington, DC: Peace Corps Information Collection and Exchange.

Gnägi, A. 1992. *Bienenhaltung im Arrondisement Ouélessébougou, Mali: Lokale Imkerei-Technologie und Möglichkeiten zur partizipativen Weiterent-wicklung*. Bern: Institut für Ethnologie.

Gutmann, B. 1922. Die Imkerei bei den Dschagga. *Archiv Anthropologie Braunschweig* 19(1), 8–35.

Harris, D.R. 1976. Traditional systems of plant food production and the origins of agriculture in West Africa. In *Origins of African plant domestication*, J.R. Harlan, J.M.J. de Wet, A.B.L. Stemler (eds), 311–56. The Hague: Mouton.

Harris, W.V. 1940. *Beeswax*. Pamphlets of the Tanganyika Department of Agriculture. No. 23. Dar-es-Salaam: Government Printer.

Himsel, H.H. 1991. Traditional bee-keeping in the Republic of Niger. *Bee World* 72, 22–8.

Irvine, F.R. 1957. Indigenous African methods of bee-keeping. *Bee World* 38(5), 113–28.

Isichei, E. 1983. *A history of Nigeria*. Harlow: Longman.

Jessen, C.F. 1967. *Bee keeping in northern Nigeria*. Extension Bulletin No. 3, Ministry of Agriculture, Kaduna, Northern Nigeria.

Kaberry, P.M. 1952. *Women of the Grassfields*. London: Colonial Office.

Keay, R.W.J. 1989. *Trees of Nigeria*. Oxford: Clarendon Press.

Kerr, W.E. 1967. The history of the introduction of African bees to Brazil. *South African Bee Journal* 39(2), 3–5.

Kuény, G. 1950. Scènes apicoles dans l'ancienne Égypte. *Journal of Near Eastern Studies* 9, 84–93.

Langstroth, L.L. 1853. *The hive and the honeybee*. Northampton, MA: Hopkins, Bridgman.

Lovejoy, P.E. 1985. The internal trade of West Africa before 1800. In *History of West Africa*, vol. I. 3rd edn, J.F.A. Ajayi & M. Crowder (eds), 648–90. Harlow: Longman.

Macgregor, Laird & R.A.K. Oldfield 1837. *Expedition into the interior of Africa by the Niger* [2 volumes]. London: Richard Bentley.

Monteiro, J. 1875. *Angola and the River Congo*. London: Macmillan.

Mutsaers, M. 1991. Bees in their natural environment in southwestern Nigeria. *Nigerian Field* **56**(1/2), 3–18.

Mutsaers, M. 1993. Honeybee husbandry in Nigeria: traditional and modern practices. *The Nigerian Field* **58**, 2–18.

Ndiaye, M. 1976. Bee-keeping development in Senegal. In *Apiculture in Tropical Climates*, E. Crane (ed.), 171–9. London: IBRA.

Nightingale, J.M. 1976. Traditional bee-keeping among Kenya tribes, and methods proposed for improvement and modernisation. In *Apiculture in Tropical Climates*, E. Crane (ed.), 15–22. London: IBRA.

Ntenga, G.M. & B.T. Mugongo 1991. Honey hunters and bee-keepers: a study of traditional bee-keeping in Babati District, Tanzania. *International Rural Development Centre Working Paper* 161. Uppsala, Sweden: Sveriges Lantbruksuniversitet.

Pager, H. 1973. Rock paintings in Southern Africa showing bees and honey hunting. *Bee World* **54**(2), 61–8.

Pager, H. 1976. Cave paintings suggest honey hunting activities in Ice Age times. *Bee World* **57**(1), 9–14.

Park, M. 1817. *Travels in the interior districts of Africa*. London: John Murray.

Petie, B. 1974. Bees or birds? *Rhodesian Prehistory* **6**(12), 2–3.

Pobéguin, H. 1906. *Essai sur la Flore de la Guinée française*. Paris: Augustin Challamel.

Portères, R. 1946. L'aire culturale du *Digitaria iburua* Stapf, céréale mineure de l'Ouest-Africain. *L'Agronomie Tropicale* **1**(11,12), 389–92.

Portères, R. 1976. African cereals: *Eleusine*, fonio, black fonio, teff, *Bracharia, Paspalum, Pennisetum*, and African rice. In *Origins of African plant domestication*, J.R. Harlan, J.M.J. de Wet, A.B.L. Stemler (eds), 409–52. The Hague: Mouton.

RIM, 1992. *Nigerian livestock resources, volume II: National synthesis*. Report to Federal Department of Livestock and Pest Control Services, Abuja, Nigeria.

Ruttner, F. 1992. *Naturgeschichte der Honigbienen*. München: Ehrenwirth.

Schweinwurth, G.A. 1873. *The heart of Africa*. London: Sampson Low, Marston, Low and Secule.

Scott, H. 1952. Journey to the Guraghé Highlands (South Ethiopia). *Proceedings of the Linnaean Society of London* **183**, 85–189.

Seyffert, C. 1930. *Biene und Honig im Volksleben der Afrikaner*. Leipzig: Voigtländer.

Sharpe, B. 1982. *Group formation and economic interrelations amongst some communities of Kauru District: Hausa, Kaivi, Rishuwa, Ruruma*. PhD thesis, University of London.

Silberrad, R.E.M. 1976. *Bee-keeping in Zambia*. Bucharest: Apimondia.

Skinner, N. 1969. *Hausa tales and traditions, vol. I*. London: Frank Cass.

Smith, F.G. 1958. Bee-keeping observations in Tanganyika 1949–1957. *Bee World* **39**(2), 29–36.

Smith, J. 1942. Bee-keeping in Cameroons Province. *Farm and Forest* **3**(3), 128–9.

Svensson, B. 1985. A short report on bee-keeping in Guinea Bissau and its possible modernisation. *Proceedings of the Third International Conference of Apiculture in Tropical Climates, Nairobi, 1984*, 27–32. London: International Bee Research Association.

Taylor, J.E. 1942. Honey and beeswax production at Zaria. *Farm and Forest* **3**(1), 11–14.

Villières, B. 1987a. *Le point sur l'apiculture en Afrique Tropicale*. Paris: GRET.

Villières, B. 1987b. L'apiculture africaine en regions tropicales et equatoriales de l'Ouest. *Bulletin Technique Apicole* **14**(4), 193–220.

Wainwright, D. 1992. From forest to supermarket: traditional bee-keeping in Zambia. In *Traditional bee management as a basis for bee-keeping in the tropics*, J. Kaal, H.H.W. Velthuis, F. Jongeleen, J. Beetsma (eds), 25–38. Bennekom: NECTAR.

Index

Note: all references are to livestock/domesticated animals, except where otherwise indicated.
Page references in **bold** indicate chapters.